Analytical Measurements and Instrumentation for Process and Pollution Control

Analytical Measurements and Instrumentation for Process and Pollution Control

Edited by

Paul N. Cheremisinoff
Associate Professor, Environmental Engineering
New Jersey Institute of Technology
Newark, New Jersey

Harlan J. Perlis
Late Professor, Electrical Engineering
New Jersey Institute of Technology
Newark, New Jersey

Contributors

W. L. Beck, Jr.
C. D. Berger
J. D. Berger
P. Carlucci
F. H. Chung
S. C. Creason
T.-Y. Fan
E. J. Farmer
D. H. Fine
R. E. Goans
F. D. Jury

A. L. Lafleur
C. J. Lind
R. L. Miller
R. L. Moore
K. Ogan
E. C. Price
G. J. Schmidt
R. Vanzetti
T. Vo-Dinh
L. J. Warren

ANN ARBOR SCIENCE
PUBLISHERS INC / THE BUTTERWORTH GROUP

Copyright © 1981 by Ann Arbor Science Publishers, Inc.
230 Collingwood, P.O. Box 1425, Ann Arbor, Michigan 48106

Library of Congress Card Catalog Number 80-70319
ISBN 0-250-40405-2

Manufactured in the United States of America

PREFACE

A prerequisite for control of processes and environmental parameters is an understanding of the analytical factors that are involved in measurement, and applying such measurements as standards of comparison and regulation. Additionally, high-sensitivity techniques for measurement of various species are required. Making measurements in the industrial and real-world environment is important for a number of reasons:

1. such analytical data may be required for designing control equipment;
2. instrumentation may be required for process control and efficiency;
3. data may be required for submission to governmental enforcement agencies to comply with source measurement requirements;
4. as legal evidence of compliance; and
5. for protection of health and property for both workers and the public at large.

Analytical Measurements is organized in 14 chapters, each written by an expert in his or her respective field. Each of these sections examines selected significant analytical methods and documents the technique from the standpoint of current practice and the case study approach. An extensive listing of information sources accompanies many sections where applicable, with the intent of stimulating further information exchange for the reader or user of this volume. The book goes beyond a state-of-the-art review by structuring the information on instrumentation/control environments, problems according to concepts and how to proceed. The aim is to provide a how-to-do-it book—as well as what the method will do and where to find it.

This book should prove useful to engineers, scientists, chemists and students concerned with a wide variety of instrumentation and measurement problems in industrial process and environmental control fields. Acknowledgment is made to the experts who gave willingly of their time and knowledge to make this volume possible.

This volume was completed and published after the untimely death of Dr. Harlan J. Perlis who was co-editor at the inception of this project. Dr. Perlis is, and will be, sorely missed for his tireless interests and efforts in the field of instrumentation and control and his ceaseless activities, pioneering spirit and research.

Paul N. Cheremisinoff

v

Contributors

W. L. Beck, Jr., Oak Ridge Associated Universities, Oak Ridge, TN

C. D. Berger, Oak Ridge National Laboratory, Oak Ridge, TN

J. D. Berger, Oak Ridge Associated Universities, Oak Ridge, TN

Patrick Carlucci, Matheson Division of Searle Medical Products U.S.A., Inc., Lyndhurst, NJ

Frank H. Chung, Sherwin-Williams Research Center, Chicago, IL

Samuel C. Creason, Beckman Instruments Inc., Process Instruments Div., Fullerton, CA

Tsai-Yi Fan, James Ford Bell Technical Center, General Mills Inc., Minneapolis, MN

Edward J. Farmer, Ed Farmer & Associates, Sacramento, CA

David H. Fine, New England Institute for Life Sciences, Waltham, MA

R. E. Goans, Waldorf, MD

Floyd D. Jury, Engineering Department, Fisher Controls Co., Marshalltown, IA

Arthur L. Lafleur, Analytical Instruments Div., Thermo Electron Corp., Waltham, MA

Carol J. Lind, U.S. Geological Survey, Menlo Park, CA

Ron L. Miller, The Perkin-Elmer Corp., Norwalk, CT

Ralph L. Moore, Engineering Dept., E.I. DuPont de Nemours & Co., Wilmington, DE

Kenneth Ogan, The Perkin-Elmer Corp., Norwalk, CT

Elizabeth C. Price, Department of Civil and Engineering, New Jersey Institute of Technology, Newark, NJ

Gary J. Schmidt, The Perkin-Elmer Corp., Norwalk, CT

Riccardo Vanzetti, Vanzetti Infrared & Computer Systems Inc., Canton, MA

Tuan Vo-Dinh, Health & Safety Div., Oak Ridge National Laboratory, Oak Ridge, TN

Leonard J. Warren, CSIRO Division of Mineral Chemistry, Port Melbourne, Australia

Contents

pH MEASUREMENT INSTRUMENTATION

Samuel C. Creason

Beckman Instruments, Inc.
Fullerton, California

ACIDS AND BASES

A common way to describe an aqueous solution of an acid or base is in terms of the weight of the material that is present in a given amount of solution. For example, specifying that one solution is 5% hydrochloric acid and that another is 8.5% boric acid describes the two solutions exactly. In this particular case, because the molecular weights of the two acids are in the ratio of 5:8.5, equal volumes of the two solutions contain the same number of molecules of acid. In spite of this, the two solutions differ markedly from each other in their properties. For example, the solution of boric acid can be used as an eyewash, whereas similar use of the solution of hydrochloric acid would be disastrous, even though the weight percents of the acids are similar. Clearly, something else must be considered to describe the solutions more precisely.

The difference between the two solutions mentioned above is related to the way in which each of the acids behaves when added to water. Adding any acid to water produces an excess of hydrogen ions, and both hydrochloric acid and boric acid do this, but to different extents. When hydrogen chloride is dissolved in water, almost all of the molecules dissociate to form hydrogen ions and chloride ions.

Since nearly all the molecules dissociate, hydrochloric acid is described as a strong acid. However, when boric acid is dissolved in water, almost all of the boric acid remains in molecular form; only a relatively small number of molecules dissociate to form hydrogen ions and the corresponding negative ions. Because of this, boric acid is described as a weak acid.

Thus, hydrochloric acid is unfit for use as an eyewash because it is a strong acid. A solution of it contains a high enough concentration of hydrogen ions to damage the eye. On the other hand, boric acid can be used as an eyewash because it is a weak acid. Even a fairly concentrated solution of it contains a relatively low concentration of hydrogen ions. What is necessary, then, to describe an acidic solution more precisely is a measure of the concentration of hydrogen ions in the solution.

For similar reasons, what is necessary to describe a solution of a base is a measure of the concentration of hydroxyl ions in the solution. Bases can be described in the same sort of terms as acids. Adding a base to water produces an excess of hydroxyl ions. Thus, sodium bicarbonate is a base. It is more properly described as a weak base because a relatively large amount of it produces only a small excess of hydroxyl ions. On the other hand, caustic soda (sodium hydroxide) is described as a strong base because it is totally dissociated in water. In fact, even water dissociates to a slight degree, forming both hydrogen and hydroxyl ions in small but equal amounts. Since the hydrogen ions and the hydroxyl ions are present in the same amount, pure water is neither acidic nor basic.

In pure water, and in any aqueous solution, the product of the concentrations of hydrogen and hydroxyl ions is a constant, provided the temperature remains constant. At 25°C, for example, the product is $10^{-14} \, mol^2/l^2$. Thus, the acidity or basicity of any aqueous solution can be described by specifying just the concentration of hydrogen ions and the temperature.

The hydrogen ion concentrations of commonly encountered solutions vary over a wide range, so that a logarithmic scale is convenient. Using such a scale introduces the concept of pH.

pH

pH may be described as the negative logarithm of the concentration of hydrogen ions in an aqueous solution. (The description is exact if the term *activity* is substituted for *concentration*.) In absolute terms, this means that pure water, which contains $10^{-7} \, mol/l$ of hydrogen ions, has a pH of 7. In relative terms, this means that a solution having a pH of 4 contains 10 times more hydrogen ions than a solution having a pH of 5. That is, a change of pH of one unit reflects a 10-fold change in hyrdogen ion concentration. A pH below 7 means that the solution is acidic. A pH above 7 means that the solution is basic. A pH of 7 means that the solution is neutral—neither acidic nor basic. The pH index is usually considered to range between pH 0 and pH 14. Values outside this range are only infrequently encountered in practice.

THE GLASS ELECTRODE—
SILVER/SILVER CHLORIDE REFERENCE ELECTRODE PAIR

If two solutions of different pH values are separated by a glass membrane of certain composition, a potential is developed across the membrane. Experimentally, it is determined that the potential is a function of the difference in pH values between the two solutions. If the pH values are the same, the potential difference will be essentially zero. As the difference in pH of the two solutions is increased, the potential across the membrane increases by about 59 mV per pH unit difference at 25°C. This property of the glass membrane may be exploited to make pH measurements. A simple configuration is illustrated in Figure 1.

A connection to each solution is necessary to measure the difference in potential. To make the connections, a silver wire coated with silver chloride is dipped into each solution. Chloride ions are added to each solution in the same amount as, say, potassium chloride. Other things being essentially equal, reproducible and equal potentials are produced between each coated wire and the solution in which it is immersed because of the formation of two nearly identical silver/silver chloride half-cells. Thus, the potential difference between the two wires depends solely on the difference in pH between the two solutions. This potential may be measured by using a high-impedance millivolt meter with characteristics that are described later.

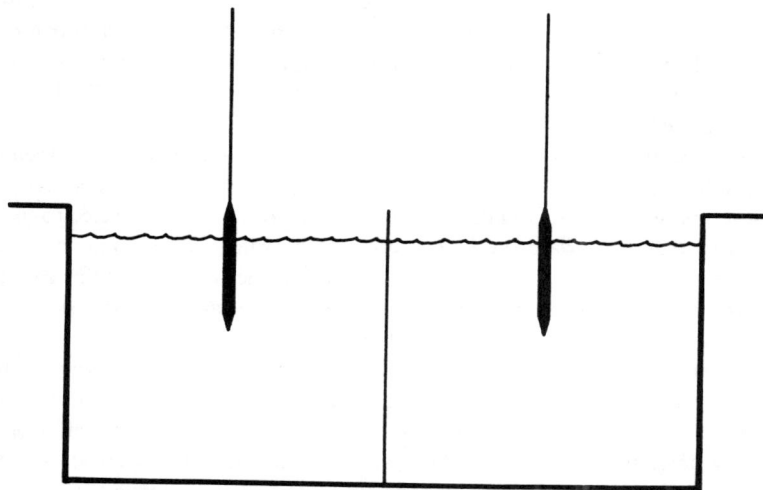

Figure 1. Cell for pH measurement showing two solutions separated by glass membrane. Probes into solutions are silver wires coated with silver chloride.

Of course, to make a useful measurement of pH, a sample of the solution to be tested is substituted for one of the solutions shown in Figure 1. Unfortunately, when this is done the potential between the coated silver wire and the solution may vary with the composition of the solution to be tested, independently of pH. If that happens, the potential difference between the two silver wires is no longer solely a function of pH. To deal with this problem, each of the coated silver wires is enclosed in an envelope so that the solution with which each wire is in contact remains of constant composition.

The envelope that encloses one of the coated silver wires is simply the glass membrane itself in the form of a thin-wall bulb. Filling the bulb with a buffered solution that contains chloride ions provides an electrode that is sensitive only to the pH of the external stream.

The remaining coated wire is enclosed within an inert envelope and that envelope filled with a solution of constant composition, which contains a fixed concentration of chloride ions. To complete the circuit, the solution within the envelope must be in electrical contact with the external stream. Classically, this is done by providing a controlled leak in a thick-wall glass bulb, so that the filling solution slowly can be forced outward into the external stream. An external, elevated reservoir is used to provide a continuous supply of filling solution to the electrode.

Maintaining a positive (but very slow) flow outward ensures that none of the external stream will enter the electrode and contaminate the filling solution. Such an electrode is properly called a reference electrode because it is essentially insensitive to changes in the pH or composition of the external stream. This is true even if the thick-wall bulb is made of a glass that is sensitive to pH because the resistance of the leak is so much less than the resistance of the bulb. The leak is a "short circuit" path.

More recently developed reference electrodes establish a junction between the filling solution and external stream by means of a porous polymeric junction. Filling solution diffuses outward through the junction of the reference electrode and the stream diffuses inward (both at extremely slow rates). The two mingle in the junction area, providing electrical contact. Since the need to force filling solution from the electrode is eliminated, the external reservoir is no longer necessary, and maintenance requirements are drastically reduced.

A refinement of the nonflowing-junction reference electrode is the double-junction reference electrode. It consists of a small, conventional, nonflowing-junction electrode and a nonflowing-junction outer body. An "inert" electrolyte, typically potassium nitrate, fills the space between the inner electrode and the outer body, providing electrical continuity. With this arrangement, the stream does not come in physical contact with the filling solution or the internals of the inner electrode. Because of this isolation, the stream may contain components that otherwise would react with the filling solution or internals, such as sulfide, silver, lead or mercury.

Industrial versions of the glass pH-sensitive electrode and silver/silver chloride electrode are available for use at temperatures as high as 130°C and pressures as high as 150 psi.

OTHER ELECTRODES

While the glass pH-sensitive electrode and silver/silver chloride reference electrode are the most widely used in online industrial applications, others are available.

The most common alternative to the glass pH-sensitive electrode is the antimony electrode. It is useful in applications that are too harsh for the glass electrode, such as pH measurements in acidic fluoride media or in some ore-extraction processes. The electrode consists of a stick of antimony metal on which a coating of antimony oxide is present. It exploits the pH dependence of the antimony/antimony oxide couple.

The antimony electrode is limited in application compared to the glass electrode. The antimony electrode is not usable at pH values less than about 2 nor greater than about 12. In comparison, the glass electrode is usable over the range of 0 to 14. Further, the antimony electrode is sensitive to the presence of oxidizing and reducing agents in the stream being measured, whereas the glass electrode is virtually interference free. Other metal/metal oxide pH-sensitive electrodes have been devised but none has gained wide acceptance by the industrial community.

The most common alternative to the silver/silver chloride reference electrode is the Calomel electrode. It is based on the mercury/mercurous chloride couple. For the vast majority of online industrial applications, the Calomel electrode provides no substantial advantage over the silver/silver chloride electrode. Further, it is limited to applications in which the temperature does not exceed about 80°C.

The remainder of this discussion is based only on the glass pH-sensitive electrode and the silver/silver chloride electrode.

THE ELECTROCHEMICAL CELL

When the reference and the glass electrodes are immersed in a stream, an electrochemical cell is formed. The potential of the cell is related to the pH of the stream. At this point it is useful to discuss that relationship and see how the potential is measured.

The output of the cell is a potential that is directly proportional not only to pH, but to temperature as well. Typically, the filling solutions of the electrodes are chosen so that the potential is within a few millivolts of zero when the pH of the stream is near 7, regardless of the temperature. In a practical sense, picking an exact pH at which the potential will be exactly zero is impossible because the properties of the electrodes will vary slightly from one set to another.

At 25°C, as the pH of the stream decreases below 7, the potential of the cell becomes more positive by about 59 mV per unit. Similarly, as the pH of the stream increases above 7, the potential becomes more negative by about 59 mV per unit. At 50°C, the behavior is similar, except that the potential changes by about 64 mV per pH unit. The temperature dependence is linear from 0°C–100°C. Finally, because of the presence of the glass membrane, the internal resistance of the cell ranges as high as 10^3 megohms.

pH ANALYZERS

These characteristics place some definite constraints on the design of a pH analyzer, although such a device is basically nothing more than a millivolt meter with a scale that is graduated in pH units. First, since essentially no current should be drawn from the cell (to avoid "loading" it), the input impedance of the analyzer must be at least 10^6 megohms. Such values are achieved in present-day analyzers through the use of field-effect transistor input amplifiers.

Further, the output of the analyzer should consistently indicate the correct value of pH regardless of the temperature of the stream. Thus, a means of compensating for the effect of temperature on the sensitivity of the electrodes must be provided. In practice, this is accomplished by placing a temperature-sensitive resistive element in thermal contact with the stream and using the resistive element to adjust the gain of the analyzer.

Finally, some means of adjusting the offset of the analyzer must be provided to compensate for minor differences between sets of electrodes. Commonly, a wide range of offset adjustment is provided in conjunction with a span control, so that an arbitrary range of pH may be selected for display. This type of system is shown in Figure 2.

For many years, this single-ended input configuration was typical of all industrial online pH analyzers. In some applications, it is a satisfactory approach. However, it tends to be sensitive to stray electrical currents in the stream. The reference electrode effectively is tied to circuit ground at the analyzer. Thus, unless the circuit ground is extremely well isolated from earth ground, an iR drop will be produced across the reference electrode in response to the stray currents within the stream being measured. The iR drop adds to or subtracts from the pH potential, thereby producing an erroneous reading. This problem is eliminated in many modern pH analyzers by measuring the potential of the cell in a different way, as shown in Figure 3. In this case, each electrode is connected to a separate

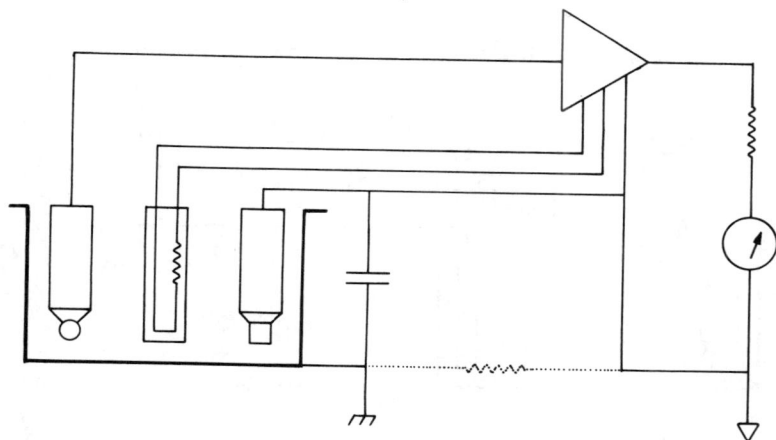

Figure 2.   Single-ended-input pH analyzer. From left to right, elements of cell are glass electrode, thermocompensator and reference electrode. Dotted path depicts leakage resistance.

high-input impedance amplifier. No current can pass through the reference electrode and no iR-drop is produced, even if circuit ground is tied directly to earth ground. The potential of each electrode is measured with respect to a metallic rod that is immersed in the stream. The potential of the metallic rod is affected by many factors, including stray currents in the stream, but this is of no consequence because of the way in which the outputs of the amplifiers are combined. The output of the first amplifier is the potential of the glass electrode with respect to the rod: $e_g - e_{rod}$. The output of the second amplifier is the potential of the reference electrode with respect to the rod: $e_g - e_{rod}$. By subtracting the second output from the first, the desired potential, $e_g - e_r$, is obtained. This potential is directly related to pH.

CALIBRATION AND MAINTENANCE

Once a pH analyzer is in service, it must be calibrated periodically. This may be accomplished in at least two ways. First, a "grab sample" of the stream being measured is taken, and the pH indicated by the online analyzer at the time of sampling is noted. The pH of the sample is determined by means of a laboratory instrument, and the online analyzer is adjusted accordingly. Care must be taken to ensure that neither the composition nor the temperature of the sample changes before it is analyzed.

A better way to calibrate a pH analyzer is by means of buffer solutions. A single buffer (with pH value as near as possible to that of the stream being

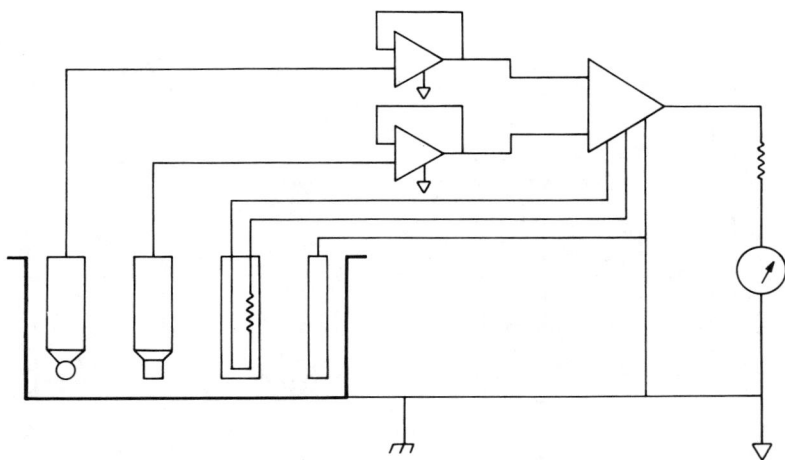

Figure 3.   Differential-input pH analyzer. From left to right, elements of cell are glass electrode, reference electrode, thermocompensator and solution ground post.

measured) serves to calibrate the analyzer. However, good practice dictates that a second buffer be used to establish that the electrodes are in proper working order.

In some applications, simply performing periodic calibration is sufficient maintenance. However, many streams contain materials that either abrade or coat the electrodes. These must be dealt with if the analyzer is to produce reliable results.

In cases of electrode abrasion, simply reducing the velocity of the stream may be helpful. If that is not possible, a somewhat crude, but often effective, solution is to slip pieces of gum-rubber tubing over the electrodes. If the tubing extends just beyond the tips of the electrodes, relatively stagnant zones are produced, which tend to reduce abrasion.

Several means exist for dealing with electrode coating. In the case of a particularly tenacious coating, boiling the tips of the electrodes in concentrated hydrochloric acid for a few minutes or letting the tips stand in hydrofluoric acid for a few seconds may be the only soluton. The latter method should be used only as a last resort, however, since the hydrofluoric acid removes a part of the already thin-glass pH-sensitive membrane. An electrode that is subjected to such treatment should be calibrated using at least two buffers to ensure that it still performs properly.

Generally, less-stringent means can be used to eliminate electrode coatings. In some cases where a submersion-style probe is employed, a water jet may be useful for periodically spray cleaning the electrodes. Conveniently automated by means of a timer and valve, this device is particularly useful if a soft, mushy coating is involved.

A second method for cleaning electrodes is an automated brush assembly, which periodically brushes the coating from the electrodes. However, if the coating is particularly abrasive, the glass pH-sensitive electrode may require more frequent replacement than would otherwise be the case.

The final method included in this discussion is the ultrasonic cleaner. While details of the device vary from manufacturer to manufacturer, each consists of an ultrasonic transducer, which is placed within the electrode chamber, and a remote exciter. Such devices are useful in a wide variety of applications involving organic or inorganic coatings, but at least some periodic manual cleaning is generally required.

REFERENCES

1.   Bates, R. G. *Determination of pH; Theory and Practice* (New York: John Wiley & Sons, Inc., 1973).
2.   Westcott, C. C. *pH Measurements* (New York: Academic Press, Inc., 1978).
3.   Shinskey, F. G. *pH and pION Control in Process and Waste Streams* (New York: John Wiley & Sons, Inc., 1973).
4.   Butler, J. N. *Solubility and pH Calculations* (Boston: Addison-Wesley Publishing Co., Inc., 1964).
5.   Creason, S. C., and B. Delettrez. "Problems and Solutions in pH Measurements," *ISA Trans.* 16(3): 67 (1977)

CHAPTER 2

AUTOMATIC CONTROL OF pH

Ralph L. Moore
Engineering Department
E. I. du Pont de Nemours & Co., Inc.
Wilmington, Delaware

HISTORICAL

The concept of pH measurement was developed early in the 20th century[1]. S. P. L. Sorenson proposed pH as a measurement of the acidity or alkalinity of a water solution. G. N. Lewis expanded the concept to the more rigorous definition that pH is actually a measure of hydrogen ion concentration $(H^+)^-$:

$$pH = -\log_{10} H^+$$

The measurement of pH is thus the measurement of the dissociation of the acid or alkali molecules into ions. Some acids dissociate readily and produce a high concentration of hydrogen ions. These are known as strong acids and include hydrochloric and sulfuric acids. Dissociation is defined by the ionization constant for the chemical[2]. The ambiguity in the use of "strong" is immediately apparent—it refers both to degree of ionization and to concentration.

The relationship between ionization and chemical concentration is known as the activity coefficient. This coefficient has the value of 1 for dilute solutions. Thus the pH usable for measurement has a range of 0 to 14, which describes chemical concentration of some 5% sulfuric acid or sodium hydroxide (caustic). The overlap in the definition of the word "strong" is of little practical significance because all chemicals to which pH applies are necessarily weak solutions.

The logarithmic relationship also applies to basic (alkaline) solutions:

$$14 - pH = \log_{10} (OH)^-$$

Thus, the relationship between ionization (concentration in mol/l) and pH is made up of two logarithmic curves, both starting at pH = 7. This point is defined as neutrality (neither acid nor base). Distilled water is neutral, or has a pH of 7 and a chemical concentration of 10^{-7} mol/l (a minute trace of chemicals).

The relationship between pH and concentration can be derived both analytically and experimentally. The resulting graph is called a titration curve and is shown in Figure 1. The curve shown is for highly ionized chemicals ("strong base—strong acid") and is typically symmetrical and very steep (highly sensitive) near neutrality. It conforms closely to the theoretical logarithmic relationship and serves as the basis for most calculations concerning neutralization. As will be seen later, titration curves for mixtures of chemicals are represented by different ionization constants for each chemical, and the curve for the mixture deviates markedly from the theoretical.

Titration curves can be developed experimentally as well as theoretically, and, in fact, experimental curves are the only ones obtainable for unknown mixtures of chemicals, such as those in the wastewater leaving a manufacturing site. The experimental titration curve is developed by placing 0.5–1 liter of wastewater in a container with a portable pH measuring device (cell). The "as found" pH can be any value on the scale, but, for simplicity, assume the sample to be acidic. A neutralizing reagent, say caustic, is chosen. Then small quantities (ml) of very weak caustic (0.1–0.01 N) are added and the pH noted after each addition. A graph of cumulative reagent addition versus pH is the titration curve.

While very weak reagents are used in the laboratory to discriminate pH values on the steep portion of the titration curve, the variable "r" (for ratio) shown in Figure 1 is more useful in an operating plant. It is simply the volumetric ratio of reagent addition to process flow and can be calculated as follows for a reagent of (for example) 20% concentration and a sample of 500 ml.

$$r = \left[\frac{(N)\,(MW)\,(100/P)}{(SG)\,(1000\ ml/l)\,(V)} \right] (ml\,N\,NaOH)$$

where N = normality of reagent = 0.1
   MW = molecular weight of reagent = 40
   P = percentage concentration of reagent = 20
   SG = specific gravity of P% reagent = 1.115
   V = volume of sample = 500 ml

$$r = \left[\frac{(0.1)\,(40)\,(100/20)}{(1.115)\,(1000)\,(500)} \right] (ml\,N\,NaOH) = (0.0000359)\,(ml\,N\,NaOH)$$

Thus, the ratio "r" is a very small number.

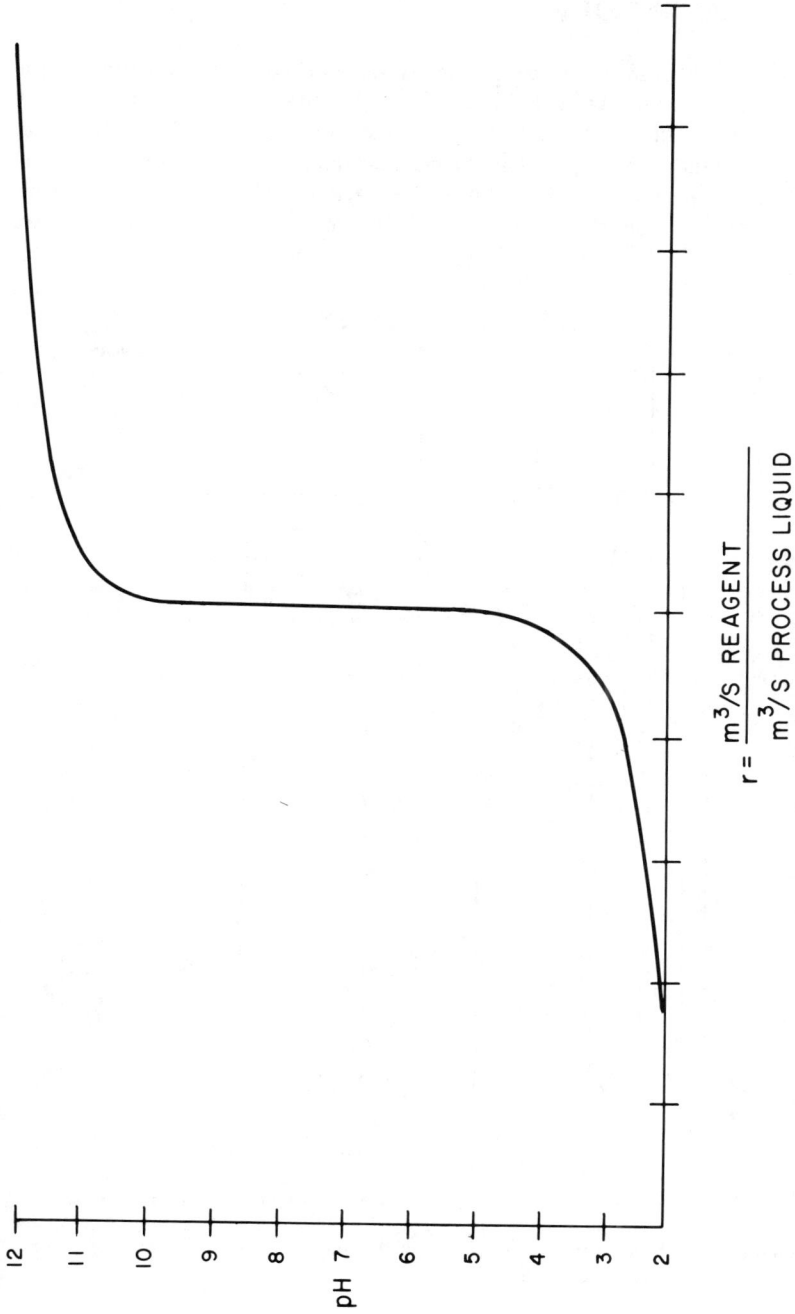

$$r = \frac{m^3/S \ REAGENT}{m^3/S \ PROCESS \ LIQUID}$$

Figure 1.   pH relationship.

MEASUREMENT OF pH

The primary means of measuring the hydrogen ion concentration of a solution is the glass electrode[3]. The glass electrode is a glass tube closed at the bottom with a specially formulated glass membrane, as shown in Figure 2. The outer surface of the membrane is in contact with the process liquid, while the inner surface contains a buffered chloride filling solution. The membrane then acts as a site for the transfer of ions from the outer surface to the inner filling solutions. A surface

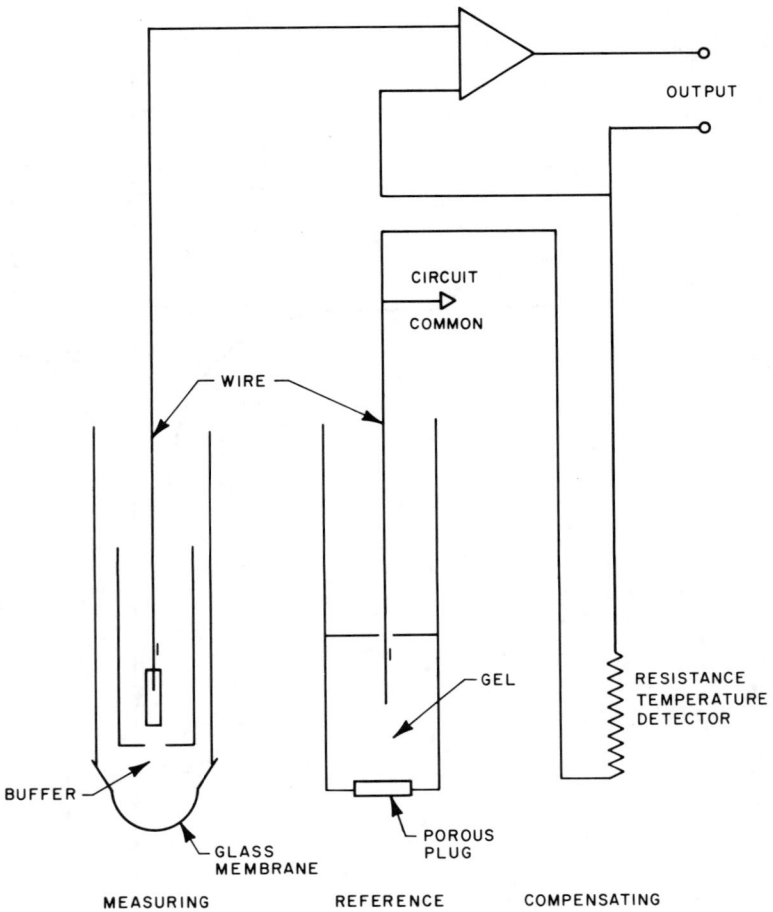

Figure 2.   pH measuring, reference and thermal compensation electrodes with simplified measuring circuit.

potential proportional to process hydrogen ion concentration is developed on the pH-sensitive membrane. A stable electrical connection is provided by the internal filling solution. A wire immersed in the filling solution then carries the current generated by the membrane potential.

Since the glass membrane functions by the exchange of ions, and ion exchange will occur among many ions having similar chemical properties, interference to the hydrogen ion selectivity of the glass membrane can take place. An example is the interference of sodium ions at high values of pH. The interference can be minimized by use of the Dow Antimony pH unit[4], which replaces the glass membrane measuring electrode with a metal (antimony) electrode.

The electrical circuit from the process through the measuring electrode is completed through the reference electrode and back to the process fluid. The internal half cells of the measuring and reference electrodes shown in Figure 2 are designed to be identical, and so cancel their voltage contributions to the circuit. The entire voltage difference in the circuit is then developed across the membrane of the measuring electrode. Since the reference electrode is designed to be insensitive to all ions, the millivoltage developed by the electrical circuit is entirely a function of the ion exchange across the membrane.

Modern reference electrodes utilize solid-state concepts. Older electrodes had a liquid fill that flowed into the process stream through a liquid junction or a wick. Maintenance difficulties included stoppage of flow due to (1) pluggage of the wick, or (2) high-process pressure. The newer reference electrodes require no electrolyte flow, but contain solid crystals sealed in the reference chamber. The junction to the process is through a porous wood, ceramic or polymer plug. Typically, the process stream diffuses through the plug and dissolves some of the crystals or gel to form a conductive path. The fill must be replaced periodically with this type of electrode. Even newer designs are completely nonflowing and require no liquid interchange between the process and the internal electrolyte.

The relationship between the process pH and the electrical voltage is shown by the simplified Nernst equation:

$$MV = \frac{2.3\ RT}{F} (7 - pH)$$

where   MV = millivoltage generated
      R = universal gas constant
      T = absolute temperature
      F = Faraday's constant

Solution of this equation shows an output of 59.1 mV per pH unit. It also shows a sensitivity of millivoltage to temperature, as well as pH. Thus, temperature compensation is required in pH applications where pH is expected to vary widely, and a compensator is usually incorporated into the pH measuring circuit, as shown in Figure 2. the inaccuracy due to temperature is small near the equalization point, however, and deletion of the thermal compensation should be considered in applications where variations are known to be small.

The dissociation of a chemical into ions in water is described by its ionization constant[2], which is itself a function of temperature. Thus, the pH of a solution changes with temperature because of the effect of temperature on ionization. This phenomenon is distinct from the temperature effect on measurement described by the Nernst equation and, quite obviously, temperature compensation affects only the measurement error.

The electrical potential developed by the measuring electrode is, at most, several hundred millivolts, while the glass membrane has a very high resistance. Amplification is required to provide a signal strong enough for transmission, recording and control. Figure 2 shows the pH electrodes and compensator combined into a unit generally designated as a pH cell, and electrical circuitry for amplification usually called a pH transmitter. The high resistance of the pH electrodes requires that the amplifier represent a very high impedance to the measuring circuit to provide an accurate transmission of the millivoltage generated.

BASIC DATA

Many types of control systems can be designed with a very minimum of basic data since adjustment within the control components are adequate to compensate for any deficiencies. Not so with pH control. Table I shows the extensive basic data required to design even a workable pH control system, where "workable" can be defined as the ability to put the controller on "automatic."

Regulating pH in a chemical process is a defined situation and thus requires somewhat fewer basic data than a wastewater application. Definition implies that pure, known chemicals are being reacted under known conditions. The lack of unknown chemical components results in a single titration curve always describing the neutralization process. Table I shows that while extensive data are required for this situation, it is considerably less than those required for the wastewater application. Conversely, however, the maximum acceptable variation in pH in a chemical process is in the order of ±0.1 pH unit, so great care is indicated in the design of the control configuration.

The neutralization of wastewater, on the other hand, is an almost completely undefined situation. The waste stream is made up of anything, in any amount, that anyone wants to discard anywhere on the plant site. Thus, the dumping of a small amount of strong acid from a laboratory bench must be accommodated along with a massive spill of organic (highly buffered) acid from a unit operation. Yet governmental regulations require that pH be between 6 and 9 as the waste

Table I.   Basic Data for Design of pH Control Configuration[4]

Process pH	Wastewater pH
Titration curve	Titration curves showing maximum and minimum buffering
	Relationship of free mineral acid to pH at strong acidities
pH recordings to show noise and variations	pH recordings to show noise and variations
Magnitude and frequency of known hydrogen ion load changes	Magnitude and frequency of known hydrogen ion load changes
Magnitude and frequency of known throughput flow changes	Magnitude and frequency of known throughput flow changes
	Define periods during which wastewater is already neutral
	Define rainstorm flows
Condition of flowing material (gummy, precipitates)	Condition of flowing material (solid objects, precipitates, gummy, sticky)
	Test results defining mixing in down-stream ditches, basins, lagoons
Define strength of neutralizing reagent and variation in strength	Define strength of neutralizing reagent and variation in strength
Required variation in pH set point	Required variation in pH set point
Maximum permissible variation in pH	Maximum permissible variation in pH

stream is discharged into a natural body of water. Extensive basic data are required to provide adequate regulation under these conditions.

The heart of the basic data is the titration curve. It defines the controller modes, the controller adjustment, valve sizes and whether multiple valves are required. Without it, quantitative design is impossible and even system stability cannot be guaranteed. Definition of the titration curve becomes more difficult in the wastewater application because the curve changes with each chemical in the stream. The neutralization process is thus described by a continuously changing spectrum of titration curves, all the way from "strong base—strong acid" to highly buffered. Extensive effort is required to define the full spectrum. Definition borders on the impossible in an as yet unbuilt plant, and titration curves can be determined only using composite samples of the anticipated waste stream. Control system revisions are almost always required after the conditions become better defined at plant startup.

The characteristics of some chemicals found in industrial reactions do not fit the concepts of ionization inherent in the pH measurement. Figure 3 shows a process titration curve that departs radically from the traditional curve shown in Figure 1. To design to Figure 1 while the process was actually following Figure 3 would result in a very poor performance. Figure 4 shows the variation in titration curve from a single plant discharging only two waste acids. Acid "A" is a strong (highly ionized) acid, while acid "B" is a highly buffered acidic compound.

Load changes in throughput flow and in hydrogen ion concentration both represent disturbances to the pH control system, and both magnitude and time history are required for the design of an adequate system. Due to the logarithmic relationship between hydrogen ion and pH, the pH measurement is more sensitive by a factor of 10 to hydrogen ion variations than to throughput flow. Definition of hydrogen ion variations therefore must be more specific than that of throughput flow. The measurement and recording of wastewater flow and pH as basic data in an existing plant is very attractive economically because it results in the minimization of the overdesign of the neutralization system. Economy is dictated since even a minimum wastewater neutralization system typically will cost hundreds of thousands of dollars.

An often overlooked fact is definition of periods when the flowing stream will be neutral and the control system inoperative. Special consideration is required because control systems do not operate well on a part-time basis, and because pH is so sensitive to loading that it can deviate wildly while the control system is coming online. Difficulties with part-time control include (1) plugging of the typically small control valve part; (2) velocity limiting of the valve plug and stem; and (3) reset windup in the controller.

CONTROL THEORY

While Figure 1 illustrates the notorious nonlinearity of the pH relationship, the concepts of linear mathematics can be used with the limited portions of the titration curve to illustrate the elements of controllability. A basic tool of linear mathematics is the block diagram, made up of blocks such as the following, which transform the incoming information through static gain and time response to the outgoing information:

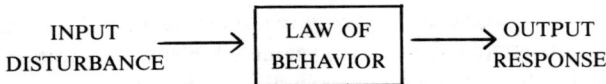

INPUT ⟶ | LAW OF | ⟶ OUTPUT
DISTURBANCE | BEHAVIOR | RESPONSE

The block diagram is used in Figure 5 as a tool for the design of the pH control configuration. The transfer functions (blocks) of Figure 5 can be related to the above generalized block as follows:

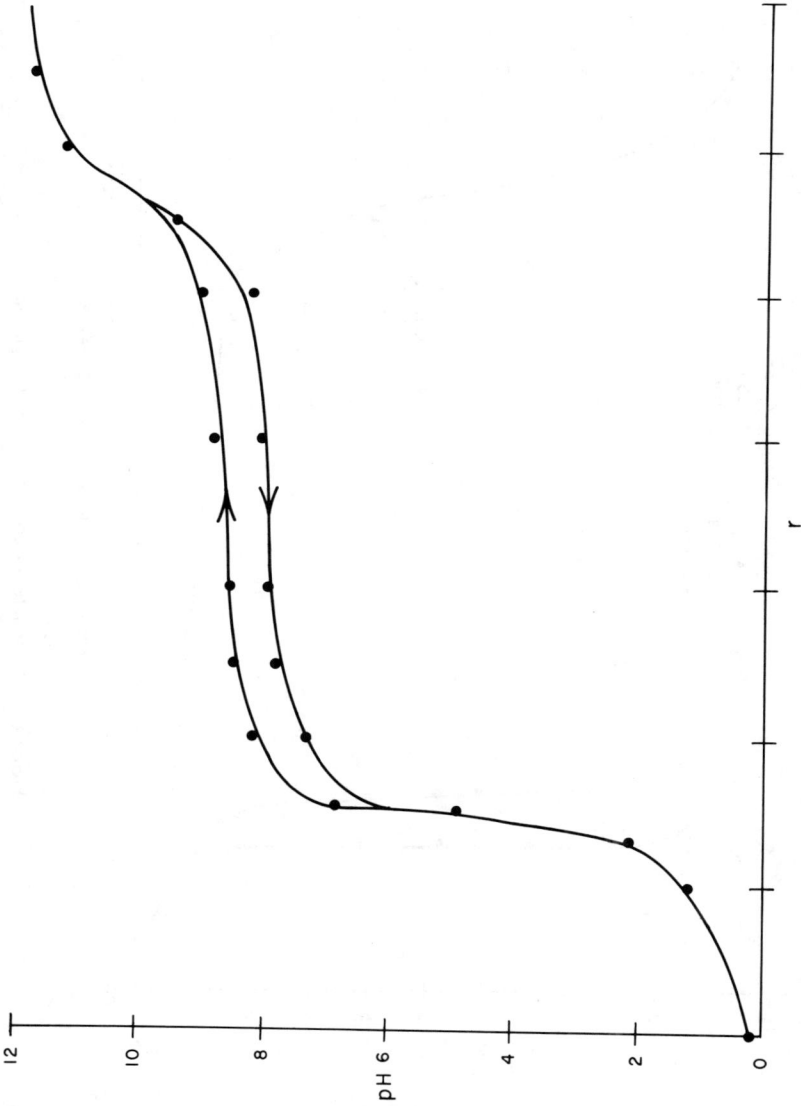

Figure 3.   Titration curve describing chemical process.

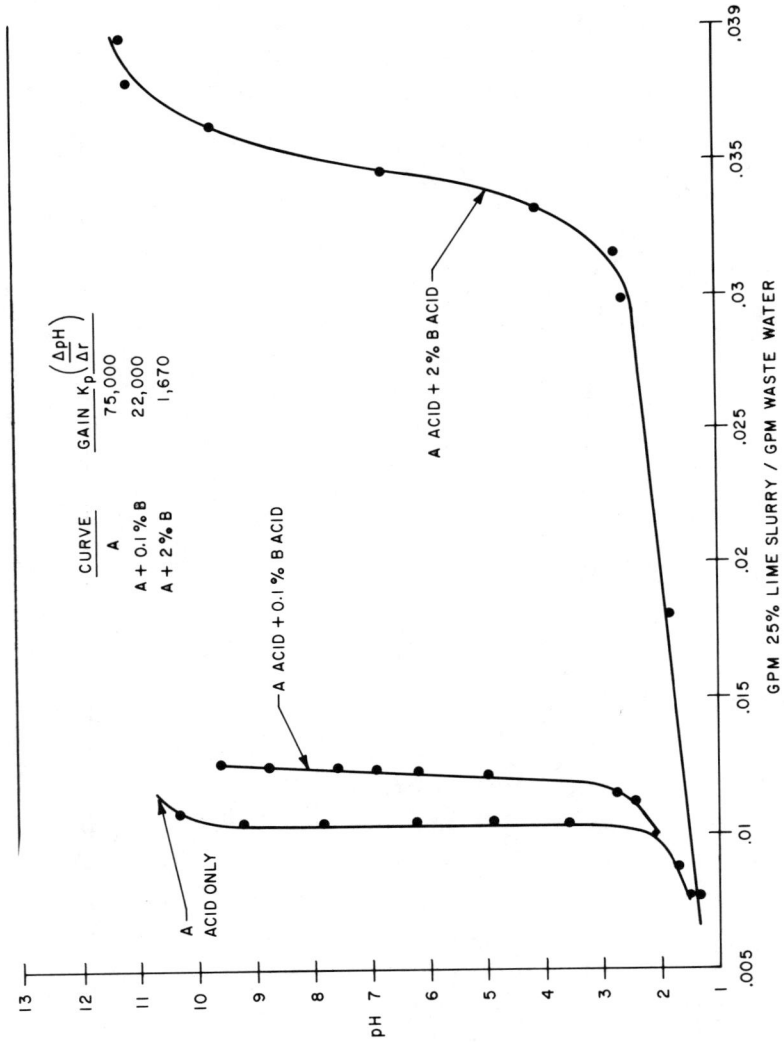

Figure 4.   Wastewater titration curves for a single plant.

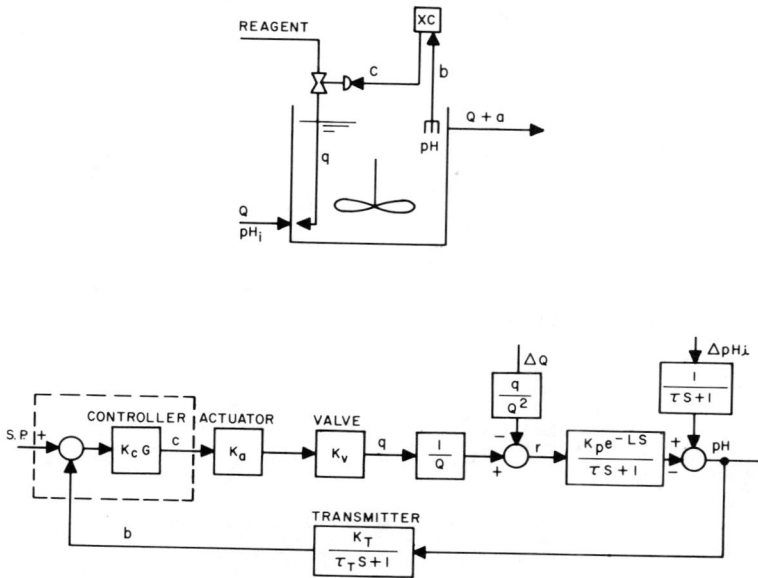

Figure 5.   Diagram of typical pH control configuration and block diagram of configuration shown above.

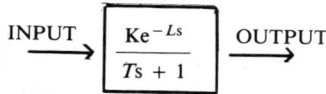

where K = gain, $\dfrac{\text{change in outlet information}}{\text{change in inlet information}}$

T = time constant (units of time)

L = dead time (units of time)

s = Laplacian operator $= \dfrac{d}{dt}$

G = dynamic controller modes, reset and rate

The controlled pH is shown at the measuring device in the instrument diagram on Figure 5 and at the right in the block diagram. Following the arrows, the pH measurement is transmitted to the controller (XC). The controller calculates the error between measured variable and set point, and operates on the error with the controller modes. The controller output (c) goes to the control valve, which sets the reagent flow (q). The ratio (r) in the block diagram is defined as previously:

$$r \doteq \frac{\text{gpm of reagent}}{\text{gpm of process flow}} = \frac{q}{Q+q}$$

And, where q is very small,

$$r = \frac{q}{Q}$$

Linearizing,

$$r = \frac{1}{Q} \left| q - \frac{q}{Q^2} \right| Q$$

The process vessel is characterized by a gain (K), dynamic response (*T*) and dead time (*L*). Dead time is defined as follows:

L = Dispersion time in tank

$$+ \ \frac{\text{Distance upstream to reagent injection (ft)}}{\text{flow velocity (ft/unit time)}}$$

$$+ \ \frac{\text{Distance downstream to pH measurement (ft)}}{\text{flow velocity (ft/unit time)}}$$

The deterioration of control performance due to dead time can be estimated using the efficiency factor "E"[5]:

$$E = e^{-L/T}$$

Since the efficiency refers to the hydrogen concentration or to the ratio "r," and since the pH is extremely sensitive to either variable, it is imperative that efficiency "E" be very high. Thus, dead time must be a small fraction of the back-mixed time constant *"T."* Reducing dead time includes minimizing the distance upstream to the reagent injection point and downstream to the measurement. These distances can be eliminated entirely by making the injection and measurement in the neutralization vessel itself.

The closed loop transfer function is determined from the block diagram:

$$\Delta pH = \frac{\dfrac{1}{Ts + 1} \, \Delta pH_i}{1 + \dfrac{K_c \, K_a \, K_v \ \ 1/Q \ \ K_p \, K_T}{(T_T s + 1) \ (Ts + 1)}}$$

For simplicity, the following assumptions have been made:

G = 1 (a proportional-only controller)
L = 0 (dead time is negligible)

Stability is determined by the denominator or the characteristic equation, which can be simplified to

$$(Ts+1)(T_Ts+1) + K_c K_a K_v \ 1/Q \ K_p K_T$$

The roots of this equation are found by the Pythagorean Theorem:

$$s_1, s_2 = \frac{-(T+T_T) \pm \sqrt{(T+T_T)^2 - 4\,T\,T_T\,(K_0+1)}}{2\,T\,T_T}$$

For overdamped or "dead beat" response,

$$(T+T_T)^2 \geq 4\,T\,T_T\,(K_0+1)$$

where $K_0 = K_c K_a K_v \ 1/Q \ K_p K_T$ = open loop gain.

Simplifying,

$$K_c K_a K_v \ 1/Q \ K_p K_T \leq \frac{T}{4\,T_T}$$

This equation governs the pH control system stability for a small region around the set point. Substantial values of the backmixed time constant (T) are required to attentuate disturbances, as shown in the block diagram in Figure 5 and to permit increasing the open loop gain, as shown in the above equation. A minimum value of 2 minutes is generally required for adequate control performance.

Figure 6 is an expanded portion of the "A acid only" titration curve shown in Figure 4. The process gain (K_p) shown as a part of the block diagram in Figure 3 comes directly from the titration curve and is defined on Figure 6 as the slope of the titration curve at the set point. For pH set points between the "knees" of the titration curve, the process gain is extremely high. Thus, disturbances are amplified, as shown in the block diagram, by a very high gain that makes pH regulation to within small deviations extremely difficult. Further, the high process gain (K_p) necessitates an extremely low controller gain (K_c) to attain stability, as shown in the open loop gain equation.

The low controller gain required for stability requires a controller that is adjustable to the required gain (wide proportional band). A pH control system with a controller that has a proportional band adjustability of 0–300% obviously will be unworkable with the typical requirement of 1000–2000% proportional band. Having attained stability with the wide proportional band, however, the low controller gain results in poor feedback control system response and emphasizes the need for well-mixed volume in the control system to increase the process time constant (T), thus permitting an increase in open loop gain and in controller gain.

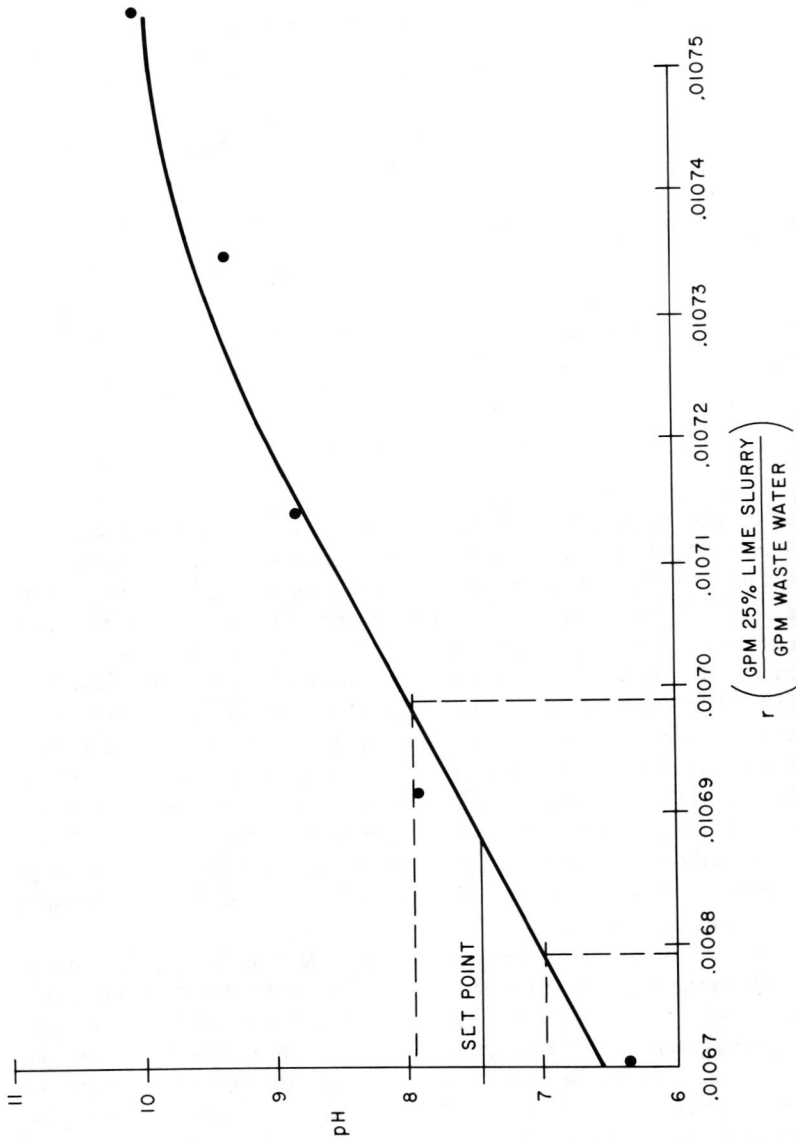

Figure 6.   Expanded portion of "A" acid titration curve shown in Figure 4.

Reference to Figure 6 shows that the strength of the reagent used works through the ratio "r" to have an effect on the process gain. Thus, process gain K_p can be reduced by reducing the strength of the reagent. However, reducing reagent strength generally will require increasing the control valve size to deliver more of the weaker reagent, which, in turn, increases the valve gain (K_v). Little advantage accrues from the tradeoff between K_p and K_v in terms of the open loop gain. The gain combination,

$$K_v K_p$$

does become quite small in a system neutralizing a very weak hydrogen ion concentration, and such systems are easily controlled.

A final reason for judiciously choosing the value gain–process gain product is to raise reagent demand sufficiently to avoid an extremely small valve that operates in a nearly closed position. The valve can be made larger by using a weaker reagent.

CONTROL VALVES

The control valve for pH control is chosen based on the criteria of resolution, rangeability and minimum change in open loop gain.

The resolution criteria is shown on Figure 6. The control valve used in regulating pH must have a positioner. Even with a positioner, however, the valve will position with an accuracy of some ±0.2% of total stroke. The uncertainty in valve positioning produces an uncertainty in the pH resulting from a given valve stem position. Suppose that an acceptable pH uncertainty is ±0.5 pH units. Then, from Figure 6,

$$\Delta r = \frac{\text{Uncertainty}}{K_p} = \frac{1}{75,000} = 0.000013$$

And for a typical wastewater flow of 5000 gpm,

$$\Delta q = Q\Delta r = (5000)(0.000013) = 0.067 \, gpm$$

And for a valve with a linear flow characteristic,

$$q = \frac{\Delta q}{0.004} = \frac{0.067}{0.004} = 16.67 \, gpm$$

The equal percentage valve characteristic is often chosen for control valves in pH control systems for rangeability considerations. Valve stem movement is much larger for small flows than the stem movement of a valve with a linear characteristic. Also, since valve gain (K_v) is proportional to reagent flow, valve gain is low at low flows, helping to compensate for the high process gain (K_p). The foregoing analysis can be undertaken using the equal percentage characteristic relationship and, with some increased complexity, the same results can be attained.

Use of the equal percentage control valve flow characteristic introduces a danger due to its nonlinearity. Referring to Figure 4, the "A acid only" curve can be moved horizontally, retaining exactly its same shape by changing the concentration of the incoming "A" acid. It is probable that sooner or later the acid concentration will be such as to require the control valve to be almost completely open. Since gain (K_v) is proportional to reagent flow, valve gain will be very high. The product of a high valve gain and a high process gain will cause the control system to become unstable and to perform poorly. For this reason, linear valve characteristics are used in pH control valves where hydrogen ion loading is unconstrained.

The foregoing calculation shows that a control valve with a maximum capacity of 16 gpm of reagent is required to meet the resolution requirements of the titration curve. However, Figure 6 shows a neutralization demand of 54 gpm to bring pH to a value of 7.5. Thus, the valve chosen on the basis of resolution does not have adequate capacity, and multiple valves are required. Similarly, a range of operating conditions sometimes can be identified that will demand a wider range of reagent flows than can be accommodated by the rangeability of a single control valve. Hoyle[6] has shown the probable necessity of multiple valves using the logarithmic pH relationship, which demands one unit of reagent to neutralize from a pH of 6 to 7, but 10,000 units to neutralize from 2 to 7.

Multiple valves can be actuated either by (1) split ranging, or (2) connecting the small valve to the pH controller and the large valve to a floating controller, which sets the position of the small valve. Split ranging has the advantage of simplicity, but the disadvantage of a discontinuity at the point where the smaller valve is fully open and the larger valve begins to open. Shinskey[7] has demonstrated a technique for matching the gains of the small- and large-split ranged valves at the switching, thus avoiding an abrupt change in open loop gain as the valves are switched. However, gain matching requires the use of equal percentage valve characteristics, with the attendent disadvantages discussed above.

Figure 7 shows the second configuration for multiple valves. The small valve is actuated by the pH controller and represents the advantage of a low valve gain (K_v) in that control loop. The floating controller has as a measured variable the signal to the small valve, or the small valve position. The floating controller is thus called a position controller. Its set point is then set at 50–70% of scale, and it will move the large valve until the small valve reaches the set point position.

The floating controller is replaced by a gap action controller if (1) the large valve cannot meet the resolution requirement of the titration curve, or (2) the vessel volume is such that the process time constant (T) is less than 1 minute. The large valve remains in a constant position while the position of the small valve is

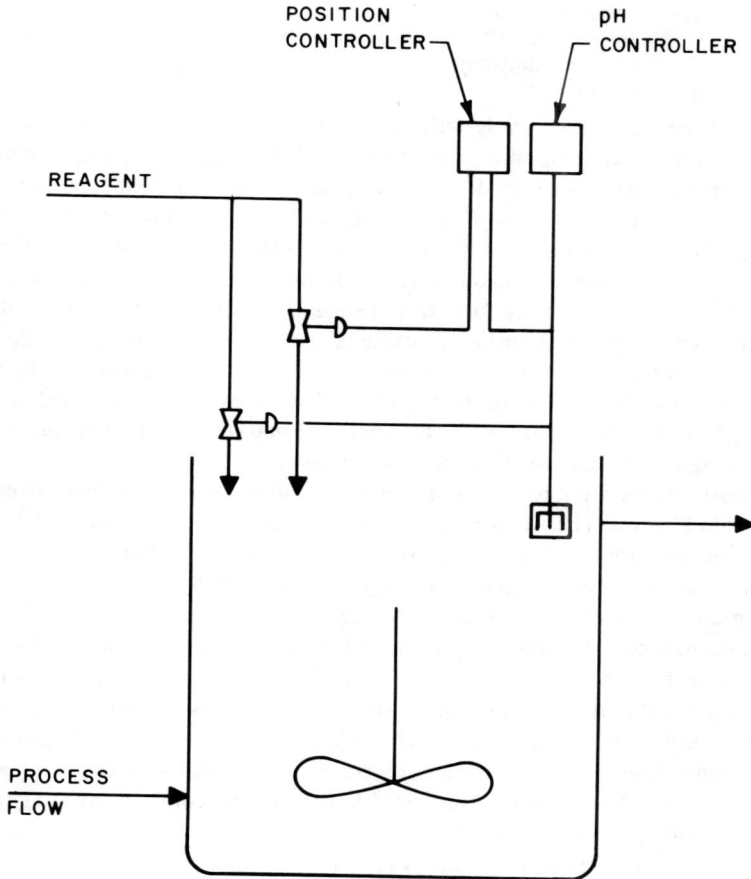

Figure 7.   Feedback control system with small control valve modulated by pH controller for resolution and large valve positioned to keep small valve in range.

within the gap of the controller. Performance has been found to deteriorate if the position controller is replaced by a ramping relay.

NEUTRALIZATION VESSEL AGITATION

The process time constant shown in the foregoing block diagram is defined:

$$T = \frac{V}{Q+q}$$

where T = time constant (min)
         V = vessel volume (m^3)
         Q = process flow (m^3/sec)
         q = reagent flow (m^3/sec)

The process time constant is derived from a mass balance on the reagent[8] and depends on the assumption of perfect mixing. Perfect mixing is unattainable in practice, thus any vessel is made up of stratified areas (which do not contribute to volume), time for dispersion of the reagent (dead time) and backmixed volume (proportional to the time constant). Energy is added to the vessel to minimize stratification and the dispersion time, and to maximize the backmixed volume. Energy can be added in the form of pressurized air for spargers, pressure drop across a mixing jet, an agitator or head loss across baffles (in a ditch or pond). Agitation in a well-dimensioned container is generally the most energy efficient. The less energy added, the less predictable will be the mixing. Excessive churning is not necessarily good mixing. Totally unmixed ponds and ditches sometimes mix surprisingly well, and sometimes do not mix at all.

Vessel configuration to accomplish both reagent dispersion and backmixing is shown in Figure 8. The aspect ratio of the cylindrical vessel is essentially "1." The ratio can decrease to 0.7, with little loss of mixing efficiency. Mixing in other configurations, e.g., very shallow or rectangular, cannot be determined analytically and must be assessed by experimental modeling.

Reagents tend to be heavy chemicals, such as 93% sulfuric acid (SG = 1.84) and 50% sodium hydroxide (SG = 1.53). Should the reagent be substantially heavier than the liquid in the vessel, it will not disperse well, but will settle to the bottom of the vessel. One technique for dispersion is to point the reagent pipe upstream, as shown in Figure 8, and force the reagent to turn around with an umbrella effect. Another is to bring the reagent to within a few inches of the rotating agitator blades. Adding the reagent to the free surface of the liquid is almost always unsuccessful since it will short circuit by following a streamline to the exit.

The incoming throughput flow ideally comes in the bottom of the tank, beneath the agitator. The agitator pumps down to avoid spray in the top of the vessel. Thus the throughput flow must flow up through the downpumping agitator. Baffles prevent rotation of the vessel contents. Placement of inlet and outlet 180° apart minimizes short circuiting around the agitator in the direction of rotation.

While mixing is predictable in a vessel such as shown in Figure 8, process conditions and economics often dictate vastly different conditions. The regulation of pH, on the other hand, depends on the dispersion and the backmixed time constant. Frequently, poor control is due to poor mixing. With an operating process, mixing efficiency must be determined experimentally. Mecklenburgh and Hartland[9] recommend the following steps in defining backmixing:

1. Refer to literature.
2. Analyze system geometry and flow patterns.
3. Form a concentration profile by steady injection of tracer and unsteady injection of tracer.

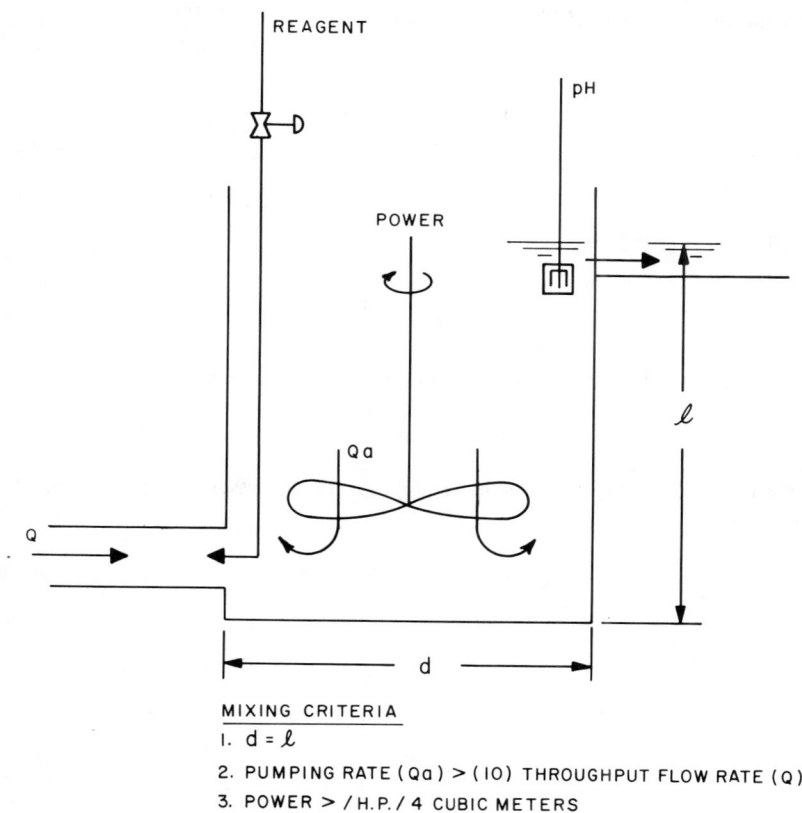

MIXING CRITERIA

1. $d = \ell$
2. PUMPING RATE (Q_a) > (10) THROUGHPUT FLOW RATE (Q)
3. POWER > / H.P. / 4 CUBIC METERS

Figure 8.   Diagam of vessel configuration showing inlet/outlet orientation, reagent injection, measurement location, agitator action and mixing guidelines.

Shinskey[2] has formulated the following relationship to describe the mixing performance of the vessel shown in Figure 8:

$$\frac{T_d}{V/Q} = \frac{1-\alpha}{2}$$

where T_d = dead time (dispersion time), min
    V = vessel volume (m³)
    Q = throughput flow (m³/min)
    Q_a = agitator pumping flow (m³/min)

and

$$\alpha = \frac{Q_a}{Q + Q_a}$$

From the guidelines of Figure 8,

$$\frac{Q_a}{Q} = 10$$

Then,

$$\alpha = 0.9$$

and

$$\frac{T_d}{V/Q} = 0.05$$

This relationship results in little loss in control efficiency[8]:

$$E = e^{-T_d/V/Q} = e^{-0.05} = 0.95$$

Since vessel throughput (Q) is usually specified, vessel volume (V) can be determined:

$$\text{Residence time} = \frac{V}{Q} = T + T_d$$

$$\frac{V}{Q} = T + .05\,\frac{V}{Q}$$

$$0.95\,\frac{V}{Q} = T$$

Thus, volume (V) can be chosen to provide an adequate backmixed time constant (T).

CONTROLLER

The process controller was discussed as a part of the block diagram of the pH control system and was shown to provide an adjustable gain and compensating dynamic functions in the system. Controllability was shown to depend largely on the titration curve. Reference to the curves in Figure 4 indicates that the difficulties in pH control include

- very high process gain (K_p) at neutrality,
- nonlinear gain throughout the span of pH,
- variable buffering, and
- wide range of reagent flows.

Control theory showed the need for wide adjustability of controller gain or proportional band. However, having chosen a proportional band that results in satisfactory response at neutrality, open loop gain will vary with the nonlinearity of the titration curve (as denoted by process gain K_p). The phenomenon of varying response is observed when pH varies from the high gain portion of the titration curve (around neutrality), past the "knee" of the curve, and to the highly insensitive "tail" of the curve near the end of the pH measurement span. As pH passes the "knee" of the curve, process gain goes from a very high value to a very low value, as shown in Figure 4. Open loop gain also becomes a very low value, and control system response becomes highly overdamped and sluggish. The low gain results in the control system being unable to bring pH back past the "knee" of the titration curve, and pH will "hang" in the low gain region for long periods of time (15 minutes to 1 hour). Thus, the system will exhibit a very high frequency response to small deviations in pH and a very low frequency response to large deviations[8].

Modern nonlinear controllers can provide both wide proportional band adjustability and compensation for the nonlinear titration curve. One nonlinear controller has the segmented gain shown in Figure 9[10-12]. The gain of each of the three segments is adjustable, as is the location of the two breakpoints between the segments. Comparing Figures 4 and 9, the breakpoints of the controller gain would be set at the "knees" of the titration curve in Figure 4. The ratio of gains at the breakpoints is set equal to the ratio of titration curve slopes at the "knees." The controller is then tuned by standard procedures while pH is in the sensitive (center) portion of the titration curve.

A more sophisticated controller uses both a continually calculated gain based on a single variable or a number of independent variables and segmented gain based on the output of the controller or a single contact[13,14]. Figure 10 shows controller gain variation as a function of a chosen external variable (process measurement, deviation from set point, or other remote input), while Figure 11 shows gain adjustability in response to an external contact closure. This controller is characterized by extreme flexibility, and compensation for the nonlinearity, sensitivity and valve rangeability in the pH application is limited only by the imagination of the user[15].

Still another technique for compensating the logarithmic pH measurement is a controller that is programmed with a series of titration curves, the most appropriate of which for compensation can be activated by a selector switch on the controller[16]. The choice of titration curves for which the controller will compensate and provide a linear measurement is shown in Figure 12. Different curves can be selected above and below neutrality, thus accomodating the unsymmetrical titration curve. This same controller unit can be used as a feedforward compensator in linearizing a feedforward pH and multiplying the pH and flow measurement signals.

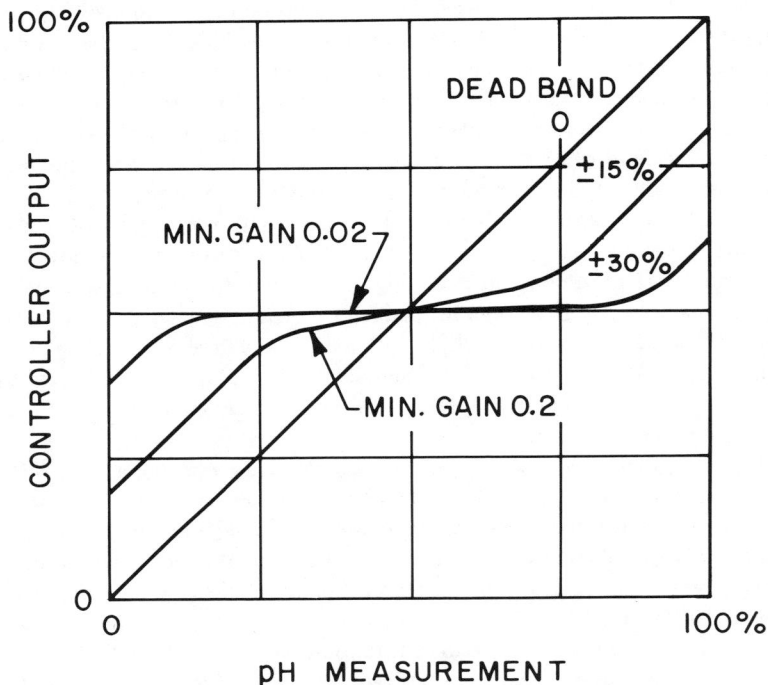

Figure 9.   Nonlinear controller with segmented gain. Breakpoint adjustment is shown as percentage measurement span. Gain within dead band is adjustable by a multiplier on controller span.

ADAPTIVE CONTROL

While the foregoing controllers offer flexibility sufficient for process pH control, wastewater pH control must cope with a continuously changing chemical mix. The result of mixing only two chemicals in a wastewater stream is shown in Figure 4. The typical petrochemical plant wastewater stream will contain many more varieties, and titration curve variation would be expected to be much more extreme. With the two chemicals in Figure 4, however, a gain variation at neutrality of 46 is observed, and a rangeability of 10 required. Open loop gain has been shown to vary directly with process gain; thus, a change in open loop gain of 46 is indicated. A change by a factor of 3 in open loop gain will cause control system response to go from oscillatory to overdamped, and performance deteriorates greatly with a factor of 46.

An adaptive controller that changes its gain with the changing titration curve is required. Concepts have been advanced[17,18], but no proven technology is avail-

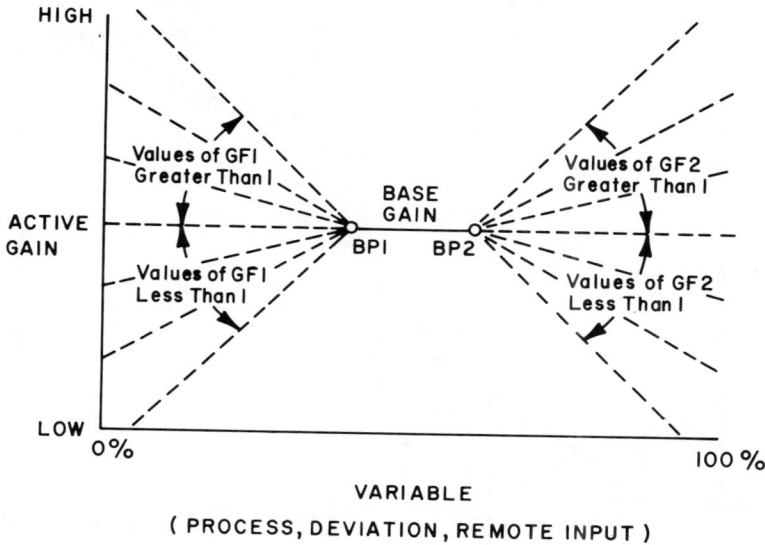

Figure 10. Nonlinear controller with segmented gain adjustable by an external signal. Note that this is a plot of gain, not input/output relationship. GF = gain factor or multiplier.

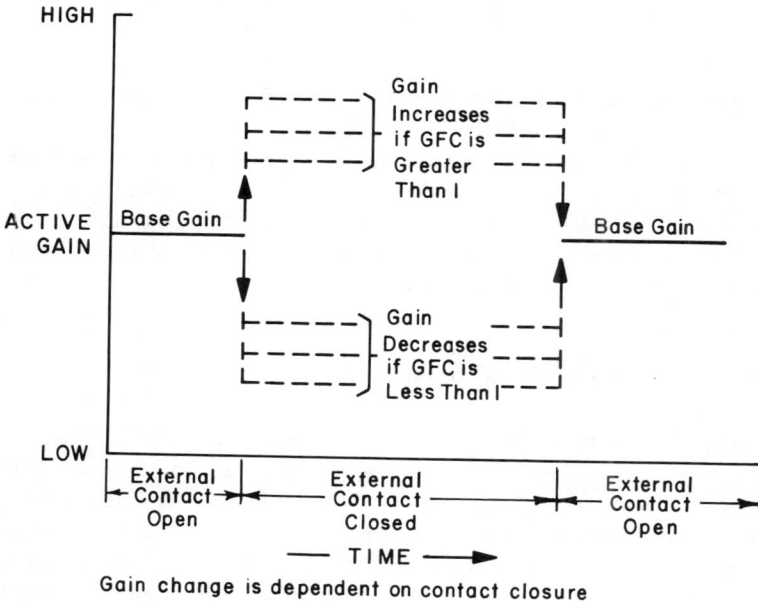

Gain change is dependent on contact closure

Figure 11. Nonlinear controller with segmented gain adjustable by an external contact closure.

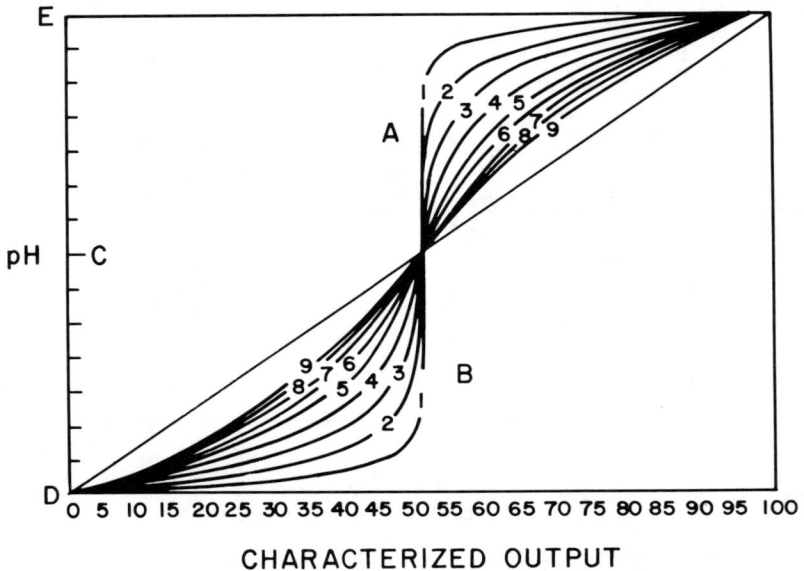

Figure 12.   Nonlinear controller with characterization, which can be chosen to compensate for titration curve nonlinearity. And curve 1 to 9 can be chosen in both A and B segments.

able to provide the adaptability. Sagacious application of the control elements described above is required for each application.

Figure 4 also illustrates that curve identification is a major stumbling block in adaptive control. Identification by measuring inlet pH is very difficult because major changes in buffering can take place (as shown), with little change in incoming pH. Given curve identification, modern microelectronics could easily provide the gain change based on the external signal.

FEEDFORWARD CONTROL

Williams[19] has delineated the advantage of feedforward, shown in Figure 13 for a pH process, as the ability of the upstream measurement to adjust the control valve before the disturbance even reaches the process vessel, thus minimizing the variation in the controlled variable in the process itself. Remarkable improvements in regulation have been attained in well-defined chemical unit operations, such as distillation. However, the nonlinearity inherent in process pH and the variation in the titration curve in wastewater neutralization represent significant difficulties in the feedforward of pH.

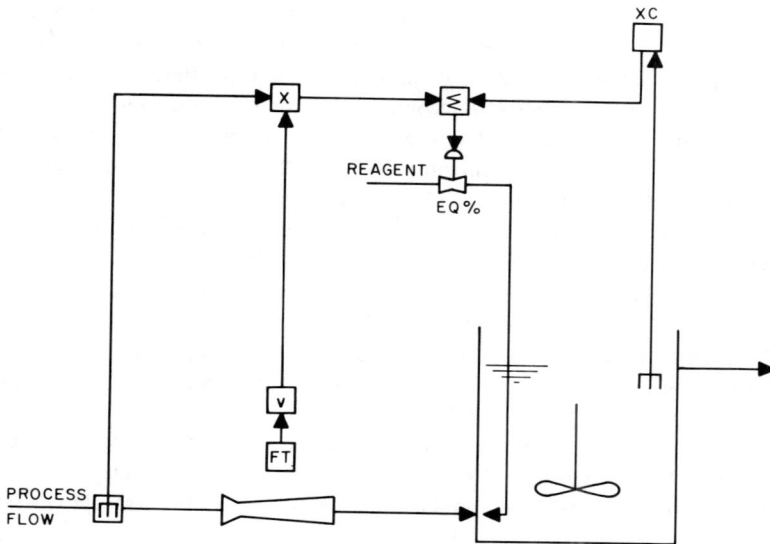

Figure 13.   Feedforward-feedback configuration for pH regulation with the logarithmic pH feedforward signal matched to the control valve equal percentage (logarithmic) characteristic, and the cubic flow signal approximately matched to the equal percentage characteristic.

Figure 13 shows a feedforward concept based on the titration curve, i.e., if process flowrate and component concentration are known, the amount of reagent required to bring the process to another pH can be calculated. The pH (concentration) is multiplied by flowrate to result in valve position (reagent addition). Reference to experimental titration curves in Figures 3 and 4 illustrates the difficulties. First, the logarithmic nature of the titration curve results in a very small change in pH at high or low values (11–13 pH units or 1–3 pH units), requiring a large change in reagent flow to maintain neutrality. Thus, inlet pH must be read very accurately. However, pH sensors have a history of drifting and fouling, which makes accuracy very difficult to maintain. The reliable measurement of small changes in inlet pH to actuate the feedforward system is a challenging task.

Second, both pH and (normally) flow are nonlinear measurements. Williams[19] has developed an instrument configuration that minimizes computational elements by matching the logarithmic titration curve with the logarithmic equal percentage valve flow characteristic. While conceptually a notable development, both the titration curve and the valve characteristic are logarithms to bases varying all the way from 3 to 10; thus, a partial match at best is attainable. Further, this configuration require the use of an equal percentage valve for which a penalty is paid in the feedback control system (as discussed previously).

An alternative to transmitting the nonlinear signals is to linearize the measurements. An antilog computing element has been used to linearize the pH-transmitted signal, but suffers from the variation in the logarithmic base described above. Linearization can also be accomplished with a function generator with a variable step size to accommodate the sensitive and insensitive portions of the titration curve. Finally, the controller shown in Figure 12 can be adapted to linearize both the flow and pH feedforward signals.

The varying titration curve describing wastewater neutralization in Figure 4 is an additional obstacle to operable feedforward. Feedforward is a calibrated system. Quite obviously, a feedforward system calibrated to the "A" acid only titration will provide for too little acid for conditions where the "A" acid + 2% "B" acid prevails. Likewise, calibration to the latter curve will provide too much reagent when the former curve describes the process. The feedforward can very possibly be detrimental to pH controllability under these conditions.

Application of pH feedforward control requires basic data (Table I) and a bit of sagacity. If the feedforward system can be shown to compensate reliably for 80% of the load (move the valve in the right direction 80% of the travel to its final position), it can be considered to be a worthwhile application.

REAGENTS

Petrochemical processes commonly use pH to bring the chemical reaction to a desired endpoint. No differentiation exists between process flow and reagent in this case. Many other processes add acidic or basic reagents to displace an unwanted molecule in the chemical structure, or to cause a precipitate to form and settle out. These processes will be controlled by the pH of the fluid, and a reagent will be added to adjust the pH. The reagents used commonly will be sulfuric acid, sodium hydroxide or caustic (which are widely available), or lime (which is most economical).

Lime is a solid taken from quarries as calcium oxide (a dry solid), which can be hydrated to calcium hydroxide (still a dry solid). It serves as an illustration of the difficulties of a reagent in this form. They are:

- difficulty of handling as a solid,
- a very weak base when dissolved in water,
- necessity of use as a slurry,
- particle settling, and
- reaction time, including mass transfer.

Lime is usually (but not always) delivered to the user as a dry solid. Since lime dissolved in water is a very weak base, it is used as a slurry for neutralization. Major equipment is required to store the dry lime and slake it to a slurry of the desired concentration. Since the equipment to handle the lime represents a significant fixed cost, the economies in the use of lime require that the stream being neutralized be large enough to justify this cost and still show a net saving on the lower purchase cost of the lime.

Lime has a significant reaction time as contrasted to other chemicals used as reagents, e.g., caustic or sulfuric acid, which react almost instantaneously. What is loosely called reaction time is made up of the actual chemical reaction rate coefficient plus the mass transfer time for the solid particles into the liquid. Neutralization process residence time must be made larger to provide time for the lime to react. The fractional proportion of the lime reacted is shown by

$$l = \frac{x_r r}{1 + T_R/T}$$

where $l = \dfrac{m^3/\text{sec reacted lime}}{m^3/\text{sec process flow}}$

x_r = concentration of lime slurry, fraction

$r = \dfrac{m^3/\text{sec lime slurry flow}}{m^3/\text{sec process flow}}$

T = residence time of process between lime injection and pH sensor, time

T_R = reaction time constant of lime, time

The equivalent lime reaction time constant (T_R) is shown in the experimental response curves of Figures 14–19. Referring to the above equation, note that only one-half of the lime will react if the process residence time is equal to the reaction time constant. Unreacted lime will settle as a sludge in downstream equipment and will continue to react, raising pH randomly through each operation. Finally, the filling of lines and vessels will require removal of the sludge. The lime will also collect on the electrodes of the pH cell, necessitating frequent cleaning. Electrode coating is difficult to detect without removal and inspection since coating slows the response and makes pH control appear much better.

The reaction time constant of the lime varies with several parameters: (1) the type of lime (dolomitic or high calcium); (2) time in the kiln (overburning); (3) the amount of energy added in grinding and slaking; and (4) the geographical location of the quarry. Figure 15 shows a time constant variation of a factor of four with slaking time. Actually, an optimum slaking-slurry storage time exists, wherein the reaction time shortens with the agitation in the slaking tank, further breaking up the lime particles, but lengthens again during storage as the particles agglomerate and grow larger.

Figure 16 shows a comparison between the reaction times of high calcium and dolomitic lime. Figure 17 shows a significantly slower reaction rate for a dolomitic lime from another section of the country. Note that the curve ending at a pH of 3.45 is entirely in the nonlinear portion of the titration curve, thus giving

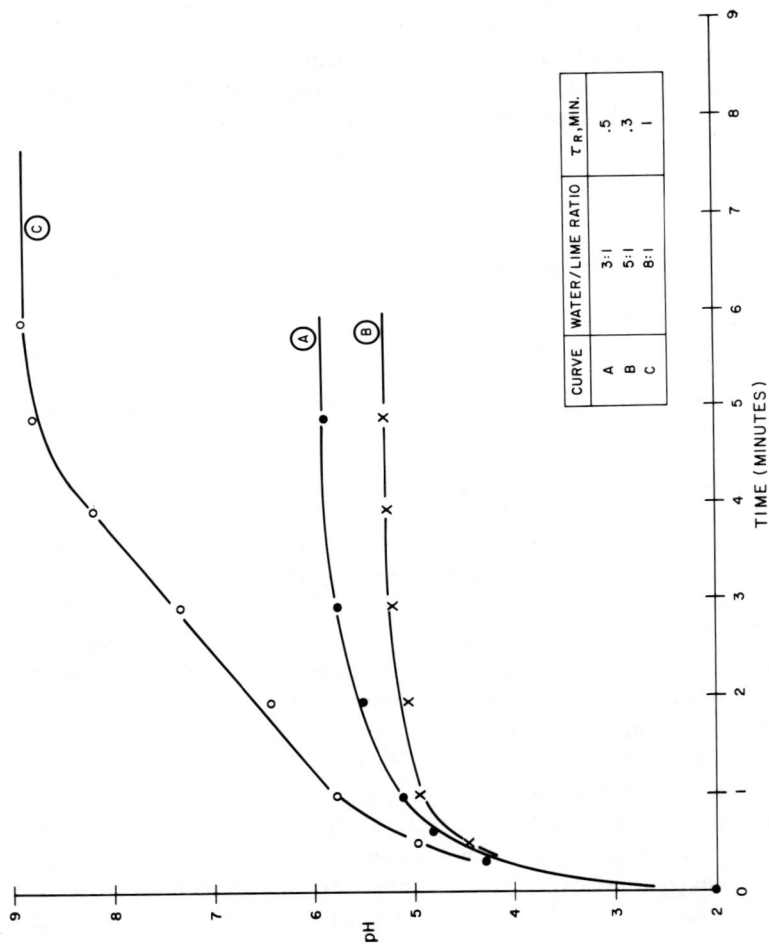

CURVE	WATER/LIME RATIO	T R,MIN.
A	3:1	.5
B	5:1	.3
C	8:1	1

Figure 14.   Variation in lime reaction time constant with slurry concentration.

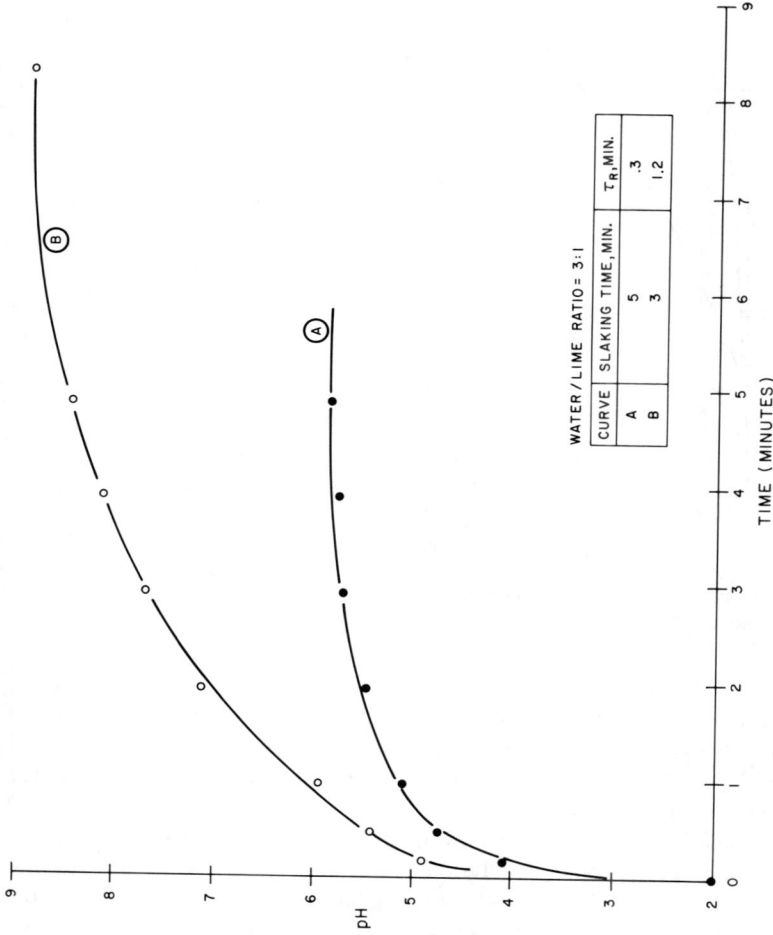

WATER / LIME RATIO = 3:1

CURVE	SLAKING TIME, MIN.	τ_R, MIN.
A	5	.3
B	3	1.2

Figure 15. Variation in lime reaction time constant with slaking time.

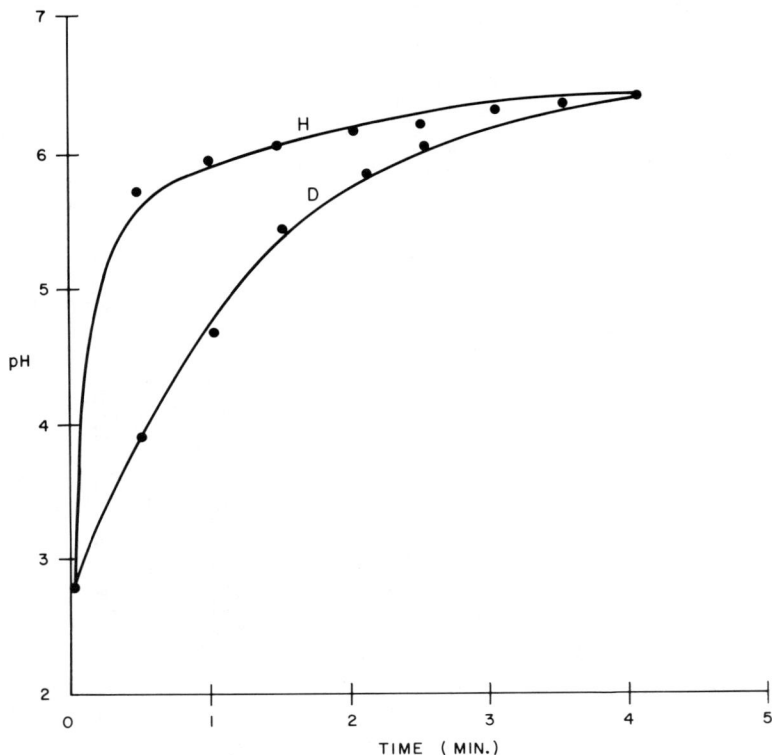

Figure 16.   Comparison of lime reaction times for high calcium lime (H) and dolomite lime (D).

misleading results when described by a time constant (a linear concept). Figure 18 illustrates the reaction variation in lime purchased with identical specifications. Finally, Figure 19 shows the reaction rate changes and different techniques for addition to the process.

The volume of a lime process must provide sufficient residence time for the lime reaction to go to completion for the slowest reaction rate. Otherwise, the pH sensor will measure a pH prior to the completion of the reaction and will call for too much lime, resulting in a final pH above set point. Since the volume must be mixed to keep particles in suspension, the volume required to accommodate dry lime addition is prohibitively expensive. Conversely, the larger volumes required for lime reaction permit improved regulation by the pH control system.

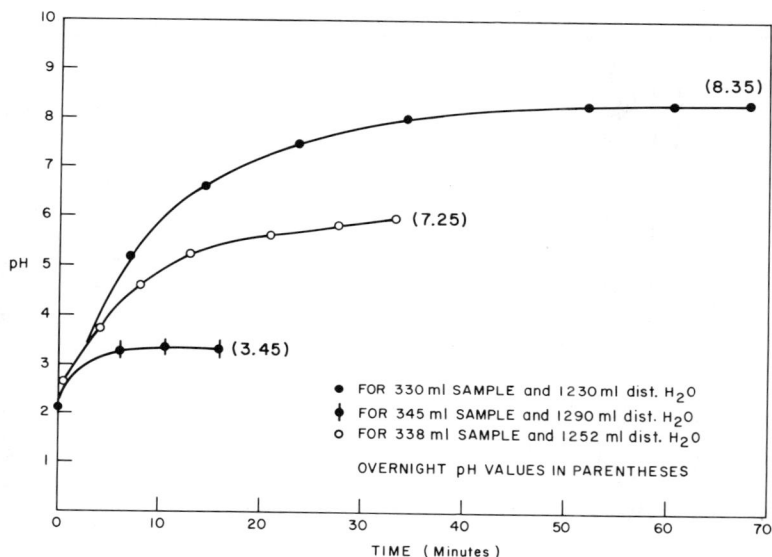

Figure 17.   Reaction rate curves using 900 mg dolomite lime and 90 ml H_2O slurried overnight and added to diluted volumes of sample.

BATCH NEUTRALIZATION

Process neutralization can be carried out in vessels that are a scaleup of the laboratory beaker and burette used in determining the titration curve. Batch neutralization is economically attractive if (1) the upstream process is operated in a batch mode, or (2) the process stream is 6 gpm or less. Since the value of the ratio "r" is very small (see Figure 3), the reagent addition equipment must add extremely small amounts for small process flows. Thus, batch neutralization is mandatory when the reagent handling equipment becomes smaller than is commercially available.

For flows larger than 6 gpm, mixed volumes for the required retention become prohibitively expensive. The batch tank must be large enough to provide time to fill, to be neutralized and to be dumped. Further, a second tank must be provided to fill while the first tank is being neutralized. Finally, a third tank is desirable for the contingency that one of the other tanks cannot be neutralized and dumped in the prescribed time.

Manual batch neutralization is extremely labor intensive. Flows have to be diverted, valves operated and small increments of reagent added at long incre-

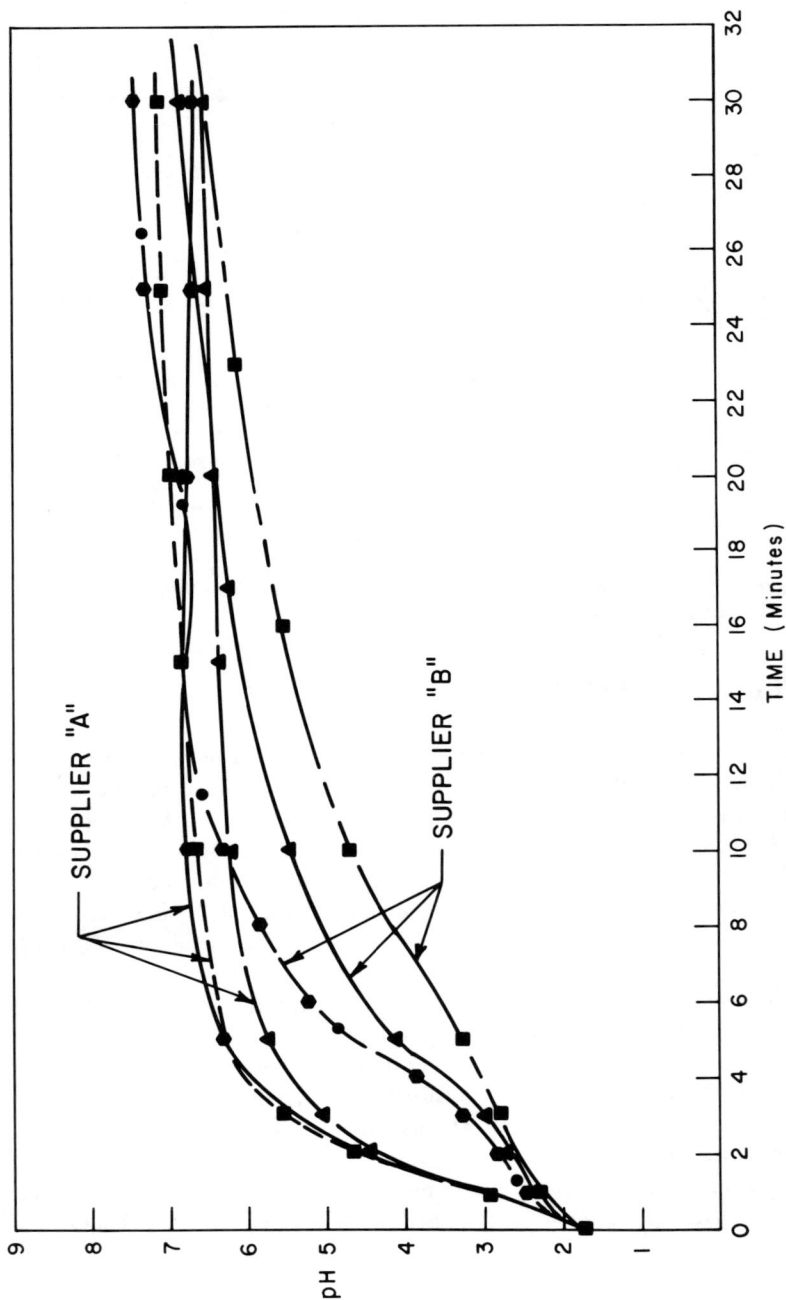

Figure 18. Comparison of reaction rates of dolomitic lime purchased under identical specifications and added to wastewater by identical procedures.

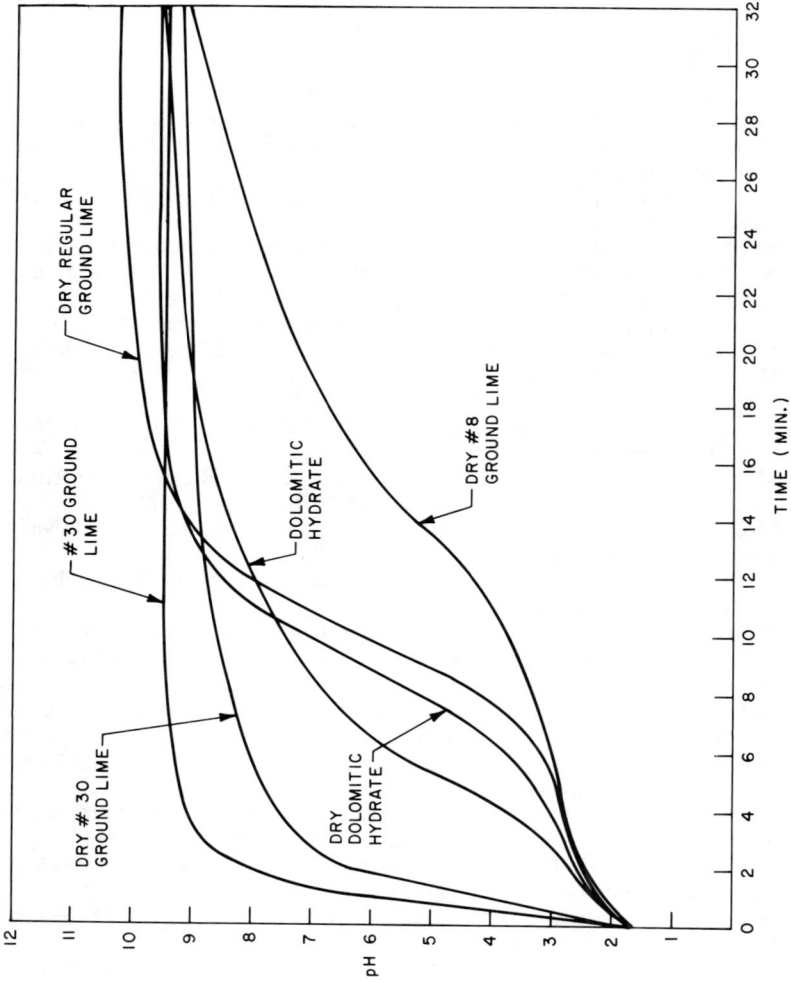

Figure 19. Comparison of reaction rates of dolomitic lime purchased from a single supplier under different specifications and added to process by different techniques.

ments of time. Thus, economics demand an automatic batch neutralization. Batch controllers are essentially automatic titrators, but have additional contacts that open and close dump valves when the chosen logic (tank full, pH at set point) is satisfied. The Luft controller[21,22] is commercially available for this service, while custom controllers can be constructed using a microprocessor[23].

REFERENCES

1. Moore, R. L. "Basic Instrumentation Lecture Notes and Study Guide—Measurement Fundamentals," Instrument Society of America, Research Triangle Park, NC (1976), p. 282.
2. Shinskey, F. G. *pH and pION Control in Process and Waste Streams* (New York: John Wiley & Sons, Inc., 1973).
3. Smith, D. E., and F. H. Zimmerli, "Electrochemical Methods of Process Analysis," Instrument Society of America, Pittsburgh, PA (1972), p. 52.
4. *Bulletin,* AMSCOR, 2512 North Velasco, Angleton, Texas.
5. Van der Grinten, P. M. E. M. "Control Effects of Instrument Accuracy and Measuring Speed," *ISA J.* 48 (December 1965); 58 (January 1966).
6. Hoyle, D. C. "The Effect of Process Design on pH and pION Control," *Proc. 18th ISA-AID Symp.,* San Francisco, CA (1972).
7. Shinskey, F. G. "Sequencing Valves to Increase Range," *Instr. Control Syst.* 9 (November 1971).
8. Moore, R. L. "The Neutralization of Wastewater by pH Control," Instrument Society of America, Research Triangle Park, NC (1978).
9. Mecklenburgh, J. C., and S. Hartland. *The Theory of Backmixing* (New York: John Wiley & Sons, Inc., 1975), p. 33.
10. Beckman Instruments, Inc. "The Beckman Non-Linear Controller," *Process Control Data Bull.* 03–012.
11. Fisher Controls. "Type TL 107 Non-Linear Controller," *Bull.* 1:1:TL 107.
12. Foxboro Co. "Non-Linear Control Unit," Instruction MI2AC–116 (November 1977).
13. Andreiev, N. "A Process Controller that Adapts to Signal and Process Conditions," *Control Eng.* (December 1977).
14. Sybron/Taylor Corp. "Instructions for Micro-Scan 1300 Indicating Controller," Instruction Manual No. 1B–11B520 (May 1978).
15. Hausman, J. F. "Adaptive Gain Control Applications," paper presented at the Texas A&M Symposium, College Station, TX, January, 1979.
16. Leeds & Northrup Co. *7081 pH Microprocessor Receiver/Controller,* Operator's Manual No. 277128 (1979).
17. Shinskey, F. G. "Adaptive pH Controller Monitors Non-Linear Process," *Control Eng.* 47 (February 1975).
18. Gupta, S. R., and D. R. Coughnowr. "An Analog Adaptive Controller For On Line Identification and Control of First-Order Process with Time-Varying Gain," *ISA Trans.* (1): 9 (1974).
19. Williams, E. T. "pH Neutralization Using Feedforward Control," *ASCE,* Chicago, IL (1969).

20. Creason, S. G. "How to Choose a pH Control System," *Instr. Control Syst.* 25 (September 1975).

21. Luft, L. "Two Set Point Automatic Control," *Instr. Control Syst.* 138: 117 (1965).

22. Luft Instruments, Inc. "Control Any Variable, Automate Any Variable, with Luft[R] Master Controller," Lincoln MA.

23. Fraade, D. J. "Use of a Microprocessor for Solving pH Control Problems," *ISA Nat. Conf. and Exhibit,* Paper No. 78–836, Philadelphia, PA, 1978.

CHAPTER 3

LUMINESCENCE SPECTROMETRY

Tuan Vo-Dinh

Health and Safety Research Division
Oak Ridge National Laboratory
Oak Ridge, Tennessee

ORIGIN AND NATURE OF LUMINESCENCE

Basic Processes

Luminescence spectroscopy is based on the detection of the electromagnetic radiation emitted from a chemical system undergoing a radiative transition between different electronic states.

The Jablonski diagram in Figure 1 schematically illustrates the basic processes. In this figure, $S_0, \ldots, S_n$ and $T_1, \ldots, T_n$ represent the electronic energy levels of a molecule (or an atom). The state of lowest energy, S_0, is called the ground state. $S_1, \ldots, S_n$ and $T_1, \ldots, T_n$ are the excited singlet and triplet states, respectively. Each state is shown with its own set of vibrational levels. Some characteristics of the singlet and triplet states are discussed in further sections. When a molecule absorbs excitation energy, it is elevated from S_0 to some vibrational level of one of the excited singlet states in the manifold $S_1, \ldots, S_n$. It deactivates rapidly to the lowest vibronic level of S_1 via internal conversion (IC), releasing thermal energy to the surrounding medium. This vibrational relaxation occurs within 10^{-13}–10^{-11} seconds.

From S_1, the molecule returns to S_0 by the following processes:

1. Radiationless deactivation to S_0 occurs by an IC process.
2. There is emission of a photon without change in spin multiplicity. This radiative transition $S_1 \rightarrow S_0$ is called "Fluorescence." The energy of the photon corresponds to the difference in energy between the excited state and the ground state of the molecule.

47

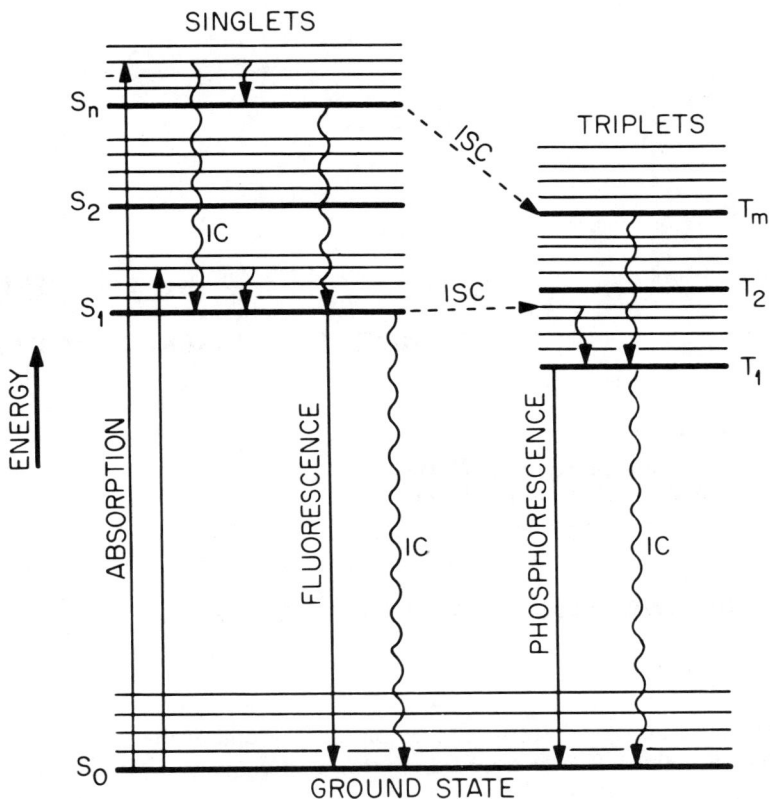

Figure 1.   Jablonski diagram. The solid lines represent electronic energy levels; the fine horizontal lines are vibrational levels. Solid arrows represent absorption of luminescence, wavy arrows are radiationless processes and dotted arrows represent intersystem crossing transitions.

3.   Transition takes place to some vibrational level of the triplet state manifold by intersystem crossing (ISC) with change in spin multiplicity. The molecule then relaxes rapidly to the lowest vibronic level of T_1 (IC process). From T_1 it returns to the ground state, S_0, by a radiationless deactivation path (ISC process) and by a radiative transition called phosphorescence.

Fluorescence and phosphorescence are the two most common luminescence processes. A less common emission process is delayed fluorescence (DF). With some chemical systems, the molecule in the triplet state, T_1, reverts back to the excited singlet state manifold. Since T_1 is always of lower energy than $S_1, \ldots, S_n$, this transition requires some additional activation energy. From S_1 delayed fluorescence may occur, exhibiting a spectrum identical to conventional fluorescence (often called prompt fluorescence), but having a longer lifetime due to the influence of the triplet state. Two types of DF processes can be differentiated: (1) a P-type (pyrene-type) DF that is produced by repopulation of S_1 by thermal activation; and (2) an E-type (eosin-type) DF that is produced when pairs of triplet state molecules interact, providing an excitation energy that is greater than, or equal to, the S_1 energy.

Nature of Fluorescence and Phosphorescence

The theory of luminescence and radiationless processes has been treated in detail in many excellent works[1-4]. This section briefly recapitulates some fundamental characteristics relevant to fluorescence and phosphorescence.

Most molecules contain an even number of electrons. In the ground state, the electrons fill up various atomic and molecular orbitals in pairs, with their spins in opposite directions in one given orbital (Pauli's exclusion principle). As a consequence, this state has no net electron spin and is called a singlet state. Whereas the ground state, S_0, is generally a singlet, the excited state can be either a singlet or a triplet, depending on the state of the spin of the electron that has been prompted to an upper orbital (Figure 2). The singlet state is diamagnetic, whereas the triplet

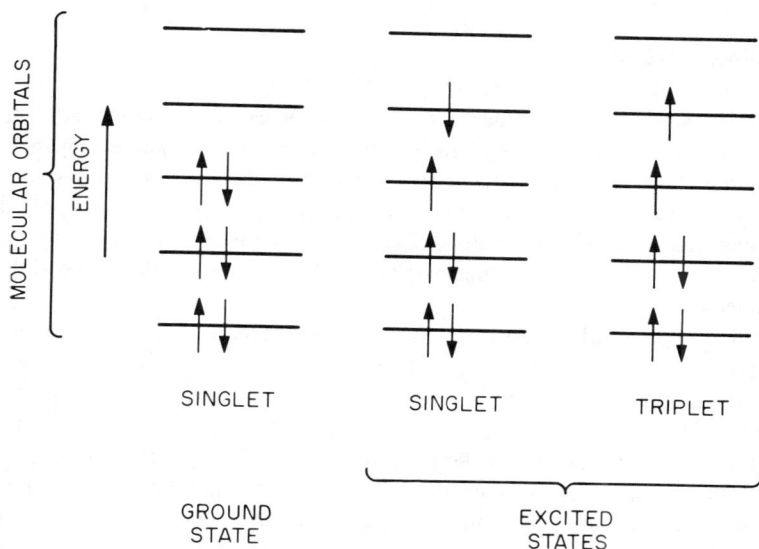

Figure 2.   Molecular orbital scheme for the ground and excited states.

state is paramagnetic. The basis of the nomenclature "singlet, triplet" is derived from multiplicity consideration of the splitting of the energy levels when the molecule is exposed to a magnetic field: under the application of an external magnetic field, the triplet state splits into three Zeeman levels, whereas the singlet state, with no spin magnetic momentum, remains unaffected.

Transitions between different excited singlet states of the same multiplicity ($S_n \leftrightarrow S_m$ or $T_n \leftrightarrow T_m$) may occur easily, but transitions between pure spin states of different multiplicity ($S_n \leftrightarrow T_n$) are quantum mechanically forbidden by the spin selection rule. Forbidden transitions do occur, however, under certain conditions. Basically, the occurrence of such transitions, i.e., singlet-triplet transitions, is made possible by the coupling of the electron spin with the orbital angular momentum (spin-orbit coupling), a nonclassical phenomenon that produces a quantum mechanical mixing of states of different multiplicities.

Experimentally, fluorescence can be differentiated from phosphorescence because of its short radiative lifetime (10^{-9}–10^{-7} seconds). Phosphorescence persists from 10^{-4} to several seconds after excitation. Due to its longer decay time, phosphorescence is more susceptible to collisional quenching processes in solution than fluorescence. The radiationless deactivation for most molecules in the triplet states is so efficient that phosphorescence is normally observed only when the molecule is frozen in a rigid glass at low temperature. Recently, intense phosphorescence from organic compounds adsorbed on solid substrates has also been observed at room temperature[5,6].

Luminescence Spectra

Measurement of luminescence intensity allows the characterization and quantitative determination of traces of many inorganic and organic species. Further insight into the factors and mechanisms involved in molecular processes can be gained by studying other spectroscopic parameters, such as the lifetime, polarization and quantum yield of the emissions. Our discussion, here and in further sections, focuses only on the instrumentation involved in investigating some of these parameters.

Two basic types of luminescence spectra can be obtained (Figure 3). When a compound is excited at a constant energy (fixed excitation wavelength), an emission spectrum is produced by recording the luminescence intensity as a function of the emission wavelength, λ_{em}. The vibrational structure in the emission spectrum gives information about the ground state, S_0. The bands in the emission spectrum correspond to vibrational energy levels of the ground state. The intensity of these bands are determined, in quantum theory, by the degree of overlap of the wavefunctions of the various vibrational levels of the ground and excited states (Franck-Condon factors).

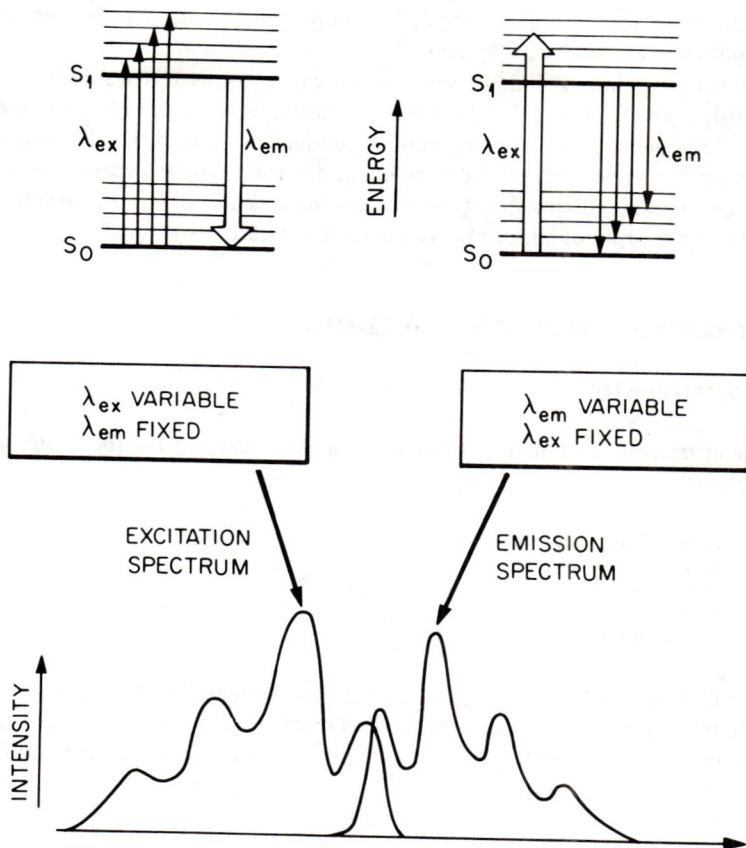

Figure 3.   Principle of excitation and emission spectrum measurement.

An excitation spectrum is obtained by varying the wavelength, λ_{ex} (or energy) of the exciting radiation while the observation of the luminescence is performed at a fixed emission wavelength. As a consequence, the vibrational structure of an excitation spectrum corresponds to the vibrational nature of the excited state, S_1 (Figure 3). If the excitation spectrum is compared to the absorption spectrum, the wavelength positions of each band in the absorption spectrum are the same as for the bands in the excitation spectrum. Excitation spectroscopy can, therefore, provide absorption characteristics of a luminescing compound, but at concentrations much too low and/or under sampling conditions not suitable to absorption spectroscopy. For example, excitation spectroscopy is necessary for the study of singlet-triplet absorptions and for the investigation of quantum efficiency in energy transfer between donor and acceptor molecules. The emission and excitation spectra of several organic polycyclic compounds (anthracene, perylene and phenanthrene) are shown in Figure 4.

A third possibility of luminescence measurement consists in scanning simultaneously both λ_{em} and λ_{ex}, while keeping a constant wavelength interval between them. This method, called synchronous luminescence, is generally used for obtaining fingerprints of real-life samples and for enhancing the selectivity in the assay of complex systems[7,8]. Figure 5 shows the synchronous and conventional spectra of a mixture of various polyaromatic hydrocarbons.

INSTRUMENTATION AND METHODOLOGY

The Spectrometer

The instrument for luminescence measurement consists of the following basic units:

- an excitation source
- a sample compartment
- dispersive devices
- a detection system
- a readout unit

Figure 6 shows a schematic diagram of the basic system. An ideal spectrometer should have an intense, monochromatic and tunable excitation source, dispersive elements with high resolution and throughput, and a detection system with high sensitivity.

Excitation Sources

Light sources are generally used for excitation in luminescence spectroscopy. They can be either of line or continuum type. They can also be used either in continuous-wave (CW) mode or in pulsed mode. Line sources exhibit sharp spectral lines, whereas continuum sources have a broad spectral intensity distribution.

Low-pressure Vapor Lamps

Low (or medium)-pressure mercury vapor lamps are often used as line sources. They are simple to use, require little power and offer intense ultraviolet (UV) radiation concentrated in a few lines (253.7 nm, 365.0/365.5/366.3 nm multiplet). They also provide excellent reference light sources for the calibration of monochromators; however, they cannot be used for recording excitation spectra because the measurement requires a continuum spectral intensity distribution.

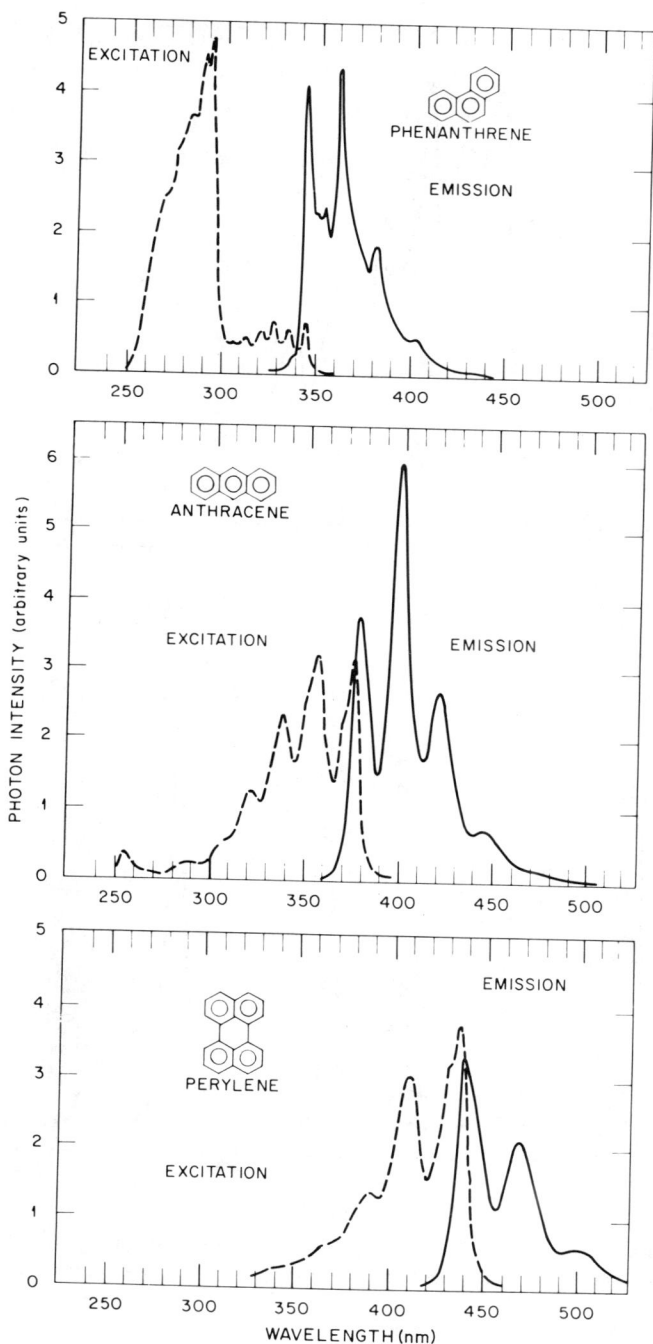

Figure 4.   Excitation and emission spectra of some polyaromatic hydrocarbons.

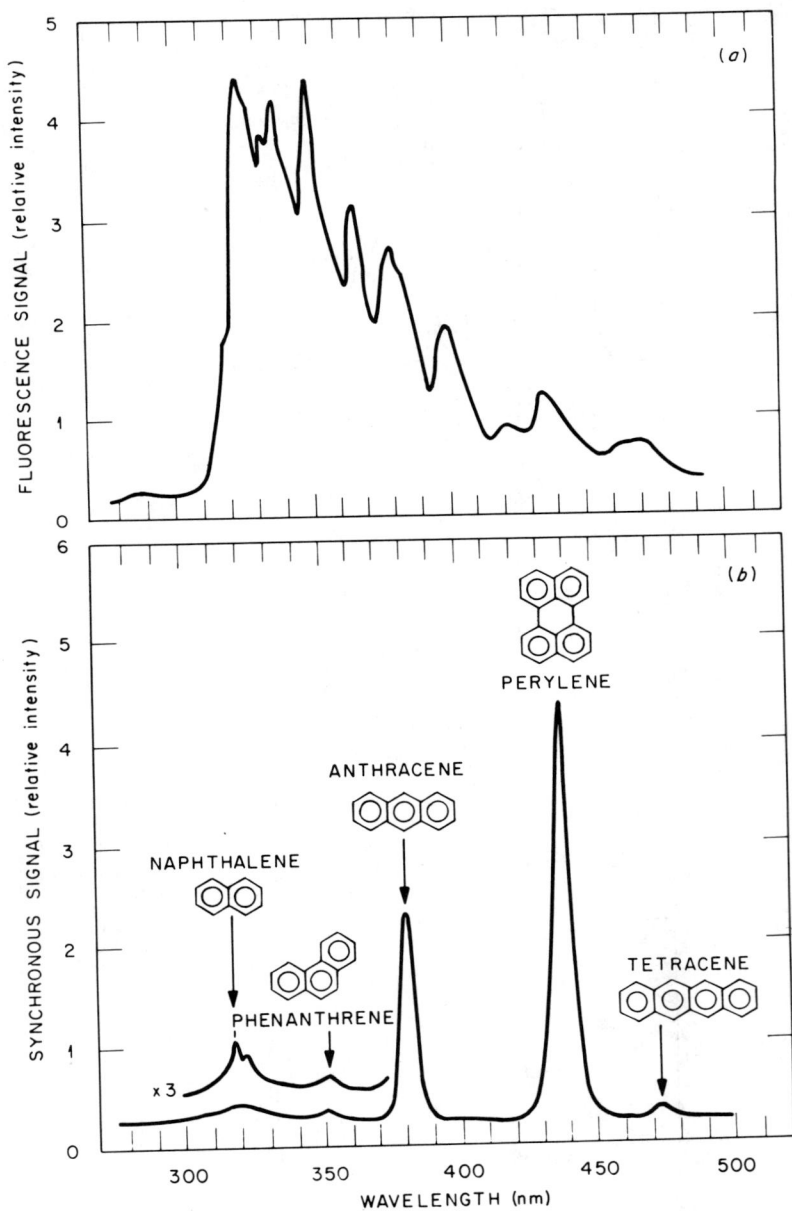

Figure 5.   (a) Fluorescence spectrum of a mixture of naphthalene, phenanthrene, anthracene, perylene and tetracene; (b) synchronous spectrum of the same mixture[8].

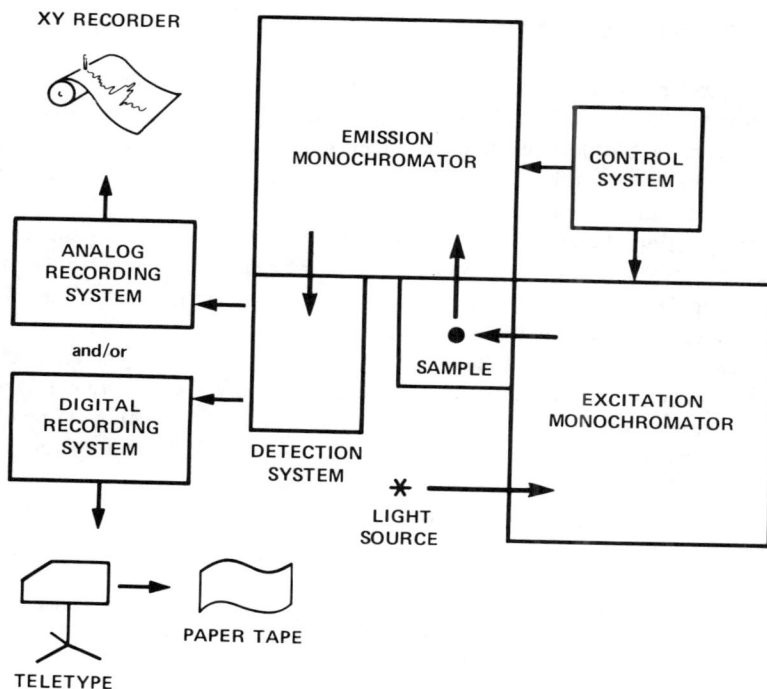

Figure 6.   Schematic diagram of a luminescence spectrometer.

Incandescent Lamps

The tungsten filament lamp is the simplest continuum source. This incandescent lamp exhibits a smooth continuous spectral profile determined by the black body radiation characteristics given by the Planck's equation:

$$S_\lambda = \frac{5.8967 \, \lambda^{-5} \Sigma_\lambda}{\exp(14388/\lambda T) - 1}$$

where S_λ = spectral radiance (watt cm^2/sr/nm)
   λ = wavelength (nm)
   T = temperature (K)
   Σ_λ = spectral emissivity of the filament material (dimensionless)

Since incandescent lamps usually have low UV output, they are seldom used as excitation sources in luminescence. Their smooth spectral profile, however, make them very suitable for intensity calibration procedures.

High-pressure Arc Lamps

The most commonly used excitation light sources are the high-pressure arc lamps (mercury, xenon or xenon-mercury arc lamps). They produce a quasicontinuum spectral profile with a few broad bands. Their intense output from the UV (200 nm) up to the near infrared (IR) is very useful for excitation spectra measurements over wide spectral ranges.

The operation of high-pressure arc lamps requires special care, such as the need to reduce stray light with a good monochromator, the use of a highly regulated dc power supply, a warmup period to minimize arc wandering and a means to remove heat generated by the output in the IR.

Modulation and pulsing of the excitation sources are achieved by coupling their power supplies to a modulation unit. Frequency stability is a very critical factor. With high-pressure lamps, modulation is usually obtained by passing the exciting light through a chopper or a device having variable optical density (Kerr cell, ultrasonic diffraction grating, electrooptical crystal).

Lasers

The development of lasers has opened new avenues to luminescence spectroscopy. The advantages offered by lasers as excitation sources include:

- monochromaticity
- high intensity
- phase coherence
- high degree of collimation
- short pulse duration (with pulsed lasers)
- polarized radiation
- low stray light

The principle of laser operation has been described in detail in many reports and manuscripts[9-17]. The laser process (stimulated emission following population inversion between different electronic-vibronic levels) is generally achieved inside a resonant optical cavity filled with an active medium that can be gas, semiconductor, dye solution, etc.

Some lasers are pulsed, others are operated in the CW mode. Various methods for producing pulses (< 10 ns) include Q-switching and mode-locking. The Q-switching method consists of first spoiling the quality factor, Q, of the laser resonator and then suddenly switching the cavity value, Q, back to a relatively high value. The shutter device in active switching can be a Kerr cell (a beam modulator in which modulation is proportional to the square of an applied electric field), a Pochels cell (a device which provides a modulation linearly proportional to the applied electric field) or simply a rotating mirror. Passive Q-switching is obtained with a saturated absorber such as a gas or a dye. The shutter is usually placed in the cavity between the laser medium and one of the reflectors. Subnanosecond pulses are obtained from lasers by the mode-locking technique, which consists of phase-locking the electromagnetic longitudinal modes inside the cavity.

The spectral output of lasers can be modified by a frequency doubling technique (second harmonic generation). The harmonic generator device usually consists of a nonlinear crystal (ADA: ammonium dihydrogen arsenate; KDP: potassium dihydrogen phosphate), which multiplies or produces a harmonic of the frequency of the laser beam. Recently, third harmonic generation has been achieved[18]. A technique that also uses nonlinear crystals generates a variable frequency laser beam by frequency mixing[19]. Stokes and anti-Stokes Raman pulses also may be used for excitation and probing experiments[20].

Gas Lasers. In a gas laser, the active medium is an atomic (helium-neon), molecular (carbon dioxide, hydrogen cyanide and water vapor) or ionized gas (argon, krypton, xenon, helium-cadmium, helium-selenium). The spectral output is available from the far infrared (FIR) to the vacuum ultraviolet (VUV), depending on the type of laser. Output powers of CW gas lasers range from milliwatts to kilowatts. Peak powers of pulsed gas lasers can reach several megawatts.

A compact, low-power gas laser, often used for alignment and for calibration, is the helium-neon laser. The most commonly used ion lasers are the argon-ion laser or argon-krypton ion laser. An ion laser that uses a metal vapor as active medium is the helium-cadmium laser.

The most commonly used pulsed laser is the nitrogen laser, which produces 1–10 ns pulses in the near UV (337 nm). Pulsed output in the UV is obtained with excimer lasers, in which the active medium is an excimer (krypton-fluoride, xenon-fluoride).

In chemical lasers, the active medium, such as hydrogen fluoride (HF) or deuterium fluoride (DR), is produced by a chemical reaction induced by electrical discharges. A CW gas laser having high power at 10 μW (up to several hundred watts) is the carbon dioxide laser.

Solid State Lasers. In solid-state lasers, the active medium is an atomic species that exists in, or can be added to, the material (chromium in ruby, neodynium in glass). Solid-state lasers can be of the pulsed type, such as the chromium-doped ruby laser, or the CW type, such as the YAG (yttrium-aluminum-garnet) laser. Spectral outputs are mostly in the red or near infrared. Although pulsed outputs are from tenths of a joule to several joules, the repetition rate of pulsed lasers is relatively low (0.1 Hz to tens of Hz). Optical pumping of pulsed lasers is usually achieved with xenon flash tube. The YAG laser, operated in the CW mode, can be mode-locked to produce pulsed output at repetition rates between kHz and gHz.

Semiconductors Lasers. Semiconductor lasers can consist of p-n junction diodes or can be of homogeneous material. Diode lasers can have either a tunable output (lead-salt type) or a fixed-wavelength output (gallium arsenide, GaAs, or gallium-aluminum, GaAl). Whereas the lead-salt lasers are used for high-resolution spectroscopy in the IR, the GaAs and GaAl lasers are usually employed in communications through fiber optic cables or in information processing.

Tunable Dye Lasers. Dye lasers offer many properties that make them close to being an ideal excitation source. Although some lasers have output that can be varied within a small spectral range, only dye lasers can be tuned across a large spectral range from the UV to the near IR with a series of different dyes. This type of laser consists of an organic dye solution that can be pumped by a variety of sources (flash lamp, nitrogen laser, ion laser). The type of output desired and the absorption properties of the dye determine the choice of the pump source. Pulsed sources usually provide peak powers of sufficiently high levels for laser action in most dyes.

Tunable lasers that are not dye lasers include parametric oscillator lasers and F-center lasers, some semiconductor lasers and some infrared lasers operated at high pressures.

Special Excitation Sources

Synchrotron Radiation. Synchrotons provide vacuum ultraviolet (VUV) and X-ray radiation that are needed for studying the spectroscopic properties of organic and inorganic materials in the vicinity of, or beyond, their ionization potential[21]. The output is a continuum spectrum ranging from the X-ray region to the FIR that is emitted by electrons or positrons accelerated in a circular accelerator. Synchrotron radiation has high intensity and low beam divergence. The beams can be modulated at very high frequencies (500 MHz). Their polarization characteristics make them unique for spectroscopic studies. At the present time, these sources are, however, used only in basic research.

Electron beams. Under certain conditions, luminescence is observed using electron beam excitation. The photophysical processes involve conduction electrons and excitons in inorganic materials. Experimentally, one noteworthy advantage of this type of excitation source is the absence of the background light that is inherent to conventional light sources. Electron beams are commonly used in electron microscopes and in cathode ray tubes (CRT)[22].

Dispersive Devices

Filters

When the variation of the excitation or emission wavelength is not needed, the simplest dispersive element is the filter. Most filters are of the absorption, birefringence or interference type.

Interference filters allow the passage of light within a narrow spectral bandpass by a process of constructive and destructive interference. Carefully controlled thickness of vacuum-deposited layers of chemicals are arranged such that light outside the spectral bandpass is eliminated by destructive interference, whereas light within the bandpass is reinforced and transmitted by constructive interference.

Absorption-type filters are the neutral density filters (constant transmission over a wide spectral range), cutoff filters (sharp transmission cutoff in the blue region) or bandpass filters.

Birefringent filters are constructed with polarizers and retardation plates[23]. They usually have a bandwidth similar to that of interference filters. Table I gives some characteristics of various types of filters.

Monochromators

Prism Monochromators. In prism monochromators, light dispersion is due to the change of refractive index of the prism material with the wavelength of the light source. The angular dispersion D is given by:

Table I.   Some Properties of Various Types of Filters

Type	Materials	Properties
Absorption	Glass Gelatin Chemical solution Chemical gas Fused silica Sapphire Alkali-halide	Broad bandpass Short wavelength cutoff Good transmission 　(up to 90%) Inexpensive Available commercially
Interference	Multilayers of alternate high 　and low refractive index 　dielectrics (ex: zinc 　sulfide and cryolite)	Narrow bandpass 　(up to 1 nm) Good transmission (30% in 　UV; 60 in visible)
Birefringence	Polarizers and retardation 　plates	Variable wavelength

$$D = \frac{d\theta}{d\lambda} = \frac{dn}{d\lambda}\frac{d\theta}{dn}$$

where θ = angular deviation
   n = refractive index of the prism materials
   λ = wavelength of the light source

Prism monochromators usually produce less stray light than grating devices and are free from overlap from multiple orders. Grating monochromators, however, are now more widely used because of their higher resolution.

Grating Monochromators. The characteristics of a grating are determined by the shape and number of its grooves. The shape of the grooves is such that the maximum of light at a given wavelength, λ, is concentrated to only one specific angle because of constructive interference.

The general diffraction grating formula is given by (Figure 7):

$$\sin\theta + \sin\theta' = k\,n\,\lambda$$

where θ = angle of incidence,
   θ' = angle of diffraction

   $n = \dfrac{1}{d}$, number of grooves per unit length,

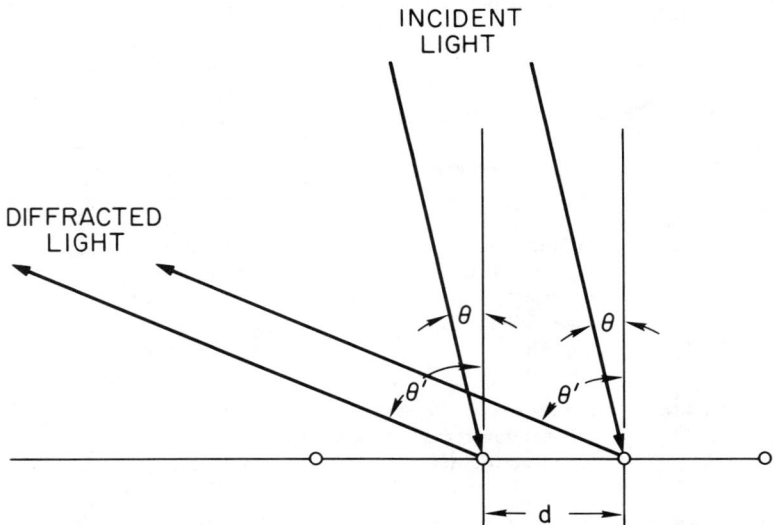

Figure 7.   The grating rule.

d = groove spacing, and
k = dispersion order.

As shown in this formula, light is dispersed by the grating because θ' depends on λ. A spectrum is obtained for each value of k. In zero-order (k = 0) there is no dispersion because all wavelengths are obtained in the same direction.

Most gratings used in modern spectrometers are of reflection type ($\theta = \theta'$). In this case, the observation is in the direction of illumination (Littrow configuration). The grating formula then becomes

$$\sin \theta' = k \, \frac{\lambda}{d}$$

Three interrelated factors, i.e., spectral dispersion, resolving power and throughput, are essential to optimizing the conditions for a given experimental situation. The spectral dispersion, D_s, is defined by

$$D_s = \frac{d\lambda}{d\ell} = \frac{1}{f} \left(\frac{d\phi}{d\lambda} \right)^{-1}$$

where $d\ell$ = distance measured across the slit
         f = focal length of the lens or mirror
         ϕ = angle of deviation (or diffraction)
         $\frac{d\phi}{d\lambda}$ = rate of change of ϕ with wavelength

The resolving power R, which measures its ability to separate adjacent spectrum lines, is defined by

$$R = \frac{\lambda}{\Delta \lambda} = \frac{w}{\lambda} \, (\sin \theta + \sin \theta')$$

where   w = width of the grating

This formula shows that R depends on the width of the grating, the working angle conditions and the wavelength.

The light throughput for various standard gratings varies between 50% and 90%. It increases with the groove density. With conventionally ruled gratings, this density has a practical limit because it induces a higher level of ghosts (spurious spectral lines caused by periodic imperfections in the grating) and stray light. The development of holography, started in the early 1960s with the appearance of lasers, has made possible the production of holographic gratings with high throughput and low stray light.

Cells and Sample Compartments

The sample cell is usually enclosed in a lighttight sample compartment (flat black paint). The two basic illumination-observation configurations are the right-angle and the front-surface configurations. For weakly absorbing samples, the right-angle configuration minimizes stray exciting light. On the other hand, front-surface configuration is preferred for strongly absorbing or opaque samples. Figure 8 shows the various sample cells for luminescence measurement at room

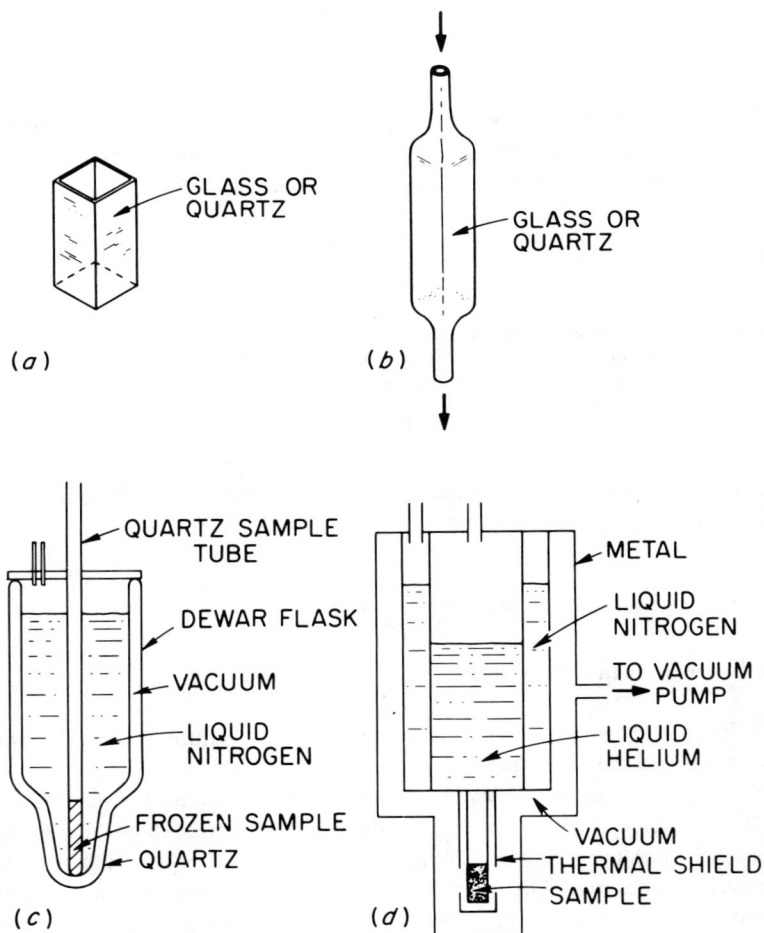

Figure 8.   Various sample cells for luminescence spectrometry.

temperature and at low temperatures. Measurements of liquid samples at room temperature are conducted with 1-cm² square cells made with Pyrex® glass, high-quality quartz or fused silica. The simplest method of cooling a sample for low-temperature studies (down to 77 K) is to immerse the sample tube directly into liquid nitrogen contained in a Dewar flask. Cooling to liquid helium temperature (4 K) requires more complex cryogenic cells and special handling procedures[24]. Figure 9 shows a typical setup for measurements at liquid helium temperature.

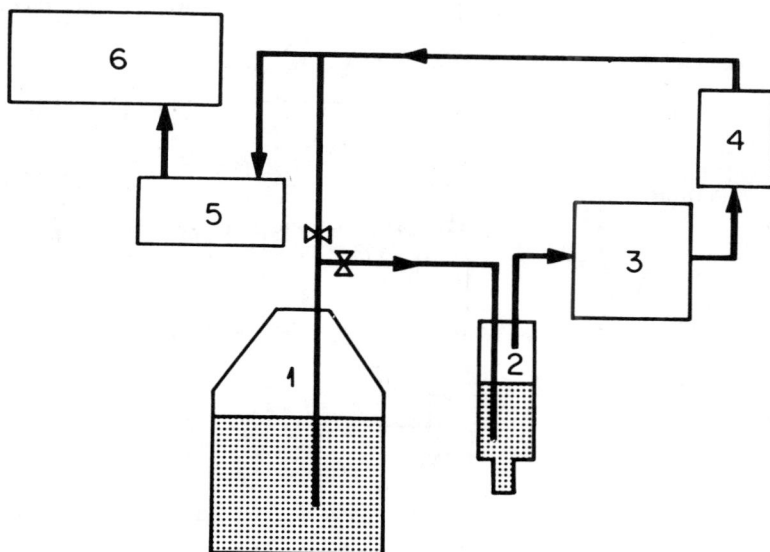

Figure 9.   Experimental setup for measurement at liquid helium temperature: (1) liquid helium vessel; (2) cryostat; (3) vacuum pump; (4) flowmeter; (5) gas meter; (6) helium gas recovery system.

Detection and Recording Systems

Photomultiplier (PM) tubes have come into widespread use as detectors in luminescence measurements. The PM is a vacuum tube that contains a highly sensitive surface, the photocathode. Within the PM, photons are converted at the photocathode into electrons, which then cascade down the dynode chain in increasing numbers due to secondary emission, which occurs at each dynode surface. The photocathode materials and the transmission characteristics of the window materials determine the spectral response of the photocathode. The long wavelength cutoff depends on the photocathode material, while the short wavelength cutoff is typically determined by the window material. One limiting factor of the PM is the dark current that may be due to leakage current (imperfect insulation), ionization of residual gases inside the tube by electrons, and thermionic emission from the cathode or the dynodes. Proper preparation of PM connections significantly reduces the dark current in photomultiplier tubes[25]. Figure 10 shows various dynode configurations and a schematic diagram of a PM tube and its associated circuitry. The PM response time depends basically on the dynode configuration. The line-focused configuration generally exhibits the faster time response (<3 ns).

Cooling the PM tube changes its performance[26]. Dark current due to thermi-

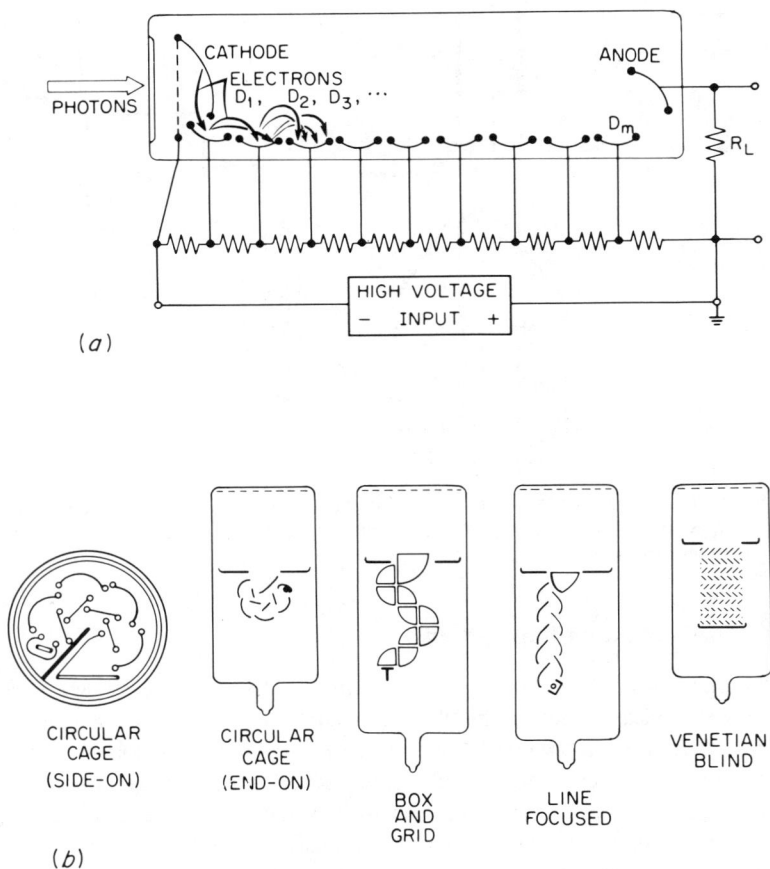

Figure 10.   (a) Schematic diagram of a photomultiplier (D_m = dynode; R_L = load resistor; E_m = dynode voltage); (b) various types of dynode configurations.

onic emission is usually decreased with lower temperatures. With phototubes having GaAs cathode, cooling causes a falloff in the red but increases the sensitivity over the rest of the visible spectrum[27].

Analog Detection

The analog (or dc) method is the oldest and simplest detection technique. Since the anode is basically a current generator, the anode current may be measured simply by a current meter. Consequently, the voltage across a load resistor is usually monitored. Typical dynode circuitries are provided by PM manufacturers. Noise in dc amplification is commonly reduced by a resistor-capacitor (RC), low-frequency bandpass filter. A too slow response of the RC filter, however, necessitates very slow scanning speeds and causes long measurement times.

Another method consists of modulating the signal and using an amplifier tuned to the modulating frequency. This technique rejects the noise outside the amplifier bandpass. A further improvement is the technique of synchronous (or lock-in) detection. Stability is improved because any drift or instability of the modulating system is locked by the amplifier. Synchronous detection, however, cannot eliminate some types of noise ("l/f noise", multiplicative noise) that are amplified along with the signal[28].

Digital data can be obtained from the dc anode current by adding a voltage-to-frequency converter[29-31].

Digital Photon-counting Technique

A true digital technique that counts discrete photon pulses is the single-photon counting technique. This method uses high-speed electronic circuitry for detecting individual photon pulses[32-35]. The block diagram of a typical photon counting system is shown in Figure 11. An incident electromagnetic light beam ejects photoelectrons from the photocathode of the photomultiplier, which are then multiplied by a cascaded secondary emission process. The number of discrete pulses of charge produced at the anode is proportional to the number of photons incident on the photocathode. This technique has proved to have several advantages over the conventional dc method, especially for low-level light detection[36,37]. First, the digital data can be processed directly in a manner suitable for further data treatment. The processing of information by digital circuitry is less susceptible to long-term drifts, which usually limit analog systems. The feature permits measurement of extremely low radiation flux by using a much longer averaging time than would be feasible with dc detection. The method also provides the ability for optimizing the signal:noise ratio by discriminating against photomultiplier dark current. With an ideal detection device, one would expect the pulses resulting from single photoelectrons to have exactly the same pulse height. Actually, the output pulses have a spectrum of pulse heights because thermionic emission of higher dynodes and a certain number of spurious pulses are generated by the photomultiplier even when the cathode is not illuminated. These pulses have a distribution in heights that will generally differ from the photoelectron pulse height distribution. Thermionic electrons have a pulse height distribution lower than photoelectrons. On the other hand, other sources of noise, such as cosmic ray muons, after pulsing, and radioactive contamination of tube materials generally produce pulses of higher amplitude than photoelectron pulses. Therefore, most of these pulses can be eliminated by discriminator units, which select pulses within a suitable range of height levels[38].

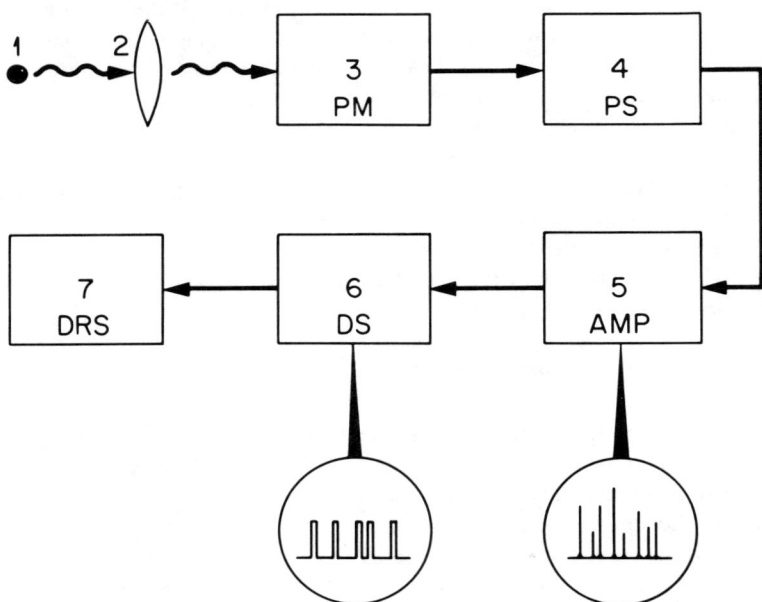

Figure 11.   Schematic diagram of the single photon counting technique: (1) light source; (2) optics; (3) photomultipler (PM); (4) pulse shaper (PS); (5) amplifier (AMP); (6) pulse height discriminator (DS); (7) digital recording system (DRS).

Spectroscopic Measurements

Measurement of fluorescence spectra under steady-state illumination is the simplest type of measurement. Phosphorescence, which has a decay time longer than that for fluorescence, is generally recorded with chopped excitation and gated detection. The exciting radiation is periodically interrupted and the emission is observed only during the off-period of the excitation. The main advantage of this method is the removal of both interfering fluorescence and stray light excitation.

Mechanical choppers are the simplest and most commonly used devices employed in phosphorimetry. The first type of mechanical choppers consists of two discs (Becquerel discs) mounted on a common axis (Figure 12). The discs are rotated by one or by two synchronous motors[39]. A more practical type of chopper is the rotating can[40], which is now available in most commercial instruments. A modified version of the cylindrical can is the rotating mirror[41]. With rotating devices, the illumination-excitation geometry in the right-angle configuration produces illumination and observation cycles that are out-of-phase with respect to each other when the device rotates. Time resolution is achieved by rotating mechanical choppers at various speeds or by electronic devices. The section entitled Time- and Phase-Resolved Spectroscopy, p. 71, discusses the various electronic devices used for time-resolution in fluorimetry as well as in phosphorimetry.

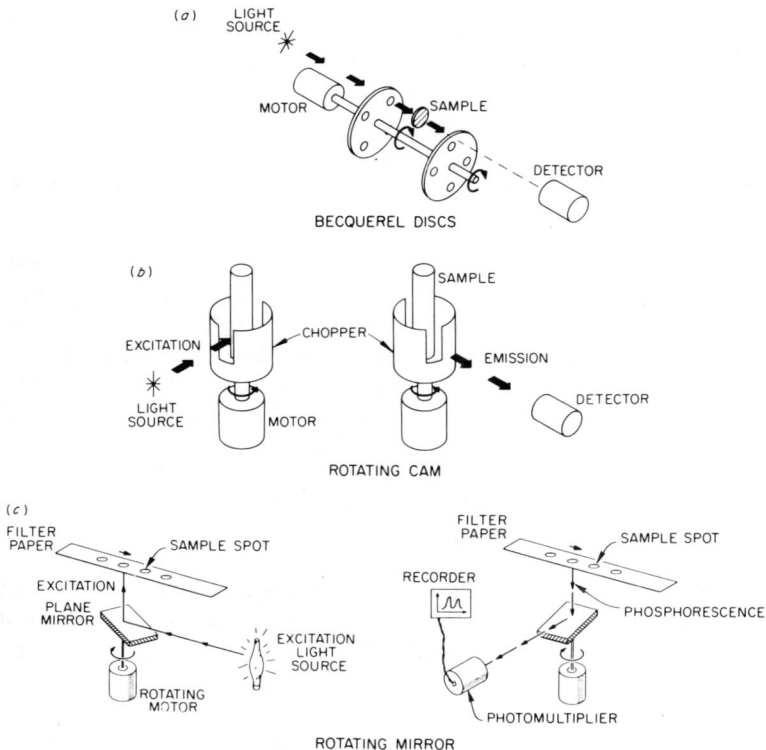

Figure 12.   Various phosphoroscopic choppers.

Lifetime Measurements

Decay times of luminescence processes have been measured by the following three methods:

1.   the pulsed-excitation analog method,
2.   the digital single-photon counting method, and
3.   the phase-shift method.

Pulsed-excitation Analog Method

In this method, the decay of the luminescence is measured after excitation by an intense and short pulse. If the emission decay time, ζ, is much greater (10-fold or more) than the exciting pulse duration, δt, it can be directly determined using a sufficiently fast oscilloscope. This method, which involves simple instrumentation, is suitable for most phosphorescence decay studies. In situations in which ζ is on the order of δt, careful derivation of the data is required. The true decay function, I_L, is derived from the following convolution function:

$$I_{obs}(t) = \int_0^t I_L(t') \, R(t-t') \, dt'$$

where I_{obs} = observed decay function
I_L = true luminescence decay function
R = response function of the instrument (exciting pulse duration; time response of detector)

One commonly used device is the box-car integrator, a gated amplifier with a low-pass filter. Variation of the delay duration between the excitation pulse and the gated detection period allows lifetime measurement (Figure 13). An instrument that performs an action similar to the box-car is the multichannel averager, in which each channel is assigned to collect a specific temporal data point of the waveform. This device has the added advantage of simultaneously collecting all the data at different delay times.

Pulsed-excitation Single-photon Counting Method (or delayed coincidence technique)

In this method, each photon signal is monitored after an excitation pulse. The time delay between the excitation pulse and the photon signal is recorded with a time-to-amplitude converter and stored in a multichannel analyzer where the counts in each channel are proportional to the probability of fluorescence emission in the time interval Δt at time $n\Delta t$ (n being the channel number and Δt the time-width of each channel). After a large number of repetitive measurements,

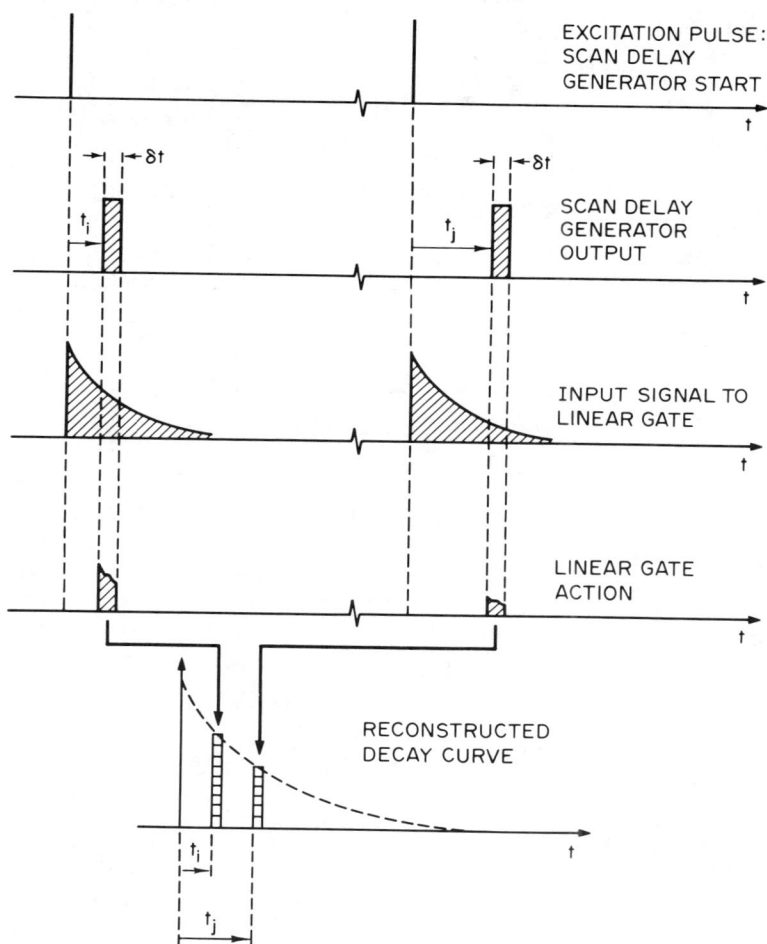

Figure 13. The box-car technique.

the multichannel analyzer memory contains a histogram of pulses that is the representation of a plot of luminescence intensity versus time (Figure 14). The single-photon counting technique requires certain strict conditions to avoid experimental errors caused by pulse overlap and detector dead time. One important parameter is the repetition rate, which must be high enough to provide sufficient data points within a reasonable measurement time. With mode-locked lasers, single-photon counting is most suitable for studies of very short lifetime (subnanosecond)[42-45].

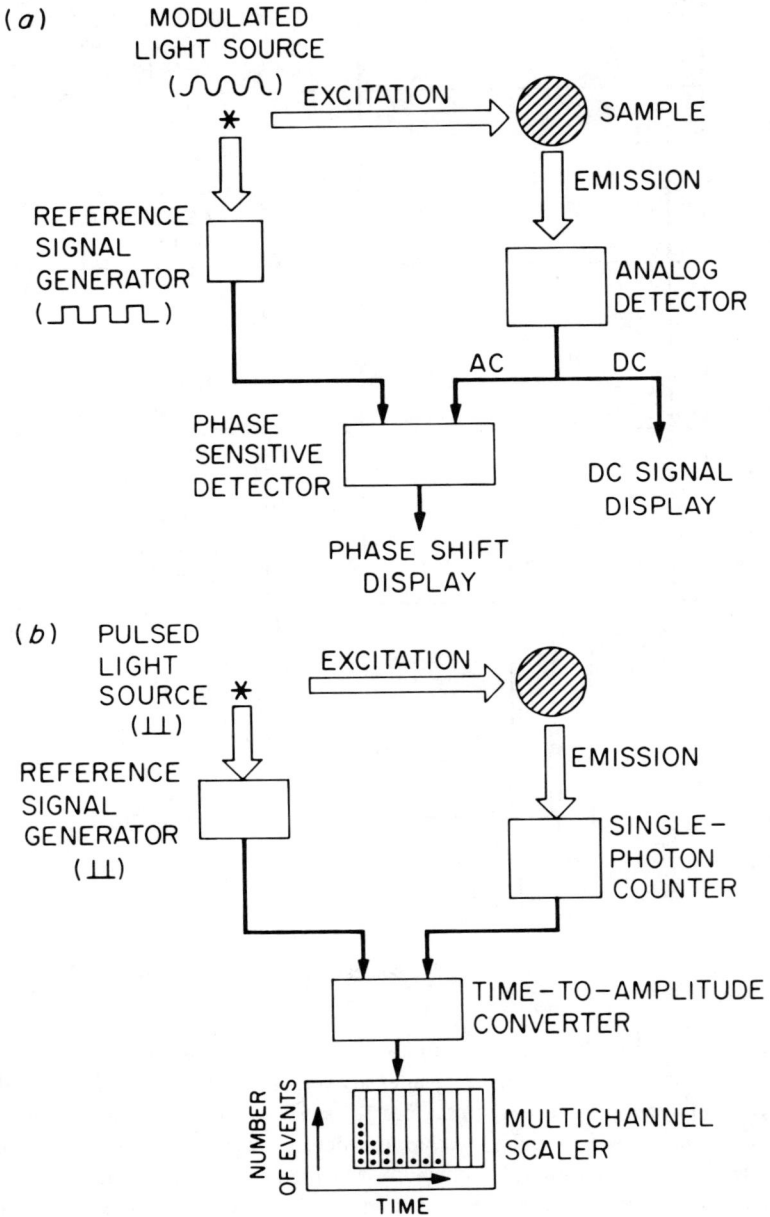

Figure 14.   Lifetime measurement: (a) the phase-shift technique; (b) the photon-counting technique.

Phase-shift Method

In the phase-shift method, the sample is excited by an intensity modulated signal[46]. The time lag between emission and excitation induces a phase-shift in the observed luminescence signal (Figure 14). For a simple exponential decay, the relationship between the lifetime, ζ, and the phase shift angle, ρ, is given by

$$\zeta = \frac{\tan \rho}{2 \pi f}$$

where   f = modulation frequency

The choice of f depends on the lifetime to be measured. For the measurement of phosphorescence decay time, f is the order of a few hundreds hertz. For fluorescence decay time, f should be in the megahertz range. Temporal resolution in the picosecond time scale has been achieved[47,48].

Pulsed-excitation methods have the advantage of producing complete decay curves that can be useful for sorting out complex multicomponent decays. The phase-shift method, on the other hand, uses simpler instrumentation and higher average excitation power that might be necessary for weak emissions.

An alternative approach to luminescence lifetime measurements involves cross-correlating the randomly fluctuating CW laser excitation with the sample emission[49,50].

SPECIAL TECHNIQUES AND APPLICATIONS

Time- and Phase-Resolved Spectroscopy

Spectroscopic measurements with time-resolution capability have been conducted using pulsed excitation and an electronic gate (box-car type), which monitors the PM output for a limited period after the excitation cutoff[51]. Another technique consists of having the PM turned off during given periods[52,53]. An instrument using the photon counting technique and an electronic gate allows simultaneous recording of total luminescence and phosphorescence spectra[54,55]. Both kinetic and spectral data can be recorded simultaneously by an online computer[56].

One practical device, previously mentioned in the section entitled Lifetime Measurements, p. 68, is the box-car integrator, which can be used for time-resolved measurements using a pulsed excitation source. Box-car integrators have been employed both in time-resolved fluorimetry and in time-resolved phosphorimetry[51,57,58].

Time-resolved measurements at low light levels usually require the single-photon counting technique. A time-resolved phosphorimeter with a pulsed dye laser and a current-to-voltage gated amplifier have been recently described for lifetime and spectroscopic measurements[59]. Subnanosecond resolution is achieved with the single-photon counting technique using an argon-ion laser as the excitation source[45].

A technique that extracts the temporal information in an indirect manner is the phase-resolution technique. The sample is illuminated by intensity-modulated light, with the detection performed by a phase-sensitive detector (lock-in amplifier). Spectral data at different phase shifts correspond to information from different components having different lifetimes. In addition to the capability of phase-resolution, modulation improves the S:N (signal:noise ratio) characteristics by reducing low-frequency additive noise (drift or "1/f noise")[60]. Phase modulation has been used for both fluorescence and phosphorescence measurement[61-64]. Phase resolution has been applied to both analog and digital detection[65, 66].

High-Resolution Quasilinear Spectroscopy

Under certain experimental conditions, luminescence spectra exhibit very sharp line structure (1–10 cm^{-1} bandwidth) at low temperatures (77°–4° K). Extensive efforts have been devoted by numerous workers to finding conditions that favor quasilinear spectral structure. One early and quite successful method is the technique or isomorphous replacement, which uses substitutional mixed crystals[67]. The idea is to select a host compound having a structural shape similar to that of the analyte, e.g., naphthalene doped in durene or anthracene doped in durene. In later years, special types of frozen solutions, the Shpoliskii matrices (normal hydrocarbons), were found to be very efficient in inducing quasilinear structure in the luminescence spectra from many polyaromatic compounds[68]. The great interest in monitoring polynuclear aromatic compounds has led to several studies and reports[69, 70]. The use of laser excitation provided a direct experimental confirmation of the hypothesis suggested by Shpolskii describing the multiplet nature of the spectra[71]. Many studies have been devoted to the understanding of the conditions causing the occurrence of quasilinear spectra[72-75]. Analytical applications of the Shpolskii effects for the identification of polynuclear aromatic compounds have also received great interest[69, 70]. Another important technique for obtaining quasilinear spectra is the matrix isolation technique at low temperatures. In the matrix isolation method, samples that are liquids or solids at room temperature are vaporized and then mixed thoroughly with a large excess of a diluent gas such as nitrogen. The resulting gaseous mixture is then deposited at low temperature as a solid onto an optical window suitable for spectroscopic measurements. Analytical applications of this technique have been reviewed recently[76]. Another method for observing highly resolved spectra involves the expansion of a gas through a supersonic nozzle. This technique is an efficient means of supercooling molecules seeded into an inert gas jet for examination by laser-induced fluorescence[77].

Laser Excitation and Nonlinear Spectroscopies

The availability of lasers with extremely narrow lines, high power output and spectral tunability has been the factor most responsible for the growth of many spectroscopic techniques in the luminescence field. These include high-resolution spectroscopy, site-selection spectroscopy and nonlinear spectroscopy[78].

Site-selection Spectroscopy

Site-selection spectroscopy is the term given to the luminescence technique in which the very narrow-bandwidth excitation lines (< 1 cm^{-1}) of lasers are used to selectively excite a limited number of molecules occupying a specific site in a crystal or a matrix[79]. Spectral lines can be broadened by two different kinds of mechanisms. A homogeneous mechanism broadens the response of all molecules equally. Lifetime broadening and thermal broadening are two examples of the homogeneous mechanism. On the other hand, the inhomogeneous broadening, which is caused by the differences in the crystal fields of molecules randomly distributed in various crystal sites, can result in broad-structure spectra with broad band excitation, even at low temperatures. Only site-selection spectroscopy can induce spectra with quasilinear structure in a variety of matrices including crystalline solids[79], Shpolskii matrices[71,80], alkane monocrystals[81] and organic glass[82].

Nonlinear Spectroscopy

Nonlinear effects are the terms used for large signal effects that cannot be described by the linearized approach and the rate equations of the atomic dynamics in spectroscopy. With the development of high-power lasers, they open horizons unknown to conventional excitation methods. One of the most significant applications of nonlinear optics is the process of multiphoton excitation (absorption), by which simultaneous absorption of several photons occurs. Although multiphoton absorption does not require the coherence properties of laser radiation, it needs the high-power output of lasers. Of particular interest is the occurrence of two-photon transitions between vibronic states of the same parity. Such transitions are dipole-forbidden in ordinary single-photon spectroscopy. Studies of these processes provide a host of additional spectroscopic information and allow a better understanding of photophysical properties[83-85]. A demonstration of certain selection rules is given in two-photon spectroscopy by controlling the polarization state of the exciting radiation[86]. Doppler-free spectroscopy is performed by absorption of one photon from each of two antiparallel beams[87].

Derivative Spectroscopy

In derivative spectroscopy, the first or some higher order derivative of the luminescence intensity with respect to wavelength is recorded. In general, differentiation offers several advantages. Weak or minor features are enhanced and thus become more easily identified; this feature is particularly useful when it comes to analyzing trace analytes whose sharp peaks are obscured by an intense and broad background. Since differentiation has an effect equivalent to that of a high-pass filter, the small minor peaks are enhanced while the broad bands are decreased. Derivative spectroscopy also decreases several types of systematic errors, such as low-frequency noise, electronic drift, light source variation and stray light. Increase in random noise, however, degrades the S:N ratio in the differentiated signal. Derivative spectroscopy can be helpful in situations where the S:N ratio is sufficiently good in the undifferentiated signal, but the spectral features are insufficiently structured for easy identification.

Derivative spectra have been obtained by several techniques, including wavelength modulation, electronic differentiation and digital data processing. If the data are readily available in digital form, then the differentiation can be accomplished by direct digital computation using the method developed by Savitzty and Golay[88].

Derivative spectra also may be recorded by wavelength modulation. This technique consists of modulating the signal by either a direct or indirect means (vibration of slits, mirrors, gratings or prisms in a monochromator, or oscillation of a refractor plate, etc.). For absorption spectrometry and flame emission spectrometry, the modulation technique is preferred because analog differentiation of the detector output may significantly degrade the S:N values of the derivative signal as compared to the original values. But for luminescence spectrometry, no significant difference was found between the S:N values of the derivative spectra obtained by differentiation or by the modulation technique.[89].

A simpler method for obtaining derivative spectra is direct electronic differentiation. In those situations in which S:N values are sufficiently high, this method is the first choice because of its simplicity and because of the availability of inexpensive commercial devices. The dc output from the spectrometer is converted into signals that are proportional to the first and second derivative with respect to time. If the wavelength scan rate $r = \dfrac{d\lambda}{dt}$ is constant, then the derivative of the intensity I with respect to wavelength $\dfrac{dI}{d\lambda}$ is a value proportional to the derivative of I with respect to time $\dfrac{dI}{dt}$:

$$\frac{d\,I}{d\,\lambda} = \frac{d\,I}{d\,t}\,\frac{d\,t}{d\,\lambda} = \frac{d\,I}{d\,\lambda}\left(\frac{d\,\lambda}{d\,t}\right)^{-1}$$

$$\frac{d\,I}{d\,\lambda} = \frac{1}{r}\,\frac{d\,I}{d\,r}$$

Videospectrometry

Developments in multichannel image detector devices have recently found many applications in luminescence spectrometry. The great potential of multi-channel, diode array, or TV-type detectors such as vidicons and photodiode arrays has been demonstrated[90-96]. Television-type multichannel detectors have been used for molecular absorption, Raman and molecular fluorescence studies.

In videospectrometry, the detector is placed at the focal plane of a mono-chromator with the exit slit removed. One primary advantage of these multi-channel detectors is that the emission at all wavelengths within the mono-chromator is detected simultaneously. The simultaneous detection of all the dispersed radiation using N spectral (or spatial) resolution elements reduces the measurement time by a factor of N in the case of an S:N ratio-limited measurement and improves the S:N ratio by a factor of N in the case of a time-limited measurement. The main disadvantage is the low sensitivity compared with PM tubes. Intensifier devices are also available for improving the sensitivity of video-detectors (silicon-intensified-target vidicon and intensified diode array). An optical multichannel analyzer with gating capability has been developed for time-resolved luminescence studies[97].

Another approach for simultaneous detection is the multiplex approach. Multi-plex spectrometric systems utilize transformation methods such as Hadamard and Fourier techniques. Whereas these systems have been successfully applied to IR, they offer limited advantage to UV-VIS, X-ray, electron and molecular system spectrometry.

Optical Microwave Double Resonance Spectroscopy

Optical microwave double resonance (OMDR) spectroscopy is based on both the magnetic and the spectroscopic properties of the triplet state. Although the degeneracy of the triplet state of aromatic molecules is removed by spin-spin interactions under certain conditions, this degeneracy is too small to be detected by direct optical methods (~ 0.1 cm^{-1}), even at very low temperatures (< 4 K) where spin-relaxation is slower than the observed lifetime[98]. Microwave radiation of resonance frequencies between triplet spin sublevels can preferentially popu-late certain sublevels. Optical detection under these conditions has provided a powerful means for investigating radiative and radiationless processes via the various spin sublevels of the triplet state[99-103].

Single Vibronic-Level Spectroscopy

The data obtained from spectroscopic measurements of gases at low pressures are of primary importance since they apply to molecules that are supposedly in their "free and isolated state." A unique luminescence technique for studying molecules in the gas phase is the single vibronic-level (SVL) excitation method. Excitation of single vibronic levels is achieved by narrow bandwidth irradiation of the individual molecular vibrational bands. An essential condition for the occurrence of SVL luminescence is the absence of collisional deactivation of the originally excited vibronic states. With an excited lifetime of 10^{-7} s and a vapor pressure of less than 0.1 torr, collisions may be neglected. Under such collision-free conditions, a molecule behaves as though it is an isolated system. Another important requirement is the absence of vibrational rearrangement while the molecule is in the excited state. The emission process occurs under conditions similar to those of resonance fluorescence in atomic spectroscopy. Unlike equilibrated luminescence spectra from a system of excited molecules under Boltzman equilibrium conditions, SVL spectra exhibit extremely well-resolved structures. The first SVL measurement with benzene used a xenon arc lamp coupled to a monochromator[104]. Using excitation from a high-pressure mercury arc, both SVL fluorescence and phosphorescence of quinoxaline have been detected. Time-resolved fluorescence and the kinetic quenching parameters of gases, such as Cl_2Cs and fluorocyclobutanone, have been investigated[105,106]. A host of photophysical processes in a wide variety of compounds were investigated by SVL spectroscopy: time-resolved emissions of quinoline, quinaxoline and cinnoline[105]; intramolecular energy transfer in styrene systems; rotational structure of tetrazine[107]; and photochemistry of formaldehyde[108].

ACKNOWLEDGMENTS

This research is sponsored by the Office of Health and Environmental Research, U.S. Department of Energy under contract W-7405-eng-26 with the Union Carbide Corporation.

REFERENCES

1. McGlynn, S. P., T. Azumi and M. Kinoshita. *The Triplet State* (Englewood Cliffs, NJ: Prentice-Hall, Inc., 1970).

2.   Bixon, M. and J. Jortner. *J. Chem. Phys.* 48: 715 (1968).
3.   Azumi, T. *Chem. Phys. Lett.* 25: 135 (1974).
4.   Siebrand, W. *Chem. Phys. Lett.* 6: 192 (1970).
5.   Schulmann, E. M., and C. Walling. *Science* 178: 53 (1972).
6.   Vo-Dinh, T., and J. D. Winefordner. *Appl. Spectros. Rev.* 13: 261 (1977).
7.   Lloyd, J. B. F. *Nature* 231: 64 (1971).
8.   Vo-Dinh, T. *Anal. Chem.* 50: 396 (1978).
9.   Schawlow, A. L. *Science* 202: 141 (1978).
10. Svelto, O. *Principles of Lasers* (New York: Plenum Publishing Corp., 1976).
11. Kaminov, I. R., and A. I. Siegman. *Laser Devices and Applications* (New York: IEEE Press, 1973).
12. Ready, J. F. *Industrial Applications of Lasers* (New York: Academic Press, Inc., 1978).
13. Letokhov, V. S., and V. P. Chebotayev. *Nonlinear Laser Spectroscopy* (New York: Springer-Verlag New York, Inc., 1977).
14. Corney, A. *Atomic and Laser Spectroscopy* (New York: Clarendon Press, 1977).
15. Beck, R., W. English and S. Gurs. *Table of Lasers Lines In Gases and Vapors* 2nd ed. (New York: Springer-Verlag New York, Inc., 1978).
16. Moore, B. *Chemical and Biological Applications of Lasers* (New York: Academic Press, Inc., 1977).
17. Pressley, R. J. *Handbook of Lasers* (Cleveland: CRC Press, Inc., 1971).
18. Bloom, D. M., G. W. Bekkers, J. F. Yound and S. E. Harris. *Appl. Phys. Lett.* 26: 687 (1975).
19. Zermicke, F., and F. E. Midwinter. *Applied Nonlinear Optics* (New York: Wiley Interscience, 1973).
20. Topp, M. R., and J. C. Orner. *Chem. Phys. Lett.* 32: 407 (1975).
21. Lopez-Delgado, R., A. Tramer and I. H. Munro. *Chem. Phys.* 5: 72 (1974).
22. Davidson, S. M., and A. J. Rasul. *J. Phys. E. Sci. Instr.* 10: 43 (1977).
23. Clarke, D., and J. F. Grainger. *Polarized Light and Optical Measurement* (Oxford: Pergamon Press, Inc., 1977).
24. Meyer, B. *Low Temperature Spectroscopy* (New York: Elsevier North-Holland, Inc., 1971).
25. Davies, W. E. R. *Rev. Sci. Instr.* 43: 556 (1972).
26. Cole, M. R., and D. V. Ryer. *Electro-Opt. Syst. Des.* 4: 6 (1972).
27. Martin, H. *Electro-Opt. Syst. Des.* 16 (1976).
28. O'Haver, T. C. *Anal. Chem.* 51: 91A (1979).
29. Fitzgerald, J. M. *Fluorescence News* 17 (1): 1 (1973).
30. Enke, C. G. *Anal. Chem.* 43: 69A (1972).
31. Ingle, J. D., Jr., and S. R. Crouch. *Anal. Chem.* 44: 285 (1972).
32. Morton, G. A. Appl. Opt. 7: 1 (1968).
33. Ford, R., R. Jones, C. T. Oliver and E. R. Pike. *Appl Opt.* 8: 1975 (1968).
34. Knight, A. E. W., and B. K. Selinger. *Austr. J. Chem.* 26: 1 (1973).
35. Weakes, F. *Electro. Opt. Syst. Des.* 6 (1977).
36. Savager, C. M., and P. Maker. *Appl. Opt.* 10: 955 (1971).
37. Malmstadt, H. V., M. L. Franklin and G. R. Horlick. *Anal. Chem.* 44: 63A (1972).
38. Gustafson, T. L., F. E. Lytle and R. S. Tobias. *Rev. Sci. Instr.* 49: 1549 (1978).

39. Parker, C. A., and S. Hatchard. *Trans. Farad. Soc.* 57: 1894 (1961).
40. Hollifield, H. C., and J. D. Winefordner. *Chem. Inst.* 1: 341 (1969).
41. Vo-Dinh, T., G. L. Walden and J. D. Winefordner. *Anal. Chem.* 49: 1126 (1977).
42. Spears, K. G., L. E. Cramer and L. D. Hoffland. *Rev. Sci. Instr.* 49: 255 (1978).
43. Koester, V. J., and R. M. Dowben. *Rev. Sci. Instr.* 49: 1186 (1978).
44. Clinelove, L. J., and L. A. Shaver. *Anal. Chem.* 48: 364A (1976).
45. Wild, U. P., A. R. Holzwarth and H. P. Good. *Rev. Sci. Instr.* 48: 1621 (1977).
46. Ware, W. In: *Creation and Detection of the Excited States,* A. A. Lamola Ed. (New York: Marcel Dekker, Inc., 1971).
47. Haar, H. P., and M. Hanser. *Rev. Sci. Instr.* 49: 632 (1978).
48. Menzel, E. R., and Z. D. Popovic. *Rev. Sci. Instr.* 49: 39 (1978).
49. Hieftje, G. M., G. R. Hungen and J. M. Ramsay. *Appl. Phys Lett.* 30: 463 (1977).
50. Dorsey, C. C., M. J. Pelletier and J. M. Harris. *Rev. Sci Instr.* 50: 333 (1979).
51. Fisher, R. P., and J. D. Winefordner. *Anal. Chem.* 44: 948 (1972).
52. Hamilton, T. D. S., and K. Razi Naqvi. *Anal. Chem.* 45: 1581 (1973).
53. Jameson, D. G., and B. D. Michael. *J. Phys. E. Sci. Instrum.* 9: 208 (1976).
54. Vo-Dinh, T., and U. P. Wild. *Appl. Opt.* 12: 1286 (1973).
55. Vo-Dinh, T., and U. P. Wild. *Appl. Opt.* 13: 2899 (1974).
56. Wilson, R. M., and T. L. Miller. *Anal. Chem.* 47: 256 (1975).
57. Badea, M. G., and S. Georghiou. *Rev. Sci. Instr.* 47: 314 (1976).
58. Matthews, T. G., and F. E. Lytle. *J. Luminescence* 1979.
59. Boutillier, G. D., and J. D. Winefordner. *Anal. Chem.* 51: 1384 (1979).
60. Benci, S., P. A. Benetti and M. Manfredi. *Rev. Sci. Instrum.* 41: 1336 (1970).
61. Gruneis, F., S. Schmeider and F. Dorr. *J. Phys. E. Sci. Instr.* 9: 1013 (1976).
62. Mousa, J. J., and J. D. Winefordner. *Anal. Chem.* 46: 1195 (1974).
63. Lytle, F. E., J. E. Eng., T. D. Harris and R. E. Santini. *Anal. Chem.* 47: 571 (1975).
64. Jessop, J., R. P. Wayne and J. T. Wayne. *J. Phys. E. Sci Instr.* 5: 638 (1972).
65. Schlag, E. W., H. L. Setzle, S. Schneider and J. G. Larsen. *Rev. Sci. Instr.* 45: 364 (1974).
66. Renkes, G. D., L. R. Thorne and W. D. Gwinn. *Rev. Sci. Instr.* 49: 994 (1978).
67. McClure, D. S. *J. Chem. Phys.* 22: 1668 (1954).
68. Shpolskii, E. V. *Sov. Phys. Usp.* 5 (3): 372 (1960); 3 (3): 522 (1962); 6 (3): 411 (1963).
69. Farocq, R., and G. F. Kirbright. *Analyst* 101: 566 (1976).
70. D'Silva, A. P., G. J. Oestreich and V. A. Fassel. *Anal. Chem.* 48: 915 (1976).
71. Vo-Dinh, T., and U. P. Wild. *J. Luminescence* 6: 296 (1973).
72. Richards, J. C., and S. A. Rice. *J. Chem. Phys.* 54: 2014 (1970).
73. Rebane, L. A., and P. M. Saari. *Sov. Phys. Sol. State* 12: 1547 (1971).

74. McCumber, D. F., and M. D. Sturge. *J. Appl. Phys.* 34: 1682 (1963).
75. Hochstrasser, R. M., and P. N. Prasad. *J. Chem. Phys.* 56: 2814 (1972).
76. Wehry, E. L., and G. Mamantov. *Anal. Chem.* 51: 643A (1979).
77. Smalley, R. E., L. Wharton and D. H. Levy. *Chem. Acc.* 10: 139 (1977).
78. Kohler, B. E. *Chemical and Biochemical Applications of Lasers,* C. Bradley, Ed. (New York: Academic Press, Inc., 1976).
79. Szabo, A. *Phys. Rev. Lett.* 27: 323 (1971).
80. Vo-Dinh, T., U. T. Kreichbich and U. P. Wild. *Chem. Phys. Lett.* 24: 352 (1974).
81. Vo-Dinh, T., U. P. Wild, M. Lamotte and A. M. Merle. *Chem. Phys. Lett.* 39: 118 (1976).
82. Alshifts, E. I., R. I. Personov and B. M. Kharmalov. *Chem. Phys. Lett.* 40: 116 (1976).
83. Hochstrasser, R. M., J. E. Weisel and H. N. Sung. *J. Chem. Phys.* 60: 31 (1974).
84. Honig, B., J. Jortner and A. Szoka. *J. Chem. Phys.* 46: 2714 (1967).
85. McClain, W. *Acc. Chem. Res.* 7: 129 (1974).
86. Bray, R. G., R. M. Hochstrasser and H. W. Sung. *Chem. Phys. Lett.* 33: 1 (1975).
87. Hansch, T. W., I. S. Shahin and A. L. Schawlow. *Phys. Rev. Lett.* 27: 707 (1971).
88. Savitzky, A., and M. Golay. *Anal. chem.* 36: 1628 (1964).
89. Green, G. L., and T. C. O'Haver. *Anal. Chem.* 46: 2191 (1974).
90. Johnson, S. A., W. M. Fairbank Jr., and L. Schawlow. Appl. Opt. 10: 10 (1971).
91. Talmi, Y. *Anal. Chem.* 47: 658A (1975); 47: 697A (1975).
92. Donati, M., G. J. Guekos, H. S. H. Gunthard, M. J. D. Strutt and U. P. Wild. *IEEE Trans. Instr. Meas.* 24: 170 (1975).
93. Winefordner, J. D., J. J. Fitzgerald and N. Omenetto. *Appl. Spectros.* 27: 369 (1975).
94. Cooney, R. P., T. Vo-Dinh, G. Walden and J. D. Winefordner. *Anal. Chem.* 49: 939 (1977).
95. Cooney, R. P., T. Vo-Dinh and J. W. Winefordner. *Anal. Chim. Acta* 89: 9 (1977).
96. Warner, I. M., J. B. Callis, E. F. Davidson, M. Gouterman and G. D. Christian. *Anal. Lett.* 8: 665 (1975).
97. Albrecht, G. F., E. Kallne and J. Meyer. *Rev. Sci. Instr.* 49: 1637 (1978).
98. Hall, L. A., A. Armstrong, W. Moomaw and M. A. El-Sayed. *J. Chem. Phys.* 48: 1395 (1968).
99. Schmidt, J., and J. H. Van der Waals. *Chem. Phys. Lett.* 2: 640 (1968).
100. Hornig, A. W., and J. S. Hyde. *Mol. Phys.* 6: 33 (1963).
101. DeGroot, M. S., I. R. M. Hesselmann and J. H. Van der Waals. *Mol. Phys.* 12: 259 (1967).
102. Tinti, D. S., M. A. El-Sayed and A. H. Maki. *Chem. Phys. Lett.* 3: 343 (1969).
103. El-Sayed, M. A. In: *Excited States,* Vol. I, E. C. Lim Ed. (New York: Academic Press, Inc., 1974).
104. Parmenter, C. S., and M. W. Schuyler. *Chem. Phys. Lett.* 6: 339 (1970).
105. McDonald, J. R., and L. E. Brus. *Chem. Phys. Lett.* 16: 587 (1972).
106. Lewis, R. S., and E. K. C. Lee. *J. Chem. Phys.* 61: 3434 (1974).

107.   Dworetsky, S. H., L. E. Brus and R. S. Horack. *J. Chem. Phys.* 61: 1581 (1974).
108.   Miller, R. G., and E. K. C. Lee. *Chem. Phys. Lett.* 33: 104 (1975).

CHAPTER 4

APPLICATIONS OF LIQUID CHROMATOGRAPHY

Kenneth Ogan, Gary J. Schmidt and Ron L. Miller
The Perkin-Elmer Corporation
Norwalk, Connecticut

INTRODUCTION

The number of applications of modern liquid chromatography (LC) is growing rapidly. It is impossible to thoroughly cover in the short space of this chapter all the different fields in which LC has been applied; therefore, we have elected to summarize the recent developments in three areas: polymers, biochemistry and environmental analysis. It is assumed that the reader is familiar with the fundamentals of LC; there are several excellent introductory books[1-3] on the subject.

There are three parts to a chromatographic analysis: (1) sample preparation, (2) chromatographic separation and detection and (3) confirmation and quantitation of the compounds of interest. Sample collection and preparation are a crucial part of an analytical method, but each sample type often requires procedures specific to it that are not necessary for other sample types. Thus, sample preparation procedures are much more a function of the specifics of the sample than of the subsequent chromatography; hence, this presentation will focus more on the details of the chromatography, detection and quantitation involved in the different applications.

In the chromatography step, the introduction of bonded-phase packings has had a significant impact on LC, permitting a much wider variety of chromatographic separations than is possible with just silica. The combination of the high efficiency of microparticulate substrates with the chromatographic flexibility afforded by bonded phases is largely responsible for the rapid growth of modern LC. Thus, microparticulate C_{18} bonded-phase silica is the most common column packing used today. Other bonded phases are available for applications in which neither silica nor C_{18}-silica provides the desired chromatographic behavior.

While the refractive index detector comes the closest of all LC detectors to being a universal detector, it suffers from a lack of sensitivity. Of the other LC detectors currently available, the ultraviolet (UV) detector offers the best sensitivity for the widest range of compounds. Several other detectors offer high sensitivity for specific types of compounds; for example, the fluorescence and electrochemical detectors. Scott[4] has given a thorough review of LC detectors.

Preliminary identification of the peaks registered by the chromatography detector is made by matching retention times with those of standards. Additional data are required or conclusive peak identification in complex chromatograms, either from additional chromatography or from spectral characterization. Reeve and Crozier[5] have described the assessment of the information content of these various additional experiments, as well as the use of the information content in defining the probability of accurate peak identification. Mass spectra obviously contain the most information for the majority of compounds, but online mass spectrometric detectors for LC have only recently been introduced commercially and are in only limited use at the present time. Optical spectra, especially UV spectra, have been used fairly extensively for peak identification, although UV spectra are not as conclusive as mass spectra. The use of highly selective detectors, such as fluorescence and electrochemical detectors, also improves the accuracy of peak identification.

APPLICATIONS IN POLYMER ANALYSES

The application of liquid chromatography and, in particular, of gel permeation chromatography, to the control of polymer processing and synthesis requires that the sampling and analysis time be short enough that the process being monitored will not become unstable or proceed too far before the analytical results can be used in a feedback loop. For batch blending processes, analyses are only needed at selected control points, typically at the end of a stage, to determine what adjustments need to be made. Batch chemical reactions require periodic monitoring of conversion levels until the desired molecular weight distribution has been reached. Continuous steady-state processes require periodic observations of conversion or composition parameters for comparison with the set points for the process.

In the product development laboratory or process development laboratory, liquid chromatography is used to measure yields, conversion levels and product characteristics, and to elucidate reaction kinetics. Additional generation of data during scaleup ensures that the same product characteristics are maintained during transition from the laboratory through pilot plant and into production scales.

This section discusses the application of liquid chromatography to the determination of polymer additives, polymer molecular weight distributions, and fingerprinting of low-molecular-weight polymers.

Polymer Additives

Various kinds of additives may be included in the formulation of a polymer product, depending on the end use of the product. These additives may be used to modify the viscoelastic properties of the polymer system at various stages of production. Plasticizers are added to lower the modulus. Tackifiers are used to develop adhesion in polymer solutions, dispersions, melts and even dried films. Diluents are added to lower the viscosities of thermoplastic melts and uncured liquid polymers. Surfactants are used to improve the wettability of latex- and emulsion-based adhesives, paints and coatings. Polymer additives may be used to affect the ability of a material to interact with its environment. Antioxidants, fire retardants and UV absorbers enhance a polymer's resistance to aging, heat and light. Dyes are sometimes added to improve marketability of a finished product. Crosslinking and vulcanizing agents may be added to improve structural integrity, and accelerators are often used to speed the cure of thermosetting systems. Also, the degree of residual monomer or solvent may be of interest.

Conventional modes of chromatographic separation, absorption, reversed-phase and ion exchange are used in the analysis of additives, but the most commonly used chromatographic mode is gel permeation chromatography (GPC), which involves spearation by molecular size on a column material having a carefully controlled pore size distribution. Molecules with hydrodynamic radius less than the pore size diffuse into the packing, while larger molecules are excluded. Thus, large molecules pass through the column unretained, medium-sized molecules enter some of the pores, and small molecules enter all the pores and are retained longest. Types of packings used in GPC include polystyrene gels, silica gels, controlled pore-size glass beads, polyvinyl acetate gels, starch-based gels and others. The polystyrene gels offer the highest number of theoretical plates, but can be used only with solvents of slight to moderate polarity, such as chlorinated hydrocarbons, toluene and tetrahydrofuran (THF), all of which swell the polymer gel. Highly polar solvents, such as water, alcohols, acetonitrile and dimethylformamide, which shrink the gel, must not be used with these columns, nor should highly nonpolar solvents such as hexane, as these tend to swell the polystyrene to too great an extent, and may even dissolve it. Silica gel and glass columns may be used with any solvent, but suffer from a lack of efficiency for very low-molecular-weight compounds.

An individual GPC column is suitable for only a specified range of effective molecular weight compounds, and several columns in series are normally used to cover the range of molecular weights of interest. The high efficiency of the newer microparticulate polystyrene gel columns has greatly reduced the number of columns needed for most analyses, often to three or fewer 25-cm columns.

Phthalate plasticizers have been analyzed in liquid epoxy resin matrices by reversed-phase chromatography after only a filtration step, and in poly(vinyl acetate) emulsions after removal by chloroform extraction. In the latter analysis, the polymer remains emulsified in the aqueous layer; the technique is applicable to other types of plasticizers as well.

Figure 1 shows a GPC separation of a urea-formaldehyde organosol. This analysis was performed using a single 25-cm column, 0.46 cm inside diameter. The technique works even better for plastisols and organosols with larger molecular weight differences between the components.

Although most common diluents in thermosetting polymer systems (e.g., hardeners for epoxy resins and urethane adhesives) are more conveniently determined by gas chromatography, liquid chromatography can be used, for example, in the determination of diisocyanates[6,7]. The wax diluents utilized in hot melt adhesives can be effectively separated from the base polymer, usually an ethylene-vinyl acetate copolymer (EVA), using high-temperature (110–135°C) GPC in a chlorinated hydrocarbon mobile phase after preliminary solvent extraction of the tackifying resin. Quantitation of the wax is difficult because differential refractometry must be used to detect the wax with any degree of sensitivity, and the refractive index of hydrocarbon waxes varies considerably. The EVA is much easier to quantitate in any case. If the wax consists mostly of n-alkanes, gas chromatography (GC), low-molecular-weight GPC[8] or differential scanning calorimetry[9] can be used to fingerprint the wax.

Methods for separating surfactants from polymers by GPC have been documented[10], but the analysis is extremely difficult when very small amounts are present in the sample, as is most often the case. In the analysis of emulsion polymers, complex wet chemical sample preparation schemes are necessary, and even then the separation scheme must be designed for a particular type of surfactant.

Antioxidants have been separated from each other by both normal- and reversed-phase chromatography. Snyder and Kirkland[10] described the separation of hindered phenols and triphenyl phosphite using a silica column. Stoveken, at Perkin-Elmer, has studied the effect of temperature on the separation of a variety of antioxidants and mold release agents using a C_{18} reversed-phase column and UV detection at 200 nm, optimum selectivity was obtained at 70°C. The separation of a mixture of BHT, UV-531, AM-340, Irganox 1010 and Irganox 1076 by reversed-phase chromatography using gradient elution with water and THF is shown in Figure 2. Ishiguro[11] described the separation of Irganox 1076 and BHT from a polypropylene extract using low-molecular-weight GPC.

Accelerators for epoxy resin systems, usually found in the hardener, have been examined by a number of techniques. Kirkland has separated chlorophenyl ureas

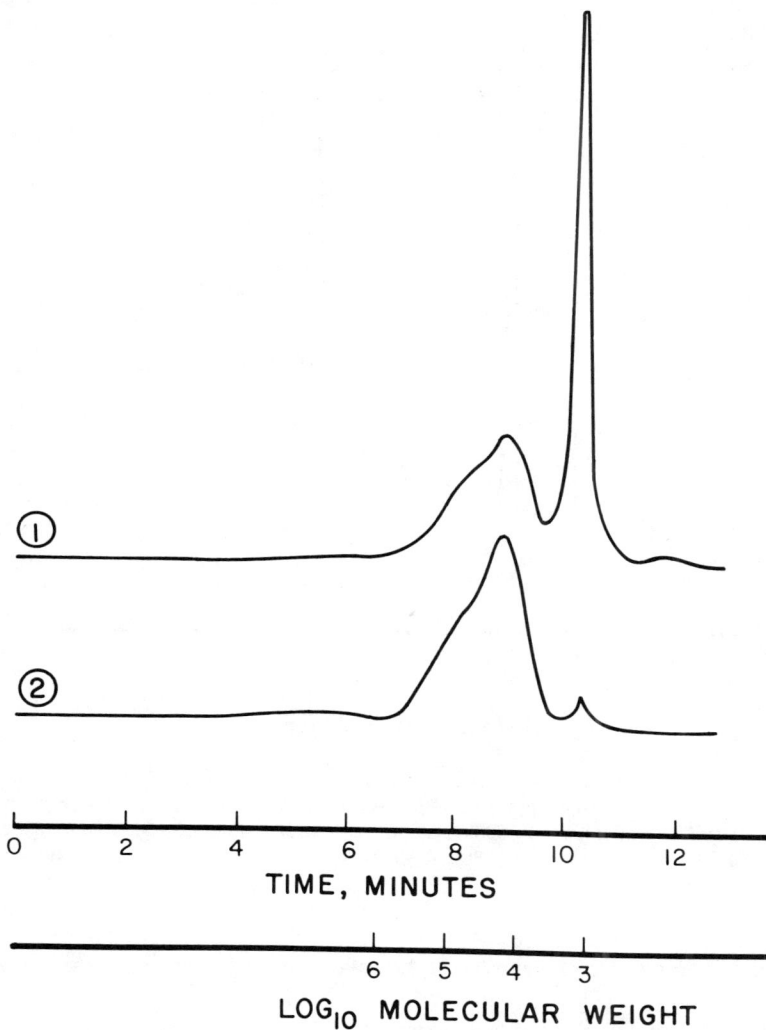

Figure 1.   Urea-formaldehyde organosol separated by GPC using a single Shodex A-80 M/S column with THF at 1.0 ml/min as the mobile phase. The higher molecular weight component is the urea-formaldehyde resin and the lower molecular weight component is a glyptal plasticizer. Chromatogram 1 resulted from an 8-μl injection of a 16.5 mg/ml solution with UV detection at 265 nm. Chromatogram 2 resulted from an 8-μl injection of 1.65 mg/ml solution with UV detection at 230 nm.

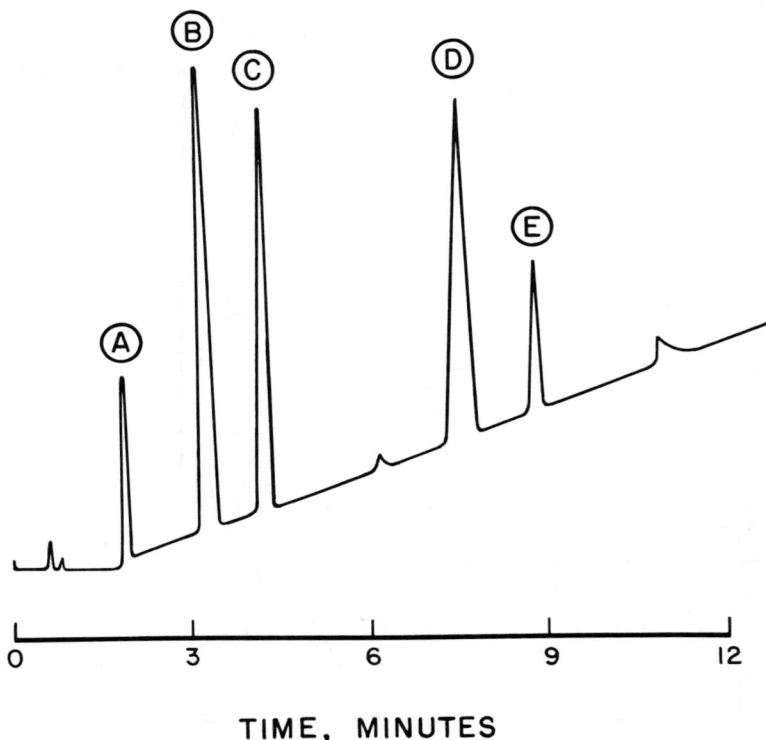

Figure 2.   Analysis of polymer additives on an HC-ODS column (25 × 0.26 cm) at a flow rate of 2.0 ml/min with UV detection at 210 nm. The mobile phase was a gradient beginning with 50% THF in water with a slope of 3.33%/min THF. Peak identities are: A: BHT; B: UV-531; C: AM-340; D: Irganox 1010; and E: Irganox 1076.

by liquid-liquid chromatography, both conventional and reversed phase[12, 13], and also by adsorption chromatography[14].

The determination of residual monomers and solvents has been carried out primarily by gas chromatography, but some LC methods exist. Normal-phase chromatography, for example, has been used to determine residual monomer in poly(methyl methacrylate)[15], and diisocyanates, monomers used in polyurethane synthesis, have been determined by normal-phase chromatography using low concentrations of acetonitrile in chloroform as the mobile phase[16]. GPC may be used to separate residual monomers and solvents from polymers and polymer films as a sample preparation step for subsequent GC analysis.

Determination of Polymer Molecular Weights

GPC can be used to characterize the range of polymer molecular weights in terms of the moments of the molecular weight distribution. The GPC column set is calibrated using a series of monodisperse polymer standards, the molecular weights of which have been well characterized by independent methods such as light scattering, intrinsic viscosity measurements and membrane osmometry. Because the separation is based on molecular size rather than molecular weight, calibration must be carried out using the same mobile phase and flow conditions to be used in the analysis of the samples, since the degree of polymer chain uncoiling in solution is dependent on the solvent.

A logarithmic plot of molecular weights vs elution volume for the standards is constructed as in Figure 3. The exclusion limit, M_e, is the molecular weight corresponding to the hydrodynamic volume of the smallest polymer chains unable to permeate any of the pores in the column packing. The associated exclusion volume, V_e (also called V_o), is the void volume of the chromatographic system. At the other extreme is the permeation volume, V_p, which includes the pore volume accessible to the molecules with the lowest molecular weight, M_p. The GPC column set can thus be used for polymer molecules, with hydrodynamic radii between those corresponding to the standard molecular weights, M_e and M_p. This calibration curve of molecular weight vs elution volume is used to translate the elution volume scale of a GPC chromatogram into a molecular weight scale. The GPC peak then represents a distribution of molecular weights. The moments of this distribution are calculated as follows. The GPC peak is divided into retention time (elution volume) slices of equal width. The molecular weight, M_i, associated with the ith slice is obtained from the calibration curve. The area of the ith slice is denoted by H_i. (If the slice width is sufficiently small, the height of the ith slice may be used instead of the area.) The moments are then given by

$$M_n = \frac{\Sigma H_i}{\Sigma (H_i/M_i)} \tag{1}$$

$$M_w = \frac{\Sigma (H_i M_i)}{\Sigma H_i} \tag{2}$$

$$M_z = \frac{\Sigma (H_i M_i^2)}{\Sigma (H_i M_i)} \tag{3}$$

The first moment, M_n, is known as the number average molecular weight and is strongly affected by changes in the low-molecular-weight end of the distribution. The second moment is called the weight average molecular weight, M_w, and

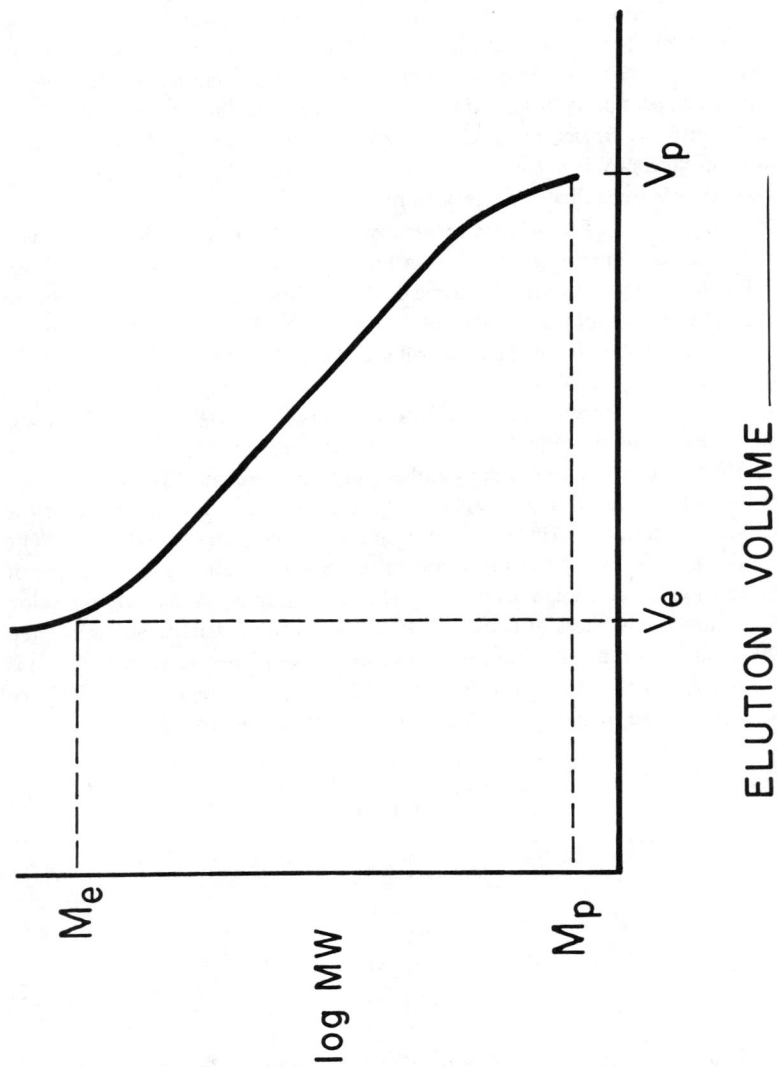

Figure 3.   Calibration curve for a GPC column set.

reflects the high-molecular-weight portion of the distribution. The third moment, or Z-average molecular weight, M_z, reflects the very high-molecular-weight portion of the curve. The ratio of the second to the first moment is the polydispersity, P, and is a measure of the breadth of the distribution; a monodisperse material has a polydispersity of unity.

Polystyrene[17,18], polyethylene[19,20], polyvinylacetate[21], polyvinylchloride[21], polydimethyl-siloxane[22,23], polymethylmethacrylate[23], polyethylene glycol[22] and various acrylic copolymers[24] have been analyzed in this manner using a differential refractometer. This detector does, however, suffer from a lack of sensitivity in some cases, but this often can be overcome by changing to an eluent with a greater difference in refractive index from the polymer.

Infrared detection has been used to analyze t-butylmethacrylate-styrene block polymers[25] and styrene-vinyl stearate copolymers[26], while ultraviolet detection has been used for styrene-butadiene copolymers[27], polyvinyl acetate[22], ethylene-vinyl acetate copolymers[22], vinyl acetate-dibutyl fumarate copolymers[22] and acrylic copolymers[24].

The computation of actual molecular weights of polymer samples chemically different from those used as standards is somewhat more complicated. One simple approach uses a parameter called the "Q factor," defined as the ratio of molecular weight to mean chain length. The Q factor is computed for the sample and standards as described by Hendrickson and Moore[28], and the moments of the molecular-weight distribution are multiplied by the ratio of the Q factor for the sample to that for the standards. This approach assumes that the degree of uncoiling of the polymer chains is the same for all samples and standards, an assumption that breaks down at high concentrations or when either of the polymeric species exhibits reduced solubility in the mobile phase. If monodisperse polystyrenes are used as standards, one can construct the calibration curve in terms of hydrodynamic radius and report the moments of the sample distribution in terms of molecular size, rather than molecular weight. This approach assumes that the Q factor for the polystyrene is independent of molecular weight, which appears to be the case for dilute solutions of polystyrene in THF and chlorinated hydrocarbons.

A more accurate approach is to use a universal calibration curve. Grubisic et al.[29] have shown that if the product of weight average molecular weight and the intrinsic viscosity $[\eta]$ is plotted against retention volume, the same calibration curve is obtained for all polymers, including highly branched polymers and copolymers. Furthermore, the intrinsic viscosity can be related to the molecular weight by

$$[\eta] = K (M_w)^\epsilon \tag{4}$$

where K and ϵ are the Mark-Houwink parameters for the polymer-solvent pair in question. The hydrodynamic volume, Z, can then be expressed as[24]

$$Z = [\eta] M = K (M)^{(\epsilon + 1)} \tag{5}$$

If the Mark-Houwink parameters are known for the polymer sample of interest, the moments of the Z distributions can be calculated from equations analogous to Equations 1, 2 and 3. Then, the actual molecular-weight moments for this case are obtained from inversion of Equation 5:

$$M_j = \left(\frac{Z_j}{K}\right) \exp \left[\frac{1}{\epsilon + 1}\right] \tag{6}$$

The analysis of the composition of copolymers, particular addition copolymers, in terms of molecular weight, introduces additional considerations that must be dealt with. The ideal analytical system would utilize a detector specific for each monomer in the copolymer, as in the analysis of a styrene-methyl acrylate-methacrylic acid terpolymer[24], for example, in which styrene is detected by a fixed-wavelength UV detector at 254 nm; methyl methacrylate by an IR detector operated at 1740 cm^{-1}; and methacrylic acid by an IR detector operated at 1700 cm^{-1}. The ideal case of a detector specific for each comonomer is not often realized in practice. Provder[24] has shown that it is sufficient if each detector (one detector for each comonomer) responds differently to each component. Provder's work utilized IR, UV and differential refractometer detection; each wavelength used in a spectroscopic detector is equivalent to a "different" detector, since the response factors of the comonomers are wavelength dependent. Following Provder[24], the response of the different detectors is given by a matrix equation; the response vector (with individual detector responses as elements) is equal to the response matrix (with the response of the ith detector to the jth component as elements) right-multipled by the weight fraction vector. Obviously, this becomes very complex for more than two or three components.

Fingerprinting of Low-Molecular-Weight Polymers

Sets of low-molecular-weight GPC columns often have sufficient resolution to separate oligomers, particularly if the monomer itself is large, as is the case with epoxy resins. Most of the same considerations discussed in the previous section still apply, plus a few others. More care must be taken with the low exclusion limit microparticulate polystyrene gel comumns, as they are extremely fragile with respect to high flowrates and pressures and physical shock. High resolution is usually needed, which often necessitates the use of more columns. Finally, since oligomers themselves are monodisperse by definition, one need not always go through the complexities of the previously outlined calculations.

As an illustration of the technique, Figure 4 shows the separation of a 1,3-butylene glycol-lactic acid condensation product. The chromatogram shows the dilactate ester (peak A), the monolactate ester (peak B) and a third peak (C) corresponding to the two monomers. Thus, the degree of conversion can be monitored for the condensation reaction; this technique has also been applied to the reaction of 1-butanol with epichlorohydrin[30].

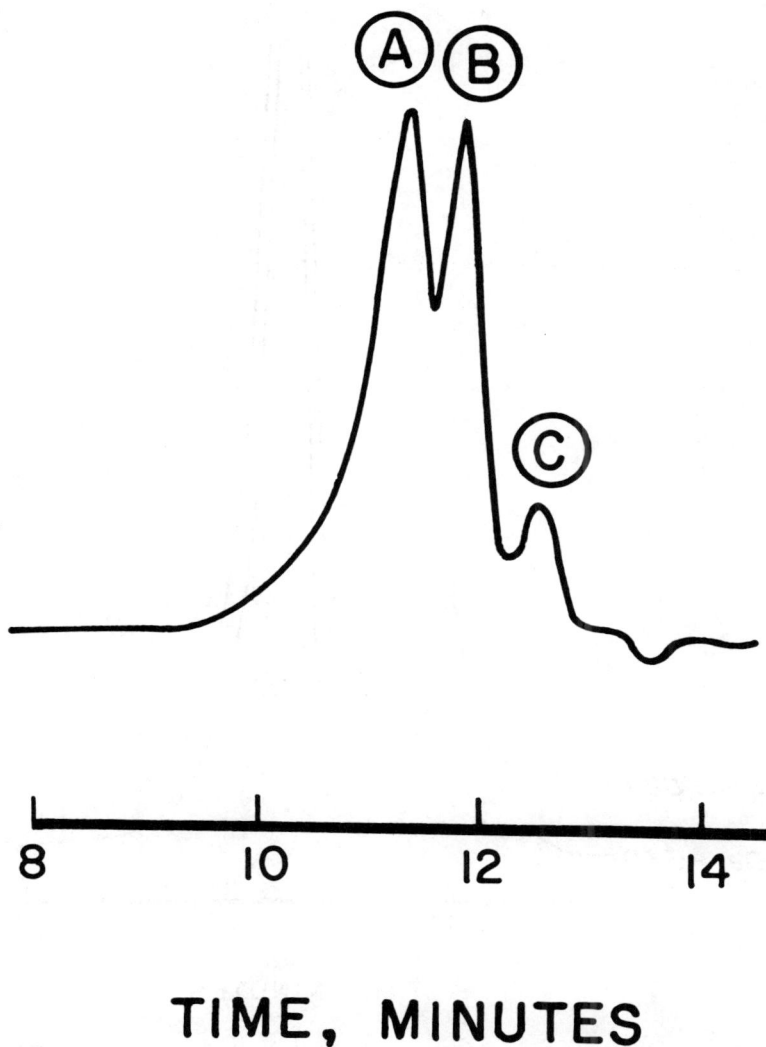

Figure 4.   Separation of condensation product of 1,3-butylene glycol and lactic acid on three Shodex A-802/S columns in THF at 2.0 ml/min; 1.5 mg were analyzed and detection was by differential refractometry.

Fingerprints of epoxy resins are shown in Figures 5 and 6. The epoxies shown in Figure 5 may be distinguished from each other by the results of the normalized integration of peaks C (trimer) and D (tetramer), as given in Table I. The integration for the chromatograms of the two high-viscosity epoxy resins shown in Figure 6 is given in Table II.

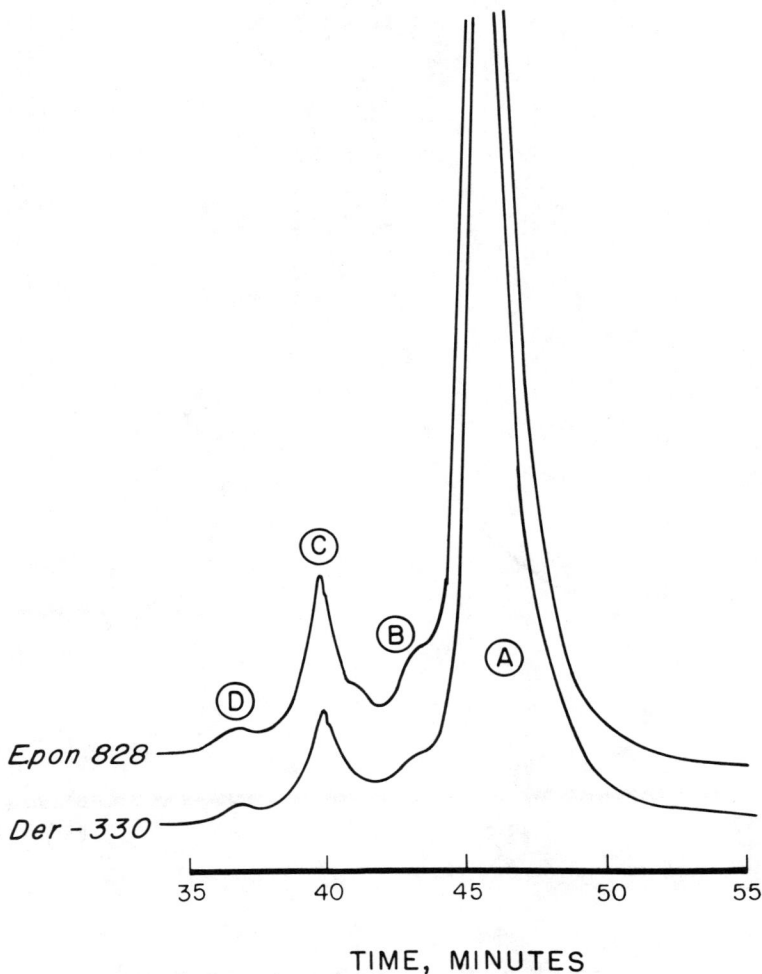

Figure 5.   GPC separation of 40 μg of epoxy resins on three Shodex A-802/S columns in THF at 0.5 ml/min. UV detection at 276 nm was used.

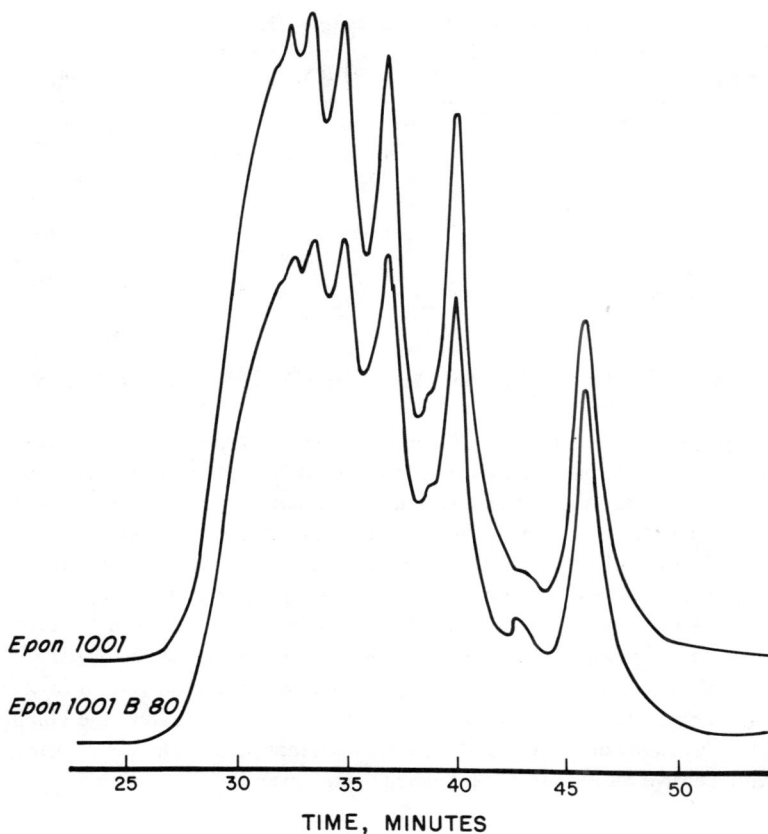

Figure 6.   GPC separation of epoxy resins, 50 μg each, on three Shodex A-802/S columns in THF at 0.5 ml/min, with UV detection at 276 nm.

Table I.   Epoxy Composition from Peak Areas

	DER-330 (Dow)	Epon 828 (Shell)
% Trimer	8.60	10.46
% Tetramer	1.02	1.45

This technique is applicable in principle to the analysis of low oligomers of any polymeric system, and has been used to analyze low-molecular-weight polystyrenes[30] and polyethylene glycols[31], diallyl phthalate prepolymers[32] and other systems.

Table II.   Peak Areas from Figure 6

Retention Time	Epon 1001-B-80	Epon 1001
32.46	28.60	31.48
33.47	10.23	11.24
34.87	12.94	13.19
36.82	15.42	15.42
39.97	17.59	18.25
43.03	2.81	–
46.00	12.41	10.42

CLINICAL APPLICATIONS OF LIQUID CHROMATOGRAPHY

The concept of chromatography is not new to the practical clinical chemist. For years, classical column chromatography systems have been used for many clinical separations, either for separating individual compounds or for fractionating mixtures into compound classes. Classical ion exchange chromatography resulted in the proliferation of amino acid analyses. The use of modern liquid chromatography has gained increasing popularity in the clinical laboratory over recent years, spurred by the rapid technological advances that have occurred in the areas of instrument design and performance, column technology and detector sensitivity.

The objective of this section is to present some general considerations to the use of liquid chromatography in clinical chemistry. Understandably, the emphasis will be toward the determination of compounds that possess clinical significance. Comprehensive reviews of this subject are available[33-35].

Sample Pretreatment

There are three commonly used sample pretreatment procedures for preparing clinical specimens for analysis by liquid chromatography. These include direct sample injection without any sample treatment preparation or modification of the sample, protein precipitation techniques and extraction procedures.

Direct injection of serum has been used only in isolated instances. The technique requires aliquoting a small volume of serum along with an aliquot of an internal standard solution into the injection syringe and injecting the sample directly onto a guard column, which is placed just prior to the analytical column in the chromatography system. A serious problem associated with this injection technique is the introduction of serum matrix material, which may precipitate onto the top of the guard column, necessitating replacement of the guard column every 20 to 40 analyses.

Protein precipitation is most usually effected by mixing approximately 50–100 μl of serum with one or two volumes of acetonitrile. This prevents the introduction of excess protein material, which might precipitate within the chromatography system causing clogging of the column and increased system backpressures.

Many of the classical solvent or adsorption type of extractions may be utilized for sample pretreatment in liquid chromatography. In those instances in which the compounds are present at sufficiently high concentrations, microextractions on serum volumes as little as 50 μl may be used successfully. Where larger samples must be processed, e.g., where compounds are present at low concentrations, larger volumes of organic extractants are often all that are required. Where interfering substances are present, classical acid-base back extractions often may be employed successfully.

Analysis of Drug Substances

There is an extensive and rapidly growing volume of literature on applying liquid chromatography to the monitoring of therapeutic drugs to control dosage and effect proper drug management programs. An excellent review by Wheals and Jane[36] references almost 360 publications dealing with drug determination.

Theophylline

A great deal of interest has been displayed in recent years in the liquid chromatography determination of theophylline in serum. Prior to the proliferation of LC methodology, the most commonly used techniques were UV spectroscopy and gas chromatography. Most UV procedures suffered from potential interference from other xanthines, including caffeine, and GC methods required time-consuming derivatization techniques.

In 1976, Adams et al.[37] described a procedure for the determination of theophylline in a 50-μl serum sample using reversed-phase chromatography on a C_{18}-silica column. The theophylline was extracted using 200 μl of a chloroform:isopropanol extractant after buffering the serum with a pH 6 phosphate buffer. The total analysis time for a single sample was about 15 minutes, and the day-to-day precision was about 5% over the therapeutic concentration range of 5–20 mg/l. A method by Peng et al.[38] also used reversed-phase chromatography, with a preliminary acetonitrile deproteinization procedure added. Theophylline concentrations of less than 1.5 mg/l could be determined in as little as 10 μl of serum and required only 7 minutes per sample. Recently, Kelley et al.[39] reported a potential interference in many theophylline procedures using protein precipitation pretreatment procedures from some cephalosporin antibiotics. These antibiotics may coelute on some reversed-phase packing materials with the theophylline. For most

LC procedures, either 8-chlorotheophylline or β-hydroxyethyl-theophylline are used as internal standards. The separation of some xanthine standards is shown in Figure 7. Theobromine and caffeine are commonly monitored to ensure that they do not interfere with the compounds of interest. Dyphylline is an antiasthmatic drug that is less frequently monitored.

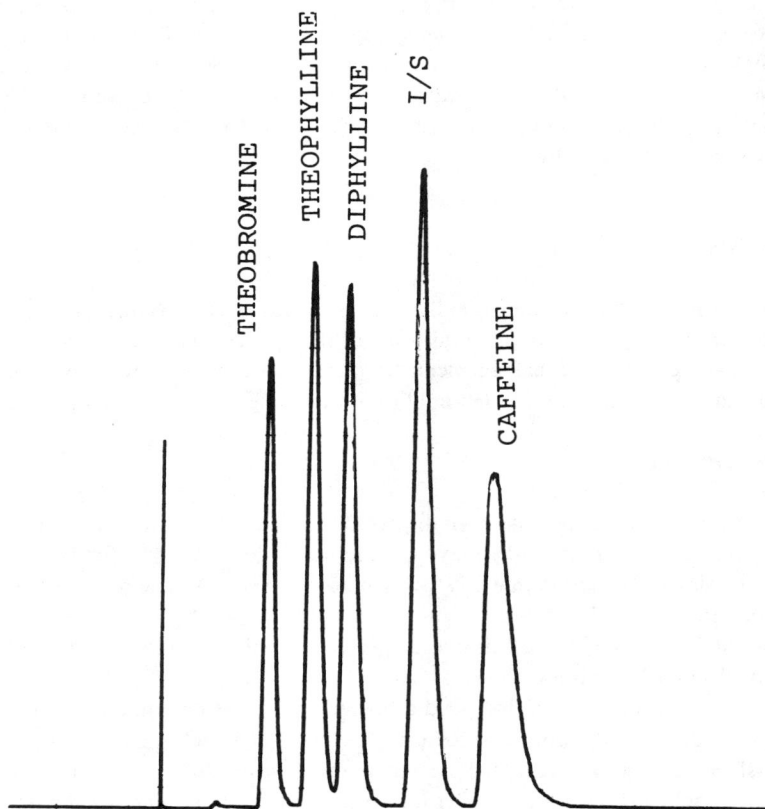

Figure 7.   Separation of xanthine standards on HC-ODS column (UV detection at 273 nm). Mobile phase was 2% acetonitrile in water at 1.5 ml/min; column temperature was 55°C. Peak identification: TB = theobromine; TP = theophylline; Dip = Diphylline; 8-Cl = 8-chlorotheophylline; Caf = caffeine.

Anticonvulsant Drugs

The use of variable-wavelength UV detection has been the foundation for many LC procedures describing the determination of the anticonvulsant drugs. In

1976, Adams and Vandemark[40] used LC to determine the five major anticonvul-sant drugs over their respective therapeutic concentration range. They used a C_{18} bonded-phase column with 17% acetonitrile in water for the mobile phase. The 0.5-ml serum sample was pretreated by adsorbing the drugs onto activated char-coal followed by their desorption using an organic solvent. A UV detector set at 195 nm was used. Several papers[41-43] describing the extension of this earlier work have since appeared in the literature. In 1977, Adams[44] also published a review article on the determination of the anticonvulsant drugs using LC, which addressed in detail the optimization of the chromatographic conditions.

Kabra and co-workers[45-47] have published extensively on the determination of anticonvulsants using LC. The most recent procedure from their laboratory[48] uses a C_{18} bonded-phase column and isocratic elution with an acetonitrile and water mobile phase. The sample pretreatment procedure they have developed consists of deproteinization of 100 μl of serum with 200 μl of acetonitrile containing 5-(p-methylphenyl)-5-phenylhydantoin (MPPH), which is used as an internal standard. Detection of the drugs was effected using a far UV detection wave-length of 195 nm. Soldin and Hill[49] also have described an LC procedure for determining the anticonvulsants; however, they used a two-wavelength detection technique to confirm the identity of the compounds eluting from the column.

A number of procedures have very recently been described for determining the anticonvulsant drug, valproic acid. Farinotti and Mahuzier[50] described a postcol-umn reaction procedure using bromocresol purple and subsequent detection at 425 nm. In 1978, Schmidt and Slavin[51] described a quick precolumn derivatiza-tion procedure using phenacyl bromide reagent and detection at 255 nm. Sutheimer et al.[52] applied these phenacyl esters to the analysis of 0.5-ml serum samples using cyclohexanecarboxylic acid as an internal standard and showed excellent quantitation throughout the entire therapeutic concentration range. Figure 8 shows the results obtained for two serum samples using this procedure. Chromatograms A and B represent valproic acid concentrations of 57 and 168 mg/l, respectively.

Tricyclic Antidepressant Drugs

Another class of drug compounds that has gained increasing interest as far as LC methods development is concerned is that of tricyclic antidepressant drugs. Since the tricyclic drugs have basic pK values, most LC procedures have utilized ion-pair techniques[53,54] or adsorption chromatography systems[55-57]. Knox and Jurand[53] described the chromatographic behavior of tricyclics paired with per-chlorate ions on silica columns. Proelss et al.[54] recently described the use of reversed-phase chromatography using pentanesulfonic acid as the ion-pairing reagent. The drugs were extracted from 2 ml of serum into a 10 ml hexane/iso-amyl alcohol extractant at pH 14. Isocratic elution with a mobile phase consisting

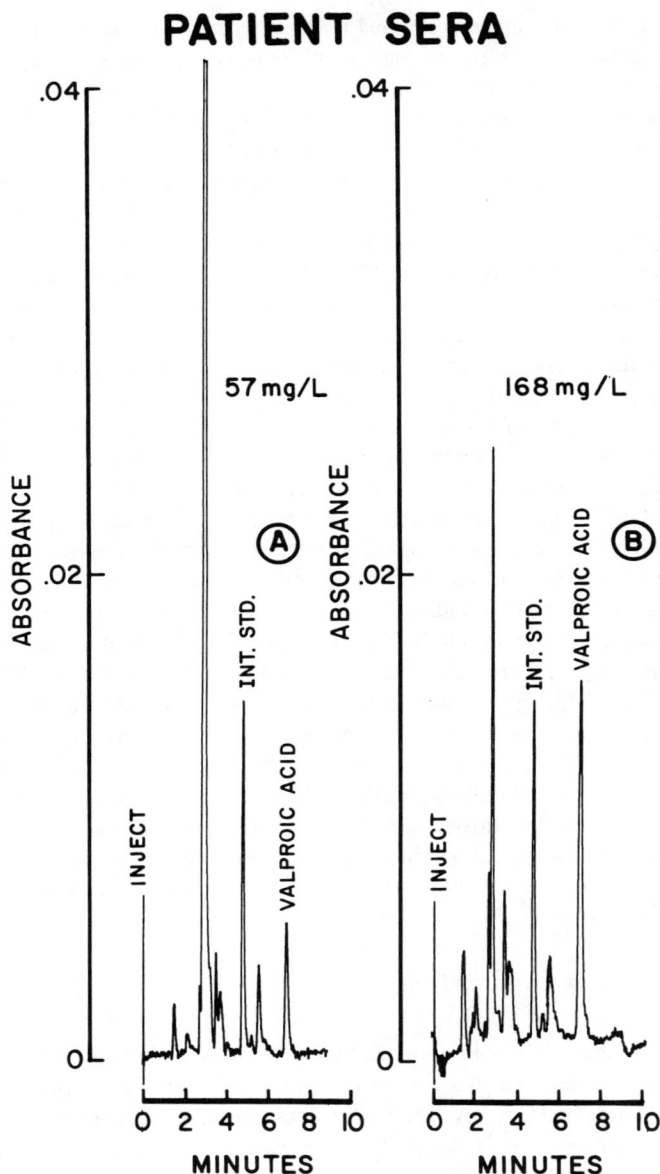

Figure 8.   Determination of valproic acid in serum. Instrumentation was a Perkin-Elmer Series 2/2, LC-55B UV detector set at 247 nm, and a 5-μm Spherisorb ODS column. The mobile phase composition was 60/40/0.2 acetonitrile/water/phosphoric acid at 2.0 ml/min and 50°C.

of methanol, acetonitrile and 0.1 mol/l phosphate buffer was used. Five of the commonly used tricyclic drugs could be determined simultaneously within 15 minutes. Using a UV detector set to 254 nm, detection limits to 2 μg/l could be obtained. Of 48 drugs tested for interference, 9 caused potential interferences with the compounds of interest.

In 1978, Vandemark et al.[56] described the use of a 5-μm silica column using an organic mobile phase for determining the tricyclics. A small amount of ammonium hydroxide was added to the mobile phase to prevent the ionization of the basic tricyclic drugs. Amitriptyline and imipramine and their pharmacologically active N-desmethyl metabolites could be determined simultaneously. Figure 9 illustrates this separation. Protriptyline, a seldom used tricyclic drug, was used as an internal standard. A variable-wavelength UV detector set to 211 nm was used. Of the drug compounds tested for interferences, doxepin and diphenhydramine interfered with the analysis of amitriptyline. A short note published by Sutheimer[57] showed that the substitution of diethylamine for the ammonium hydroxide in the mobile phase permitted the separation of doxepin from amitriptyline.

Analgesics

Several papers[58-61] have appeared in the literature describing LC methods for determining the analgesic drug, acetaminophen. Blair and Rumack[61] described the use of cation exchange chromatography using Aminex A-5 resin. Serum samples were analyzed by aspirating a 1.5-μl sample into the injection syringe along with 1.5 μl of N-butyryl-p-aminophenol solution, which was used as an internal standard, and injecting into the LC. Since the procedure involved direct serum injection, a precolumn was used, which was replaced after approximately 40 injections. The total chromatography time was about 1 hour per sample.

Horwitz and Jatlow[59] described the use of a C_{18} reversed-phase chromatography system and isocratic elution with 5% acetonitrile. A 0.5-ml serum sample was extracted with ethyl acetate, and an aliquot of the reconstituted residue was injected onto the column. A variable-wavelength UV detector set at 250 nm was used. The concentration of the drug was quantitated using the peak height ratio method, with N-propionyl-p-aminophenol as an internal standard. Hydrochlorothiazide was found to interfere with acetaminophen, and sulfadiazine had retention characteristics similar to those of the internal standard. Rosano et al.[58] and Gotelli et al.[60] have published similar procedures.

Cardiac Drugs

A number of excellent LC procedures have been developed for determining the cardiac drugs. Adams et al.[62] described a reversed-phase procedure for determin-

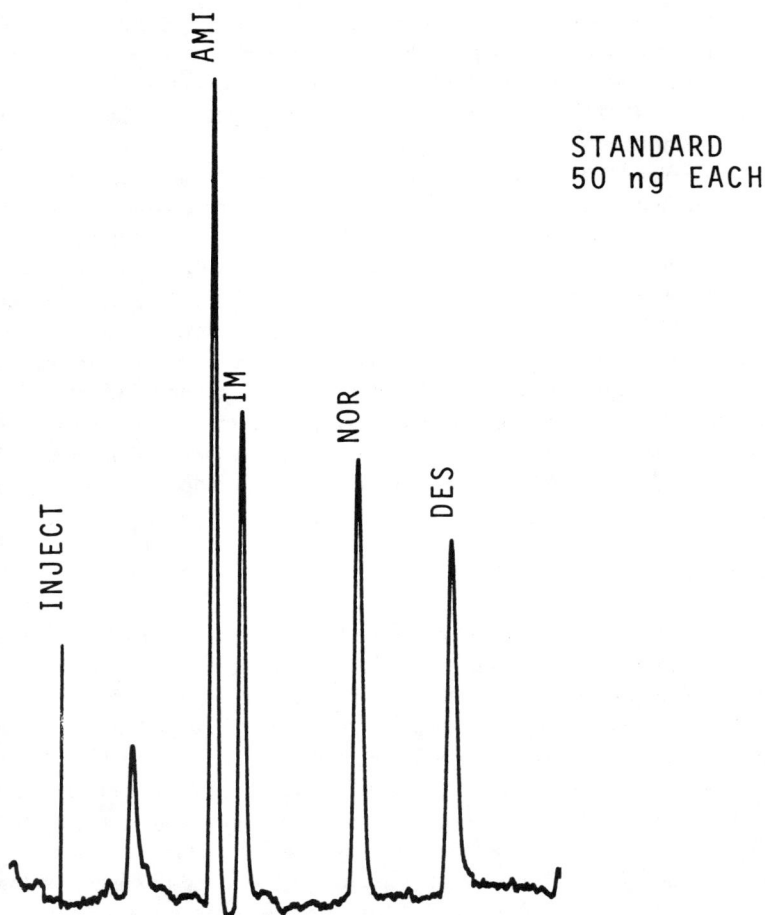

Figure 9. Separation of the major tricyclic antidepressant drugs on a Silica B/5 column. Mobile phase was 89.8/10/0.2 acetonitrile/isopropyl alcohol/ammonium hydroxide at 1.5 ml/min flowrate and 65°C column temperature. Peak identification: Ami = amitriptyline; Im = imipramine; Nor = nortriptyline; and Dep = desipramine.

ing procainamide and lidocaine using 10% acetonitrile in a 200-mM phosphate buffer and detection at 205 nm. The drugs were extracted from serum by adsorption onto activated charcoal, followed by desorption with an organic solvent. A more recent publication[63] from this laboratory described a modification of this sample pretreatment procedure, which included a microextraction of the drugs from 50-μl serum samples. Shukur et al.[64] described a procedure for determining procainamide and its acetylated metabolite by detecting at 280 nm, although no provision for using an internal standard was made. Of the drugs tested for interference, only sulfathiazole interfered with the analysis of N-acetylprocainamide. Rocco et al.[65] described a similar reversed-phase procedure and also incorporated the use of N-proprionyl procainamide as an internal standard. Dutcher and Strong[66] also described a procedure for determining some of the cardiac drugs using adsorption chromatography.

Several procedures have been described for determining the β-adrenergic blocker drug, propranolol[67-69]. Jatlow et al.[68] used reverse-phase chromatography on a 10-μm C_{18} column, with fluorescence detection at excitation and emission wavelengths of 285 nm and 350 nm. Serum samples of 2 ml were extracted into an organic solvent. After evaporation of the extractant, a portion of the redissolved residue was injected into the LC. The total chromatography time was about 7 minutes, and the linearity of the procedure was demonstrated over the range of 6 to 400 μg/l in plasma.

Analysis of Endogenous Compounds

Polyamines

A considerable interest and rapidly developing body of literature exists for determining the polyamines using LC. Ion exchange[70] and reversed-phase[71] chromatography have been used for the separation of fluorescamine derivatives, and a reversed-phase procedure has utilized UV detection of the tosyl derivatives[72]. The most sensitive LC methods available to date are those that make use of the dansyl derivatives. Dansyl derivatization followed by adsorption chromatography on silica[73,74], normal-phase gradient elution[75] and reversed-phase gradient elution chromatography[76,77] have been described.

Fluorescamine derivatization followed by ion exchange chromatography provides very nice separation of putrescine, spermine and spermidine in 50 minues[70]. This work was not applied, however, to the analysis of biological samples. The reversed-phase procedure of Samejima et al.[71] described the analysis of fluorescamine derivatives, which required two sets of analytical conditions to obtain results for the three major polyamines. Ultraviolet detection of the tosyl derivatives was used, with good chromatographic separations obtained in approximately 20 minutes, although the procedure required a two-step sample cleanup procedure prior to derivatization[72]. Seiler et al.[76] recently reported the analysis of dansylated polyamines on a C_8 bonded reversed-phase column. The analysis time was 40 minutes using a multistep gradient elution program and a postderivatization microcolumn chromatography cleanup procedure. Vandemark et al.[77]

studied the chromatography of the dansyl derivatives using several reversed-phase columns using both isocratic and gradient elution. Results are also given on determining normal and elevated polyamine concentrations in urine samples.

Estrogens

Many publications dealing with the determination of estrogenic steroids have been described[78-85]. Most of these publications have used reversed-phase chromatography. Gotelli et al.[83] described the use of a reversed-phase column for determining urinary estriol concentrations with a variable-wavelength UV detector set to 280 nm. A 1-ml volume of a 24-hour urine specimen was enzymatically hydrolyzed with β-glucuronidase to free the conjugated estrogen followed by extraction into ether. The total chromatography time was about 10 minutes, and carbamazepine was used as an internal standard. The use of the enzymatic hydrolysis reduced the potential interference commonly observed when common acidic hydrolytic procedures are used due to elevated sugar concentrations. Schmidt et al.,[84] used precolumn derivatization to the dansyl derivative and measured urinary estriol concentrations with fluorescence detection. Taylor et al.[85] recently described fluorescence detection of the free estriol using a far UV excitation wavelength of 220 nm. They showed that a considerable increase in both sensitivity and selectivity could be obtained when compared to UV detection.

Amino Acids

The determination of amino acids has often been performed using classical ion exchange liquid chromatography. These procedures generally utilize cation exchange resins and postcolumn reaction with ninhydrin or a fluorescent derivatizing reagent. In 1976, Ersser[86] used cation exchange chromatography to study disorders in amino acid metabolism. Ninhydrin was reacted postcolumn with the eluted amino acids using an air-segmentation flow system to minimize band-broadening. Voelter and Zech[87] also used cation exchange resins for separating many of the protein amino acids. They used a citrate buffer gradient and did an online postcolumn reaction with fluorescamine. The use of a fluorescence detector provided two to three orders of magnitude greater sensitivity as compared to reactions involving ninhydrin. An excellent review of the use of ion exchange systems for determining amino acids has been published[88].

A number of precolumn derivatizing reagents have been described for chemically modifying the amino acids to make them chromatographically more suitable for reversed- or normal-phase chromatography systems. Several of these have been discussed by Deyl[89]. Of particular interest has been the use of dansyl chloride. Bayer et al.[90] described both reversed- and normal-phase chromatography of the dansyl amino acids. Using a fluorescence detector, less than 0.1 pmole of the didansyl lysine derivative could be detected. In 1977, Adams et al.[91] described a C_8 reversed-phase procedure for determining ϵ-aminocaproic acid, which included the formation of the dansyl derivative in a 15-minute precolumn procedure. A procedure developed by Hsu and Currie[92] described the use of a C_{18} bonded phase for separating the dansylated protein amino acids and detected them using a fluorescence detector. Wilkinson[93] used a C_{18} column with an acetonitrile and sodium phosphate buffer at neutral pH for separating 20 protein amino acids with UV detection at 250 nm. The limit of sensitivity was about 100 pmol. Olson et al.[94] recently applied LC with precolumn dansylation for the determination of amino acids in 10 μl of serum. The serum was deproteinized with ethanol and an aliquot derivatized in 2 minutes using an elevated reaction temperature of 100°C. A multistep, nonlinear gradient and a C_8 bonded-phase column was used, and the compounds were detected using a fluorescence detector with excitation and emission wavelengths of 298 and 545 nm, respectively. The chromatography was complete in about 40 minutes. A chromatogram showing the separation of amino acid standards using this procedure is shown in Figure 10.

ENVIRONMENTAL APPLICATIONS

Polycyclic Aromatic Hydrocarbons

The presence of polycyclic aromatic hydrocarbons (PAH) in our environment has been of considerable concern since many members of this class of compounds have long been known or suspected to be carcinogens. PAH are ubiquitous in the environment, occurring in many oil- or coal-based commercial products and being generated by combustion processes. These compounds are well suited to determination by LC, being retained on both normal- and reversed-phase columns, and they are easily detected with a UV detector. Also, many of the PAH of interest fluoresce strongly and, in these cases, LC with fluorescence detection provides the most sensitive means of monitoring these compounds.

There are many references in the literature to the application of LC to the determination of PAH in environmental samples, but it has only been since the introduction of microparticulate packings that there have appeared analyses for more than three or four PAH[95-106]. These particular reports describe the analysis of air particulate extracts[95-98,100,101], dust[104], automobile exhaust[105], wastewater extracts[100,106], oysters[102], and coal, tar or asphalt-related extracts[99,100,103]. Table III identifies most of the PAH for which these samples were analyzed. All but two[99,104] of these analyses used reversed-phase chromatography on a C_{18} bonded-phase column. The first 16 PAH in Table III are those PAH on the U.S. Environmental Protection Agency (EPA) Priority Pollutant list. The chromatogram in Figure 11 shows the separation of 15 of these 16 PAH, plus benzo[e]pyrene, on an HC-ODS column. Figure 12 shows the separation of these 16 PAH plus 10 others.

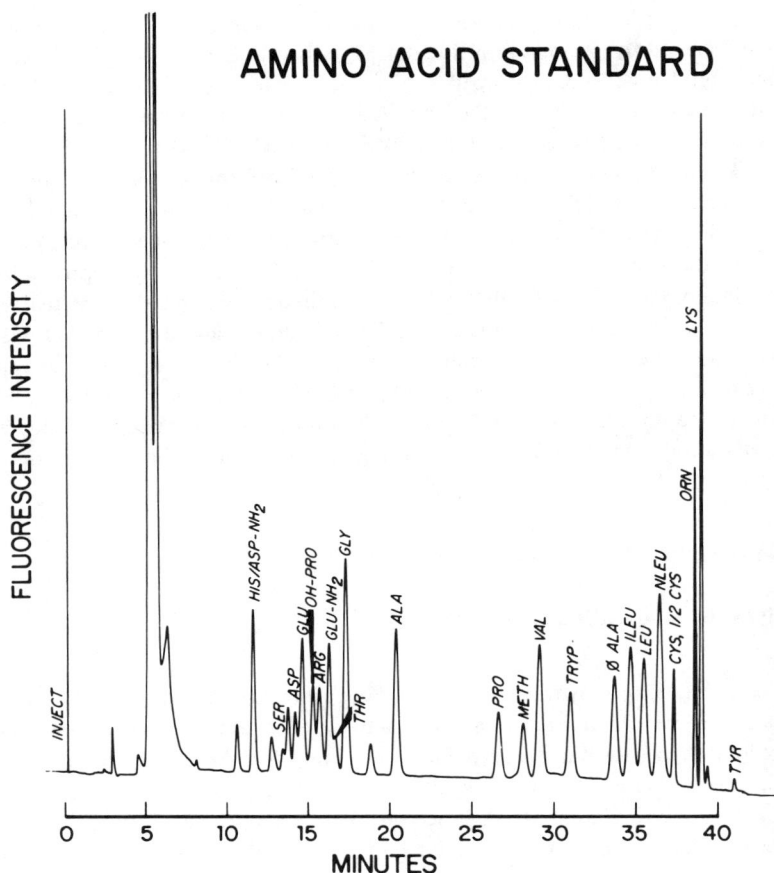

Figure 10.   Separation of dansylated amino acids on RP-8 column. A nonlinear gradient of acetonitrile in water using a Perkin-Elmer Series 3 was used, with fluorescence detection using a Perkin-Elmer Model 650-10 LC[94].

The compounds in Table III range from two to six fused rings and, consequently, have different spectral characteristics, which affects their detection. Absorbance and/or fluorescence detection were used in all of these analyses. In several of these analyses, the detector wavelengths were changed during the analysis, either during development of the chromatogram or between repeated injections of the same sample. Wavelength changes during an analysis serve several purposes:

  1. The choice of detector wavelengths corresponding to absorbance maxima or

Table III.   Summary of PAH Studied

PAH	Reference Number:										
	95	96	97	99	100	101	102	103	104	105	106
Naphthalene	x		x					x			x
Acenaphthylene											
Acenaphthene											x
Fluorene								x			
Phenanthrene	x		x					x		x	x
Anthracene	x			x				x		x	x
Fluoranthene	x	x	x	x	x	x	x	x	x		x
Pyrene	x	x	x	x		x	x	x	x	x	x
Benz[a]anthracene	x	x		x	x	x	x	x	x	x	x
Chrysene	x	x	x	x	x	x	x	x	x	x	x
Benzo[b]fluoranthene	x			x			x		x		x
Benzo[k]fluoranthene	x	x		x	x	x	x		x	x	x
Benzo[a]pyrene	x	x	x	x		x	x	x	x	x	x
Dibenz[a,h]anthracene	x	x		x	x	x	x	x	x	x	x
Benzo[ghi]perylene	x	x			x	x	x		x	x	x
Indeno[1,2,3-cd]pyrene	x			x					x	x	x
Benzo[e]pyrene	x	x		x	x	x	x		x	x	x
Perylene	x		x	x	x	x	x		x	x	
Triphenylene	x						x		x	x	
Coronene	x		x				x		x	x	
Dibenz[a,c]anthracene									x	x	x
Benzo[j]fluoranthene	x			x						x	

fluorescence maxima of compounds of interest will enhance their detectability in complex samples. Smillie et al.[101] and Hanus et al.[102] varied the UV detector wavelength, and Ogan et al.[106] varied the fluorescence excitation and emission wavelengths during the development of chromatograms to increase the sensitivity for selected compounds. Grant and Meiris[99] used different absorbance wavelengths, and Das and Thomas[100] used different fluorescence excitation wavelengths and emission cutoff filters in sequential analyses of the same sample to selectively enhance the response of different compounds. The high sensitivity possible with fluorescence detection is demonstrated in two of these studies[100,106], in which picogram and subpicogram detection limits are reported (see Appendix A). These amounts correspond to minimum detectable concentration levels as low as 0.03 ng/ml.

2. Detection at different wavelengths often can be used to achieve spectral resolution of compounds that are incompletely resolved by the chromatography. Das and Thomas[100] and Smillie et al.[101] observed incomplete resolution of

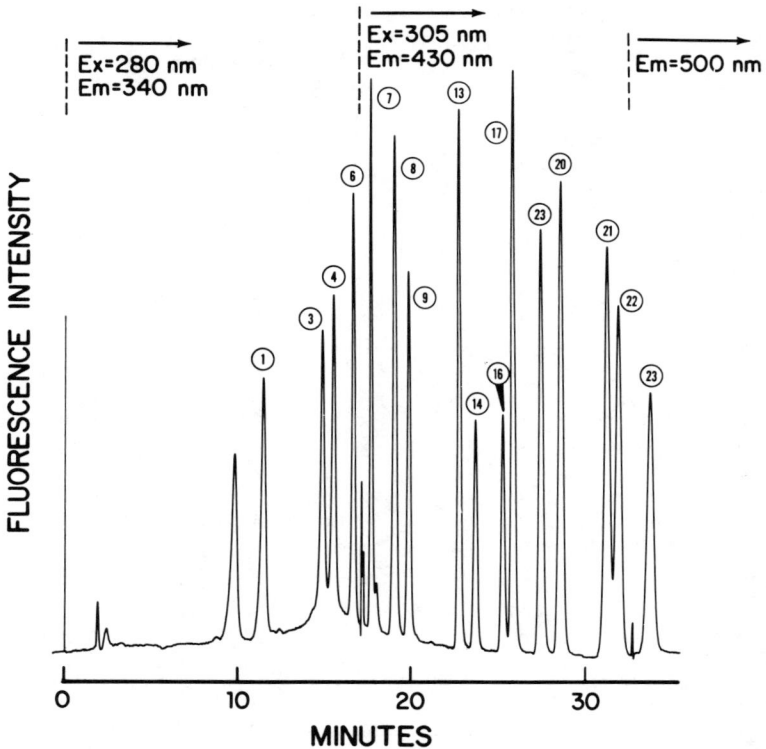

Figure 11.   Chromatogram of 16 PAH standards on an HC-ODS column using 0.5 ml/min flowrate. The mobile phase was acetonitrile in water, starting at 40% acetonitrile and going to 100% acetonitrile in 15 minutes. The mobile phase composition was then held at 100% acetonitrile for 22 minutes. A Perkin-Elmer Model 650-10S fluorescence detector was used, with excitation and emission wavelengths changed to the values indicated during the chromatogram. For peak identification, see Table IV.

benz[a]anthracene and chrysene. These two compounds have different fluorescence spectra, so these groups used the data from chromatograms recorded with different detector wavelengths to quantitate the amount of each compound present. Smillie et al.[101] also applied this technique to the quantitation of benzo[e]pyrene and perylene.

3. After preliminary identification of peaks in a complex chromatogram by matching retention times with those of standards, further confirmation of peak identity can be achieved by measuring the ratio of peak height (or area) from detection at two (or more) wavelengths. These values are then compared with the values of the peak height (area) ratios found for the standards. Krstulovic et al.[97]

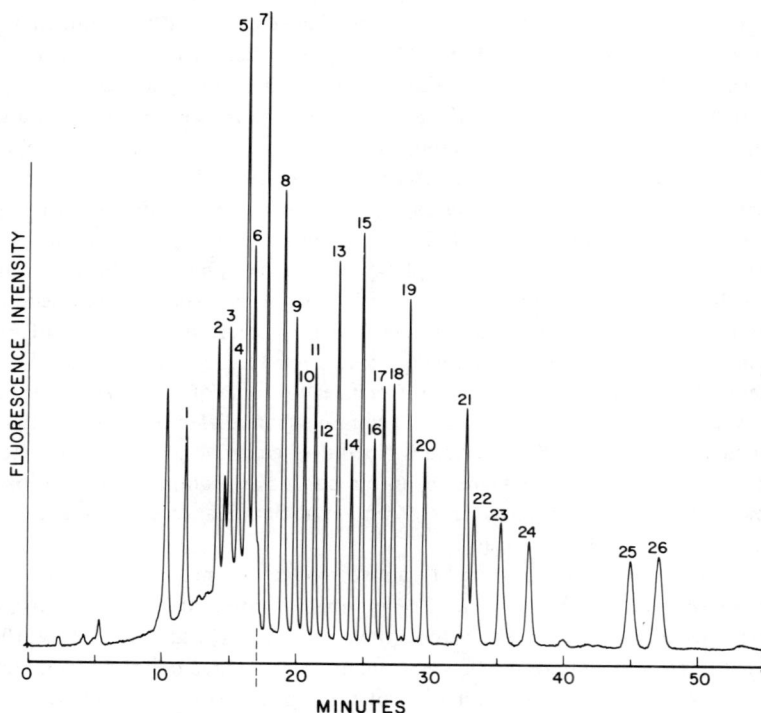

Figure 12.   Separation of 26 PAH on an HC-ODS column using the same gradient as in Figure 11, but with 35 minutes at 100% acetonitrile. A Perkin-Elmer Model 650-10S fluorescence detector was used, with Ex = 280 nm and Em = 340 nm at the start of the run. These wavelengths were changed to Ex = 305 nm and Em = 430 nm at the point indicated by the dotted line. For peak identification see Table IV.

Table IV.   Peak Identification

1. Naphthalene	14. Chrysene
2. 1-Methylnaphthalene	15. 7,12-Dimethylbenz[a]anthracene
3. Acenaphthene	16. Benzo[e]pyrene
4. Fluorene	17. Benzo[b]fluoranthene
5. 1,4-Dimethylnaphthalene	18. Dibenz[a,c]anthracene
6. Phenanthrene	19. Benzo[k]fluoranthene
7. Anthracene	20. Benzo[a]pyrene
8. Fluoranthene	21. Dibenz[a,h]anthracene
9. Pyrene	22. Benzo[ghi]perylene
10. 9,10-Dimethylanthracene	23. Indeno[1,2,3-cd]pyrene
11. 2-Methylanthracene	24. Dibenzo[a,e]pyrene
12. Benzo[a]fluorene	25. Picene
13. Benz[a]anthracene	26. p-Quaterphenyl

used peak area ratios at several different absorbance wavelengths to confirm peak identification, while Das and Thomas[100] used peak height ratios at different fluorescence wavelength settings. This technique is essentially the quantitative comparison of two (or more) points in the absorbance or fluorescence spectra that one would obtain from collected fractions or from stop-flow scanning. The benefits of absorbance-ratioing have been described by others[107-108].

Lankmayr and Muller[104] compared the retention and selectivity characteristics of a C_{18} silica (LiChrosorb RP-18) in reversed-phase chromatography and a NH_2 silica (LiChrosorb NH_2) in normal-phase chromatography with those of a new bonded-phase sorbent, NO_2 silica (Nucleosil NO_2), used in a normal-phase mode. They found that this latter material gave the best separation for their set of PAH standards. Differences in selectivity are not restricted to bonded-phase silicas with different bonded groups. Selectivity differences for some of these PAH have been reported[101, 106, 109] for C_{18} bonded-phase columns from different manufacturers.

Dong and co-workers[110-112] have described the determination of several heterocyclic PAH, specifically aza-arenes, in air particulate samples and in tobacco smoke. This class of compounds is of potential significance since several of its members are known carcinogens.

Several of the analyses referred to above yielded chromatograms with a large number of peaks, reflecting the large number of compounds present at trace levels. Severe overlap of peaks reduces the accuracy of the analysis; hence, there is great interest in simplifying these chromatograms. Three different approaches can be taken; additional fractionation of the sample prior to chromatography, higher efficiency columns and selective detection. Chromatographic fractionation of samples by TLC has been the most widely used method for sample treatment. Extraction followed by back extraction with dimethyl sulfoxide[113] has been reported to give a pure PAH extract. Another promising approach is the use of an NH_2 bonded-phase LC column[114] to fractionate the sample on the basis of the number of fused rings. Columns with greatly improved efficiency will reduce the number of overlapping peaks. Experimental columns having much higher efficiencies than are available commercially have been reported by Scott and Kucera[115] and by Novotny Tsuda and co-workers[116, 117]. The use of selective detectors[118, 119] can also be used in cases when coeluting compounds have different spectral (or other) characteristics, as was described earlier.

Pesticides

The development of analytical methods for pesticides has received a great deal of attention, resulting in a large number of articles being added to the literature. Gas chromatography is the major analytical tool for pesticide analyses, but there is an increasing number of analyses based on liquid chromatography. Lawrence and Turton[120] recently summarized the liquid chromatography data for 166 pesticides. Sparacino and Hines[121] have made an extensive investigation of the retention characteristics of 30 carbamates on various normal- and reversed-phase columns. Both of these studies have also detailed the UV characteristics of many of these compounds pertinent to their detection. Table V, taken from Lawrence and Turton[120], is a summary of the UV response of various pesticide classes. In this table, detection at 254 nm is compared to detection at a more optimum wavelength for several pesticide classes and provides a dramatic demonstration of the advantages of using a variable-wavelength UV detector in trace analyses for these compounds.

Table V.   UV Absorption Sensitivities of Pesticides

Pesticide Class	Sensitivity (μg) [a]	Wavelength (nm)
Phenyl Ureas	0.03 – 0.06	254
Phenyl Carbamates	0.05 – 0.08	254
Phenyl Carbamates	0.005– 0.01	207
Methyl Carbamates	0.5 –20	254
Methyl Carbamates	0.2 – 2	200–206
Triazines	0.1 –10	254
Triazines	0.01 – 0.02	220
Organophosphates	0.5 –16	254
Phenoxy Acids	0.05 – 0.35	280
Phenoxy Esters	15	254
Organochlorines (DDT type)	1 –15	254
Anilides	0.05	254
Nitrophenols	0.02 – 0.05	254
Uracils	0.2	254
Uracils	0.06 – 0.07	270–280
Thiocarbamates	0.07 – 0.60	205
Thiocarbamates	0.3 –10	254

[a] Amount injected to produce a response to 0.01 absorbance units.

Many of the newer pesticides, for example, the carbamate-derived insecticides and urea-derived herbicides, are thermally labile, which presents problems in analyses using GC. Pribyl and Herzel[122] have evaluated the liquid chromatographic determination of several herbicidal urea derivatives. Lawrence[123] has reported an analytical method for several phenylurea herbicides in food, based on LC with a 5-μm silica column. Farrington et al.[124] have described an analytical method for determining many of the same herbicides in grain soil and river water, using LC on a 5-μm C_{18} bonded-phase silica column. The latter method achieved nearly baseline resolution for eight phenylurea herbicides. There is also interest and a need for analytical methods for pesticide metabolites, and Glad et al.[125] have described the determination by LC of several metabolites of linuron, one of the phenylureas included in these other studies.

Methyl and ethyl parathion are organophosphorus insecticides frequently used in place of more persistent organochlorine compounds. Paschal et al.[126] have reported on the analysis of runoff water for these two compounds, using an XAD-2 column for sample collection. Potential interferences from several other pesticides were investigated. Phosphorothioate insecticides can be oxidized to their oxygen analogs, oxons, which are much more potent cholinesterase inhibitors than the parent compounds. Kvalvåg et al.[127] have described the determination of azinphos methyl oxon in foliar dislodgeable residues and in soil surface dusts from orange groves treated with azinphos methyl.

Skelly et al.[128] and Stevens et al.[129] have described a procedure for separating the popular herbicide 2,4-dichlorophenoxyacetic acid (2,4-D) from its isomers by reversed-phase chromatography, with a saponification step to convert 2,4-D esters to the parent 2,4-D. The separation of 2,4-D and several of the hydroxy derivatives suspected to be formed in the metabolism of 2,4-D is reported by Drinkwine et al.[130]. The chromatography of 2,4-D and several of its amino acid conjugates has been given by Arjmand et al.[131].

Pesticide analyses using LC have used the UV detector almost exclusively. Other more specific and sensitive detectors would be of great value in this field. Two such candidates finding increased use in other application areas are the fluorescence detector and the electrochemical detector. There are a few examples of the use of these detectors in pesticide determinations. Ott[132] has evaluated the use of an electrochemical detector in the determination of 2-phenylphenol residues in orange rind. There are several published applications using fluorescence detection. A few have monitored native fluorescence, for example, that from naphthaleneacetic acid, a growth regulator applied to fruit[133,134]. Most applications using fluorescence detectors have utilized the chemical derivatization of the compound(s) of interest with a fluorophore. Derivatization has been applied most frequently to the carbamate class of insecticides[135-139]. Lawrence et al.[140] have also described the derivatization of the phenolic breakdown products of organophosphate pesticides using dansyl chloride.

Derivatization is usually employed to improve the detectability of a compound of interest. Thus, Lawrence and Leduc[141] reported a detection limit of 0.05 ppm for the direct determination of carbofuran in crops, with UV detection at 280 nm, while the detection limits for the dansyl[138] and o-phthalaldehyde[136] derivatives of carbofuran are 0.01 ppm. For carbaryl, Lawrence and Leduc[142] reported that fluorescence detection of the dansyl derivative (10 ppb detection limit) is eight times more sensitive than the direct detection of the parent compound at 254 nm.

As these latter authors demonstrate, derivatization also can be used for additional confirmation of a compound identification. The derivatization can be done online, on the effluent flowing from the column. Krause[143] has described a post-column reaction detector in which carbamates are hydrolyzed and the amine product labeled with o-phthalaldehyde for fluorescence detection. He obtained subnanogram detection limits for cabaryl, carbofuran and methomyl using this detector. In their work on the determination of oxons, Kvalvåg et al.[127] also used a postcolumn chemical reaction detector following the UV detector for increased specificity as well as sensitivity.

The next few years should see a proliferation of applications of liquid chromatography in pesticide analyses as modern microparticulate columns and sensitive detectors become more common in laboratories engaged in pesticide analyses.

Nitrosamines

The toxicity and carcinogenicity reported for several pesticides will have to be reassessed as a result of recent reports of the presence of nitrosamines in the pesticide formulations[144-146]. Almost all of the N-nitroso compounds tested have proven to be carcinogenic, several of them being very potent carcinogens. Dimethylnitrosamine and diethylnitrosamine have received the most attention, in part because they are volatile and, hence, readily determined using GC. Studies of nonvolatile nitrosamines have, until recently, been hampered by the lack of suitable methods having sufficient sensitivity. Among the nonvolatile N-nitroso compounds of potential interest are N-nitroso derivatives of pesticides, nitroso metabolites of volatile nitrosamines and nitrosation reaction products resulting from reactions among the ingredients of commercial products. Sensitive analytical methods are needed for these compounds.

The N-nitroso group has a characteristic UV absorption band between 230 and 240 nm, so nonvolatile nitrosamines can be directly determined using LC with UV detection. This absorption band is not exceptionally strong, so detection limits using this approach are in the low ppm region, while detection at the ppb level is desirable in most cases because of the high activity of these compounds. Thus, Bontoyan et al.[146], in their analysis of pesticide formulations for nitrosamines, were able to use UV detection only for formulations having high concentrations of N-nitroso compounds.

Fine and co-workers[147] have adapted the thermal energy analyzer (TEA) for use as a detector for liquid chromatography. This detector is a nitrosyl-specific detector that has been used primarily with GC. In this detector, the column effluent is rapidly volatilized and passed over a catalyst, which cleaves the N–N bond of nitrosamines (Figure 13). The nitrosyl radical reacts with ozone to produce excited state NO_2, which then relaxes to the ground state with emission of a photon. Detection of this photon generates the detector signal. Most published applications using LC in determinations of nitrosamines have used the TEA detector because of its sensitivity and selectivity. Using LC with a TEA detector, Fan et al.[148] found significant levels of N-nitrodiethanolamine in several cosmetic products—levels as high as 48 ppm in one case. Peak identification was done by retention time matching with a standard sample, using chromatography on three different types of columns for further confirmation. Fan et al.[149] also found parts per hundred levels of N-nitrosodiethanolamine in synthetic cutting fluids. Extensive skin contact is involved with both of these product categories, so these findings are of great concern. Baker and Ma[150] surveyed several cured meats for the presence of N-nitrosoproline, a suspected precurser for the highly carcinogenic N-nitrosopyrrolidine.

Klimisch and Ambrosius[151] and Wolfram et al.[152] have described the use of NBD-Cl (7-chloro-4-nitrobenzo-2-oxa-1,3-diazole) to put a fluorescent label on the amine generated by denitrosation of N-nitrosoproline with HBr in glacial acetic acid. The fluorescent amine derivative was then determined using LC with fluorescence detection. Wolfram et al.[152] reported this approach to be ten times more sensitive than the direct determination of the nitrosamine using gas chromatography-mass spectrometry (GC-MS), or TLC, with a detection limit of 3.5 pmole (0.5 ng) of N-nitrosoproline.

CONCLUSIONS

Although a large number of applications of modern liquid chromatography are referenced here in just three fields, this is still just a survey, not a comprehensive

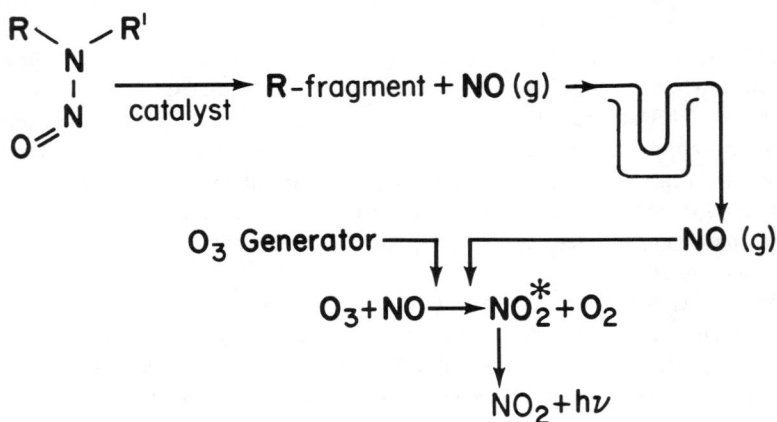

Figure 13.   Schematic description of the principle of operation of the Thermal Energy Analyzer (TEA).

review. Many other references could be cited in these and other fields. The impact of the new technology in LC is only beginning to appear. We expect LC applications to be a very active area in the early 1980s. It should be noted that while comparisons are often made between LC and GC methods, the two techniques tend to be complementary, rather than competitive. As its name implies, the carrier gas in LC serves a transport function, with the separation of various compounds occurring as a result of their interaction with the liquid coating on the column support material. However, in LC, the interaction of the sample molecules with the liquid mobile phase plays a central role in the separation process. Compounds that weakly interact with liquids and are, therefore, volatile, are usually better determined using GC, while selection and adjustment of an appropriate mobile phase in LC provides the most direct means of separating compounds that interact more strongly with their environment. Both techniques are very powerful analytical tools.

APPENDIX A

Great care must be exercised in comparing detection limits reported as absolute amounts injected, i.e., as ng or pg. Most LC detectors are concentration detectors, so the output signal reflects the concentration of a solute in the detector flow cell. With everything else being constant, i.e., injection volume, mobile phase linear flow velocity, etc., an increase in the inner diameter of the column results in an increase in the peak volumes. Consequently, there is a decrease in the solute concentration in an eluting peak and a corresponding decrease in the detector signal. The injection volume can be increased in proportion to the increased column cross-sectional area, which increases the total amount injected by the same proportion, and the solute concentrations in the eluting peak are again identical for the two columns. However, note that while the detection limits expressed in terms of the *concentration* of solute in the injected sample are now identical for both columns, the detection limit expressed in terms of the total *amount* of solute injected is lower for the smaller diameter column. Thus, for purposes of liquid chromatography, the detection limit is more meaningful when expressed as the minimal detectable concentration of injected sample.

REFERENCES

1. Bristow, P. A. *Liquid Chromatography in Practice* (Cheshire, England: hetp Press, 1976).
2. Snyder, L. R., and J. J. Kirkland. *Introduction to Modern Liquid Chromatography* (New York: Wiley-Interscience, 1974).
3. Scott, R. P. W. *Contemporary Liquid Chromatography* (New York: John Wiley & Sons, Inc., 1976).

4. Scott, R. P. *Liquid Chromatography Detectors* (New York: Elsevier North-Holland, Inc., 1977).
5. Reeve, D. R., and A. Crozier. "Molecular and Sub-cellular Aspects of Hormonal Regulation in Plants," in *Encyclopeadia of Plant Physiology— New Series,* Vol. 1, J. MacMillan, Ed., (Berlin: Springer-Verlag, 1979).
6. Cox, G. B., and K. Sugden. *Anal. Chim. Acta* 91: 365–368 (1977).
7. Dunlap, K. L., R. L. Sandridge and J. Keller. *Anal. Chem.* 48: 497–499 (1976).
8. Ishiguro, S., et al. *11th Japan Applied Spectrometry Tokyo Symposium Abstracts,* 111 (1975).
9. Miller, R. L., and G. Dawson. "Characterization of Hydrocarbon Waxes and Polyethylenes by DSC," paper presented at the 1979 Conference of the North American Thermal Analysis Society.
10. Snyder, L. R., and J. J. Kirkland. *Modern Liquid Chromatography,* (Washington, DC: American Chemical Society, 1973).
11. Ishiguro, S., et al. "Shodex Liquid Chromatography Data No. 75-1," Showa Denko, Japan (1975); reproduced in Perkin-Elmer Technical Note No. 89 (1978).
12. Kirkland, J. J. *J. Chromatog. Sci.* 7: 7–12 (1969).
13. Kirkland, J. J. *Anal. Chem.* 43(12): 36A–48A (1971).
14. Kirkland, J. J. Paper presented at the Ninth International Symposium on Chromatography, Switzerland, 1972.
15. Aitzetmuller, K., and W. R. Eckert. *J. Chromatog.* 155: 203–205 (1978).
16. Wronksi, J., and S. Wroblewski, National Loss Control Labs. Private communication to R. L. Miller (1979).
17. Fetters, L. J. *J. Appl Polymer Sci.* 20: 3437–3442 (1976).
18. Janca, J. *J. Chromatog.* 134: 263–272 (1977).
19. Nakano, S., and Y. Goto. *J. Appl Polymer Sci.* 20: 3313–3320 (1976).
20. Bassett, D. C., B. A. Khalifa and R. H. Olley. *J. Polymer Sci. Polymer Phys. Ed.* 15: 995–1009 (1977).
21. Janca, J., and M. Kolinsky. *J. Appl. Polymer. Sci.* 21: 83–90 (1977).
22. Miller, R. L. Unpublished results (1979).
23. Campos, A., L. Borque and J. E. Figueruelo. *J. Chromatog.* 140: 219–227 (1977).
24. Provder, T. "Analysis of Copolymers by GPC," *Proc. 58th Can. Chem. Conf.* 189 (1975).
25. Dawkins, J. V., and M. Hemming. *J. Appl. Polymer Sci.* 19: 3107–3118 (1975).
26. Mirabella, F. M., E. M. Barrall, E. F. Jordan and J. F. Johnson. *J. Appl. Polymer Sci.* 20: 581–589 (1976).
27. Runyon, J. R., D. E. Barnes, J. F. Rudd and L. H. Tung. *J. Appl. Polymer Sci.* 13: 2359–2369 (1969).
28. Hendrickson, J. G., and J. C. Moore. *J. Polymer Sci.* 4(A1): 167–188 (1966).
29. Grubisic, Z., P. Rempp and H. Benoit. *J. Polymer Sci.* 5 (Part B): 753 (1967).
30. Mori, S. *J. Chromatog.* 156: 111–120 (1978).
31. Mori, S. *Anal. Chem.* 50: 1639–1643 (1978).
32. Bledzki, A., and M. I. Prusinska. *J. Polymer Sci. Polymer Chem. Ed.* 16: 107–114 (1978).

33. Dixon, P. F., M. S. Stoll and D. K. Lim. *Ann. Clin. Biochem.* 13 (Pt. 4): 409–432 (1976).
34. Dixon, P. F., C. H. Gray, C. K. Lim and M. S. Stoll, Eds. *High Pressure Liquid Chromatography in Clinical Chemistry,* (London: Academic Press, Inc., 1976).
35. Hawk, G. L., P. B. Champlin, H. C. Jordi and D. Wenke, Eds. *Biological/Biomedical Applications of Liquid Chromatography,* (New York: Marcel Dekker, Inc., 1979).
36. Wheals, B. B., and I. Jane. *Analyst* 102: 625–644 (1977).
37. Adams, R. F., F. L. Vandemark and G. J. Schmidt. *Clin. Chem.* 22: 1903–1906 (1976).
38. Peng, G. W., M. A. F. Gadalla and W. L. Chiou. *Clin. Chem.* 24: 357–360 (1978).
39. Kelley, R. C., D. E. Prentice and G. M. Hearne. *Clin. Chem.* 24: 838–839 (1978).
40. Adams, R. F., and F. L. Vandemark. *Clin. Chem.* 22: 25–31 (1976).
41. Adams, R. F., G. J. Schmidt and F. L. Vandemark. *Chromatog. Newslett.* 5: 11–13 (1977).
42. Adams, R. F., and G. J. Schmidt. *Chromatog. Newslett.* 4: 8–10 (1976).
43. Adams, R. F., G. J. Schmidt and F. L. Vandemark. *J. Chromatog.* 145: 275–284 (1978).
44. Adams, R. F. In: *Advances in Chromatography,* Vol. 15, J. C. Giddings, E. Grushka, J. Cazes, P. R. Brown, Eds. (New York: Marcel Dekker, Inc., 1977).
45. Kabra, P. M., G. Gotelli, R. Stanfell and L. J. Marton. *Clin. Chem.* 22: 824–827 (1976).
46. Kabra, P. M., and L. J. Marton. *Clin. Chem.* 22: 1070–1072 (1976).
47. Kabra, P. M., G. Gotelli, R. Stanfill and L. J. Marton. *Clin. Chem.* 22: 1672–1674 (1976).
48. Kabra, P. M., D. M. McDonald and L. J. Marton. *J. Anal. Toxicol.* 2: 127–133 (1978).
49. Soldin, S. J., and J. G. Hill. *Clin. Chem.* 23: 2352–2353 (1977).
50. Farinotti, B., and G. Mahuzier. *J. Liq. Chromatog.* 2: 345–364 (1979).
51. Schmidt, G. J., and W. Slavin. *Chromatog. Newslett.* 6: 22–24 (1978).
52. Sutheimer, C., D. Fretthold and I. Sunshine. *Chromatog. Newslett.* 7: 1–4 (1979).
53. Knox, J. H., and J. Jurand. *J. Chromatog.* 103: 311–326 (1975).
54. Proeless, H. F., H. J. Lohmann and D. G. Miles. *Clin. Chem.* 24: 1948–1953 (1978).
55. Watson, I. D., and M. J. Stuart. *J. Chromatog.* 110: 389–392 (1975).
56. Vandemark, F. L., R. F. Adams and G. J. Schmidt. *Clin. Chem.* 24: 87–91 (1978).
57. Sutheimer, C. *Chromatog. Newslett.* 7: 38–39 (1979).
58. Rosano, T. G., C. A. Brito and J. M. Meola. *Chromatog. Newslett.* 6: 1–3 (1978).
59. Horwitz, R. A., and P. I. Jatlow. *Clin. Chem.* 23: 1596–1598 (1977).
60. Gotelli, G. R., P. M. Kabra and L. J. Marton. *Clin. Chem.* 23: 957–959 (1977).
61. Blair, D., and B. H. Rumack. *Clin. Chem.* 23: 743–745 (1977).

62. Adams, R. F., F. L. Vandemark and G. J. Schmidt. *Clin. Chem. Acta.* 69: 515–524 (1976).
63. Schmidt, G. J., F. L. Vandemark and R. F. Adams. *Chromatog. Newslett.* 4: 32–35 (1976).
64. Shukur, L. R., J. L. Powers, R. A. Marques, M. E. Winter and W. Sadee. *Clin. Chem.* 23: 636–638 (1977).
65. Rocco, R. M., D. C. Abbott, R. W. Giese and B. L. Karger. *Clin. Chem.* 23: 705–708 (1977).
66. Dutcher, J. S., and J. M. Strong. *Clin. Chem.* 23: 1318–1320 (1977).
67. Schmidt, G. J., and F. L. Vandemark. *Chromatog. Newslett.* 5: 42–44 (1977).
68. Jatlow, P., W. Bush and H. Hochster. *Clin. Chem.* 25: 777–779 (1979).
69. Nation, R. L., G. W. Peng and W. L. Chiou. *J. Chromatog.* 145: 429–436 (1978).
70. Radhakrishnan, A. N., S. Stein, A. Licht, K. A. Gruber and S. Udenfriend. *J. Chromatog.* 132: 552–555 (1977).
71. Samejima, K., M. Kawase, S. Sakamoto, M. Okada and Y. Endo. *Anal. Biochem.* 76: 392–406 (1976).
72. Hayashi, T., T. Sugiura, S. Kawai and T. Ohno. *J. Chromatog.* 145: 141–146 (1978).
73. Abdel-Monem, M. M., and K. Ohno. *J. Chromatog.* 107: 416–419 (1975).
74. Abedl-Monem, M. M., and K. Ohno. *J. Pharm. Sci.* 66: 1195–1197 (1977).
75. Newton, N. E., K. Ohno and M. M. Abdel-Monem. *J. Chromatog.* 124: 277–285 (1976).
76. Seiler, N., B. Knodgen and F. Eisenbeiss. *J. Chromatog.* 145: 29–39 (1978).
77. Vandemark, F. L., G. J. Schmidt and W. Slavin. *J. Chromatog. Sci.* 16: 465–469 (1978).
78. Huber, J. F. K., J. A. R. J. Hulsman and A. Meijers. *J. Chromatog.* 62: 79–91 (1971).
79. Butterfield, A. G., B. A. Lodge and N. J. Pound. *J. Chromatog. Sci.* 11: 401–405 (1973).
80. Dolphin, R. J. *J. Chromatog.* 83: 421–429 (1973).
81. Dolphin, R. J., and P. J. Pergande. *J. Chromatog.* 143: 267–274 (1977).
82. Fitzpatrick, F. A., and S. Siggia. *Anal. Chem.* 45: 2310–2314 (1973).
83. Gotelli, G. R., J. H. Wall, P. M. Kabra and L. J. Marton. *Clin. Chem.* 24: 2132–2134 (1978).
84. Schmidt, G. J., F. L. Vandemark and W. Slavin. *Anal. Biochem.* 91: 636–645 (1978).
85. Taylor, J. T., J. G. Knotts and G. J. Schmidt. *Clin. Chem.* (in press).
86. Ersser, R. S. In: *High Pressure Liquid Chromatography in Clinical Chemistry,* P. F. Dixon, C. H. Gray, C. K. Lim and M. S. Stoll Eds. (London: Academic Press, Inc., 1976).
87. Voelter, W., and K. Zech. *J. Chromatog.* 112: 643–649 (1975).
88. Benson, J. V., Jr., and J. A. Patterson. In: *New Techniques in Amino Acid, Peptide, and Protein Analysis,* A. Niederwieser and G. Pataki, Eds. (Ann Arbor, MI: Ann Arbor Science Publishers, Inc., 1971).
89. Deyl, Z. *J. Chromatog.* 127: 91 (1976).

90. Bayer, E., E. Grom, B. Kaltenegger and R. Uhmann. *Anal. Chem.* 48: 1106–1109 (1976).
91. Adams, R. F., G. J. Schmidt and F. L. Vandemark. *Clin. Chem.* 23: 1226–1229 (1977).
92. Hsu, K., and B. L. Currie. *J. Chromatog.* 166: 555 (1978).
93. Wilkinson, J. M. *J. Chromatog. Sci.* 16: 547–552 (1978).
94. Olson, D. C., G. J. Schmidt and W. Slavin. *Chromatog. Newslett.* 7: 22 (1979).
95. Dong, M., D. C. Locke and E. Ferrand. *Anal. Chem.* 48: 368–372 (1976).
96. Fox, M.A., and S. W. Staley. *Anal. Chem.* 48: 992–998 (1976).
97. Krstulovic, A. M., C. M. Rosie and P. R. Brown. *Anal. Chem.* 48: 1383–1386 (1976).
98. Dong, M. W., D. C. Locke and D. Hoffman. *Environ. Sci. Technol.* 11: 612–618 (1977).
99. Grant, D. W., and R. B. Meiris. *J. Chromatog.* 142: 339–351 (1977).
100. Das, B. S., and G. H. Thomas. *Anal. Chem.* 50: 967–973 (1978).
101. Smillie, R. D., D. T. Wang and O. Meresz. *J. Environ. Sci. Health* A13: 47–59 (1978).
102. Hanus, J. P., H. Guerrero, E. R. Biehl and C. T. Kenner. *J. Assoc. Off. Anal. Chem.* 62: 29–35 (1979).
103. Rietz, E. B. *Anal. Lett.* 12: 143–153 (1979).
104. Lankmayr, E. P., and K. Muller. *J. Chromatog.* 170: 139–146 (1979).
105. Nielsen, T. *J. Chromatog.* 170: 147–156 (1979).
106. Ogan, K. E. Katz and W. Slavin. *Anal. Chem.* 51: 1315–1320 (1979).
107. Berg, R. G., C. Y. Ko, J. M. Clemons and H. M. McNair. *Anal. Chem.* 47: 2480–2482 (1975).
108. Yost, R., J. Stoveken and W. MacLean. *J. Chromatog.* 134: 73–82 (1977).
109. Ogan, K., E. Katz and T. J. Porro. *J. Chromatog.* 188: 115–127 (1980).
110. Dong, M., and D. C. Locke. *J. Chromatog. Sci.* 15: 32–35 (1977).
111. Dong, M., D. C. Locke and D. Hoffman. *Environ. Sci. Technol.* 11: 612–618 (1977).
112. Dong, M., I. Schmeltz, E. Jacobs and D. Hoffmann. *J. Anal. Toxicol.* 2: 21–25 (1978).
113. Natusch, D. F. S., and B. A. Tomkins. *Anal. Chem.* 50: 1429–1434 (1978).
114. Wise, S. A., S. N. Chesler, H. S. Hertz, L. R. Hilpert and W. E. May. *Anal. Chem.* 49: 2306–3210 (1977).
115. Scott, R. P. W., and P. Kurcera. *J. Chromatog.* 169: 51–72 (1979).
116. Tsuda, T., and M. Novotny. *Anal. Chem.* 50: 271–275 (1978).
117. Hirata, Y., M. Novotny, T. Tsuda and D. Ishii. *Anal. Chem.* 51: 1807–1809 (1979).
118. Ettre, L. *J. Chromatog. Sci.* 16: 396–417 (1978).
119. Ogan, K., E. Katz and T. J. Porro. *J. Chromatog. Sci.* 17: 597–600 (1979).
120. Lawrence, J. F., and D. Turton. *J. Chromatog.* 159: 207–226 (1978).
121. Sparacino, C. M., and J. W. Hines. *J. Chromatog. Sci.* 14: 549–556 (1976).
122. Pribyl, J., and F. Herzel. *J. Chromatog.* 125: 487–494 (1976).
123. Lawrence, J. F. *J. Assoc. Off. Anal. Chem.* 56: 1066–1070 (1976).

124.   Farrington, D. C., R. G. Hopkins and J. H. A. Ruzicka. *Analyst* 102: 377–381 (1977).
125.   Glad, G., and O. Theander. *J. Chromatog. Sci.* 16: 118–122 (1978).
126.   Paschal, D. C., R. Bicknell and D. Dresbach. *Anal. Chem.* 49: 1551–1554 (1977).
127.   Kvalvåg, J., D. E. Ott and F. A. Gunther. *J. Assoc. Off. Anal. Chem.* 60: 911–917 (1977).
128.   Skelly, N. E., T. S. Stevens and D. A. Mapes. *J. Assoc. Off. Anal. Chem.* 60: 868–872 (1977).
129.   Stevens, T. S., N. E. Skelly and R. B. Grorud. *J. Assoc. Off. Anal. Chem.* 61: 1163–1165 (1978).
130.   Drinkwine, A. D., D. W. Bristol and J. R. Fleeker. *J. Chromatog.* 174: 264–268 (1979).
131.   Arjmand, M., R. H. Hamilton and R. O. Mumma. *J. Agric. Food Chem.* 26: 971–973 (1978).
132.   Ott, D. E. *J. Assoc. Off. Anal. Chem.* 61: 1465–1468 (1978).
133.   Cochrane, W. P., and M. Lanouette. *J. Assoc. Off. Anal. Chem.* 62: 100–105 (1979).
134.   Moye, H. A., and T. A. Wheaton. *J. Agric. Food Chem.* 27: 291–294 (1979).
135.   Frei, R. W., J. F. Lawrence, J. Hope and R. M. Cassidy. *J. Chromatog. Sci.* 12: 40–44 (1974).
136.   Moye, H. A., S. J. Scherer and P. A. St. John. *Anal. Lett.* 10: 1049–1073 (1977).
137.   Krause, R. T. *J. Chromatog. Sci.* 16: 281–288 (1978).
138.   Lawrence, J. F., and R. Leduc. *J. Chromatog.* 152: 507–513 (1978).
139.   Lawrence, J. F., and R. Leduc. *J. Assoc. Off. Anal. Chem.* 61: 495–499 (1978).
140.   Lawrence, J. F., C. Renault and R. W. Frei. *J. Chromatog.* 121: 343–351 (1976).
141.   Lawrence, J. F., and R. Leduc. *J. Agric. Food Chem.* 25: 1362–1365 (1977).
142.   Lawrence, J. F., and R. Leduc. *J. Assoc. Off. Anal. Chem.* 61: 872–876 (1978).
143.   Krause, R. T. *J. Chromatog. Sci.* 16: 281–288 (1978).
144.   Ross, R., T. Morrison, D. R. Rounbehler, S. Fan and D. H. Fan. *J. Agric. Food Chem.* 25: 1416–1418 (1977).
145.   Fine, D. H., D. P. Rounbehler, T. Fan and R. Ross. *Cold Spring Harbor Conf. on Cell Proliferation* 4: 293–307 (1978).
146.   Bontoyan, W. B., M. W. Law and D. P. Wright. *J. Agric. Food Chem.* 27: 631–636 (1979).
147.   Oettinger, P. E., F. Huffman, D. H. Fine and D. Lieb. *Anal. Lett.* 8: 411–414 (1975).
148.   Fan, T. Y., U. Fogg, L. Song, D. H. Fine, G. P. Arsenault and K. Biemann. *Food Cosmet. Toxicol.* 15: 423–430 (1977).
149.   Fan, T. Y., J. Morrison, D. P. Rounbehler, R. Ross, D. H. Fine, W. Miles and N. P. Sen. *Science* 196: 70–71 (1977).
150.   Baker, J. K., and C.-U. Ma. *J. Agric. Food Chem.* 26: 1253–1255 (1978).
151.   Klimisch, H.-J., and D. Ambrosius. *J. Chromatog.* 121: 93–95 (1976).
152.   Wolfram, J. H., J. I. Feinberg, R. C. Doeer and W. Fiddler. *J. Chromatog.* 132: 37–43 (1977).

CHAPTER 5

GC/MS IN TRACE ORGANIC ANALYSIS

Elizabeth C. Price

Department of Civil & Environmental Engineering
New Jersey Institute of Technology
Newark, New Jersey

INTRODUCTION

For years, the old routine methods of analysis for organics included the standard biochemical oxygen demand (BOD) and chemical oxygen demand (COD) tests. These methods have optimum sensitivities of approximately 1 mg/l. The standard nonspecific 4-amino-antipyrene derivative test for phenols is somewhat more sensitive; however, none of these provide definitive information as to the nature of the organics present or to the toxicity, origin, cause, carcinogenicity or ultimate fate of the organic[1]. It has been estimated that four of five occurrences of cancer in man are due to environmental effects such as carcinogens in air and water[2]. Therefore, the health and environmental implications of organic compounds in water and air depend on the nature of the compound. For example, 1 mg/l of endrin in water would be totally unacceptable since it is toxic to fish at 1/1000 of this concentration, and 1 mg/l of geosmin in 10,000 times the odor threshhold of that compound.

The reason that even minute concentrations of recalcitrant compounds can be harmful is the phenomenon known as biomagnification, whereby a concentration of toxic chemicals by organisms occurs not only in a series of steps up the food chain, but also by direct uptake from water[3]. Compounds such as dichloro-diphenyl-trichloro-ethane (DDT), polychlorinated biphenyls (PCB) and aromatic hydrocarbons are recalcitrant and biomagnified. For example, the highest level observed for the pesticide dieldrin in drinking water is 14 ng/l (ng = 10^{-9}g). One would have to drink 2 l/day of water for 17 years to ingest the same amount of dieldrin as from eating a 0.5 kg catfish grown in the same water[4].

The GC/MS system, a gas chromatograph coupled with a mass spectrometer, can be used for separation followed by either identification or quantification of organic compounds present in the nanogram range. In fact, it is the only system that can both identify and quantify. It is for this reason that essentially all of the discoveries of organic pollutants in environmental samples in the last 15 years have been based on GC/MS information.

SAMPLE PREPARATION

Since it is inadvisable to inject more than a few microliters of solution into a GC column, a certain amount of sample preparation is usually required to effect concentration of the organics prior to injection. In some cases, elimination of interfering substances is also necessary; however, for reasons that will become apparent, much less "cleanup" is required when a mass spectrometer is used as the GC detector, than for conventional GC work.

Direct Aqueous Injection

Direct aqueous injection is only suitable when relatively high concentrations of organics are present, as, for example, in industrial wastes. Minimum detectable concentrations are from 1–5 mg/l, depending on the compound.[5].

It is possible to lower the minimum detectable quantity to about 1–10 μg/l (μg $= 10^{-6}$ g) by concentration of the sample. This can be accomplished by distillation and sometimes redistillation to concentrate low boiling organics, and by evaporation of the water to concentrate organics with boiling points too high for steam distillation. Of course, direct aqueous injection must be used with a compatible column, some of which are listed in Table I.

Membrane separations have been used to introduce volatile contaminants from water directly into a mass spectrometer without chromatographic separation. A membrane is chosen that is permeable to the compounds of interest, but not to the water or undesirable matrix components. A silicone membrane has been used, one side of which was in contact with the aqueous solution while the other surface was open to the ion source[6]. Obviously, with this method there is no separation of compounds, and single ion monitoring must be relied on for identification and quantification. Lower detection limits of response are listed in Table II.

Results must be interpreted with care; for example, although the m/e 83 ions are attributed in Table II to chloroform, any other compound that would produce a $CHCl_2^+$ fragment would result in a signal that would be counted as chloroform. The sensitivity of this method is increased by heating, which, at the same time, renders the silicone membrane more permeable to water, decreasing the life-

Table I.   Representation GC Columns for Direct Aqueous Injection[5]

Column Packing	Application
Chromosorb 101 80/100 mesh	General for low molecular-weight compounds up to MW 200; water elutes before organics
4% FFAP on Chromosorb W 60/80 mesh	Phenols, acids
0.4% Carbowax 1500 on Carbopack A	Low molecular-weight acids
Tenax GC 60/80 mesh	Amines
Carbowax 400 on Porasil C 100/200 mesh	General for low-molecular weight compounds

Table II.   Lower Detection Limits of Membrane/Mass Spectrometer Responses[6]

Compound	Detection Limit (ppb)	Ion Monitored
Benzene	0.5	78
Methyl Salicylate	2	120
Chloroform	4	83
Methylene Chloride	4	49
Acetone	50	58
Isopropanol	50	45
Phenol	600	94

time of the mass spec filaments. In general, equilibrium, the time needed to establish constant migration through the membrane and transfer line, can be reached in about 20 seconds. Mieure et al.[6] are investigating the use of a membrane probe, which inserts through the mass spectrometer solid probe vacuum lock.

Headspace Analysis—Static and Dynamic

Static headspace analysis involves direct sampling and analysis of the equilibrium atmosphere in a closed container. Of course, it is limited to those compounds that have a relatively high vapor pressure over a water sample, and detection limits are restricted by the equilibrium gas-liquid partition of the solutes, as well as the limited amount of headspace gas that can be conveniently sampled and analyzed. Sample temperature is obviously significant.

An improvement over the "static" headspace method is "dynamic" headspace sampling, whereby the low boiling compounds are sparged from the water sample with an inert gas and concentrated by adsorption on a resin column for later analysis. Such "purge and trap" procedures have in common the following features:

1. The inert gas (helium or nitrogen) is forced through a glass frit or other diffusing device to produce tiny bubbles, which rise through the water absorbing the volatile organics.
2. The eluted gas passes through a porous polymer trap, which adsorbs the organics while the inert gas and the water vapor are vented to the air.
3. Sparging gas flows are generally on the order of 20–40 ml/min for a duration of 10–20 minutes.
4. The organics are then desorbed from the polymer by heating the trap and backflushing into the chromatographic system with carrier gas.

A schematic of a typical sample concentrator (Tekmar's LSC-1) is shown in Figure 1[7].

Tenax GC (60/80 mesh) is an effective trap absorbant for compounds boiling above 30°C, and is often used in combination with silica gel (grade-15) to absorb the lower boiling compounds.

Recovery efficiencies differ between systems, as well as between compounds. Factors affecting recoveries are purge gas diffuser characteristics, purge gas flow and duration, trap heater characteristics, desorb carrier gas flow and sample characteristics such as temperature, pH, ionic strength and sample matrix. It is therefore essential that blanks, standards and spiked samples be run daily along with the unknown samples[5,6,8].

Purging at elevated temperatures enhances the recoveries of some compounds, but may also promote chemical reactions that would alter the composition of the organics, particularly in the case of chlorinated water or effluents. Researchers are not in agreement as to whether the advantages outweigh the difficulties involved in purging at elevated temperatures[5,9].

Liquid-Liquid Extraction

Liquid-liquid extraction of organic compounds from water using an organic solvent has been used successfully with many variations. Methods based on batch extraction with a low boiling solvent (hexane or methylene chloride are commonly used) followed by concentration by evaporation of the solvent are described in the literature[10,11]. Disadvantages of this method are that low boiling compounds are lost and solvent impurities as well as sample extracts are concentrated.

Liquid-liquid extraction also may be done with a high-boiling, late-eulting sol-

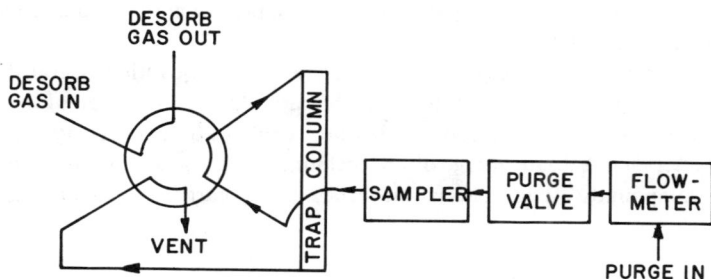

PURGE MADE: He IS BUBBLED THROUGH SAMPLE, PARTITIONING VOLATILE ORGANICS INTO GAS PHASE FOR CONCENTRATION BY TRAP COLUMN.

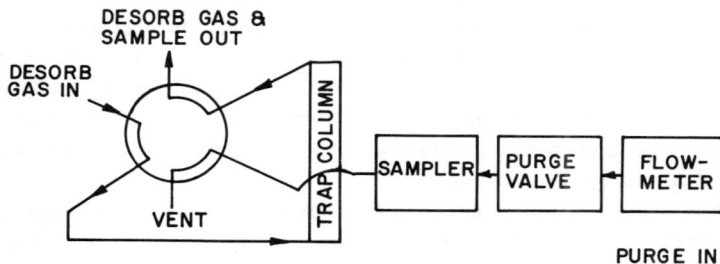

DESORB MADE: TRAP COLUMN IS HEATED & BACKFLUSHED WITH He TO TRANSFER CONCENTRATED SAMPLE TO THE GC.

Figure 1.   Volatile organic sample concentrator (courtesy of Tekmar Co.).

vent (i.e., hexadecane). The advantage of this method is the retention of volatile compounds usually masked or lost when low-boiling solvents are used. A large amount of sample is usually extracted with a small amount of solvent, and the extract is analyzed without further concentration[5,12].

Liquid-liquid extraction also may be continuous, allowing greater volumes of sample to be extracted and extremely low concentrations of organics to be detected.[13,14].

Dialysis into Solvent

Another technique that has been used to extract and concentrate trace organics is dialysis through a thin polymer membrane into a solvent, a process that can selectively concentrate the organics in a sample. It can further be used to transfer organics to water-miscible solvents and can be adapted to automated or continuous sampling and analysis[6].

Studies show that it takes about 24 hours to effect a quantitative transfer of organics through the membrane; however, by carefully controlling all of the variables, including time, fairly reproducible values could be obtained using one-hour dialysis intervals. A disadvantage of this technique is that it is only useful for dissolved compounds; any material adsorbed onto particulate matter is not recovered.

Adsorption Methods

These methods consist of passing the sample of water or air through a resin or activated charcoal filter trap, followed by desorption with the application of heat or solvents. An activated carbon filter method (CAM) was introduced by the U.S. Public Health Service (USPHS) in 1951[15]. It is useful for some organopesticides, but undependable for others due to lack of control of the adsorption and desorption characteristics of the carbon filter bed. Other disadvantages include bacterial breakdown of adsorbed organics during the sampling period and relative difficulty of desorption[13].

Recently, the use of polymer resins has become popular for the isolation and concentration of lipophilic organics in aqueous samples[4,16-18]. A low-polarity styrene-divinyl benzene copolymer (Amberlite XAD-2) has been widely used. Its advantages include optimum pesticide recovery with minimum background, ease of elution with a solvent and the ability to extract sample sizes up to several hundred liters.

Air samples are usually collected by passing a measured amount of air through an absorbant trap and desorbing thermally into the chromatograph. Tenax GC (2,6-diphenyl-*p*-phenyleneoxide polymer) is an effective sorbent for nonpolar and slightly polar compounds, and Spherocarb is often used for highly polar compounds.[5,19].

Sample collection has been called the weakest link in the analytical chain for the determination of airborne organics[20]. One problem is the formation of artifacts by chemical or biochemical reactions on the adsorbant. The characterization of particulate-bond organics presents another set of problems; for example, variations in the individual compounds identified with the type of filter and the solvent used have not been well defined. Of course, if the organic concentrations are high enough, direct injection into the chromatographic system is feasible. Methods have also been developed for the determination of organics in sediment and biological samples[5,11].

GAS CHROMATOGRAPHY

The function of the gas chromatographic portion of the instrument is the separation of the individual organic compounds contained in the sample. A small amount of sample extract is vaporized on introduction into a heated injection port. The chromatographic process occurs as a result of repeated sorption-desorption acts as the sample compounds move along the column. The gaseous fraction is carried through the column by an inert carrier gas (usually helium for a GC/MS system) and interacts with a stationary phase along the way. The stationary phase may either be distributed as a thin film on the particulate surfaces of an inert column packing or applied to the walls of an open tubular or capillary column.

The speed with which the majority of the molecules of a particular compound move along the stationary phase depends on the attraction of the compound to the stationary phase. On a nonpolar phase, the solutes emerge in the order of their boiling points, while the retention on a polar phase depends on the interaction between the active groups in the solute molecule and those of the phase. The affinity of stationary phases for the various classes of organic compounds are characterized by McReynolds Constants[21,22].

The separation column is placed in a temperature-controlled oven that is either maintained at a constant temperature or programmed to increase with time. Factors affecting separation, resolution and retention time are carrier gas pressure and flowrate, column length, sample loading and temperature, as well as the particular characteristics of the stationary phase and sample compounds. Generally, the packed glass columns utilized for environmental samples are 1.8 meters (6 ft) long, 2–4 mm i.d., with helium carrier gas flowing at 20–40 ml/min. The packing is inert solid supports of 80/100 mesh coated with stationary-phase material. Although hundreds of stationary phases are available, four characteristic and commonly used materials are listed in Table III with their general applications.

In recent years, there has been considerable interest in "capillary" or "open tubular" columns. These are open columns, usually glass, 45.7 meters (150 ft) or longer, with an i.d. of 0.2–0.4 mm. The stationary phase is either applied as a thin film on the inside surface of the tube (wall-coated open tubular or WCOT), or the inside wall is first covered with a layer of very fine porous particles before the stationary phase coating (support-coated open tubular, or SCOT). These columns utilize a flowrate of 1–5 ml/min and require a smaller sample injection size. Their advantage is that they can often effect better resolution in a shorter time as compared to packed columns. This is significant in the analysis of very complex substances, such as some industrial wastes, as well as for difficult-to-separate but environmentally significant components such as PCB. Conventional open tubular columns are not recommended for:

Table III.   Representative Stationary Phases and Their Applications

Stationary Phase	Polarity	Applications
Methyl Silicone	Nonpolar	Gasoline, fuel oil, hydro-carbons, essential oils
Methyl/Phenyl Silicone	Moderate	Chlorinated hydrocarbons, PCB, pesticides, aromatic solvents
Carbowax (polyglycol)	High	Alcohols, ketones, esters, fatty acids
Cyano-Silicone	Very high	Mercaptans, aromatics, heterocyclics

- analysis of inorganic gases,
- analysis of thermally unstable samples that might decompose on contact with hot metal surfaces (i.e., some pesticides), or
- analysis of very high-boiling substances.

It should be kept in mind that high-resolution GC performance is not as necessary when a mass spectrometer is online as when a conventional GC detector is used. Indeed, the "mass spec people" tend to think of the GC portion as an inlet to their spectrometers, while the chromatographers view the mass spectrometer as a fancy detector for their chromatographic systems[5,23-26].

GC-MS INTERFACE

The unique combination of the separation abilities of the gas chromatograph and the identification capabilities of the mass spectrometer forms a powerful analytical tool. Interfacing the two instruments, however, involves the transition from a pressure system (GC) to a high vacuum system (MS). Ideally, an interface device would remove as much of the carrier gas from a GC peak as possible, while transporting a maximum amount of the remaining organic material into the mass spectrometer ion source. The success with which a given device performs these functions can be expressed by the yield (y), the fraction of sample that reaches the MS ion source and the enrichment factor (N), indicating the amount of carrier gas removed.

These parameters are calculated as follows:

$$\text{yield (or efficiency)}, y = \frac{Q_{ms}}{Q_{qc}}$$

$$\text{enrichment factor (or separation factor), } N = \frac{(Q_{ms})/(He_{ms})}{(Q_{qc})/(He_{gc})}$$

where Q_{qc} = amount of sample entering separator
         Q_{ms} = amount of sample entering spectrometer
         He_{gc} = amount of carrier gas entering separator
         He_{ms} = amount of carrier gas entering spectrometer

Although many designs of interface devices have been proposed, they fall into four general classes, as shown in Figure 2[27].

Direct coupled interfaces use a section of narrow bore tubing to carry the column effluent to the ion source. The pressure drop is effected by splitter valves, adjustment of GC flows or auxiliary pumping. Direct coupled interfaces are commonly used in connection with capillary columns.

In the effusive type of interface, carrier gas molecules move selectively through the walls of a fritted tube housed in an evacuation chamber. In the jet orifice separator, the momentum of the heavier organic molecules carries them through the interface, while the lighter helium molecules are deflected and removed by the vacuum pumps. In the permeable membrane interface, the organic molecules diffuse selectively through a membrane that resists the conductance of helium. High yields and efficiencies are possible with this interface.

MASS SPECTROMETER

Mass spectrometry is the separation, or sorting, of charged molecular species according to their mass to charge ratios (m/e), and a mass spectrum is a record of the relative abundance of these species. The most common way of producing ions for organic mass spectrometry is by electron bombardment, or electron impact (EI), of the gaseous sample. When an electron collides with an organic molecule, enough energy may be transferred to remove an electron from the molecule producing a positively charged molecular ion (M^+). Or, if excess energy is transferred into the impact, the molecule may break apart to form lower massed fragments. The resulting collection of molecular ion and fragment ions is characteristic of the original compound and, when the ions are sorted and quantified in a mass spectrum, they form a "fingerprint" of the original molecule.

Electron impact spectra of the trihalomethanes chloroform and dibromochloromethane and the pesticides lindane and methoxychlor are shown in Figures 3 and 4. The fragmentation patterns for each compound appear to be somewhat predictable from the structural formulae (inserts in figures). Note that when the halogens are involved, there are clusters of m/e values for a single fragment, representing the various combinations of isotopes, as indicated above the spectra.

Fragmentation patterns for any compound may vary slightly as to the relative

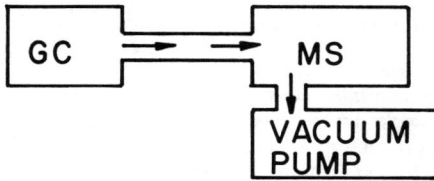

DIRECT COUPLED

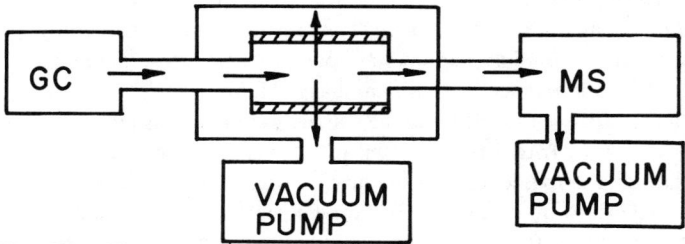

EFFUSIVE

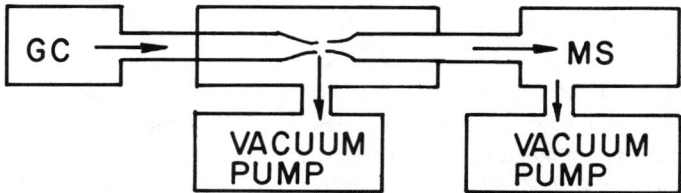

JET ORIFICE

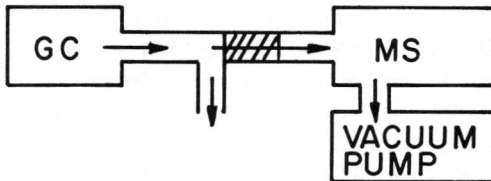

PERMEABLE BARRIER

Figure 2.   GC/MS interfaces[27].

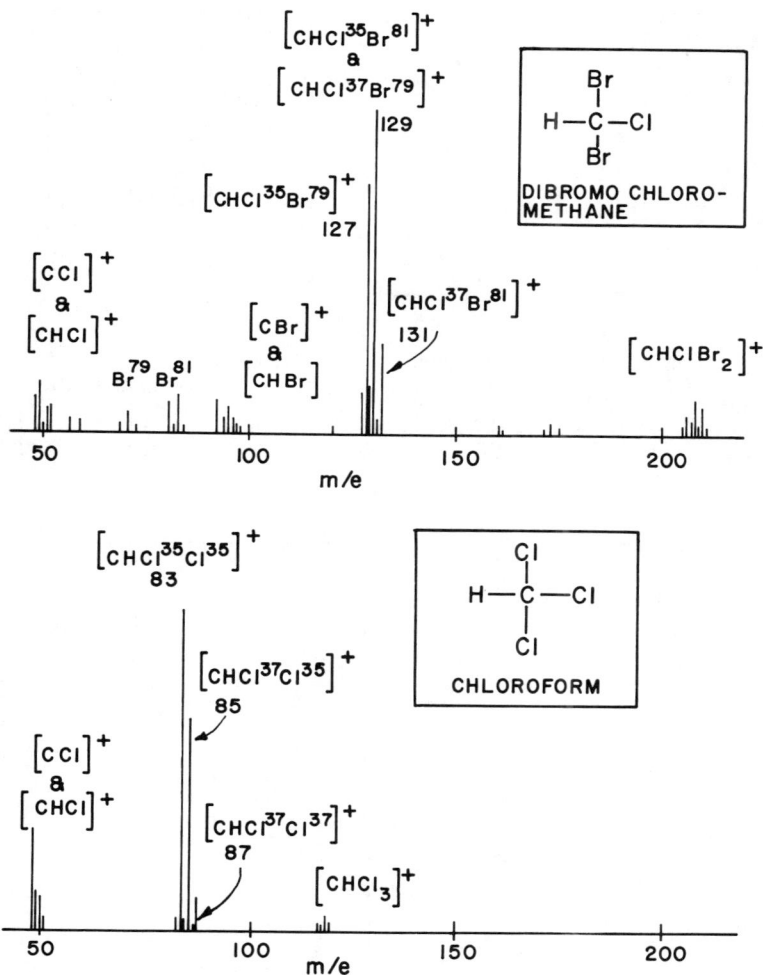

Figure 3.   Mass spectra of $CHCl_3$ plus $CHBr_2Cl$.

abundance of species from instrument to instrument, but the variations are usually minor. There are certain "rules of thumb" for these patterns. For example, for hydrocarbons:

1.   The parent, or molecular ion, peak is largest on the least-branched hydro-carbons.
2.   A single-carbon fragment is not usually lost from a straight chain; a peak corresponding to a methyl group generally represents a side chain.

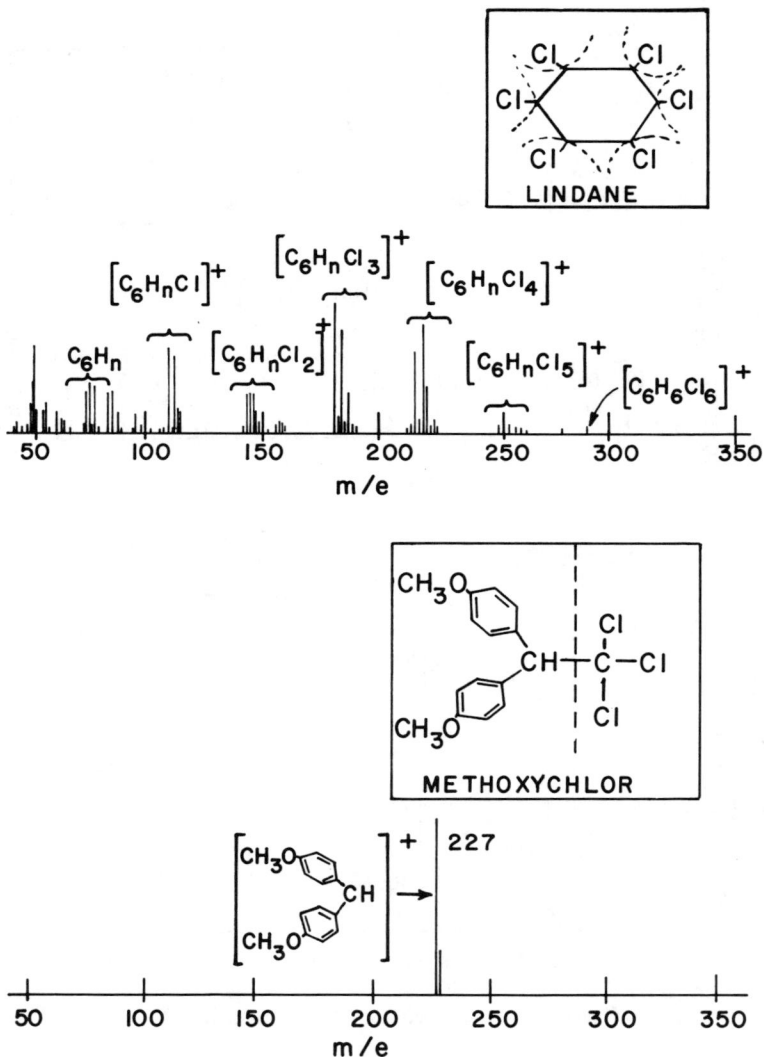

Figure 4. Mass spectra of lindane and methoxychlor (courtesy of New Jersey Institute of Technology Environmental Lab).

3. Breaks are most likely at the most highly branched chain.
4. Odd mass ions are more frequent than even mass ions.
5. If prominent even peaks are observed, compound must be highly branched (have broken two bonds).

All of the lines in a given spectra are not necessarily predictable; certain re-arrangements and metastable ions are characteristic of particular compounds. Identification is usually confirmed by comparison with a known spectra. This comparison may be conducted manually with a search of a spectral index[28], or electronically, if the MS system is equipped with library search capabilities.

Other methods of producing ions are chemical ionization (CI) and field ioniza-tion (FI). Chemical ionization is a process in which a reagent gas is ionized and then allowed to react with the sample. This "softer" or lower energy ionization results in less fragmentation of the molecule and a larger percentage of molecular ions. Figure 5 shows a comparison of the EI and CI mass spectra of methyl para-thion[29]. The most commonly used reagent gas is methane, although hydrogen, water, ammonia, nitric oxide and low-molecular-weight aliphatic hydrocarbons have been used. A typical reaction when methane is used is

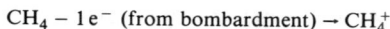

$$CH_4 - 1e^- \text{ (from bombardment)} \rightarrow CH_4^+$$

$$CH_4^+ + CH_4 \rightarrow CH_5^+ + CH_3$$

$$CH_5^+ + M \rightarrow MH^+ + CH_4$$

where M is the sample molecule and MH^+ is the molecular ion plus one (M + 1).

The use of different gasses as reagents varies the fragmentation pattern of a molecule. Water and ammonia are relatively mild protonating agents, while pro-pane is intermediate between methane and isobutane.

Field ionization (FI) is another "soft" or low-energy method of ionization. The sample is subjected to a high electric field (potential gradients of 10^7 to 10^8 V/cm), and outer shell electrons are "picked off" the neutral molecule. FI mass spectra are therefore characterized by prominent molecular ion peaks and only a few fragment ion peaks. For this reason, it is useful in determining exact molecular weights and for the analysis of multicomponent mixtures without pre-separation.

Types of Mass Spectrometers

The sorting of ions according to their mass to charge ratio (m/e) is usually done by one of three basic types of mass spectrometers: magnetic, quadrupole or time-of-flight.

Magnetic Sector

The mass analysis in this type of instrument is a wedge-shaped, or sector, mag-netic field (Figure 6), which disperses the total ion beam into discrete ion beams

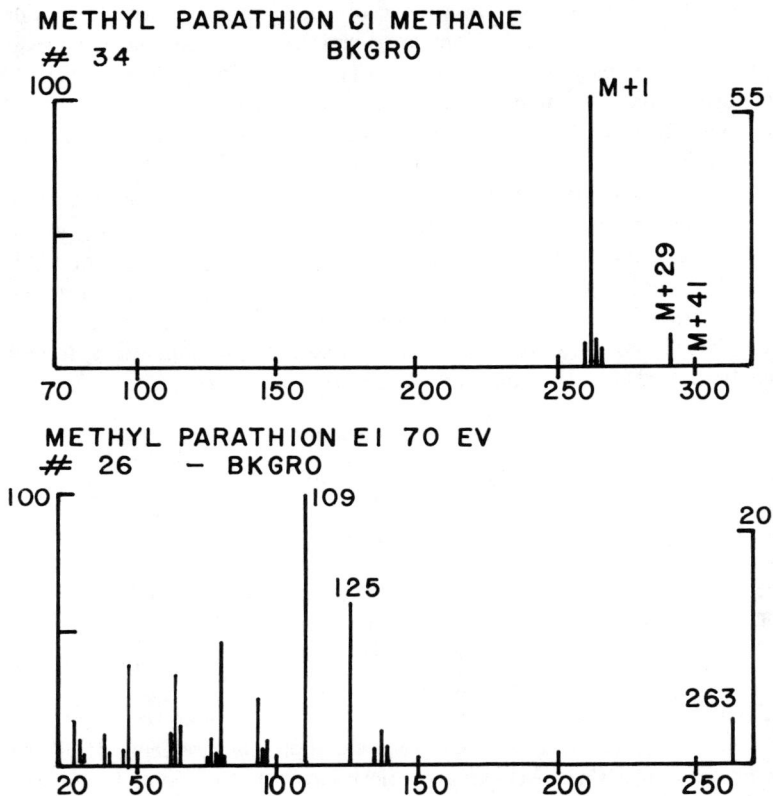

Figure 5.   EI and CI specta of methyl parathion[29].

of individual m/e values[30]. Only when the centrifugal and centripetal forces are balanced ($HeV = mv^2/R$) will a given mass reach the detector; ions of other masses will collide with the walls. To detect ions of differing m/e values, either the voltage or the magnetic field can be varied; however, magnetic scanning using electromagnets is the usual procedure.

Quadrupole

A schematic of quadrupole design is shown in Figure 7[30]. The four rods are alternately connected to radiofrequency (rf) and dc voltage generators. For any given level of rf/dc voltage, only ions of a specific m/e avoid collision with the rods and successfully traverse the quadrupole to reach the detector. Scanning is accomplished by varying the rf frequency.

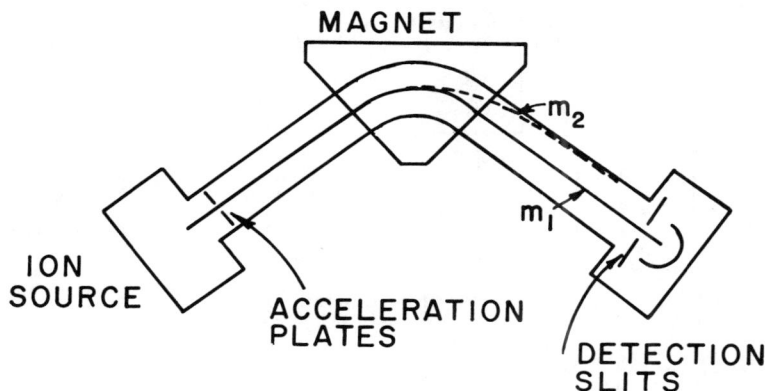

Figure 6.   Schematic of a sector magnet mass spectrometer[30].

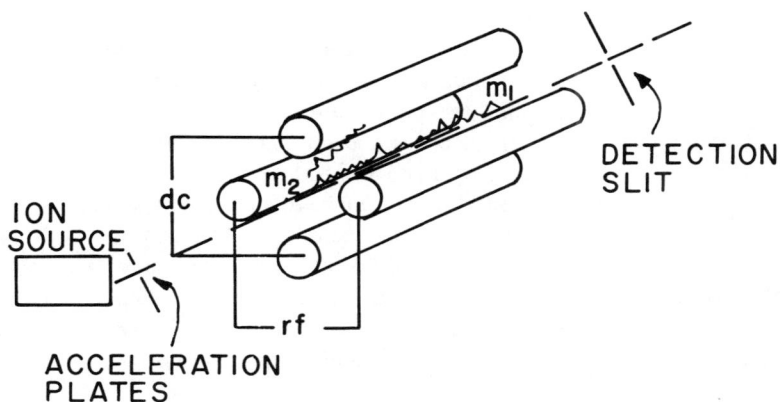

Figure 7.   Schematic of a quadruple design mass spectrometer[30].

Time-of-Flight

The operating principle of the time-of-flight instruments involves producing a discrete "bunch" of ions and accelerating them with the same kinetic energy toward a detector one to two meters away, as shown in Figure 8. Ions of low mass will have a higher velocity and will reach the detector before those of higher mass, and the m/e values are determined by the time of arrival at the detector. This type of mass spectrometer is the least common of the three types described.

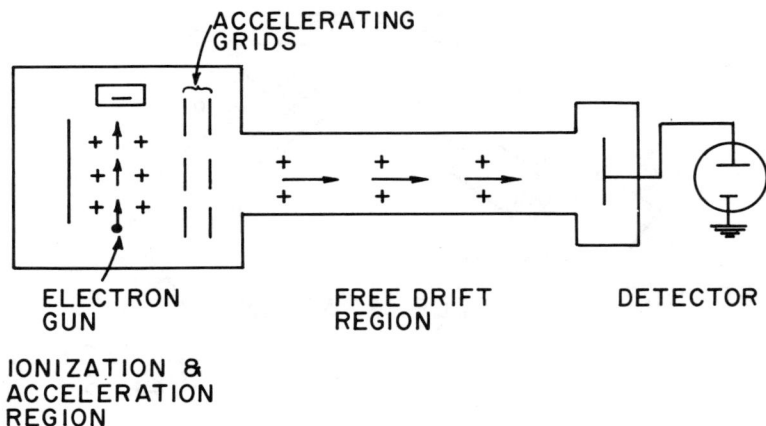

Figure 8.   Schematic of a time-of-flight mass spectrometer.

Resolution

The levels of resolution in mass spectroscopy are generally referred to as low, medium and high. Low resolution usually implies unit resolution or a resolving power of 100 to 1000; medium resolution is a resolving power of 2000 to 10,000; and high resolution is a resolving power above 10,000. Resolution also can be expressed by the ratio $M:\Delta M$ when M and $M + \Delta M$ are the m/e values of two adjacent peaks. For low-resolution data, ΔM will be equal to one mass unit (unit resolution), and for high-resolution, ΔM may be 0.01 mass units.

Exact mass measurements with accuracies of 0.01 m/e units are extremely valuable in the identification of unknown organics. This type of resolution requires a double focusing instrument to eliminate velocity dispersion[30]. A certain amount of sensitivity must be sacrificed for each increase in resolution.

Detectors

The original photographic method of detection is used occasionally in high resolution work, but it has largely been replaced by electron multipliers or Faraday cup detectors. Electron multipliers use the principle of secondary electron emission to effect amplification gains of 10^6. In addition to high sensitivities, they also have a fast response time. However, they do exhibit some mass discrimination and are less stable than electrometric (i.e., Faraday cup) detectors. Faraday cup detectors are simple, inexpensive, rugged and reliable. The principal disadvantage is the relatively long delay in the amplification system, making them inappropriate for use with rapidly scanning instruments.

Modes of Operation—Scan vs Selected Ion Monitoring

The mass spectrometer may be programmed to monitor m/e values as required by the analysis. For example, all of the m/e values between 50 and 250 may be scanned at a rate of, say, 50 scans per minute (or 1.2 seconds per scan). This is done by varying appropriately the rf frequency (quadrupole instruments) or magnetic field (magnetic instruments). The total ion current within these or other specified m/e limits may also be plotted vs time. In this scan mode, the mass spectra for identification of each compound is displayed and/or recorded as it elutes from the GC column; and the relative amounts of the various compounds are related to the magnitude of the total ion current generated as they pass through the mass spectrometer. Thus, in the scan mode, we are given the spectra, or fingerprint, of each of the compounds separated by the column, as well as an indication of its relative concentration.

On the other hand, if the instrument is directed to monitor and quantify only a few m/e values of interest, a great deal of sensitivity as well as specificity is obtained. This mode of operation is referred to as "selected ion monitoring" (SIM) and has the following advantages as a quantitative tool:

1. Sensitivity is increased since the mass spectrometer is only looking at a limited number of m/e values (instead of every one in a whole spectral range), and thus may spend more time on each value.
2. When individual characteristic ions of a substance are monitored and the abundance plotted with time, much less interference in the form of baseline noise or coeluting of overlapping compounds is experienced.

This increase in sensitivity and decrease in background interferences is illustrated in Figure 9, in which a total ion chromatograph of 20 ng of chlordane in urine is compared with a SIM chromatograph at m/e 410[29].

Generally, the base peak and molecular ion are the most diagnostic ions of any compound and are normally used as long as there are no evident interferences. The ions selected for identification and quantification of any particular compound must be found to be present in the appropriate relative intensities and must elute from the GC column after the proper retention time. Obviously, using the SIM mode, only a limited number of preselected compounds can be detected and/or quantified.

Quantification is based on the integrated area from the specific ion plot, as compared to the signal produced by an internal standard. The response of the internal standard compared to that of the material in question in the appropriate concentration range has been determined. An ideal internal standard is achieved by the synthesis of the compound in question with a stable isotope label, effectively shifting the mass of the standard beyond the interference of the naturally occuring analog[29,31,32].

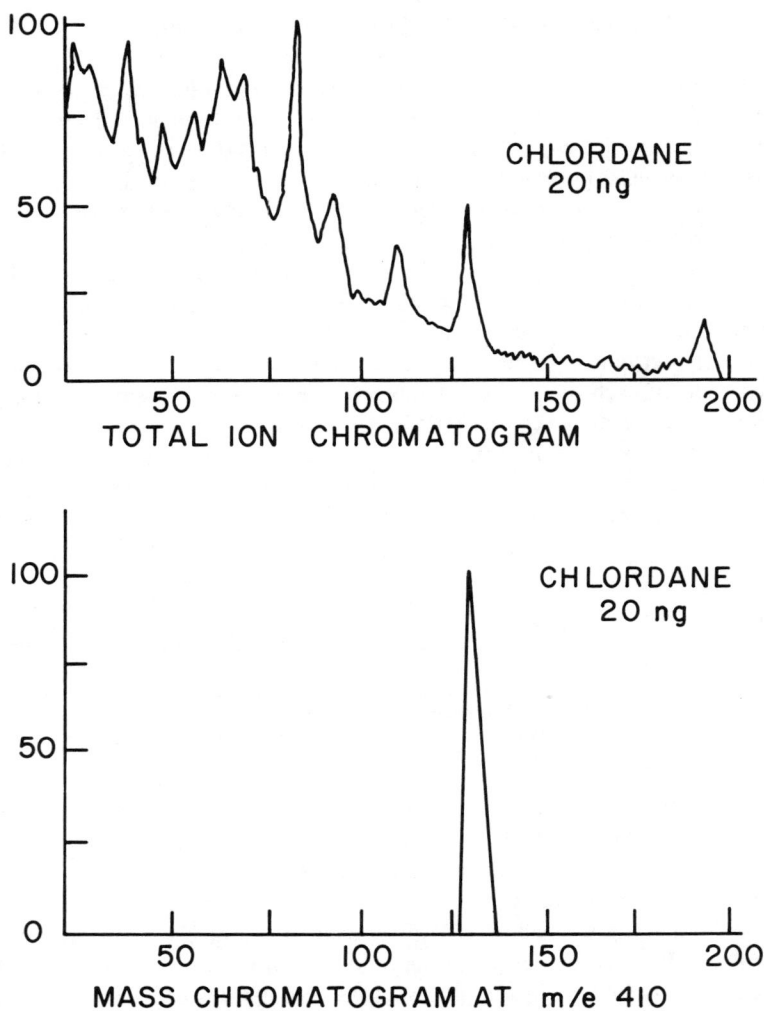

Figure 9.   Total ion chromatogram and SIM chromatogram of 20 ng chlordane in urine[29].

SELECTION OF ANALYTICAL METHOD

There are two general goals that affect the selection process. The "target compound" goal is one in which a list of compounds to be searched for and/or quantified is submitted or proscribed. In this case, detailed procedures for separation and isolation from interferences are usually designated. Systems for optimizing recovery as well as internal standards and controls for quantification are usually employed.

By contrast to the "target compound" approach is the "broad spectrum" approach, in which one is seeking to obtain information on all the compounds present in a sample. Sample preparation methods are as simple as possible to preclude losses of components and minimize contamination. The development of computerized mass spectral search systems has made identification simpler and faster.

Target Compound Approach

The priority pollutant program of the U.S. Environmental Protection Agency (EPA) is an example of the target compound approach. During 1976, the EPA was required to establish effluent limitations for a group of 129 "priority pollutants" selected on the basis of known human or animal toxic and carcinogenic effects. These 129 products include 106 specific organic compounds and 9 product formulations that are mixtures of organic compounds.

Since this large number of target compounds covers a broad range of physical and chemical properties, they were divided into five general classes:

1. A group of 45 compounds isolated from a pH 11 adjusted sample by extraction with methylene chloride, followed by drying and concentration of the extract forms the first class. Due to the broad range of boiling points, a general purpose column with high-temperature stability is necessary for the chromatographic separation. EPA has suggested a methyl-phenyl silicone stationary phase (i.e., Supelco's SP-2250) with temperature programming from 50–260°C at 8°C/min. The base-neutral compounds in this fraction, as well as their characteristic most prominent EI ions, are listed in Table IV.

2. A group of 30 volatile compounds isolated by the inert gas purge and trap method are shown in Table V. Suggested chromatographic column packings have been 0.2% carbowax 1500 on 80/100 or 60/80 carbopac C, and SP1000 (carbowax and substituted terephthalic acid) on carbopac B. The GC oven is held at room temperature during trap desorption, then rapidly heated to 60°C and held there for 4 minutes followed by a programmed increase of 8°C/min to 170°C.

3. A group of 11 compounds extracted from a sample adjusted to pH 2 comprises the third group. After acidification, the organics are extracted from the aqueous fraction with methylene chloride. The dried and concentrated methylene chloride fraction is injected into a GC column, which uses tenax or SP-1240 DA, a specially deactivated ester/acid polyester, as the stationary phase. Compounds of the acid fraction are listed in Table VI.

4. A group of 24 compounds and product formulations, the pesticide fraction,

Table IV.   Compounds of the Base-Neutral Fraction

Compound	Characteristic EI Ions (relative intensity)		
1,3-Dichlorobenzene	146(100)	148(64)	113(12)
1,4-Dichlorobenzene	146(100)	148(64)	113(11)
1,2-Dichlorobenzene	146(100)	148(64)	113(11)
Hexachloroethane	117(100)	199(61)	201(99)
bis(2-Chloroethyl)ether	93(100)	63(99)	95(31)
bis(2-Chloroisopropyl)ether	45(100)	77(19)	79(12)
N-Nitrosodi-n-propylamine	130(22)	42(64)	101(12)
Isophorone	82(100)	95(14)	138(18)
Nitrobenzene	77(100)	123(50)	65(15)
Hexachlorobutadiene	225(100)	223(63)	227(65)
1,2,4-Trichlorobenzene	74(100)	109(80)	145(52)
Naphthalene	128(100)	127(10)	129(11)
bis(2-Chloroethoxy)methane	93(100)	95(32)	123(21)
Hexachlorocyclopentadiene	237(100)	235(63)	272(12)
2-Chloronaphthalene	162(100)	164(32)	127(31)
Acenaphthylene	152(100)	153(16)	151(17)
2,6-Dinitrotoluene	165(100)	63(72)	121(23)
Acenaphthene	154(100)	153(95)	152(53)
Dimethylphthalate	163(100)	164(10)	194(11)
Fluorene	166(100)	165(80)	167(14)
2-Chlorophenyl phenyl ether	204(100)	206(34)	141(29)
2,4-Dinitrotoluene	165(100)	63(72)	121(23)
1,2-Diphenylhydrazine	77(100)	93(58)	105(28)
Diethylphthalate	149(100)	178(25)	150(10)
N-Nitrosodiphenylamine	169(100)	168(71)	167(50)
Hexachlorobenzene	284(100)	142(30)	249(24)
4-Bromophenyl phenyl ether	248(100)	250(99)	141(45)
Phenanthrene	178(100)	179(16)	176(15)
Anthracene	178(100)	179(16)	176(15)
Di-n-butylphthalate	149(100)	150(27)	104(10)
Fluoranthene	202(100)	101(23)	100(14)
Pyrene	202(100)	101(26)	100(17)
Benzidine	184(100)	92(24)	185(13)
Butylbenzylphthalate	149(100)	91(50)	
bis(2-Ethylhexyl)phthalate	149(100)	167(31)	279(26)
Chrysene	228(100)	229(19)	226(23)
Benzo(a)anthracene	228(100)	229(19)	226(19)
Benzo(b)fluoranthene	252(100)	253(23)	125(15)
Benzo(k)fluoranthene	252(100)	253(23)	125(16)
3,3-Dichlorobenzidine	252(100)	254(66)	126(16)
Benzo(a)pyrene	252(100)	253(23)	125(21)
Indeno(1,2,3-cd)pyrene	276(100)	138(28)	277(27)
Dibenzo(a,h)anthracene	278(100)	139(24)	279(24)
Benzo(g,h,i)perylene	276(100)	138(37)	277(25)
Nitrosodimethylamine	42(100)	74(88)	44(21)
Deuterated anthracene (int. std.)	188(100)	94(19)	80(18)

Table V. Compounds of the Purgable Fraction

Compound	EI Ions (relative intensity)	Ion Used to Quantify
Chloromethane	50(100); 52(33)	50
Dichlorodifluoromethane	85(100); 87(33); 101(13); 103(9)	101
Bromomethane	94(100); 96(94)	94
Vinyl Chloride	62(100); 64(33)	62
Chloroethane	64(100); 66(33)	64
Methylene Chloride	49(100); 51(33); 84(86); 86(55)	84
Trichlorofluoromethane	101(100); 103(66)	101
1,1-Dichloroethylene	61(100); 96(80); 98(53)	96
Bromochloromethane(IS)	49(100); 130(88); 128(70); 51(33)	128
1,1-Dichloroethane	63(100); 65(33); 83(13); 85(8); 98(7); 100(4)	63
trans-1,2-Dichloroethylene	61(100); 96(90); 98(57)	96
Chloroform	83(100); 85(66)	83
1,2-Dichloroethane	62(100); 64(33); 98(23); 100(15)	98
1,1,1-Trichloroethane	98(100); 99(66); 117(17); 119(16)	97
Carbon Tetrachloride	117(100); 119(96); 121(30)	117
Bromodichloromethane	83(100); 85(66); 127(13); 129(17)	127
bis-Chloromethyl ether	79(100); 81(33)	79
1,2-Dichloropropane	63(100); 65(33); 112(4); 114(3)	112
trans-1,3-Dichloropropene	75(100); 77(33)	75
Trichloroethylene	95(100); 97(66); 130(90); 132(85)	130
Dibromochloromethane	129(100); 127(78); 208(13); 206(10)	127
cis-1,3-Dichloropropene	75(100); 77(33)	75
1,1,2-Trichloroethane	83(95); 85(60); 97(100); 99(63); 132(9); 134(8)	97
Benzene	78(100)	78
2-Chloroethylvinyl ether	63(95); 65(32); 106(18)	106
2-Bromo-1-chloropropane(IS)	77(100); 79(33); 156(5)	77
Bromoform	171(5); 173(100); 175(50); 250(4); 252(11); 254(11); 256(4)	173

Table V, continued

Compound	EI Ions (relative intensity)	Ion Used to Quantify
1,1,2,2-Tetrachloroethene	129(64); 131(62); 164(78); 166(100)	164
1,1,2,2-Tetrachloroethane	83(100); 85(66); 131(7); 133(7); 166(5); 168(6)	168
1,4-Dichlorobutane(IS)	55(100); 90(30); 92(10)	55
Toluene	91(100); 92(78)	92
Chlorobenzene	112(100); 114(33)	112
Ethylbenzene	91(100); 106(33)	106

Table VI.   Compounds of the Acid Fraction

Compound	Characteristic EI Ions (relative intensity)		
Phenol	94(100)	65(17)	66(19)
2-Chlorophenol	128(100)	64(54)	130(31)
2-Nitrophenol	139(100)	65(35)	109(8)
2,4-Dimethylphenol	122(100)	107(90)	121(55)
2,4-Dichlorophenol	162(100)	164(58)	98(61)
p-Chloro-m-cresol	142(100)	107(80)	144(32)
2,4,6-Trichlorophenol	196(100)	198(92)	200(26)
2,4-Dinitrophenol	184(100)	63(59)	154(53)
4-Nitrophenol	65(100)	139(45)	109(72)
4,6-Dinitro-o-cresol	198(100)	182(35)	77(28)
Pentachlorophenol	266(100)	264(62)	268(63)

are listed in Table VII. These compounds are extracted at ambient pH with 15% methylene chloride in hexane. GC separation is effected with a methyl/phenyl silicone stationary phase alone, or in combination with a fluoropropyl silicone. EPA recommends that these compounds be analyzed using gas chromatography and an electron capture detector, with GC/MS used only for confirmation.

5. Finally, two compounds, acrolein and acrylonitrile, are very water soluble and not easily isolated from aqueous samples. These are identified by direct aqueous injection.

These "target compounds" are quantified using selected ion monitoring, and the internal standards bromochloromethane, 2,bromo-1-chloropropane, 1,4-dichlorobutane and deuterated anthracene, as indicated in Tables IV and V[32-34].

Table VII.   Compounds of the Pesticide Fraction

Compound	Characteristic EI Ions (relative intensity)		
β-Endosulfan	201(100)	283(48)	278(30)
α-Benzenehexachloride	183(100)	109(86)	181(91)
γ-Benzenehexachloride	183(100)	109(86)	181(91)
β-Benzenehexachloride	181(100)	183(93)	109(62)
Aldrin	66(100)	220(11)	263(73)
Heptachlor	100(100)	272(60)	274(46)
Heptachlor Epoxide	355(100)	353(79)	351(60)
δ-Endosulfan	201(100)	283(48)	278(30)
Dieldrin	79(100)	263(28)	279(22)
4,4'-DDE	246(100)	248(64)	176(65)
4,4'-DDD	235(100)	237(76)	165(93)
4,4'-DDT	235(100)	237(72)	165(59)
Endrin	81(100)	82(61)	263(70)
Endosulfan Sulfate	272(100)	387(75)	422(25)
δ-Benzenehexachloride	183(100)	109(86)	181(90)
Chlordane	373(19)	375(17)	377(10)
Toxaphene	a		
Aroclor-1242	a		
Aroclor-1254	a		
Aroclor-1221	a		
Aroclor-1232	a		
Aroclor-1248	a		
Aroclor-1260	a		
Aroclor-1016	a		

[a] These preparations are mixtures of compounds.

Broad Spectrum Approach

An example of the "broad spectrum" approach is the National Organics Reconnaissance Survey (NORS), undertaken by EPA after President Ford signed into the law "The Safe Drinking Water Act" in December 1974. This program was an attempt to determine the identity, concentrations and potential effects of organics in the water supplies of representative cities across the nation. More than 100 compounds have been isolated and identified using a variety of separation and concentration techniques, including purge and trap, continuous liquid-liquid extraction and adsorption on activated carbon and macroreticular resins[35-37]. Recovery efficiencies vary among individual compounds, as well as with the methods employed, and the techniques of concentration and separation continue to be refined.

Mass spectra of the compounds eluting from the GC column are stored by the software of the mass spectrometer as it operates on a "scan" mode. Chromatographic separation and resolution are considerably more critical in the broad base search and identification procedure than when the instrument is operating in a quantitative SIM mode, since coeluting or overlapping compounds would not produce an easily identified spectra. Many investigators recommend capillary columns for this type of work. CI spectra have been found to be complimentary to EI spectra and, in some cases, necessary for positive identification.

MASS SPECTRA IDENTIFICATION

Any search and match system must have a data base, or organized collection of reference data, as well as a system or approach to that data base. Manual searching of printed data bases have been attempted, and elaborate index systems have been developed to aid the searcher. Manual systems are, however, time consuming and tedious.

Computerized search systems have been explored since 1971, when the EPA developed a system whereby spectral data were transmitted via telephone lines from a minicomputer coupled with the mass spectrometer to a program running in a large-scale remote time-sharing computer. The remote computer conducts a search of the data base and transmits the names of the compounds with "best fit" spectra back to the user's terminal. The degree of similarity of each spectra to the unknown is reported as a similarity index (SI) calculated on a scale of 0 to 1. SI values must be used with caution because the computer weights each of the m/e values equally and cannot make value judgments concerning extraneous ions caused by background or overlapping spectra. Another criterion for identification is the quality index (QI), as a weighting factor for the SI. Using this system, the presence of the molecular ion (M^+) and its naturally occuring isotopes ($M+1^+$) in appropriate abundancies far outweighs most other ions.

Further developments and refinements of computerized search systems have led to a consolidation of systems into an international mass spectral search system (MSSS). This system offers a number of search options: spectra can be retrieved based on molecular weights, partial or complete molecular formulae, mass losses from the molecular ion, Mass Spectrometry Data Center classification codes and combinations of all of these—29 options altogether. In addition, the data base may be searched from a keyboard/printer terminal, which is not connected to a GC/MS minicomputer, by typing in mass and abundance data, one pair at a time.

Individual GC/MS systems may extract and retain from the data base a library of 500–2000 spectra in their particular area of concern. With the further development of GC/MS software as well as the MSSS, it is expected that the effectiveness of computerized GC/MS should improve and the cost per identification should decrease[37-40].

SPECIALIZED INSTRUMENTS

Specialized, or dedicated, instruments are available for both air and water analysis. A portable vapor detector system was developed for the U.S. Army, the idea being that hidden personnel could be detected by selected ion monitoring of masses indicative of organic compounds emitted by the human body. Adaptations of this system also have been mounted in aircraft to detect vapor plumes from explosives manufacturing facilities, and in a microbus for monitoring chemical-industrial vapors.

A portable instrument package has been housed in two heavy-duty aluminum suitcases, one of which contains the control panel and data display, the main power system and a miniaturized computer, while the quadrupole mass spectrometer, its vacuum system and the specialized inlet system are housed in the other case. During operation, the two cases are interconnected and operated via switch commands from the control panel (Figure 10)[41]. The inlet system is composed of three membranes of dimethyl silicone, which is more permeable to organic molecules than to the air gasses. Pumping requirements for each stage of the separation are met by evacuated canisters, and sample enrichment is, typically, about 10^6.

The quadrupole analyzer structure is floated within the vacuum housing so that the rods maintain their alignment within 0.001 inch. Power for the ion pump is supplied by the main power supply or external source during operation and by batteries during transit. Three modes of operation are possible: full scan from mass 10 to 346, partial scan and selected ion monitoring. Further, sensitivies of 10^{-8} g/sec in full-scan mode and 10^{-11} g/sec in SIM mode are possible.

In 1979 The Finnigan Corporation introduced an Organics-in-Water Analyzer, a GC/MS designed to sell for considerably less than the research model, yet provide all the essential features required for priority pollutant analysis. In this system, a Perkin-Elmer Sigma series microprocessor-controlled GC coupled with an MS with quadrupole analyzer, turbomolecular pump, jet separator, 9-track magnetic tape drive and a scan range of 4 to 800 amu.

COST-EFFECTIVENESS

A cost analysis of GC vs GC/MS has been reported by Robert E. Finnigan[42]. Specifically, costs per sample were compiled of performing a priority pollutant

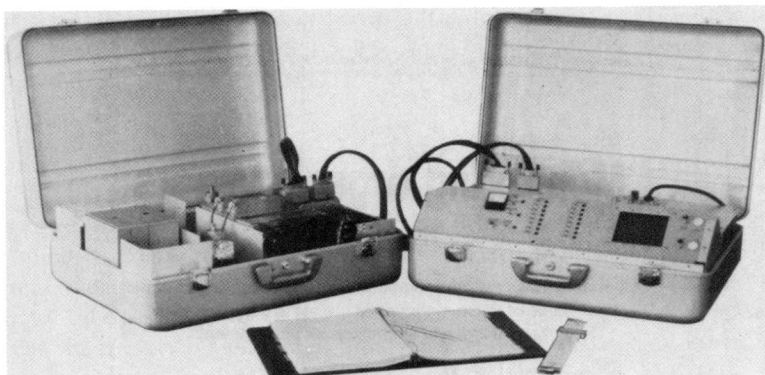

Figure 10.   Portable mass spectrometer for sampling air[41].

analysis by GC alone and by GC/MS. For this analysis, costs were divided into two categories: capital costs and operator time and costs (1979). Comparative capital costs are shown in Table VIII and operator costs in Table IX. Although capital costs are greater for the GC/MS system, operator costs are considerably lower. One reason for this is that much simpler extraction and cleanup procedures are indicated for GC/MS than for GC alone. A summary of total charges per analysis is shown in Figure 11. It should be noted, however, that the instrument charges per analysis are based on usages of 1500 hr/yr for GC/MS and 1800 hr/yr for each GC, estimates that may be quite high for many laboratories. If the

Table VIII.   Capital Costs of GC and GC/MS[42]

GC		GC/MS/DS	
Item	**$**	**Item**	**$**
4 GCs with Multiple Detectors	40 K	1 GC/MS/DS (includes 9-track magnetic tape)	140 K
1 HPLC	15 K	Installation Training Room Modules	2.5K
4 Integrators	12 K	1 GC with HD and FID	10 K
1 Lab Data System	30 K		
Installation Training Room Modules	4.5K		
Total	101.5K	Total	152.5K

Table IX.   Operator Time and Cost

Phase	Operation	GC		GC/MS	
		Time (hr)	Cost ($)	Time (hr)	Cost ($)
Sample Extraction	Extraction	6	28.02	3	14.01
	Concentration	2	9.34	1	4.67
	Purge and trap	0.5	3.50	0.5	4.67
	Pentane extract	0.5	3.50	–	–
Sample Cleanup	Column chromatograph	3	21.00	0.5	4.67
	Concentration	2	9.34	0.5	2.34
Calibration	Calibrate instrument (1 per 15 samples)	–	–	0.3 ÷ 15	0.19
Qualitative Analysis	Sample analysis	5	35.00	2.5	23.33
	Std and sample analysis	5	35.00	–	–
Quantitative Analysis (time per 15 samples)	Std analysis and std curve (3pts)	15 ÷ 15	7.00	7.5 ÷ 15	4.67
	Validate curves	10 ÷ 15	4.67	5 ÷ 15	3.11
Direct operator cost per sample:			$156.37		$ 61.66
Overhead At 126% of direct labor			197.03		77.69
General and Administrative At 14% of direct labor and overhead			49.48		19.50
Fee at 8% of Total Cost			32.23		12.71
Total labor related			$435.11		$171.56

instrument costs per hour increase as the result of less usage, the difference between the cost per sample for the two methods will decrease[42]. Further, if priority pollutant analysis is the mission of the instrument, a less expensive and more specialized Organics-in-Water Analyzer type may be the instrument of choice.

Of course, one cannot do a cost comparison on the abilities of separation and compound identification belonging uniquely to GC/MS systems.

CONCLUSIONS

The growing concern over the many potentially hazardous chemicals being released into the environment, as well as recent legislation establishing the need

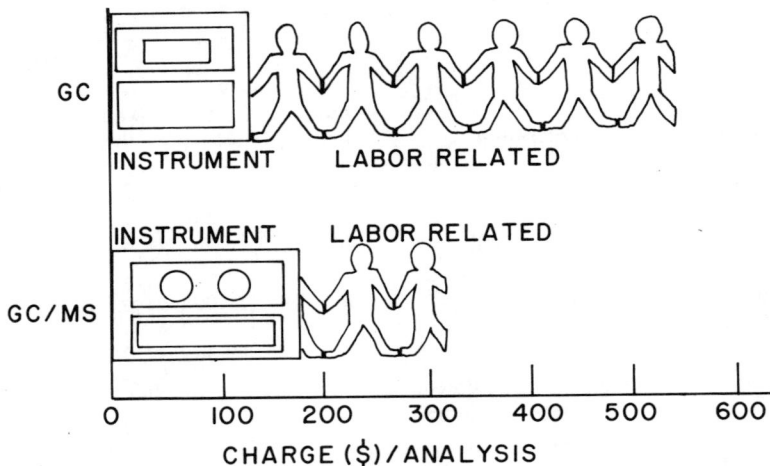

Figure 11.   Summary of total charges.

Figure 12.   Varian MAT 44 GC/MS at New Jersey Institute of Technology, Newark, New Jersey.

for extensive measurements of the presence and concentration of a broad variety of organic compounds, has spurred the development of specific and sensitive methods for their analysis. The GC/MS system represents one of the most powerful instruments for the identification and quantification of trace organic compounds, and it is becoming useful and popular as both a research and routine analytical tool.

The mass spectra of a compound contains a large amount of structural information about the molecules, and a GC/MS can therefore be used as a research tool to delineate structure of unknown organic compounds. In more recent

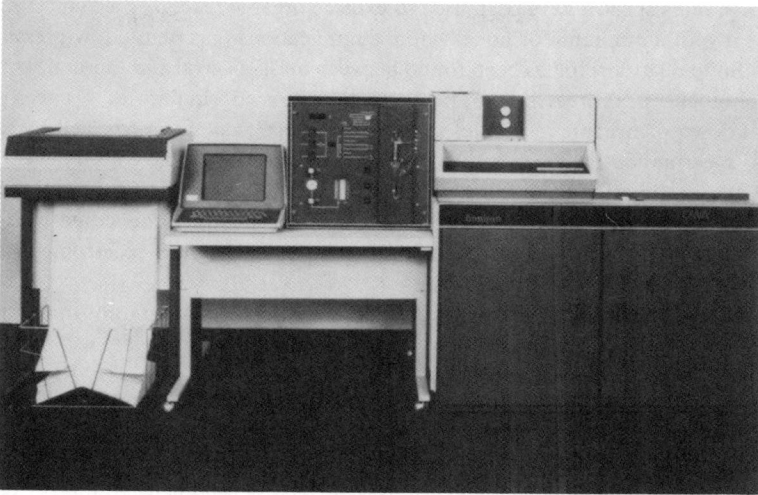

Figure 13. Finnigan organics-in-water analyzer GC/MS system.

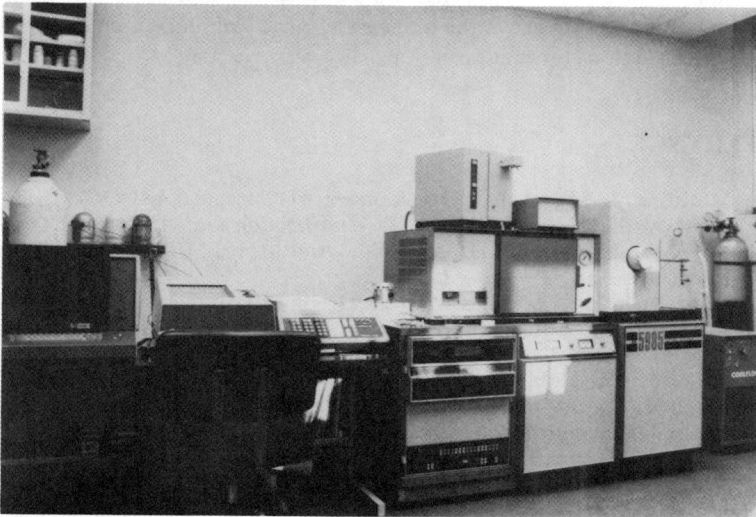

Figure 14. Hewlett-Packard 5985 GC/MS system at Passaic Valley Water Commission, Little Falls, New Jersey.

148   ANALYTICAL MEASUREMENTS

work[43], mass spectra have been used to explore organic reaction mechanisms in a system with a minimum of nonessential complications (gas phase, low pressure). A definite relationship has been found between mass spectral and thermal, as well as photochemical, reactions. There are pronounced similarities between the breakdown of organic molecules in the mass spectrometer, as revealed by their mass spectra, and their behavior on pyrolysis. Correlations have been developed that allow estimation from the mass spectrum of a gas oil of the catalytic cracking yields of dry gas, butane and gasoline. Radiation chemists, concerned with the effects of ionizing radiation on materials, have accepted mass spectrometry as a legitimate branch of their discipline.

There is little doubt but that the field of gas chromatography/mass spectroscopy is growing and expanding. Future developments probably will include:

1. improvements and refinements of the instrument in general, to the end that sensitivity, reliability and ease of repair are improved;
2. development of improvements in computer hardware and software, resulting in better MS minicomputer design;
3. development of a wide variety of specialized instruments designed for specific applications and selling for considerably less than more versatile "research models"; and
4. marketing of a host of proprietary "secondary appliances" to be used for sampling, injecting, extracting, concentrating, separating, etc., organics from matrix material.

The unique ability of the GC/MS system to separate, identify and quantify organic compounds present in the monogram range will make it invaluable as a diagnostic, as well as research, tool for many years to come.

REFERENCES

1. McGuire, J. M., et al. "Development of Computerized GC/MS Techniques within the U.S.-E.P.A.," *Identification and Analysis of Organic Pollutants in Water,* L. H. Keith, Ed. (Ann Arbor, MI: Ann Arbor Science Publishers, Inc., 1976).
2. Hase, A., and R. A. Hites. "On the Origin of Polycyclic Aromatic Hydrocarbons in the Aqueous Environment," in *Identification and Analysis of Organic Pollutants in Water,* L. H. Keith, Ed. (Ann Arbor, MI: Ann Arbor Science Publishers, Inc., 1976).
3. Mitchell, R. *Introduction to Environmental Microbiology* (Englewood Cliffs, NJ: Prentice-Hall, Inc., 1974).
4. Junk, G. A., et al. "Resin Sorption Methods for Monitoring Selected Contaminants in Water," in *Identification and Analysis of Organic Pollutants in Water,* L. H. Keith, Ed. (Ann Arbor, MI: Ann Arbor Science Publishers, Inc., 1976).
5. Budde, W. L., and J. W. Eichelberger. *Organics Analysis Using Gas Chromatography–Mass Spectrometry* (Ann Arbor, MI: Ann Arbor Science Publishers, Inc., 1979).

6. Mieure, J. P., et al. "Separation of Trace Organic Compounds from Water," in *Identification and Analysis of Organic Pollutants in Water,* L. H. Keith, Ed. (Ann Arbor, MI: Ann Arbor Science Publishers, Inc., 1976).

7. *Tekmar Scientific Apparatus,* Tekmar Company, Cincinnati, OH (1979).

8. Bellar, T. A., and J. J. Lichtenberg. "Determining Volatile Organics at Microgram-per-Liter Levels by Gas Chromatography," *J. Am. Water Works Assoc.* 66: 739 (1974).

9. Kopfler, F., et al. "GC/MS Determination of Volatiles for the National Organics Reconnaissance Survey (NORS) of Drinking Water," in *Identification and Analysis of Organic Pollutants in Water,* L. H. Keith, Ed. (Ann Arbor, MI: Ann Arbor Science Publishers, Inc., 1976).

10. *Standard Methods for Examination of Water and Wastewater* 14th ed. (New York: Americal Public Health Association, 1976).

11. U.S. Environmental Protection Agency. *Manual of Analytical Methods for the Analysis of Pesticide Residues in Human and Environmental Samples,* Research Triangle Park, NC (1977).

12. Henderson, J. E., et al. "Liquid Extraction Method for the Determination of Halomethanes in Water at the PPB Level," in *Identification and Analysis of Organic Pollutants in Water,* L. H. Keith, Ed. (Ann Arbor, MI: Ann Arbor Science Publishers, Inc., 1976).

13. Brodlmann, N. V., Jr. "Quantitation of Chlorinated Pesticides—A Comparison of Methods," *J. Am. Water Works Assoc.* (October 1975).

14. Kahn. L., and C. H. Waymon. *Anal. Chem.* 36: 1340 (1964).

15. Braus, H., et al. *Anal. Chem.* 23: 1160 (1951).

16. Junk, G. A., et al. *J. Chromatog.* 99: 745 (1974).

17. Musty, P. R., and G. Nickless. *J. Chromatogr.* 89: 185 (1974).

18. McNeil, E. E., and R. Otson. "Determination of Chlorinated Pesticides in Potable Water," *J. Chromatog.* 132: 277 (1977).

19. Mertz, R. "The Air Around Us," *The Alumnus* 33(1) (1979).

20. L. R. E. "Sampling and Analyzing Airborne Organics," *Environ. Sci. Technol.* 13(10) (1979).

21. McReynolds, W. O. "Characteristics of Some Liquid Phases," *J. Chromatog. Sci.* 8: 685–691 (1970).

22. *Supelco Chromatography Catalogue No. 15,* Supelco, Inc., Bellefonte, PA (1979).

23. Ettre, L. S. *Open Tubular Columns—An Introduction,* Perkin-Elmer Corp., Norwalk, CT (1973).

24. Ettre, L. S. *Basic Relationships of Gas Chromatography,* Perkin-Elmer Corp., Norwalk, CT (1977).

25. Applied Science Laboratories, Inc. "Technical Bulletin No. 6, Practical Chromatographic Theory," State College, PA.

26. Grab, K., and G. Grab. "Glass Capillary Gas Chromatography in Water Analysis: How to Initiate Use of the Method," in *Identification and Analysis of Organic Pollutants in Water,* L. H. Keith, Ed. (Ann Arbor, MI: Ann Arbor Science Publishers, Inc., 1976).

27. Gurtzinowicz, B. J., et al. "Integrated GC-MS Analytical Systems," *Chem. Instr.* 8(4) (1977).

28. American Society for Testing and Materials. *Index of Mass Spectral Data,* Philadelphia, PA.

29.   Velde, G. V., and J. F. Ryan. "Gas Chromatography–Mass Spectrometry as Applied to Pesticide Analysis," *J. Chromatog. Sci.* 13 (July 1975).
30.   Watson, J. T. *Introduction to Mass Spectrometry: Biomedical, Environmental and Forensic Applications,* (New York: Raven Press, 1976).
31.   Bonelli, E. J., et al. "Mass Fragmentography GC/MS in the Analysis of Hazardous Environmental Chemicals," *Am. Lab.* (July 1975).
32.   U.S. Environmental Protection Agency. "Sampling and Analysis Procedure for Screening of Industrial Effluents for Priority Pollutants," Cincinnati, OH (1977).
33.   Budde, W. L., and J. W. Eichelburger. "Organics in the Environment," *Anal. Chem.* 5(6) (1979).
34.   Supelco, Inc. "Water Pollution Analysis and Standards," Bulletin 775, Supelco, Inc., Bellefonte, PA (1978).
35.   Coleman, W. E., et al. "The Occurrence of Volatile Organics in Five Drinking Water Supplies Using GC/MS," in *Identification and Analysis of Organic Pollutants in Water,* L. H. Keith, Ed. (Ann Arbor, MI: Ann Arbor Science Publishers, Inc., 1976).
36.   Keith, L. H., et al. "Identification of Organic Compounds in Drinking Water from Thirteen U.S. Cities," in *Identification and Analysis of Organic Pollutants in Water,* L. H. Keith, Ed. (Ann Arbor, MI: Ann Arbor Science Publishers, Inc., 1976).
37.   Suffet, I. H. "GS/MS Identification of Trace Organic Compounds in Philadelphia Waters," in *Identification and Analysis of Organic Pollutants in Water,* L. H. Keith, Ed. (Ann Arbor, MI: Ann Arbor Science Publishers, Inc., 1976).
38.   Heller, R., et al. "Trace Organics in GC/MS," *Environ. Sci. Technol.* 9 (March 1975).
39.   Finnigan, R. E., and J. B. Knight. "The Use of GC/MS in the Analysis of Unusual Environmental Chemicals," Finnigan Corp., Sunnyvale, CA.
40.   Budde, W. L., and J. W. Eichelberger. "Development of Methods for Organic Analysis for Routine Application in Environmental Monitoring Laboratories," in *Identification and Analysis of Organic Pollutants in Water,* L. H. Keith, Ed. (Ann Arbor, MI: Ann Arbor Science Publishers, Inc., 1976).
41.   Evans, J. E., and J. T. Arnold. "Monitoring Organic Vapors," *Environ. Sci. Technol.* 9 (December 1975).
42.   Finnigan, R. E., et al. "Priority Pollutants II, Cost-Effective Analysis," *Environ. Sci. Technol.* 13 (May 1979).
43.   Meyerson, S. "Mass Spectrometry and Real-Life Chemistry," *Chemtech* (September 1979).

CHAPTER 6

X-RAY DIFFRACTION
TECHNIQUES AND INSTRUMENTATION

Frank H. Chung

Sherwin-Williams Research Center
Chicago, Illinois

INTRODUCTION

The stroke of luck greatly favors the prepared mind. Roentgen discovered X-rays in 1895[1]. It heralded the advent of modern physical sciences. Four branches of science have grown directly from X-rays: (1) X-ray crystallography for crystal structure determination; (2) X-ray spectroscopy for elemental analysis; (3) X-ray radiography for medical diagnosis or defect detection; and (4) X-ray therapy for treatment of disease, the extensions of which are radioisotope therapy and particle therapy.

Von Laue demonstrated the diffraction of X-rays by crystals in 1912. His experiment proved the wave nature of X-rays and the periodicity of atomic packing within a crystal. The success of X-ray diffraction led to the development of electron diffraction and neutron diffraction techniques for the study of matter. The theory and practice of these two techniques are nearly the same as that of X-ray diffraction. However, each has its own peculiarities in interacting with matter. Electron diffraction is useful for studying the structure of surfaces and films, while neutron diffraction is powerful for examining the magnetic structure and/or locating light atoms in crystals.

Laue's account of X-ray diffraction was read with great interest by two English physicists, W. H. Bragg and his son W. L. Bragg, who successfully interpreted Laue's data on zinc blende (ZnS) with the now well-known Bragg law and determined the crystal structure of the rocksalt series (NaCl, KCl, KBr and KI) the next year in 1913.

Before 1912, mineralogists and crystallographers had accumulated a great deal of knowledge on crystals concerning symmetry systems, group theory, morphological, chemical, physical and optical properties, but nothing about their internal structure. This knowledge we now call optical or classic crystallography. The use of X-ray diffraction to determine the internal structure of crystals pioneered by Laue and Bragg is called X-ray crystallography.

More than 95% of the solid materials on earth are crystalline. These crystalline materials are in the form of single crystal, powder or polycrystalline aggregate. Both powder and polycrystalline materials can be studied by powder diffraction techniques. Pauling's early work on the structure of minerals and Zachariasen's studies on thorides and actinides were done with powder diffraction data. However, except in rare cases, crystal structure determination is usually done by single-crystal techniques.

Three sets of experimental data can be obtained from X-ray diffraction of crystals: position, intensity and profile of the diffracted X-rays. These data carry the information of the specimen concerning its symmetry, atomic packing, crystallite size, lattice distortion, phase composition, stress and texture, etc. Various X-ray diffraction techniques and instrumentation are designed to extract this information from diffraction data. Among them, powder diffraction is the most frequently used technique in the realm of material science, and chemical analysis of unknown materials is the most popular application of powder diffraction in industry.

INSTRUMENTATION

All the X-ray diffraction instruments are of two types—camera or diffractometer. The first powder diffraction camera was designed by Debye and Scherrer in 1916. It was quickly accepted and improved on by many scientists. The major improvement is the use of electronic counters in place of the photographic film. Counter diffractometers appeared in the late 1950s using parafocusing optics. Although the camera techniques are still preferable in certain cases, powder diffractometer techniques are most popular in industry. The advantages of the diffractometer over the camera techniques are high intensity, linear response, better accuracy, simpler absorption effect and suitability for automation. There are in excess of 20,000 X-ray diffractometers in use in the world today[2]. These diffractometers have been manufactured by a number of companies during the past 30 years. Some companies offer complete X-ray diffraction systems. Others specialize in X-ray sources, detectors, cameras or other accessories. Among them are Philips, General Electric, Siemens, Rigaku, Diano, Supper, Rank, Syntex and many others. The International Union of Crystallography publishes and updates periodically an Index of Crystallographic Supplies, which lists various X-ray instruments, their manufacturers and suppliers. The many available techniques of using these instruments for structural or chemical analyses have been described by Bunn[3], Buerger[4], Klug and Alexander[5], Cullity[6], Azaroff[7] and Kaelble[8]. Almost all the publications pertaining to crystallography are available through Polycrystal Book Service, Pittsburgh, Pennsylvania.

X-Ray Sources

The wavelength of X-rays spans from 0.1 to 1000 Å (1 Å = 10^{-8} cm) in the electromagnetic spectrum. X-rays of about 1 Å wavelength (0.2–2.5 Å) are used for diffraction analysis because they have adequate penetration power yet are not destructive, and the interatomic distances in crystals have the same order of magnitude.

Radioisotopes emit X-rays in their decay process. For instance, the decay of Fe-55 to Mn-55 emits X-rays of 2.1 Å wavelength. Whenever matter is bombarded by high-energy particles, X-ray excitation takes place. The practical source of X-rays for diffraction work is from an X-ray tube, where electrons are accelerated by an electric field to a high velocity in vacuum. These electrons then collide with a metal target and thus excite the characteristic X-rays of the target. The target metals for X-ray diffraction tubes and their applications are cited in Table I. The X-ray tube and its electrical apparatus is called an X-ray generator and includes power supply, cooling system, safety device and control panel. Modern X-ray diffraction tubes have up to four windows, hence four pieces of X-ray apparatus can be set up simultaneously using a single X-ray source.

X-Ray Detectors

There are three ways to detect the presence of X-rays: fluorescent screen, photographic film and electronic counters. In X-ray diffraction work, a fluorescence screen is widely used to locate the position of the primary X-ray beam when adjusting the instrument. Photographic film is used for all camera techniques, and counters are used for all diffractometer techniques. The advantages of photographic film are simplicity, permanent record, and displaying all the diffracted beams with their relative disposition and intensity. Its major limitations are lower accuracy and less sensitivity. For single-crystal analysis, the advantages of films far outweigh their limitations, so that their use remains widespread. However, for chemical analysis, powder diffractometers are far superior to camera techniques.

X-ray counters used in diffractometry may be classified into three types: gas ionization counters[9,10], scintillation counters[11,12] and semiconductor counters[13,14]. Early Geiger-Muller gas ionization counters have been replaced by

Table I.   Target Metals for X-Ray Diffraction Tubes

Target	Wavelength (Å)	Filter	Applications
Tungsten	0.211	—	Very intense white radiation, well suited for Laue exposure and microradiography.
Silver	0.561	Pd Rh	High penetration, permits transmission exposure for determination of texture in transformer sheets. The (100) plane of iron has an "easy" direction of magnetization [001] parallel to the magnetic field flux.
Molybdenum	0.711	Nb Zr	Short wavelength, high penetration and low resolution. Suited for inorganic compounds that usually have high absorption coefficient and/or small unit cell. Used for austenite determination and single-crystal analysis.
Copper	1.542	Ni	Best suited for most general diffraction applications. Standard tube for all kinds of powders, metals, ceramic and organic materials.
Nickel	1.659	Co	Measurement of stress in metal, such as the (331) plane in brass (cartridge) at $2\theta = 158°$.
Cobalt	1.790	Fe	Cobalt and iron X-rays do not excite iron fluorescence radiation. Suited for all kinds of steel analyses, especially by means of film techniques in which PHA discrimination is not possible.
Iron	1.937	Mn	Same as cobalt X-rays.
Chromium	2.291	V	Long wavelength, low penetration and high resolution. Suited for organic compounds that usually have low absorption and/or large unit cell. Chromium X-rays do not excite iron fluorescence radiation. Used for steel analyses and stress analysis such as the (211) plane in iron at $2\theta = 156°$.

proportional gas ionization counters, which afford higher wavelength slectivity and much lower resolving time. Subsequently, scintillation counters now rival proportional counters for many applications and currently dominate the X-ray diffraction field. Semiconductor counters, originally developed for nuclear spectroscopy, were adapted for X-ray emission spectroscopy in the 1960s, and recently modified for X-ray diffractometry[15]. They offer exciting prospects for future applications such as fast data collection for dynamic systems[16]. Potentially, semiconductor counters could have superior resolving power due to efficacy of pulse-height discrimination and might require less stringent stability of the instrument[17]. Detectors of current use in diffractometers are all counters, the output of which is in digital form suitable for computer data processing. The feature of these counters are compared in Table II.

Table II.   Performance Data of X-Ray Counters

	Geiger-Muller	Proportional	Scintillation
Detection Process	Gas ionization	Gas ionization	Scintillation in NaI (Tl) crystal
Linear Count Rate (cps)	10^3	10^5	10^6
Background (cps)	1	0.5	10
Dead Time (μsec)	$50 \sim 3000$	< 1	< 1
Wavelength Region (Å)	0.3–4	Sealed 0.3–4 Flow 0.7–10	0.1–4
Energy Resolution	—	Excellent	Good
Pulse Size	Equal size	Proportional to X-ray energy	Proportional to X-ray energy
Pulse-Height Analysis	Cannot use PHA	Permits PHA	Permits PHA
Rating for Diffraction Use	Poor	Good	Good

X-Ray Cameras

X-ray diffraction instruments of any design are based on the Bragg law, which prescribes the geometrical conditions necessary to achieve X-ray diffraction. Bragg explained the X-ray diffraction effect in terms of reflection from a set of parallel and equispaced atomic planes. He conceived the following equation in 1912 now known as the Bragg law:

$$n\lambda = 2d\sin\theta \tag{1}$$

where   n = an integer denoting the order of reflection
   λ = the wavelength of the incident X-rays
   d = the interplanar spacing
   θ = the angle between the incident beam and the crystal planes

Note that the phenomenon is not a surface reflection, as with visible light. X-ray reflection occurs only at the precise Bragg angle θ, which satisfies Equation 1, a condition of constructive interference. No X-ray reflection occurs at other angles due to destructive interference. Reflection of visible light by a polished surface occurs over a continuous angular range.

Perhaps the simplest way to obtain a diffraction pattern is to use the powder camera technique. When the sample is extremely small (a few micrograms), powder camera technique is the only way to obtain its diffraction pattern. Besides crystal structure, a powder photograph reveals other properties of the sample, including particle size, texture and crystallinity.

Note that the film in a camera intercepts simultaneously all the X-ray beams diffracted by a crystal, whereas the counter on a diffractometer detects only one diffracted beam at a time. Therefore, the power stability is not so critical for camera techniques as that for diffractometer techniques. Besides chemical analysis, various powder cameras can be used for structural analysis of polycrystalline materials including minerals, polymers, ceramics, metals and alloys. For structural analysis of single crystals, camera and diffractometer techniques are complementary to one another. Diverse kinds of X-ray cameras for powder or single-crystal analyses are summarized in Table III.

X-Ray Diffractometers

An improvement over the film technique is X-ray diffractometry. This approach is the one most widely used in modern analytical laboratories and provides for replacement of the film and camera by a goniometer-mounted X-ray detector and a strip chart readout system. In this application, the sample in solid or powder form is placed in a flat specimen holder in the goniometer. A fixed source of X-rays impinges on the sample, which diffracts the X-rays at discrete angles. By rotating the sample at half the angular velocity of the detector, diffraction maxima as a function of the Bragg angle are recorded. A full pattern recording angular positions and relative intensities of diffraction maxima may be obtained in about 30 minutes by this technique. The data obtained are more accurate and easier to interpret than the data produced by the film technique. A schematic representation of a powder diffractometer, which includes a curved crystal monochromator, is shown in Figure 1[18]. The crystal can be either LiF or graphite and is essential for elimination of undesired K_β radiation and excess background radiation. Prior to the utilization of the curved crystal monochromator, a nickel filter was used for this purpose. While adequate, the nickel filter does not provide the high signal-to-noise advantage of the monochromator (1.4/1 for Ni, 11.0/1 for LiF, ~25/1 for graphite).

Single-crystal diffractometers are of two basic designs: Equi-inclination geometry and Eulerian geometry. Equi-inclination geometry design is the analog

Table III.   X-Ray Cameras and X-Ray Diffractometers

Diffraction Instruments	X-Rays	Crystals
Debye-Scherrer Camera	Monochromatic	Powder
Parafocusing		
Seemann-Bohlin camera	Monochromatic	Powder
Guinier camera	Monochromatic	Powder
Guinier-deWolff camera	Monochromatic	Powder
Pinhole Camera	Monochromatic	Powder
Powder Diffractometer	Monochromatic	Powder
(Analog of Debye-Scherrer camera)		
Laue Cameras		
Front-reflection camera	White	Single-crystal
Back-reflection camera	White	Single-crystal
Weissenberg Camera (U.S. & Europe)	Monochromatic	Single-crystal
deJong & Bouman Camera (Europe)	Monochromatic	Single-crystal
(i.e., Retigraph)		
Buerger Precession Camera (U.S.)	Monochromatic	Single-crystal
Equi-inclination Diffractometer	Monochromatic	Single-crystal
(Analog of Weissenberg camera)		
Four-Circle Diffractometer	Monochromatic	Single-crystal
(i.e., Eulerian geometry type)		

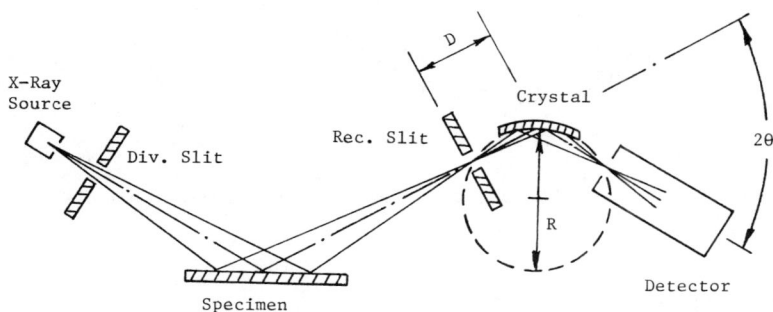

Figure 1.   Schematic of an X-ray diffractometer for powder analysis. The crystal monochromator and the receiving slit travel with the detector in alignment.

of Weissenberg camera with film replaced by a counter. In the Eulerian geometry design, the detector and the primary beam always lie in a horizontal plane. The crystal is carried in an Eulerian cradle, which permits rotation about each of the Eulerian axes: omega (ω), chi (χ), and phi (ϕ). The intensity of the diffracted

X-rays from a set of planes in the crystal can be measured by setting the ω, χ, ϕ and θ such that the Bragg condition is satisfied. A schematic diagram of the Eulerian geometry diffractometer is shown in Figure 2. The incident X-ray beam is fixed. The counter is constrained to move in the horizontal plane. The crystal is rotated until the normal to the set of diffracting planes lies in the horizontal plane. The angle between the incident beam and the set of diffracting planes in the crystal is made equal to $\theta = \sin^{-1} (\lambda/2d)$. The counter is then moved to 2θ position to measure the intensity of the diffracted beam.

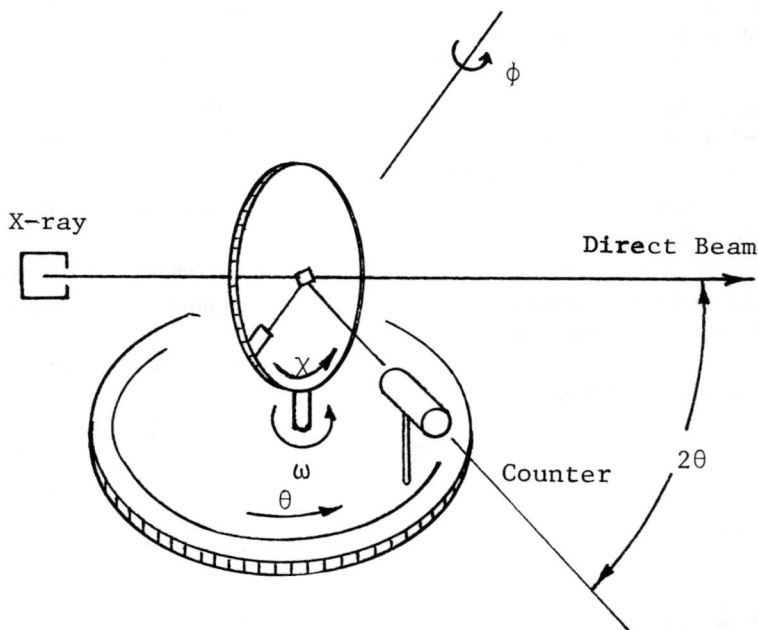

Figure 2.   Schematic of a four-circle diffractometer for structure analysis of single crystals.

Current Trend

Many advances have been made in the basic instrumentation used to obtain X-ray diffraction patterns. Computer-controlled powder diffractometers are now commercially available that enable the operator to search and match[19] patterns in the Powder Diffraction File[20], which is maintained by the Joint Committee on Powder Diffraction Standards (JCPDS). The completely automated data acquisition and search-match system depends on the quality of the X-ray diffraction pattern of the sample and the standard X-ray diffraction pattern in the File. An over-all upgrading of the File is needed. One advantage of the new instrumentation will be in comparing known in-house materials with unknowns.

Advances in accessory equipment include a primary beam monochromator[21], which is inserted between the X-ray tube and the specimen. This monochromator, mounted at the tube stand, is used when radiation-sensitive specimens are examined. Also available is a low- and medium-temperature attachment[22], which will hold the specimen in a range of about $-180° - +300°C$. This device is used for analysis of crystal structures, determination of temperature dependence of lattice parameters and examination of phase transitions.

New developments have been reported for three areas. The first area is in the use of synchrotron radiation[23], wherein the extreme energy intensity available and tunable wavelength feature provide an opportunity for performing experiments that were not possible in the past or, at best, were very difficult to carry out. This technique has been used to study high-temperature, high-pressure problems in geological systems[24]. The second area is in the use of energy-dispersive X-ray diffractometry[25], which is tied into the availability of the synchrotron source and the recent advances in electronics. This technique is especially suited to low- or high-temperature and/or high-pressure research. The third area is in the design or redesign of position-sensitive time-resolved detector systems. One such advance is the multiwire proportional counter system[26], which should speed up data gathering and has been used to obtain a powder pattern in one minute.

CHEMICAL ANALYSIS OF POWDERS

Qualitative Analysis

Each crystalline compound produces a characteristic X-ray diffraction pattern, whether in the pure state or as a component in a mixture. The identification of components in an unknown mixture is simply matching the diffraction pattern of the unknown to that of standards. Therefore, for chemical analysis by the X-ray diffraction method, a library of standard patterns is essential. To serve this purpose, the Joint Committee on Powder Diffraction Standards (JCPDS) publishes a Powder Diffraction File (PDF) annually. Over the years, the file has grown in size and improved in quality. Currently, it contains about 33,000 standard patterns of crystalline organic and inorganic compounds.

Powder Diffraction File

The Powder Diffraction File is available on 3×5-inch cards. A single X-ray diffraction pattern is registered on each card. A typical PDF card, as shown in

Figure 3, lists the interplanar spacing (d value) and the relative intensity (I/I_1) for every diffraction line of the compound. The intensity of the strongest line is assigned an arbitrary number of 100. The three most intense lines and the largest d value are sorted out and repeated on the upper left corner. Besides the X-ray diffraction pattern, the card contains the chemical name and formula of the compound, experimental conditions used for obtaining the pattern, optical and crystal data, source of the standard sample and references, when available. To effect a fast search-match for identification, three indexes accompany the Powder Diffraction File.

d	3.04	2.29	2.10	3.86	$CaCO_3$		
I/I_1	100	18	18	12	Calcium Carbonate (Calcite) ★		

Rad.CuKα₁ λ 1.5405 Filter Ni
Dia.   Cut off   Coll.
I/I_1 G.C. Diffractometer d corr. abs.?
Ref. Swanson and Fuyat, NBS Circular 539,
         Vol. II, 51 (1953)

Sys. Hexagonal   S.G. $D_{3D}^6 - R\bar{3}_c$
a₀ 4.989 b₀   c₀ 17.062 A   C 3.420
a   β   γ   Z 6
Ref. Ibid.

$ε_a$ nωβ1.659 $ε_γ$ 1.487   Sign -
2V D_x 2.711mp   Color
Ref. Ibid.

Sample from Mallinckrodt Chem. Works Spect.
Anal.:<0.1% Sr;<0.01% Ba;<0.001% Al, B, Cs,
Cu, K, Mg, Na, Si, Sn;<0.0001% Ag, Cr, Fe,
Li, Mn,
X-Ray Pattern at 26°C

Replaces 1-0837, 2-0623, 2-0629, 3-0569,
         3-0593, 3-0596, 3-0612, 4-0636,
         4-0637

d Å	I/I_1	hkl	d Å	I/I_1	hkl
3.86	12	102	1.297	2	218
3.035	100	104	1.284	1	306
2.845	3	006	1.247	1	220
2.495	14	110	1.235	2	1.1.12
2.285	18	113	1.1795	3	2.1.10
2.095	18	202	1.1538	3	314
1.927	5	204	1.1425	1	226
1.913	17	108	1.1244	<1	2.1.11
1.875	17	116	1.0613	1	2.0.14
1.626	4	211	1.0473	3	404
1.604	8	212	1.0447	4	138
1.587	2	1.0.10			{0.1.16
1.525	5	214	1.0352	2	{1.1.15
1.518	4	208	1.0234	<1	1.2.13
1.510	3	119	1.0118	2	3.0.12
1.473	2	215	0.9895	<1	231
1.440	5	300	.9846	1	322
1.422	3	0.0.12	.9782	1	1.0.17
1.356	1	217	.9767	3	2.1.14
1.339	2	2.0.10	.9655	2	234

Figure 3.   A data card on $CaCO_3$ from the Powder Diffraction File.

Davey (Alphabetical) Index. It lists alphabetically the names of compounds in the PDF. All entries are fully cross-indexed. For example, $CaCl_2 \cdot H_2O$ has three entries: calcium chloride hydrate; chloride hydrate, calcium; and hydrate, calcium chloride. Each compound name is followed by its chemical formula, its three most intense diffraction lines with their relative intensities indicated by subscripts (g for greater than 100, x for 100, 6 for 60, 2 for 20, etc.), then its card number and the microfiche number. The alphabetical index is useful when some information about the sample is available from elemental analysis, previous experience or educated guess.

Hanawalt (Numerical) Index. It lists the d values of the eight most intense lines of the pattern in order of decreasing intensities. Each d value has a subscript indicating its relative intensity. To cope with the possible uncertainty of relative intensities, each pattern is given three entries by permuting the three strongest lines. The patterns are divided into 45 groups according to the first d value of the entry. In each group the patterns are arranged according to the second d value in decreasing sequence. The corresponding compound name and PDF card number follow the d value listing. Some patterns are preceded with an "*" indicating high accuracy, "0" indicating less reliable and "c" indicating calculated pattern. The Hanawalt Index is used when there is little information about the sample.

Fink (Numerical) Index. It lists the d values of the eight most intense lines of the pattern in order of decreasing d values. Each pattern has four entries according to its four most intense lines. The format of the Fink Index is similar to that of the Hanawalt Index. It was known that both X-ray and electron diffraction patterns have the same groups of lines with high relative intensities although their intensity ranking may differ considerably. The Fink Index was developed mostly for the identification of electron diffraction patterns. It can be used for the identification of X-ray diffraction patterns where the intensity data are not reliable due to superimposed lines, preferred orientation or minor constituents.

Cases of Applications

Positive identification of the components in a mixture often leads to the solution of a technical problem, the mechanism of a reaction or the quality of a product. A few industrial cases are illustrated below:

Tin Plate. A steel mill makes sheet metal for "tin" cans where the tin coating is applied by "hot-dipping." A thin film of tin adheres to the surface of the steel and protects it from corrosion. It was found that certain steel sheets could not be wetted by the molten tin[6]. X-ray diffraction analysis of these faulty steel sheets indicated that a finely divided graphite was deposited on the surface. The presence of graphite prevented the contact between steel and molten tin.

Martensite. Many hardened steels consist of martensite and retained austenite. Austenite tends to transform and increase in volume, thereby leading to an increase in residual stress in the steel. Therefore, the presence of even a few percent retained austenite is undesirable in some applications. The presence and amount of austenite in hardened steel can be determined by microscopic examination when the austenite content is high. X-ray diffraction analysis must be used when the austenite content is below about 15%[27].

Fertilizer. After an application of composite fertilizer, the grass became withered instead of flourishing. Elemental analysis of the fertilizer could not find the cause. X-ray diffraction analysis indicated the presence of a potassium pyrosulfate, while a potassium sulfate was expected. Potassium pyrosulfate is highly acidic in water, hence killing the grass.

Paint. The sheet metal of an appliance was originally coated green. Its green color changed to blue after a brief period in service. X-ray diffraction analysis indicated that the pigment in the original paint was composed of lead chromate and iron blue; the exposed paint contained an extra component lead chloride. This finding led to the conclusion that the green color is a composite of yellow (lead chromate) and blue (iron blue). HCl fumes in that particular environment attacked the lead chromate forming lead chloride. The loss of a yellow component of the green color ended in a blue color[18].

Molybdate. Conventional corrosion-inhibiting pigments contain lead or chromium, which are toxic. A nontoxic pigment, Molywhite®, is being used for corrosion inhibitors. It is a basic zinc molybdate made from zinc oxide and molybdenum oxide[28]. The sequence of chemical reactions in its manufacturing process were followed by X-ray diffraction techniques. Therefore, the optimum reaction conditions for its production could be studied.

Limitations

In theory, the Hanawalt (Index) Search Manual should lead to the positive identification of any component in a mixture, and it does in most cases. However, it is not uncommon that a match of patterns may not be obtained for the following reasons:

1. The pattern of the unknown compound is not listed in the Powder Diffraction File.
2. The relative intensities of the diffraction lines changed drastically due to preferred orientation of the specimen.
3. There are superimposed lines from different components.
4. The intensities of the lines are too weak due to extreme absorption or very low concentration.
5. Some patterns have only one strong line, such as $CaCO_3$ (PDF 5-586) and β-SiO_2 (PDF 11-252).
6. The X-ray patterns of the unknown and the standard in the PDF were made with different instruments. For diffractometers, the absorption factor is independent of the Bragg angle, whereas in a Debye-Scherrer camera, the absorption is larger at lower Bragg angles. Therefore, low angle lines appear stronger relative to high angle lines on a diffractometer chart than on a Debye-Scherrer photograph.
7. The pattern obtained nearly matches several patterns in the PDF due to isomorphism, such as the numerous spinel-type compounds[29].
8. Clays, minerals and other natural products contain random substitution and/or irregular stacking, thus defy exact match[30]. Specialized procedures are often necessary for their identification[31].

9. Organic compounds usually have complex patterns due to low symmetry and large unit cell of the crystals[32]. The pattern of an organic mixture usually has too many overlapped lines to unravel[33,34].

Quantitative Analysis

In a powder mixture, each component produces its characteristic X-ray diffraction pattern independent of others. As already discussed, the identification of different components is achieved by unscrambling the superimposed patterns. Moreover, the intensity of the pattern of a component is proportional to the amount present. Unfortunately, the absorption effect of the total sample complicates the situation. The total absorption effect has been called the matrix effect. It depends on the nature of all other materials surrounding the component sought besides the component itself. To extract the weight fractions from intensity data, the matrix effect must be suppressed or circumvented. To this end, Klug and Alexander[5,35] developed an internal standard method in 1948. It involves the empirical construction of a calibration curve from standards, which is rather tedious, especially for multicomponent analysis. In 1974, Chung[36-38] conceived a Matrix-Flushing method, which avoided the calibration curve procedure, hence greatly simplifying the X-ray diffraction techniques for quantitative chemical analysis.

Matrix-Flushing Theory

The Bragg equation neatly prescribes the position of a diffraction line, but says nothing about its intensity. There are two theories dealing with the intensity of diffraction—Darwin's kinematical theory[39] and Ewald's dynamical theory[40,41]. Both theories lead to the same intensity expression for powder diffraction. For a single-phase powder with all crystallites randomly oriented, the intensity of diffracted X-rays in a diffractometer is given by Equation 2:

$$I(h\,k\,\ell) = \left(\frac{I_0\, e^4\, \lambda^3\, d}{32\,\pi\, m^2\, c^4\, r} \right) (N^2 p\, F^2) \left(\frac{1 + \cos^2 2\theta}{\sin^2 \theta \cos \theta} \right) T A V \qquad (2)$$

where
I = intensity of X-rays diffracted by (hkℓ) plane,
I_0 = intensity of primary X-rays,
e, m = charge and mass of electron,
λ = X-ray wavelength,
d = slit width of detector,
c = velocity of light,
r = specimen-to-detector distance,
N = number of unit cells per unit volume,
p = multiplicity,

F = structure factor,
θ = Bragg angle,
T = temperature factor,
A = absorption factor, and
V = volume of powder in the beam.

For a multiphase powder, the first four factors in Equation 2 can be kept constant. A indicates the absorption factor of the total mixture, and V indicates the volume fraction of the component sought. Let μ_t and s be the linear absorption coefficient and thickness of the total sample, and let ρ_i and X_i be the density and weight fraction of component i. Since the absorption of X-rays, just like the absorption of visible light, follows the well-known exponential law, for infinite specimen thickness we have

$$A = \int_0^\infty \exp\left(-\mu_t s\right) ds = \frac{1}{\mu_t} \tag{3}$$

$$I_i = \frac{K_i}{\rho_i} \cdot \frac{X_i}{\mu_t} = k_i X_i \tag{4}$$

Note that K_i/ρ_i is a characteristic constant of component i, μ_t is a function of X_i, and k_i is a factor containing the mass absorption coefficient of the total sample. At this moment, k_i is constant for a very small variation of X_i. Later on, it will be shown that k_i in the ratio k_i/k_j is constant for any X_i. For the quantitative X-ray diffraction analysis of a mixture of n components, we have n unknowns (X_i, i = 1 to n), which must satisfy the following equation:

$$\begin{bmatrix} k_1 & 0 & 0 & ...0 \\ 0 & k_2 & 0 & ...0 \\ 0 & 0 & k_3 & ...0 \\ \multicolumn{4}{c}{....................} \\ 0 & 0 & 0 & ...k_n \\ 1 & 1 & 1 & ...1 \end{bmatrix} \begin{bmatrix} X_1 \\ X_2 \\ X_3 \\ \cdot \\ \cdot \\ X_n \end{bmatrix} = \begin{bmatrix} I_1 \\ I_2 \\ I_3 \\ \cdot \\ \cdot \\ I_n \\ 1 \end{bmatrix} \tag{5}$$

The necessary and sufficient conditions for the existence of a unique solution of Equation 5 is that all its characteristic determinants vanish, which gives

$$X_i = \left(\frac{k_i}{I_i} \sum_{i=1}^n \frac{I_i}{k_i}\right)^{-1} \tag{6}$$

The unusual property of this unique solution is that the weight fraction of any component in a multiphase system is expressed in terms of ratios like I_i/I_j and

k_i/k_j (i, j = 1, 2, ..., n). The use of an intensity ratio makes it immune to many sources of errors in intensity measurement. The use of k-ratio makes it free from matrix effect because all the absorption factors are exactly cancelled out. Furthermore, it can be shown[36] that the k_i's in this unique solution are the corresponding Reference Intensity Ratios defined and published in the Powder Diffraction File. Therefore, the X-ray diffraction patterns of mixtures can be interpreted quantitatively and directly, without using any internal standard.

When some components in the mixture are amorphous and/or unidentified, an internal standard, such as corundum α-Al_2O_3, must be added into the sample. Let the subscript c and o stand for corundum and original sample respectively. Then we have:

$$X_c + X_o = X_c + \sum_i^n X_i = 1 \tag{7}$$

$$X_i = \left(\frac{X_c}{k_i}\right)\left(\frac{I_i}{I_c}\right) \tag{8}$$

Note that the weight fractions in Equations 7 and 8 are referring to the composite mixture because X-rays faithfully reveal what is actually observed, but cannot differentiate an internal standard from an original component. Equation 8 prescribes the slope of a calibration curve for every component in the sample and thus greatly simplifies the internal standard method. It is the working equation for quantitative multicomponent X-ray diffraction analysis.

Usually X-ray diffraction gives little information about the amorphous component in sample. An interesting feature of the Matrix-Flushing theory is that it provides a way of detecting the presence of amorphous materials in a sample, and the amorphous content can be determined when necessary. The following discriminant Equation 9 can be derived from equations 7 and 8:

$$\sum_1^n \frac{I_i}{k_i} \overset{>}{\underset{<}{=}} I_c \cdot \frac{X_o}{X_c} \tag{9}$$

where   > indicates wrong data
         = indicates all components are crystalline
         < indicates the presence of amorphous materials

Note that neither assumption nor approximation was made in the derivation of this theory. All absorption factors are exactly cancelled. Equations 6 and 8 are to be used for quantitative multiphase analysis. Equation 9 is to be used for testing the presence of amorphous components.

Applications

Practically all cases of quantitative chemical analysis by X-ray diffraction techniques can be classified into two categories. First, all components in a sample are crystalline and identified. Second, some components in a sample are amorphous and/or unidentified. The applications of Equations 6, 8 and 9 to these cases are illustrated below. The precision of routine analysis by this method is about 8% relative or better. Better precision is expected when both sample and standard have the same physical properties, such as particle size and lattice imperfection. The usual precautions are assumed such as homogeneity of samples, purity of standards, counting statistics, instrumental conditions and other sources of error. However, a set of Reference Intensity Ratios, k_i, has to be determined first.

Simultaneous k_i Determination. The JCPDS defined Reference Intensity Ratio, k_i, as the intensity ratio (I_i/I_c) of the most intense lines from a binary mixture made with a pure compound and synthetic corundum by one-to-one weight ratio. Before theoretically calculated k_i can be made more dependable[42,43], experimentally determined k_i should be used. A set of k_i of interest can be determined simultaneously by use of Equation 8, as shown in Table IV, where the simultaneously determined k_i is compared with the individually determined k_i[44].

Table IV.   Simultaneous k_i Determination

Component	Composition		Intensity		k_i	
	(g)	(wt %)	(hkl)	(cps)	Calculated	50/50 Mixture
ZnO	0.4183	16.29	101	1564	4.28	4.35
CdO	0.3562	13.87	111	2418	7.76	7.62
LiF	0.6302	24.54	200	729	1.32	1.32
CaF_2	0.5395	21.01	220	651	1.38	1.41
Al_2O_3	0.6242	24.30	113	546	1.00	1.00

Direct Interpretation. When all components in a sample are crystalline and identified, the X-ray diffraction pattern of the original sample can be interpreted quantitatively and directly. The weight fraction of every component in the sample can be easily calculated with Equation 6. When amorphous components are present, the results thus obtained represent the composition of the crystalline portion of the sample. The data of two examples are given in Table V, where all intensity data refer to the strongest reflections of corresponding components.

Quite often, only the relative amounts of concerned components in a sample are sought. The information also can be obtained with Equation 6 by considering these concerned components only and ignoring all other crystalline or amorphous components.

Note that a special case in this category is binary systems, where n = 2. An unusual binary system is partially crystalline polymers[45], which consist of a crystal-

Table V.   Direct Interpretation of X-Ray Diffraction Patterns

	Composition (g)	Intensity I_i (cps)	Reference Intensity k_i	% Composition Known	% Composition Found
ZnO	0.2236	610	4.35	9.87	9.2
NiO	o.5454	1412	3.81	24.06	24.5
CdO	0.6588	3303	7.62	29.07	28.7
KCl	0.8386	2207	3.87	37.00	37.7
ZnO	0.6759	2408	4.35	24.38	25.3
TiO_2	0.4317	931	2.62	15.57	16.2
$CaCO_3$	1.1309	2558	2.98	40.79	39.2
Al_2O_3	0.5341	420	1.00	19.26	19.2

line component and a noncrystalline component. The slope of the straight line ($k = 2.83$) in this reference verifies Equation 6 remarkably well.

Internal Standard. In many practical situations, the composition of a sample is only partially identified. The weight fractions of the identified components can be determined simultaneously with the addition of an internal standard such as corundum. The composite sample should be ground into a homogeneous fine powder before the X-ray diffraction analysis. The intensity and composition data of two examples are shown in Table VI. Equation 8 was used to calculate the weight fractions. The discriminant Equation 9 was used to indicate whether an amorphous component was present. The large intensity imbalance (1842 vs 2312) implies the presence of amorphous materials, which scatter X-rays incoherently,

Table VI.   X-Ray Diffraction Analysis with Internal Standard

	Composition (g)	Intensity I_i (cps)	Reference Intensity k_i	% Composition Known	% Composition Found	$\Sigma \frac{I_i}{k_i}$	$I_c \cdot \frac{X_o}{X_c}$
ZnO	1.8901	5968	4.35	41.49	41.1		
KCl	1.0128	2845	3.87	22.23	22.0		
LiF	0.8348	810	1.32	18.32	18.4	2721	2736
Al_2O_3	0.8181	599	1.00	17.96	—		
ZnO	0.9037	4661	4.35	34.43	36.4		
$CaCO_3$	0.7351	2298	2.98	28.00	26.2		
SiO_2 Gel	0.4234	0	—	16.13	15.9		
Al_2O_3	0.5629	631	1.00	21.44	—	1842	2312

increasing the background. The nearly equal intensity balance (2721 vs 2736) implies that all components are crystalline where the use of Equation 6 should give the same results. Note that the weight fraction of total amorphous materials was calculated from material balance.

When all the peaks in the X-ray diffraction pattern are accounted for, one is still not sure whether there is any amorphous material present. A normal scan of the last example in Table VI did not indicate the presence of amorphous materials, although the sample contained 16% silica gel. Therefore, it is safe to check with an internal standard for the following reasons: First, the true values of X_c and k_c of the internal standard are absolutely known. Secondly, the weight fraction, X_i, is dependent on I_i and k_i only, independent of any X_j, I_j and k_j, where $j \neq i$. Thirdly, the weight fraction of a component sought is independent of the presence or absence of any other component, crystalline or amorphous.

Factors Vital to Precision

The basic intensity Equation 2 was derived from an "ideally imperfect" mosaic crystal model. All mosaic blocks are small ($0.1 \sim 1$ μm size) and nonparallel (but with no more than $0.2 \sim 0.5°$ misalignment)[46]. All lattice planes are given every opportunity to diffract X-rays by rotating the crystal. For powder diffraction analysis, it implies that the crystallites have optimum particle size and proper amount of imperfection. The orientation of all crystallites is statistically random. These conditions are not always fulfilled. Besides, the sample preparation and the intensity measurement are equally important to attain reproducibility. These factors are discussed below.

Particle Size. The effective crystalline size of powder affects both the magnitude and the spread of intensity. To achieve 1% standard deviation, the crystallite size should not exceed 5 μm. If $2 \sim 3\%$ standard deviation is sufficient, the crystallite size can be as large as 10 μm. For a spinning sample, the crystallite size can be 2–3 times larger.

When the crystallite size falls below 0.1 μm, the peak height of the diffraction line decreases and peak width increases. The amount of line broadening provides a rather accurate measurement of particle size in the colloidal range (<0.1 μm). When the crystallite size is above 20 μm, the intensity of diffraction lines fluctuate sharply as a function of the orientation of the larger crystals. Powder of >50 μm size gives spotty lines in the Debye-Scherrer photograph. In essence, when the particle size of a powder is reduced to a few microns, say $1 \sim 10$ μm, several sources of errors can be minimized, including line-broadening[5], preferred orientation[47], extinction[48] and microabsorption[49] due to either the constant nature or trivial level of these effects. In many chemical precipitates and mineral clays, each particle may consist of a large number of crystallites, even though the apparent particle size is 50 μm.

Lattice Imperfection. Most crystals are neither perfect nor "ideally imperfect." Various imperfections such as dislocations, interstitials, stacking faults and random layers may exist. The profile of diffraction lines reveals the average of all sorts of lattice imperfection. Highly perfect crystals give lower intensity due to extinction. Highly imperfect crystals give lower intensity due to line broadening. For free-flowing powders of 1 to ~50-μm particle size, the extinction and line broadening are insignificantly small. Iodine, iodide and many organic compounds are very soft and malleable. Extended grinding may cause lattice distortion. In such cases, strain-free powder can be obtained by grinding the sample with liquid nitrogen or dry ice.

Preferred Orientation. Plate- or needle-like crystallites tend to set in certain preferred orientation. Preferred orientation due to particle shape is called shape texture. Talc and asbestos powders show extreme shape texture. Preferred orientation due to particle shape is called shape texture. Talc and asbestos powders show extreme shape texture. Preferred orientation due to mode of creation is called orientation texture. Fiber and film of polymers, wire and sheet of metals show orientation texture. Shape texture is important in sample preparation for chemical analysis of powders. Orientation texture is important in structural analysis of polycrystalline aggregates. Because of preferred orientations, some lines of the diffraction pattern are intensified and others weakened. When serious discrepancies arise in relative intensity between the pattern obtained and the pattern in the PDF, preferred orientation should be suspected and some correction steps be taken. Size reduction by grinding, dilution with amorphous powder, embedding in collodion and roughening the top surface can reduce the preferred orientation and make the intensity data more reproducible.

Sample Preparation. For any meaningful chemical analysis, the sample must be representative and homogeneous. It is extremely difficult to attain a truly homogeneous powder mixture. Mixing and grinding are usually done by use of an automatic mortar-and-pestle grinder, a ballmill or a Wig-L-Bug. The grinding time ranges from a few minutes to several hours, depending on nature and amount of sample. Sometimes wet grinding in a nonsolvent liquid can speed up the dispersion and size reduction. The powder sample should be free flowing into the sample holder from a side opening to avoid preferred orientation and induced packing[50].

Intensity Measurement. Integrated intensity rather than peak height should be used for serious quantitative X-ray diffraction analysis, since the peak height varies due to particle size and lattice distortion. Observe counting statistics and accumulate enough counts, N, such that the standard deviation $N^{-\frac{1}{2}}$ is within expected limits. The integrated intensity may be measured from the area under the line profile or by recording the total counts, while the counter scans the line profile. It is essential that a normal scan of the total sample be made to observe intensity level, background and possible interferences.

STRUCTURAL ANALYSIS OF SINGLE CRYSTALS

Many volumes have been written on the determination of single-crystal structures by X-ray diffraction techniques[3,4,46,51]. The short description here presents a concise concept for those with casual interest in this field about its goal, work and problems.

Single crystals are a type of laboratory curiosity. However, much of our understanding of the properties of polycrystalline materials such as metals, plastics and minerals has been gained by studies of isolated single crystals. Single-crystal structure determination is quite an involved science. The primary goal of this particular science, X-ray crystallography, is to obtain a set of coordinates of various atoms in the repeating unit (Unit cell) of the crystal. Thereupon, a detailed three-dimensional picture of atomic positions and molecular packings within the crystal can be visualized as through a super-power microscope.

The size and symmetry of the unit cell in the crystal are deduced from the positions of the diffraction lines, and the coordinates of atoms in the unit cell are calculated from the intensities of the diffraction lines. The main work for single-crystal structure analysis is twofold: (1) experimental, which includes the selection of a single crystal of 0.2 to ~ 3-mm size, the determination of its density, the indexing of reflections and the collection of a set of intensity data from diffraction experiments by use of Weissenberg/Precession camera or diffractometers; and (2) computation, which includes the derivation and establishment of a set of most probable coordinates of atoms, thermal ellipsoids, bond lengths and bond angles in the crystal based on the set of intensity data collected from diffraction experiments.

The computation is far from straightforward. Given the position of atoms in the unit cell, the direction and intensity of diffracted X-rays can be calculated precisely. However, given a set of direction and intensity data, the position of atoms in the unit cell cannot be calculated straightforwardly because the phase angle (out of alignment of wavelets) in the Fourier series is unknown. Hence, the central task of structure analysis is to deduce the missing phase angles and to derive a trial structure. The Patterson map and the Direct Method of Karle-Hauptman are two widely used approaches to the phase problem. But neither method will always result in a trial structure. Chemical information, broad knowledge, experience and imagination all play important roles.

A trial structure must be chemically plausible and give good agreement between observed and calculated intensities. This trial structure is then refined by least-squares process or Fourier synthesis to obtain a set of most probable coordinates of atoms in the crystal.

The discrepancy index, R, denotes how well the derived structure fits the observed intensity data. The lower the R, the greater the confidence in the derived structure. $R = 0.2 \sim 0.3$ indicates good trial structures. $R = 0.03 \sim 0.08$ indicates most reliably determined structures.

CONCLUSIONS

X-ray diffraction techniques are widely used in industry for chemical analysis of powders, and elegantly handled in academia for structure analysis of single crystals. Powder analysis is rather simple and routine, while structure analysis is filled with puzzles and excitement. Both are well established and yield quantitative results. Other applications include the analysis of particle size and lattice imperfection from the line profiles, the crystallinity and orientation of polymers from the total diffraction pattern, the analysis of stress in metal or alloy from the total diffraction pattern, the analysis of stress in metal or alloy from small changes in d-spacing, the texture of metal or alloy from pole figures, and the phase diagram from the composition of quenched specimens. Various models and mathematical expressions have been developed for each case. However, both expertise and diligence are required to attain reliable data. The use of simplified approaches gives only simplified pictures. Nevertheless, the derived results are practically useful and academically interesting.

An X-ray powder diffraction pattern is not necessarily made from a dab of finely pulverized solid particles. It can be made from any polycrystalline aggregate, such as metal wire or sheet, polymeric fiber or film, plastic plate or block, etc. The term particle size may indicate the size of crystallites (small single-crystals) in powders, the size of grains in metals, the size of crystalline domains in polymers, or the size of micelles in organic fibers. With X-ray diffraction techniques, the size, perfection, orientation and composition of crystallites in a specimen can be studied. Their effects on properties and performance can be correlated. Therefore, X-ray diffraction has been an important tool for research in materials science.

Besides the composition and performance of materials, physicists and physical chemists are seriously concerned about the electronic configuration of atoms, the lattice and atomic vibrations, the structure of biopolymers, and the mechanism of enzyme actions, etc. They need either the most accurate data or a huge collection of data to solve these problems. With the advent of modern electronics and computers, X-ray diffraction instrumentation has progressed from simple manual cameras to monstrous automated diffractometers handling all data collection and processing. However, the effectiveness of this automated instrumentation is yet to be fully realized. The conventional X-ray sources and detectors have prevented automated systems from a real breakthrough. New developments[52,53] in this direction include synchrotron X-ray sources, laser plasma X-ray sources, position-sensitive detectors, energy-dispersive detectors, holographic image processing and three-wavelength method of phase determination. A real breakthrough could be expected at the turn of the century.

REFERENCES

1. Ewald, P. P., Ed. "Fifty Years of X-Ray Diffraction," International Union of Crystallography, Utrecht, Netherlands (1962).
2. Jenkins, R., Y. Hahm, S. Pearlman and W. N. Schreiner. *Norelco Rept.* 26(1): 1 (1979).
3. Bunn, C. W. *Chemical Crystallography* (New York: Oxford University Press, 1961).
4. Buerger, M. J. *Crystal Structure Analysis* (New York: John Wiley & Sons, Inc., 1960).
5. Klug, H. P., and L. E. Alexander. *X-Ray Diffraction Procedures* (New York: John Wiley & Sons, Inc., 1974).
6. Cullity, B. D. *Elements of X-Ray Diffraction* (Reading, MA: Addison-Wesley Publishing Co., Inc., 1956).
7. Azaroff, L. V., and M. J. Buerger. *The Powder Method* (New York: McGraw-Hill Book Co., 1958).
8. Kaelble, E. F. *Handbook of X-Rays* (New York: McGraw-Hill Book Co., 1967).
9. Parrish, W., and T. R. Kohler. *Rev. Sci. Instr.* 27: 795 (1956).
10. Lang, A. R. *J. Sci. Instrum.* 33: 96 (1956).
11. Nuffield, E. W. *X-Ray Diffraction Methods,* Part 2 (New York: John Wiley & Sons, Inc., 1966).
12. Arndt, U. W., and B. T. M. Willis. *Single Crystal Diffractometry,* Chapter 4 (New York: Cambridge University Press, 1966).
13. Muller, R. H. *Anal. Chem.* 38: 155 (1966).
14. Birks, J. B. *The Theory and Practice of Scintillation Counting* (London: Pergamon Press, Inc., 1964).
15. Giessen, B. C., and G. E. Gordon. *Norelco Rept.* 17(2): 17 (1970).
16. Tanner, B. K. *Prog. Crystallog. Growth Charact.* 1: 23 (1977).
17. Frankel, R. S., and D. W. Aitken. *Appl. Spectrosc.* 24: 557 (1970).
18. Scott, R. W. *Treatise on Coatings,* Vol. 2, Part II (New York: Marcel Dekker, Inc., 1976), p. 597.
19. Johnson, G. G. *Comput. Chem. Instr.* 5: 45 (1977).
20. JCPDS. "Powder Diffraction File," Swarthmore, PA (1979).
21. Denne, W. A. *J. Appl. Crystallog.* 9: 510 (1976).
22. Rudman, R. *Low Temperature X-Ray Diffraction* (New York: Plenum Publishing Corp., 1976).
23. Koch, E. E, C. Kunz and E. W. Weiner. *Optik (Stuttgart)* 45: 395 (1976).
24. Akimoto, S., and Y. Takenchi. *Nippon Kessho Gakkaishi* 18: 163 (1976).
25. Mantler, M. *Mikrochim. Acta* Suppl. 7: 555 (1977).
26. Hashizume, H., Y. Memiya, K. Kohra, T. Izumi and K. Mase. *J. Appl. Phys.* 15: 2211 (1976).
27. Averbach, B. L., and M. cohen. *Trans. AIME* 176: 401 (1948).

28. Kirkpatrick, T., and J. J. Nilles. U.S. Patent 3,677,783 (1972).
29. Frevel, L. K. *Anal. Chem.* 44: 1850 (1972).
30. Brown, G., Ed. *X-Ray Identification and Crystal Structures of Clay Minerals*, 2d ed. (London: Mineralogy Society, 1961).
31. Walker, G. F. *Nature* 164:577 (1949).
32. Matthews, F. W., and J. H. Michell. *Ind. Eng. Chem. Anal. Ed.* 18: 662 (1946).
33. Williams, P. P. *Anal. Chem.* 31: 140 (1959).
34. Garska, K. J. *Appl. Spectros.* 30: 204 (1976).
35. Alexander, L., and H. P. Klug. *Anal. Chem.* 20: 886 (1948).
36. Chung, F. H. *J. Appl. Crystallog.* 7: 519 (1974).
37. Chung, F. H. *J. Appl. Crystallog.* 7: 526 (1974).
38. Chung, F. H. *Adv. X-Ray Anal.* 17: 106 (1974).
39. Darwin, C. G. *Phil. Mag.* 27: 315, 675 (1914).
40. Ewald, P. P. *Ann. Phys.* 49: 1, 117 (1916).
41. Ewald, P. P. *Acta Crystallog.* 11: 888 (1958).
42. Hubbard, C. R., E. H. Evans and D. K. Smith. *J. Appl. Crystallog.* 9: 169 (1976).
43. Hubbard, C. R., and D. K. Smith. *Adv. X-Ray Anal.* 20: 27 (1977).
44. Chung, F. H. *J. Appl. Cryst.* 8: 17 (1975).
45. Chung, F. H., and R. W. Scott. *J. Appl. Cryst.* 6: 225 (1973).
46. Stout, G. H., and L. H. Jensen. *X-Ray Structure Determination* (New York: Macmillan Publishing Co., Inc., 1968).
47. Sturm, E., and W. Lodding. *Acta Crystallog.* A24: 650 (1968).
48. Darwin, C. G. *Phil. Mag.* 43: 800 (1922).
49. Bindley, G. W. *Phil. Mag.* 36: 347 (1945).
50. *Standard X-Ray Diffraction Powder Patterns,* NBS Monograph 25, U.S. Government Printing Office, Washington, DC (1971).
51. Alexander, L. E. *X-Ray Diffraction Methods in Polymer Science* (New York: John Wiley & Sons, Inc., 1969).
52. Cole, H., Ed. "Instrumentation for Tomorrow's Crystallography," *ACA Trans.* 12 (1976).
53. Abrahams, S. C., and J. B. Cohen. *Phys. Today* 29: 34 (1976).

CHAPTER 7

A REVIEW OF POLAROGRAPHIC DETERMINATION*

Carol J. Lind
United States Geological Survey
Menlo Park, California

INTRODUCTION

Polarography measures electrochemical parameters of reactions occurring at an electrode-solution interface. The resulting data provide a means of evaluating concentrations of dissolved chemical species and of interpreting physical-chemical processes. Modifications of the basic measurement method separate and amplify the measured parameters and permit the solution of problems that many modern analytical tools cannot resolve. Indiscriminate application of polarography can lead to false conclusions, even though the data are reproducible and consistent.

Polarography is a form of voltammetry. Voltammetry measures the effect of an applied electrode potential on the current flowing through a cell. Historically, polarography is the measurement of the current flowing between a dropping mercury electrode (DME) and a reference electrode during the application of a known potential. (Appendix I lists the abbreviations used in this chapter and their definitions.) The current is measured as the voltage is continually and linearly changed from about 200 mV positive of the potential for the pertinent half-reaction ($E_{1/2}$) to at least 200 mV negative of this potential. The half-reaction is expressed as

---

*The use of brand names in this chapter is for identification purposes only and does not imply endorsement by the U.S. Geological Survey.

$$Ox + ne \rightleftarrows Red \qquad\qquad (1)$$

where n is the number of electrons, e, involved. (The $E_{1/2}$ value is the potential at which the activities of the oxidized (Ox) and the reduced (Red) forms of the pertinent chemical species (depolarizer) are equal.) The $E_{1/2}$ value depends on the depolarizer identity, the degree and type of depolarizer complexing, the support electrolyte identity and concentration, and the electrode pair. (The support electrolyte is an electrochemically inert salt that comprises the major ionic composition of the test solution.) $E_{1/2}$ values are tabulated by Vydra et al.[1], pp. 200–201, Meites[2], pp. 615–711, Sawyer and Roberts[3], pp. 358–382, Meites[4], pp. 2364–2371, and others[5-8].

A graph of the current-voltage relationship, usually made by a recorder, is called a polarogram and is shown as the solid line in Figure 1.

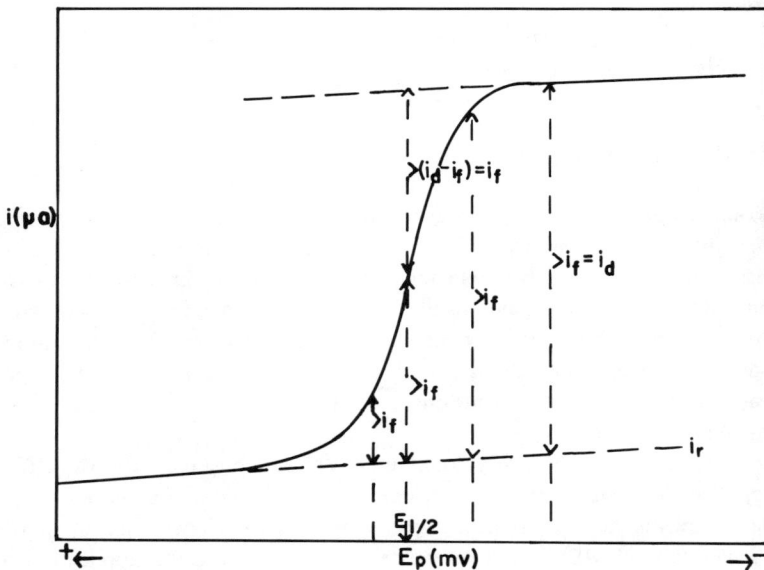

Figure 1.   Polarogram with current contributions delineated.

The creation of new working electrode designs, specialized highly sensitive techniques and modern instrumentation obscure more and the more the distinction between polarography and other forms of voltammetry. The popular, highly sensitive current amplification techniques discussed are not really polarography, but are included because they are closely associated with it. A limited amount of information on the advantages, disadvantages, special restrictions and mathe-

matical relationships accompanies the described technique modifications and refinements of classical dc polarography. The accompanying referenced material
elaborates on this information. The mathematical relationship of current to
voltage for the most popular, highly sensitive technique—anodic stripping voltammetry (ASV)—when using the hanging mercury drop electrode (HMDE) or the
mercury film electrode (MFE), are presented in the section entitled Current-
Voltage Equations for Popular Analysis Methods, p. 203.

Modifications of classical dc polarography extend detection limits from $10^{-5} M$
down to $10^{-10} M$. This chapter briefly describes the more common modifications
of classical polarography, but the books by Vydra et al.[1], Meites[2], Sawyer and
Roberts[3] and Galus[9] describe these modifications in greater detail. Roe and Eggi-
mann[10] and Roe[11] review the theory and instrumentation of more sophisticated
dynamic techniques of analytical electrochemistry through 1977. A Bibliography
updates much of this literature to early 1979.

RESTRICTIONS REQUIRED FOR DATA INTERPRETATION

The conclusions drawn from the analytical data relate to reaction reversibility
at the DME surface. The reaction in question, the other solution components and
the technique used all govern the degree of reversibility. When the forward, back-
ward and any other associated reactions are fast enough to exert no control on the
current measurement, the reaction is reversible. When one or more of these reactions is the major control, the reaction is irreversible. During irreversible reactions, the rate of the electron transfer process and the influence of the potential
on this process control the relation between current and potential. Polarography
requires an electron transfer rate of 2×10^{-2} cps or greater for reversible reactions
and an electron transfer rate of less than 2×10^{-5} cps for totally irreversible reactions. Data from reactions that are in between (neither reversible nor irreversible)
are difficult to interpret.

For a reaction to be reversible, Equation 2, the Nernst equation, must be
obeyed over the entire potential range considered:

$$E_p = E_s^\circ - \frac{RT}{nF} \ln \frac{a_{Red}}{a_{Ox}} \tag{2}$$

where E_p = the potential at the (DME)
E_s° = the standard potential of the half-reaction
R = the gas constant
T = the temperature
n = the number of electrons
F = the Faraday constant
a = the activity

The activity equals the molar concentration (C) multiplied by the activity coefficient (γ):

$$a = \gamma C \tag{3}$$

Thus, Equation 2 may be expressed in concentration units as shown in Equation 4. (Activity coefficients will be discussed further under the section entitled Formation Constants.)

$$E_p = E_s^o - \frac{R\,T}{n\,F} \ln \frac{\gamma_{Red}\,C_{Red}}{\gamma_{Ox}\,C_{Ox}} \tag{4}$$

When both the oxidized and reduced species are soluble in the solution, the concentration terms refer to their respective concentrations in the solution. When the reduced form (such as free metal or simple or complexed metal ion) is soluble in the mercury, the concentration term for the reduced species refers to the reduced-species concentration in the mercury. When this amalgam exists, the free energy of the reduced-species solution in the mercury is included in the standard potential and must be accounted for in the Nernst equation. Since the amalgam formed is dilute, assuming that the amalgam activity equals pure mercury metal activity (unity) and entering this activity value in the divisor portion of the ln terms accounts for the added free energy. Thus, Equation 4 is still applicable regardless of whether the reduced species is soluble in the mercury or in the solution.

Interpretation of polarographic data requires that measured reactions be diffusion controlled. Diffusion control means that the measured depolarizer current value is proportional to the depolarizer diffusion rate to and from the electrode.

Simple mathematical manipulations convert the Nernst equation to a form relating potentials to measurable polarographic currents for a diffusion-controlled reversible reaction. The conversion is possible because, for a diffusion-controlled reversible reaction, both the current produced by the oxidized species and the current produced by the reduced species are proportional to their respective species activities. The final expression as shown in Equation 5 is for experimental conditions (e.g., ionic strength and temperature) and relates current to concentrations:

$$E_p = E_{1/2} - \frac{2.303\,R\,T}{n\,F} \log \frac{i_f}{i_d - i_f} \tag{5}$$

The current caused by the depolarizer is the faradaic current, i_f. When the E_p value is so negative that all the measurable current is contributed by the depolarizer in the reduced form, the i_f is the diffusion-limiting current, i_d. The residual current, i_r, is the current that occurs in the absence of the depolarizer. The i_r is caused primarily by the supporting electrolyte, but other causes, such as charging phenomena, make their contribution. On the polarogram, the straight portion measured at voltages positive of $E_{1/2}$ shows the i_r. The i_f at a given E_p is determined by subtracting the i_r value, when extrapolated to the E_p of the i_f to be measured, from the total current at that E_p. Figure 1 illustrates the current terms and $E_{1/2}$[12].

For a reversible reaction, a plot of the electrode potential against the log term in Equation 5 is linear, but a linear slope does not prove reversibility. For a reversible reaction, the linear plot must have a slope of $-0.05915/n \pm 0.003$ V at 25°C. An irreversible reaction will have a greater slope. When the log term equals zero, $E_{1/2}$ equals E_p[2].

For the many technique variations applied to reversible and irreversible reactions, there is extensive literature on the electrochemical-mathematical relationships and their derivations. The books by Vydra et al.[1], Meites[2] and Galus[9] contain good, detailed summaries of this literature, and the Bibliography lists many of the more recent articles about these relationships.

COMMON APPLICATIONS

Polarography is most frequently used to determine ionic concentration of free metals, complex metals and organic species; to define reactions; and to determine formation constants. For much of this work, reversible reactions are required. Polarographic work with irreversible reactions provides information on reaction rates and processes and on surface active properties of the solution and the solution components. Experimental objectives determine whether the test reactions are cathodic or anodic.

Ionic Concentration

The first-row transition metal elements and the heavy metals (e.g., Pb) exhibit measurable i_f. These metals are common in natural water but their concentrations are low. Modern polarographic methods eliminate preconcentration steps and make the analysis for these metals a popular polarographic application.

Polarography is also applicable to the analysis of complex organic systems and the identification of complex organic substances. Application of polarography to organics is possible because only certain organic groups are electroactive at specific potentials when in particular solvent combinations.

When a ballistic galvanometer is used for recording current, the Ilkovič Equation 6 defines the average i_f during a drop life (i_d) as

$$i_d = 607.0\,nD^{1/2}\,C_b m^{2/3}\,t^{1/6} \tag{6}$$

where D = the diffusion coefficient (cm^2/sec)
   C_b = the concentration in the bulk solution (mmol/1000 cm^3)

m = the mass of the mercury drop at time t (g)
t = the age of the drop (sec)

Most modern instrumentation uses constant speed-drive motors for the recording pen, and then Equation 7 applies and i_d is defined by the "envelope of the maxima of the polarographic current oscillations."

$$i_d = 706 \, nD^{1/2} C_b m^{2/3} t^{1/6} \tag{7}$$

Regardless, i_d is proportional to C_b for a given set of electrode specifications. For most determinations of minor constituents, the assumption in the Ilkovič equation that the drop is planar gives only 2–3% error in the value of C_b[1,2].

Sawyer and Roberts[3] and Meites[2] discuss the meaning and application of I (defined in the section entitled Formation Constants, p. 182), of i_d and of D (Sawyers and Roberts,[3] pp. 331–338 and Meites[2], pp. 95–202).

Comparison of the i_f displayed by the test solution with those displayed by a set of standards, or the application of the method of standard additions, defines the depolarizer concentration. For both methods, all comparisons must be at the same E_p, pH and temperature, in the same kind and concentration of support electrolyte, and with electrodes having the same specifications. The method of standard additions measures the test solution, i_f, first and then the i_f after each of several added increments of depolarizer. Interpretation requires the following: (1) sufficient data to establish a well-defined i_f versus concentration relationship; (2) linearity for even the highest concentration; and (3) no slope change with increased depolarizer concentration. An i_f versus concentration plot, extrapolated to zero i_f, determines the test solution depolarizer concentration. Figure 2 illustrates the plot used in the method of standard additions. Franke and de Zeeuw[13] describe the evaluation and optimization of the standard addition method.

Analysis of natural water is coupled with handicaps that must be addressed if valid conclusions are to be drawn. For example, the method of additions applied to metal analysis presumes that the added metal reacts rapidly with all ligands present and that, in the process, the metal assumes the same complexation degree and complexation distribution as in the original test solution. During total metal analysis, acidification may not free all complexed metal (e.g., metal-macromolecule complexes) and can increase current-voltage response interferences by organics[14]. The separation mode specifications (e.g., filter size, centrifugation intensity, degree of destruction of organic matter and of particulate matter, resin-column separation technique, and chemical controls on the solution) govern the meaning of "dissolved," "bound" and "sorbed" species concentration values.

Practical Applications are described by Vydra et al.[1], pp. 193–276, Galus[9], pp. 498–501 and others[13–16].

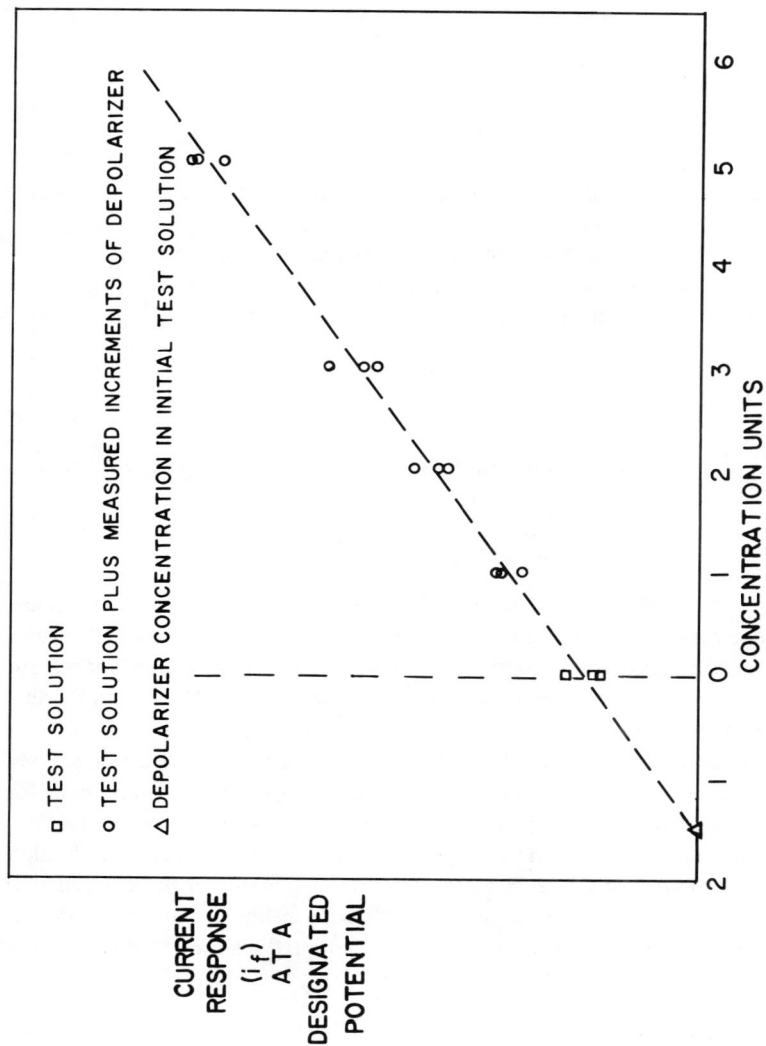

Figure 2.   Determination of depolarizer concentration by method of standard additions.

Formation Constants

A formation constant, β_p, is a mathematical term that defines the equilibrium activity relationships for the formation of a complexed molecule. For example, for the reaction expressed in Equation 8, β_p, is defined by Equation 9:

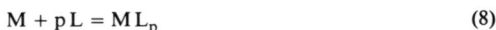

$$M + pL = ML_p \tag{8}$$

$$\beta_p = \frac{a_{ML_p}}{a_M\,(a_L)^p} \tag{9}$$

where M denotes metal, L denotes ligand and p denotes the ratio of ligand-activity units to metal-activity units.

Substitution of concentration units for activity units in Equation 9 permits calculation of the apparent formation constant $\beta_p{}'$. Application of γ_s to the calculated value of $\beta_p{}'$ permits computation of β_p, Equation 10.

$$\beta_p = \frac{\beta_p{}'\,\gamma_{ML_p}}{\gamma_M\,(\gamma_L)^p} \tag{10}$$

(The value of $-\log\gamma$ varies with the square root of the ionic strength and with several other parameters. Equation 11 defines ionic strength.

$$\text{Ionic Strength} = \tfrac{1}{2}\sum Cz^2 \tag{11}$$

where C is the molar concentration and z is the charge of the ion. The support electrolyte makes the major contribution to the ionic strength, which decreases with support-electrolyte concentration. As the total ionic concentration approaches zero, the γ value generally approaches unity.) Both Hem[17] and Butler[18] describe the estimation of activity coefficient values.

Once β_p and all but one of the molecular species activities for a reaction are known, the remaining activity can be calculated. This calculation can be made for a wide range of activities of the reacting species. Formation constants are particularly valuable for the calculation of species activities, solubilities and distribution of free and complexed species in solutions containing many constituents, such as natural waters. Formation constants, equilibrium constants, solubility products, ionic strength and activity coefficients are discussed in greater detail in the literature[17,18].

Single Complexes

The formation constant for Reaction 12 can be defined by substituting the diffusion current constant (I) and Equations 3 and 9 in the ln term of the Nernst equation and rearranging the resulting expression to give Equation 13:

$$ML^{(n-b)+} + ne + Hg \rightleftharpoons M\ (Hg) + L^{-b} \tag{12}$$

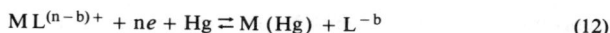

$$\text{antilog} \left[\frac{0.434\ n\ F}{R\ T} (E_{\frac{1}{2}M} - E_{\frac{1}{2}ML}) + \log \frac{I_M}{I_{ML}} \right] = \frac{\gamma_M\ \gamma_L\ \beta\ C_L}{\gamma_{ML}} \tag{13}$$

where C = the concentration
M = the free oxidized ion
ML = the complexed oxidized species
L = the ligand

(The term I is defined by

$$I = \frac{i_d}{C_b\ m^{2/3}\ t^{1/6}} \tag{14}$$

From the Ilkovič equation,

$$I = 607\ n\ D^{1/2} \tag{15}$$

when using a ballistic galvanometer, and

$$I = 706\ n\ D^{1/2} \tag{16}$$

when using modern instrumentation[1,2]). The polarogram defines the $E_{\frac{1}{2}M}$ value when only the uncomplexed form of the free oxidized ion is present, and the $E_{\frac{1}{2}ML}$ value when the complexed oxidized species is present. The term $I_M : I_{ML}$ is proportional to $i_{f_M} : i_{f_{ML}}$ when both the free and complexed species are measured at the same concentrations and at the same place on their respective polarographic waves and with the same cell characteristics present. The ligand concentration is either measured with each set of data or is made high enough that the ligand concentration does not measurably decrease when the free depolarizer ion is complexed. For less precise work, the ratio $I_M : I_{ML}$ and the activity coefficients are all assumed to equal one.

Multiple Complexes

When the free oxidized ion complexes two or more ligands, Equations 17 and 18 apply. The method of DeFord and Hume[19] extracts the differing β_ps from the data, Equations 19a–d:

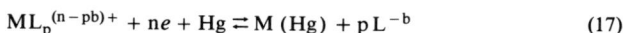

$$ML_p^{(n-pb)+} + ne + Hg \rightleftharpoons M\ (Hg) + pL^{-b} \tag{17}$$

$$F_0\ (L) = \text{antilog} \left[\frac{0.434\ n\ F}{R\ T} (E_{\frac{1}{2}M} - E_{\frac{1}{2}ML}) + \log \frac{I_M}{I_{ML}} \right] = \gamma_M \sum_0^p \frac{\beta_p\ C_L{}^p\ \gamma_L{}^p}{\gamma_{ML_p}} \tag{18}$$

Ignoring activity coefficients, application of the measured data to Equation 18 to get $F_0(L)$ and then application of Equations 19a–d permits calculation of apparent formation constants, $\beta_p{}'$.

$$F_0\,(L) = 1 + C_L\beta_1{}' + C_L{}^2\beta_2{}' + C_L{}^3\beta_3{}' \ldots C_L{}^p\beta_p{}' \tag{19a}$$

$$F_1\,(L) = \frac{F_0\,(L) - 1}{C_L} = \beta_1{}' + C_L\beta_2{}' + C_L{}^2\beta_3{}' + \ldots C_L{}^{(p-1)}\beta_p{}' \tag{19b}$$

The y intercept of a plot of $F_0(L) - 1$ versus C_L defines $\beta_1{}'$, and the y intercept of a plot of $F_1(L) - \beta_1{}'$ against C_L defines $\beta_2{}'$, etc.

$$F_2\,(L) = \frac{F_1\,(L) - \beta_1{}'}{C_L} = \beta_2{}' + C_L\beta_3{}' + C_L{}^{(p-2)}\beta_p{}' \tag{19c}$$

$$F_3\,(L) = \frac{F_2\,(L) - \beta_2{}'}{C_L} = \beta_3{}' + \ldots C_L{}^{(p-3)}\beta_p{}' \tag{19d}$$

The relation of formation constants to measurable polarographic parameters is described by Meites[2], pp. 267–284, Meites[4], pp. 360–381, DeFord and Hume[19] and Lind[20].

Other Applications

The book by Vydra et al.[1] contains a long chapter describing practical applications of a sensitive analysis method—stripping voltammetry. The Bibliography at the end of this chapter lists articles (up to early 1979) on polarographic applications to inorganic depolarizers and similar articles applying to organic depolarizers. Heineman and Kissinger[21] review recent methodologies and applications of dynamic techniques of analytical electrochemistry through 1977. The *Journal of Electroanalytical Chemistry and Interfacial Electrochemistry* is an international journal devoted to all aspects of electrode kinetics, of interfacial structure, of properties of electrolytes, and of colloid and biological electrochemistry.

Adsorption in ac polarography is discussed by Galus[9], pp. 501–504 and Jehring[22]. Electrode kinetics are discussed by Galus[9], pp. 491–498, Erdey-Gruz[23], Vielstich and Schmickler[24] and Albery[25]. Surface studies are discussed by Conway[26] and Conway and Angerstein-Kozlowska[27].

CLASSICAL DC POLAROGRAPHY

Classical dc polarography employs a DME as a working electrode and a calomel or silver-silver chloride electrode as a reference electrode for the current measurement during an applied voltage ramp. (The working electrode is the electrode at whose surface the measured redox reaction occurs.) The voltage scan rate is usually between 1 and 50 mV/sec, commonly 2–5 mV/sec. The scan rate is set so that the potential does not change much over the life of a drop.

Electrode Systems

Two-Electrode System

The two-electrode system consists of a working electrode (e.g., DME) and a reference electrode. The two-electrode system introduces errors by a voltage drop when current passes through high-resistance solutions and by possible potential changes caused by the current flowing through the reference electrode. A high-support electrolyte concentration lowers solution resistance. A low internal resistance reference electrode, capable of passing up to 20 μa with no significant potential change, decreases the possible potential changes in the reference electrode. Commercial reference electrodes usually do not meet these specifications, so satisfactory electrodes must be made. Saturated calomel (SCE), normal calomel (NCE), mercurous sulfate and silver chloride electrodes with large, active surface areas are generally used. These electrodes are immersed in salt solutions that contact the solution via sintered glass filters (and in some cases a layer of agar-agar), Vycor plugs or tubing with an ultrafine tip opening.

Three-Electrode System

A modification, the three-electrode system, employs a operational amplifier control loop and a large, active surface area auxiliary electrode. The auxiliary electrode is sometimes called a counter electrode. The electrode is most commonly of platinum, but may be of carbon. In the three-electrode system, the auxiliary electrode and the operational amplifier control loop use a negative feedback circuit to adjust the potential difference between the working electrode and the reference electrode to the desired control potential. The reference electrode is held at a constant potential. The reference electrode, or the reference electrode bridge contact with the solution, is placed as close as possible to the working electrode. There is no flow of current through the reference electrode and no voltage drop due to solution resistance. The current read is that flowing through the auxiliary and working electrodes[28,29]. Figure 3 diagrams the basic elements of a three-electrode cell.

The three-electrode system is discussed by Vydra et al.[1], pp. 157–158, and by Ellis[28] and Flato[29].

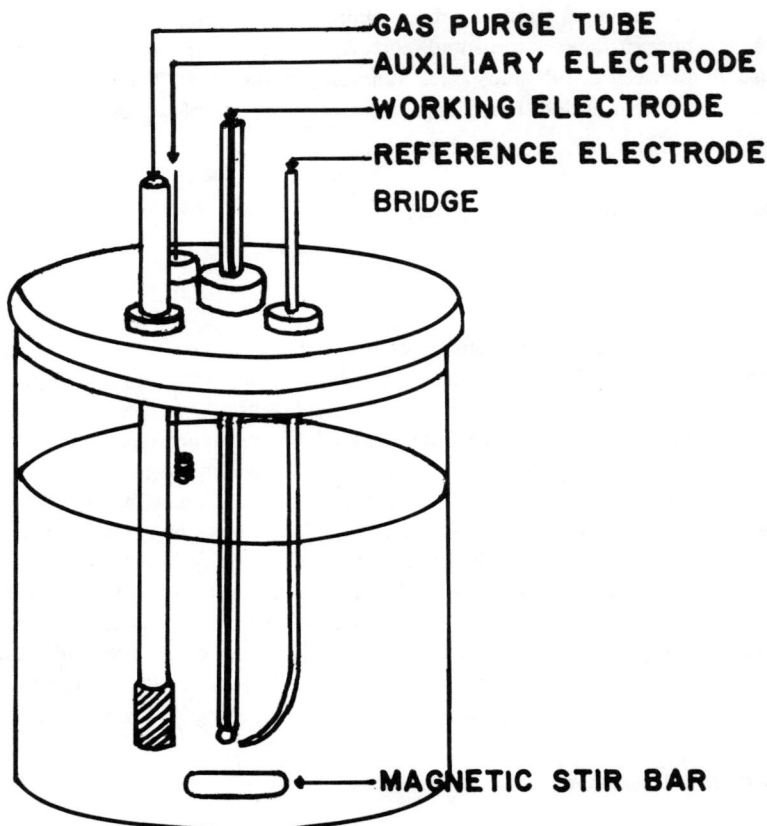

Figure 3.   Elements of a three-electrode cell.

Working Electrode—The Dropping Mercury Electrode

The dropping mercury electrode system is an unbroken mercury-filled system. The mercury passes from a reservoir, fills a column to the height of the reservoir, flows through a capillary located below the column, and forms a drop at the capillary tip. The drop dislodges at timed, regular intervals and passes through the test solution. A wire connects the external instrumental electronics with the mercury, and the growing mercury drop contacts the test solution. A measuring gauge attached to the column indicates the height of the mercury above the capillary tip. The capillary is 10–20 cm long and has an inside diameter of about 0.05–0.08 mm.

The drop time is proportional to the capillary length, inversely proportional to the mercury column height, and varies with the applied potential and with the solution composition. Modern instrumentation makes possible electronically controlled drop time. Regardless of the control mechanism, the drop time must be reproducible. The repeated cycle of increasing drop size and the drop fall creates a stirring action and causes renewal of the solute concentration exposed to the electrode[2,3]. The DME is discussed by Meites[2], pp. 73–82 and Sawyer and Roberts[3], pp. 82–85.

Impure mercury in the DME makes anodic waves difficult to interpret and causes decreased hydrogen-evolution overpotential. (In a negative voltage scan, overpotential is the potential range negative of $E_{1/2}$, where the measurable depolarizer is in the reduced form. The current caused by the hydrogen-evolution overpotential masks currents caused by the test depolarizer when both currents are measured at the same potential. When the voltage at which the masking occurs is made less negative, there is a shorter voltage span available for display of the i_d plateau of the depolarizer being tested.) Dirty mercury makes the capillary useless by soiling the capillary walls[3]. Mercury is either purchased pure or is purified. Mercury purification is described by Meites[2], p. 83 and Sawyer and Roberts[3], pp. 79–81.

Reference Electrode

Most polarographic $E_{1/2}$ listings refer to SCE. The silver-silver chloride electrode has replaced the SCE in popularity. Next to the hydrogen electrode, the silver-silver chloride electrode is the most reproducible and reliable electrode. Compared with the SCE, the silver-silver chloride electrode is less sensitive to oxygen and has smaller temperature hysteresis. Further, the electrode is easier to prepare and use. The silver-silver chloride electrode is sensitive to bromine and is undesirable for organic solvents. Substitution of 3.5 F KCl for the saturated KCl decreases possible precipitation caused by temperature changes to lower than 25° C. Substitution with a mercury-mercurous sulfate electrode alleviates chloride interference[3].

Salt Bridge

A low-leak salt bridge containing an ultrapure, noninterfering salt solution isolates the reference and/or the auxiliary electrode from the test solution and lessens the contamination from these electrodes. Electrode isolation is described by Vydra et al.[1], p. 157 and Sawyer and Roberts[3], pp. 17–34.

Support Electrolyte

A support electrolyte is necessary to conduct the current through the solution to dissipate the electrostatic attraction forces. This electrolyte is an indifferent electrolyte with a concentration at least 100 times that of the depolarizer. The support electrolyte influences the double-layer structure and can also buffer the solution, complex the depolarizer or form ion pair or micellular aggregates with the depolarizer. The redox properties of the support electrolyte can restrict the usable voltage range[3]. Support electrolytes are discussed by Vydra et al.[1], p. 33, Sawyer and Roberts[3], p. 187 and Meyer et al.[30].

Cell and Solution Preparation

A lowered i_r improves the detection limit and permits use of very dilute support electrolytes. The i_r is composed of background current, i_b, and charging current, i_c. The i_b is caused by impurity redox reactions, electrolyte decomposition and noise from the electrode system or instrument. The i_c is caused by the charging of the double-layer capacitance at the working electrode-solution interface. A high i_r swamps out the i_f when the depolarizer concentration is low.

Oxygen Removal

Dissolved oxygen causes a large and poorly reproducible increase in i_r, and possible amalgam oxidation and hydroxyl ion formation. Oxygen is reducible at the working electrode over most of the usable polarographic potential range and is chemically reactive with many solutions. Amalgam oxidation decreases reduction efficiency, and hydroxyl ion production can change unbuffered solution pH's when near neutral. Oxygen removal greatly lowers the i_r and diminishes these other undesirable oxygen-related problems. Flushing the test solution with an inert gas removes the oxygen, and blanketing the solution surface with the gas keeps the oxygen content low. Nitrogen, argon and hydrogen are the purging gases used most commonly. Prior to entering the polarographic cell, the gases are freed of trace oxygen by passing them through a purification unit. Purification units are prepared in the laboratory or commercially[1]. Gas purification is described by Vydra et al.[1], p. 174, Meites[2], pp. 87-91 and Sawyer and Roberts[3], pp. 134-137.

Contaminant Removal

For trace element analysis, the choice of sample storage container, the proper container precleaning and the cell components have a direct bearing on the reliability of the final results. Also, if the sample is filtered, filter composition and prewashing are matters for consideration. Where possible, replacing glass cell components with acid-leached quartz or with plastic, like Teflon,®* reduces contamination from and adsorption by the cell components[1]. Container selection and cleaning are discussed by Vydra et al.[1], p. 173, Moody and Lindstrom[31] and Branica et al.[32]. Trace metals in filters are discussed by Watling and Watling[33].

Reagent purification and double distillation of water in glass (even better in quartz) lowers the i_b. The high support-electrolyte concentration, as compared to that of the depolarizer, necessitates removal of the correspondingly concentrated impurities. Heavy metals and transition metals are typical support-electrolyte impurities. Electrolytic purification is often satisfactory when solutions are not highly acidic and when a large-area mercury electrode is used. Distillation purifies many acids and bases and isothermal distillation purifies many volatile substances[1]. Reagent purification is described by Vydra et al.[1], pp. 175-177, Meites[2], p. 380, Sawyer and Roberts[3], pp. 210-212, Branica et al.[32] and Riley[34].

Maxima Suppressors

A very small amount of surfactant (such as Triton X-100) suppresses the polarographic maxima (peaks formed on various portions of the polarographic wave). Test solution concentrations of Triton X-100 generally are 0.002-0.001% or lower, but rare conditions may require concentrations as high as 0.004%[2]. Maxima suppressors are only used when necessary because they can cause interferences that decrease the i_f[14]. Maxima suppressors are discussed by Meites[2], pp. 319-331 and Brezonik et al.[14].

CURRENT-DIFFERENTIATING MODIFICATIONS

In dc polarography, the changes in i_f and i_c do not coincide during a drop life period. The i_f increases in proportion to the increasing drop area and the i_c rapidly decreases after the initial drop formation. A low-pass filter averages the current changes during drop life. Special instrumentation measures the current when either the faradaic current alone, or the charging current alone, is predominant, or when one is out of phase with the other. Figure 4 illustrates the current-voltage-time relationships for dc polarography. Galus[9], pp. 472-513, presents an overview of current differentiating modifications.

---

*Registered trademark of E. I. du Pont de Nemours and Co., Inc., Wilmington, Delaware.

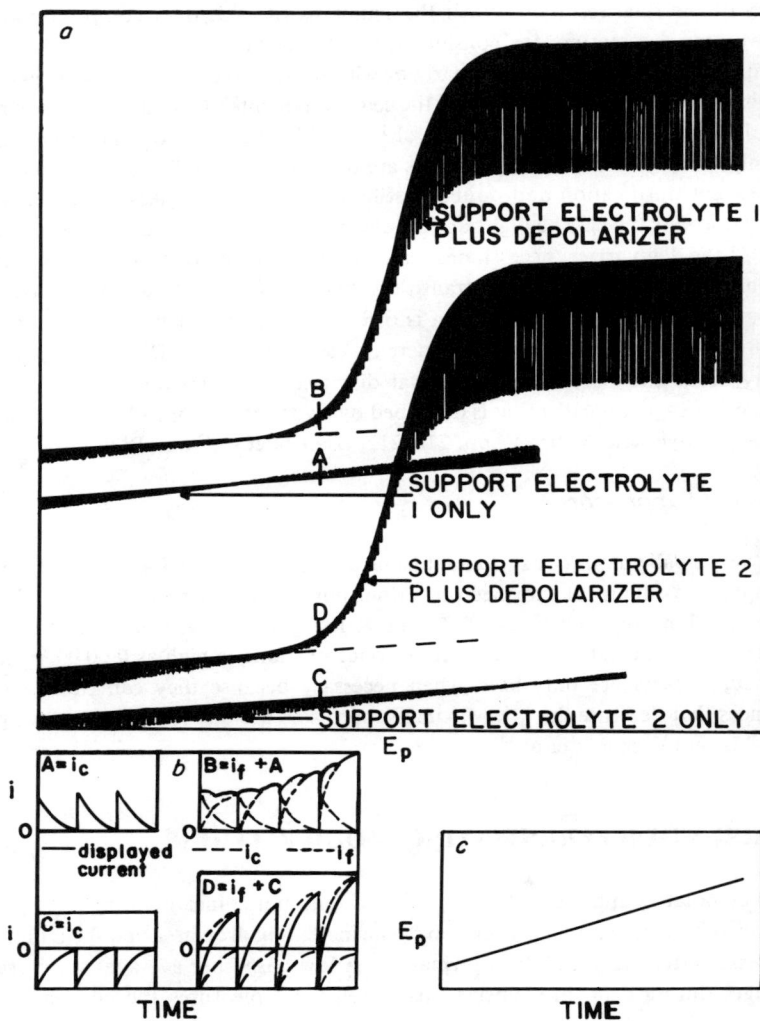

Figure 4.   Current-voltage time relationships for dc polarography[45]: (a) current-voltage relationships at A and C for two support-electrolytes with different E_p's for the crossover and i_c contribution variation, and the current-voltage relationships at B and D for these electrolytes with the same depolarizer added; (b) the i_c, i_f and total current variations over successive drop lives for conditions A, B, C and D; and (c) voltage-time relationships.

Current-Sampled dc Polarography

Current-sample dc polarography measures the current for a short interval just before the drop is dislodged. For example, with a drop life of 0.5 seconds or more, the sampling time may be 16.7 msec. Current measurement near the end of the drop life gives a value with a maximized i_f and minimized i_c. The short measuring time eliminates the need to average out the current change during the drop life. The polarogram displays this measured current until the current value of the following drop is measured and displayed[35]. Figure 5 illustrates the current-voltage relationships for current-sampled dc polarography.

Pulse Polarography with a Linearly Changing Voltage Ramp

Barker[36] and Barker and Gardner[37] originated pulse polarography by modifying Baker's square-wave methods. Pulse polarography applies a pulse during the last portion of the drop life and, as in current-sampled polarography, measures the i_f at the maximum value, just before the end of the drop life[29]. Pulse polarography is reviewed extensively in the literature[29,38,39].

Differential or Derivative Pulse (DPP)

In addition to the linearly changing voltage ramp of dc polarography, the potential in DPP includes a small-amplitude fixed pulse of 5–100 mV. The polarogram displays the difference in the current just before the pulse application and the current just before the end of the drop life[28]. The current just before the pulse is the ordinary polarographic current and the current during the pulse, at the end of the drop life, includes also the potential-pulse current when i_f is maximized and i_c is minimized. The resulting curve is a peak presentation similar to the first derivative of the usual polarogram[29]. The drop life can be 0.5–4 seconds, the pulse applied during the last 100 msec of drop life, and the current integrated over the last 33.5 msec of drop life. The potential sweep rate is low (e.g., 5–10 mV/sec), so that the potential will change only slightly during each pulse period[39].

The pickup of significant amounts of power line signal by reference electrodes and by high-resistance organic solutions causes a noise in the output. Synchronizing the current measuring time with the power line frequency and adjusting the sampling time to a known multiple of the power line frequency minimizes this noise[29,39]. For example, the current measuring time may cover two complete cycles of 60-Hz line frequency. Panel controls on some instruments regulate the compensation for i_r and, to some extent, the compensation for capillary response caused by wetting of the capillary-tip inside wall[39].

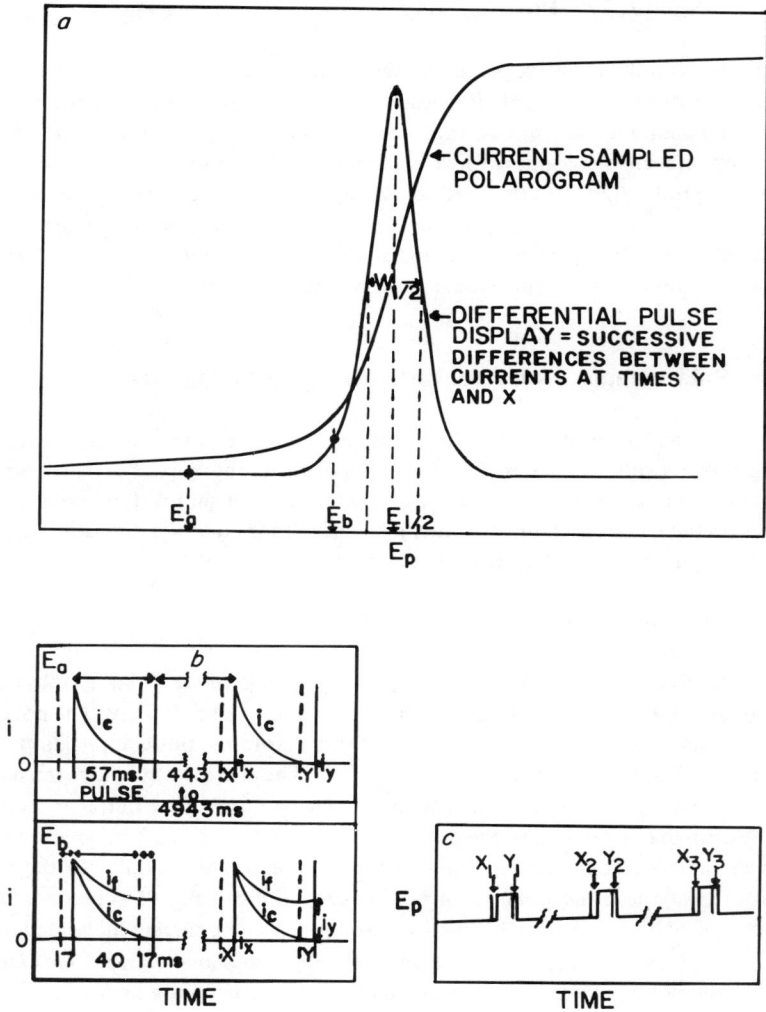

Figure 5. Current-voltage-time relationships of i_c-minimizing-and-i_f-maximizing dc polarography modifications[41]: (a) current-voltage relationship for current-sampled polarography overlaid by current-voltage relationship for DPP, W½ and E½ indicated; (b) current contribution changes at E_a and at E_b during pulse application and current measured just before the pulse (x) and just before the end of the pulse (y); and (c) DPP voltage-time relationship, measuring times x and y indicated.

The signal magnitude is proportional to the pulse amplitude and the system gain, but the smaller the pulse, the better the resolution[35]. DPP has several advantages. The method

1. has far greater sensitivity since it displays essentially only the i_f;
2. has the advantages presented with the use of the DME, but permits the application of other working electrodes;
3. permits analysis of solutions with low support-electrolyte concentrations;
4. except in extreme cases, requires no surfactant;
5. has baselines with less slope than for dc polarography;
6. distinguishes closely spaced waves; and
7. measures both large and small peaks in the same sample[29].

Figure 5 illustrates the current-voltage-time relationships for DPP.

AC Polarography

Ac polarography incorporates a small-amplitude sinusoidal modulation with a linearly changing-potential ramp. Only the ac components of the signal are presented. In special instrumentation, these signal components relate to a preselected phase relationship with the applied potential modulation. The ac portion presents the existing current difference between the minimum and maximum potentials during a modulation period. The i_f is 45° out of phase with the applied ac potential wave, and the i_c is 90° out of phase with this wave. Depending on the experimental objectives, a phase-sensitive detector measures either the i_f or the i_c. Because the sensitivity and charge transfer rate decrease together, only fast reactions are detectable with ac polarography.

The electrode reaction controls the peak position, height and shape. Frequency modulation variation moves the peak position and can transpose the peak positions. The frequency controls reaction reversibility. Ac polarography, especially phase-sensitive and second-harmonic presentations, makes interpretation possible when many different electroactive substances are present. In a mixture of electroactive species, a depolarizer can be reversible with differential pulse and not reversible for ac polarography, so a combination of methods facilitates analysis of some complex mixtures[29]. Figure 6 illustrates the voltage-time relationships for ac polarography.

Pulse Polarography with a Linearly Changing Pulse Voltage

True derivative-pulse polarography and normal-pulse polarography (NPP) employ a linearly changing amplitude potential pulse and, between pulses, maintain the initial potential.

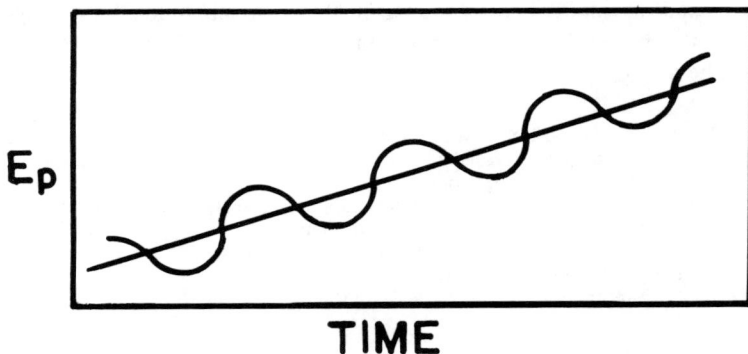

E_p

TIME

Figure 6.   Voltage-time relationship for ac polarography.

True Derivative-Pulse Polarography

In true derivative-pulse polarography, the difference displayed between the maximized i_f for successive drops resembles a first-derivative curve. For true derivative-pulse polarography, the pulse-amplitude rate of change and the time between pulses determine the sensitivity[39].

Normal- or Integral-Pulse Polarography (NPP)

In NPP, the polarogram displays each measured i_f until the i_f of the following drop is measured and displayed. This method produces a polarographic wave form.

The sensitivity of NPP is 2½–5 times that of dc polarography but is not as great as in DPP. The increased sensitivity is the result of current measurement over only a short time span during each drop life. The brief measurement causes a much thinner diffusion layer and a higher depolarizer concentration near the electrode surface. The increased sensitivity applies to diffusion-controlled, reversible and irreversible reactions[35]. The method is good for concentration and reaction kinetic studies[39].

Maintaining the initial potential for the relatively long periods between pulses has a cleaning effect on the working electrode. For a reversible reaction, the reduced depolarizer is reoxidized during this period and essentially returns the solution concentration near the electrode to the original value. The return to the initial potential also often removes adsorbed products and leaves the electrode fresh and clean for the next pulse. Like DPP, working electrodes other than DME are applicable and no surfactant is necessary[35]. Figure 7 illustrates the voltage-

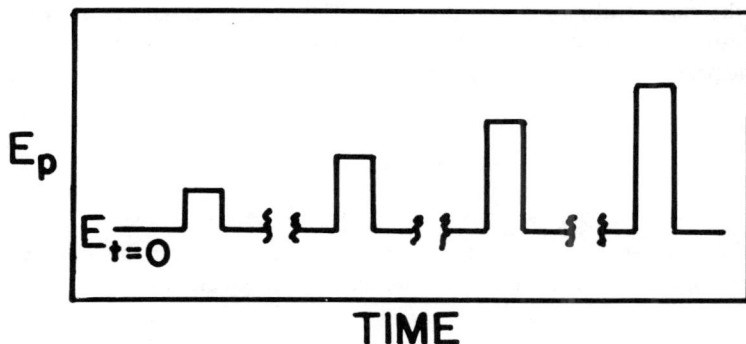

Figure 7.   Voltage-time relationships for true-derivative pulse polarography and for NPP.

time relationships for true derivative-pulse polarography and for NPP. Instrumental circuitries and special instrumentation are described in the literature[1-3,35,40-42].

CURRENT AMPLIFICATION TECHNIQUES

Current amplification techniques require either constant surface parameters for working electrodes during analysis, or, with DME, sufficiently short analysis time to eliminate effects of increasing drop size.

Linear-Sweep Voltammetry

Voltammetry is a broad term defining the study of current-potential relationships that result from the potential applied to a working electrode. Polarography is a voltammetric form. Voltammetric measurements in stirred solutions resemble polarography[35].

Linear Potential-Sweep Voltammetry or Single-Sweep Voltammetry

Single-sweep peak voltammetry applies a linearly increasing voltage ramp at 10 mV/sec or greater to an unstirred solution and produces a peak-shaped curve. Only diffusion controls the mass transport. As the voltage is changed, the faradaic current increases. When the potential is such that the reaction rate is as fast as the diffusion rate, the concentration near the electrode begins decreasing and the current displayed also decreases. The peak potential occurs when the increased reaction rate is equal to the diffusion layer depletion effect.

The peak, E_p, for a reversible reaction at 25° C is defined by the equation

$$E_p \text{(peak)} = E_{1/2} - \frac{0.029}{n} \text{ V} \tag{20}$$

Also, for a reversible reaction, the potential difference between the peak, E_p, and the one-half-the peak current, E_p, is defined by the equation

$$\left| E_p \text{(peak)} - E_p \text{(one-half the peak)} \right| = \frac{0.057}{n} \text{ V} \tag{21}$$

For an irreversible reaction, the difference in E_p is defined by the equation

$$\left| E_p \text{(peak)} - E_p \text{(one-half the peak)} \right| = \frac{0.048}{\alpha n} \text{ V} \tag{22}$$

where α is the electron transfer coefficient. For a reversible reaction, E_p (peak) and E_p (one-half peak) are independent of the scan rate, while for an irreversible reaction, these potentials change by $0.3/\alpha n$ for each tenfold sweep-rate change[35].

Fast Linear-Sweep Voltammetry

The dc polarographic scan rates are only 1–50 mV/sec, while the fast linear-sweep voltammetric scan rates are 100 mV/sec and more. The rapid voltage scan rates require an oscilloscope to display the current-votage output. Some newer recorders have response speeds that correspond to the fast voltage scan rates. These recorders are applicable to fast linear-sweep voltammetry. With such a rapid scan of the linearly changing voltage ramp, the diffusion layer thickness does not have time to increase greatly. The effective concentration near the electrode surface diminishes only slightly and, as a result, produces a greater current output compared with slow scans.

Fast linear scan has the advantage of rapid analysis, of increased sensitivity and of low-cost, unsophisticated instrumentation. The rapid scan applied to a single DME drop eliminates the drop oscillations superimposed on the polarographic curve. Increasing i_c and i_f with drop size causes a sloping baseline. The replacement of the DME with a constant area electrode or the introduction of a dc differentiator overcomes the baseline slope[29].

Stripping Voltammetry

Stripping voltammetry increases the current output by depolarizer preconcentration on the working electrode and subsequent depolarizer stripping simultaneous with i_f measurements. The stripping current is proportional to the depolarizer concentration and is frequently related to the depolarizer $E_{1/2}$[35]. Metal analysis is a popular stripping voltammetry application. ASV can analyze for several metals in one stripping scan. In aqueous solutions, the stripping potential range is 0.2 to -2.5 V for mercury electrodes and 1.5 to -0.7 for graphite electrodes. Precipitates and complexes formed by reactions of solution components with the mercury shift the mercury dissolution potential negatively. The degree of shifting depends on the character and concentration of the product formed[1]. Mercury dissolution potential is discussed by Vydra et al.[1], pp. 66, 134–135. Stripping voltammetry is also described in the literature[1,28,35].

Anodic Stripping Voltammetry (ASV)

During the controlled metal electrolysis, the potential remains 200 mV negative of the most negative $E_{1/2}$ of the metals being tested. (A low overvoltage can produce multiple or broad peaks caused by the presence of more than one film layer.) If a mercury electrode is used, the reduced metals amalgamate with the mercury electrode. The controlled electrolysis continues for a sufficiently long preset time period to give measurable signals during the stripping period[1]. All of each polarizer can be plated out from a small solution volume, but for good reproducibility, removal of at most 2% of a depolarizer from a large solution volume usually is preferred[35]. The cell conditions must be the same for the unknowns and the standards. The solutions can be unstirred or they can be stirred at a set rate. The stirring must be gentle for hanging mercury drop electrode (HMDE).

After deposition, there is at least a 30-second rest period during which the solution is unstirred. The rest period allows the depolarizer flux to the electrode to decrease, the current magnitude to drop to that of the diffusion current, and, in the case of the HMDE, the amalgam concentration to become more homogeneous.

After the rest period, changing the potential in a positive direction strips the metal depolarizer from the working electrode. The solution remains unstirred. Depending on sensitivity requirements, one of the previously described techniques with a linearly changing voltage ramp is used to measure the resulting current-voltage relationship[28].

Dc anodic-stripping voltammetry (DCASV) applies an increasing potential. Because noise factors for DCASV are amplified along with the signal, long deposition times of a half-hour or more are required. A scan rate fast enough to produce a peak form gives the highest sensitivity.

Differential-pulse anodic-stripping voltammetry (DPASV) essentially differentiates the dc stripping waveform. DPASV minimizes noise by electronic signal processing. The greater sensitivity and signal-processing capability permit use of greater gains, shorten deposition times to only a few minutes and produce a maximum sensitivity. The shorter deposition times reduce diffusion into the electrode body and provide greater reproducibility because electrode instabilities are less of a problem. At a scan rate of 5 mV/sec, the electrode and system remains in equilibrium[29].

(When deposition of more than a monolayer may occur or when the peaks are small, distorted, broad or flat, the area under the curve defines the current value rather than the peak height[1].) Peak height measurement is illustrated and discussed by Vydra et al.[1], pp. 185–187 and Meites[2], pp. 418–419, 576.

First-derivative stripping also has several advantages over DCASV. The first-derivative method minimizes i_b when it changes more slowly than i_f, minimizes i_c and increases the effect of sweep rate (v) on peak current. With an HMDE, the peak current is proportional to $v^{1/2}$ in DCASV, to $v^{3/2}$ for the first derivative, to $v^{5/2}$ for second derivative and to $v^{7/2}$ for third derivative[43].

The Two Identical HMDE Method of DPASV[44] applies both the plating and stripping processes to one electrode and only the stripping process to the other. The method cancels nearly all the i_r, and the current difference between the electrodes is almost all i_f.

The rotating ring-disk method employs a rotating ring-disk electrode (RRDE), which can be either a solid or mercury film electrode. The RRDE consists of a rotating-disk electrode (RDE) encircled by another electrode in the form of a concentric ring. The depolarizer is deposited on the disk, stripped from the disk by a linear-sweep positive voltage ramp, and plated on the ring at a controlled potential. For a split-ring electrode, separate ring segments are at different potentials and more than one depolarizer plate is out at once. During the ring deposition, the RRDE has an i_c equal to zero.

The anodic stripping sensitivity is limited by the stripping technique used, by the duration of electrode system and instrumentation stability, by the deposition controls and reproducibility (like timing or stirring rate control) and by the operator's patience. For dilute solutions (10^{-11} M), the time it takes for depolarizer adsorption onto container walls can be another limiting factor[29]. The adsorption properties of the cell material, the capacity of the depolarizer to be adsorbed, the degree of depolarizer competition with the supporting electrolyte for adsorption sites, and the time required for analysis versus adsorption rate all determine the severity of the adsorption interference. The depolarizer concentration during the ASV preelectrolysis step accentuates the i_f decrease caused by adsorption on the electrode surface[45].

Cathodic-Stripping Voltammetry (CSV)

Cathodic-stripping voltammetry is similar to ASV except that the charges on the electrodes are reversed and the depolarizer is oxidized during the preconcentration process. CSV is applicable to the determination of suspension surface areas, humic substance concentrations in water[45] and concentrations of certain anions. Barendrecht[46] covers cathodic stripping in his review of stripping voltammetry.

Reverse Scan

Reverse scan is the application of DPP while scanning towards negative potentials and then in the opposite direction. The peaks from the reverse scan shift to more negative potentials than those presented by the negative scan. Application of reverse scan can be used as a quick test for whether ASV can be applied, since the method produces a larger peak on the reverse scan for metals to which stripping analysis is applicable[35]. Reverse scan is described in the literature[35].

WORKING ELECTRODES OTHER THAN DME

Several types of working electrodes are applicable in the current differentiating modifications and the current amplification techniques discussed. When applied to the same electroactive systems as the DME, several other mercury-type electrodes and solid electrodes offer increased sensitivity[4]. Each electrode type has its advantages and disadvantages that, along with analysis requirements and current-voltage measurement mode, govern the choice of the working electrode type. Among the analysis requirements are the form and redox potential of the depolarizer deposition. Depending on the depolarizer, the deposition can be anodic or cathodic. The depolarizer deposition can be accomplished by amalgam formation with the electrode material, by film formation on the electrode, or by collection of a sparingly soluble material produced by the depolarizer reaction with the electrode or with a solution component[1]. Depolarizer deposition is described by Vydra et al.[1], pp. 20–22.

The mercury pool electrode is presented here because, historically, it is a step toward greater sensitivity and, although no longer used much for analysis, is an alternative that can be considered and utilized when purifying reagents. The MFE and the solid electrode are discussed in terms of their application to ASV because their applications are so commonly associated with ASV. The general description and properties of the MFE and solid electrodes are applicable to their use in other electrochemical methods also.

Mercury Pool Electrode

A mercury pool electrode is a mercury pool in the bottom of the cell. The pool contacts the external electronics via a platinum wire protuding through and sealed into the bottom of the cell. Within the cell, the wire contacts only the mercury, not the cell solution. Use of as large a volume of test solution as practical, and of fast scan rates, gives best results.

The advantages of the quiet mercury pool electrode are: (1) much lower detection limits than for the DME; (2) i_f detection earlier on the polarogram; (3) separation of waves that overlap with a DME; and (4) more reliable i_r correction. The i_c is nearly flat. The $E_{1/2}$ measured with a mercury-pool electrode can shift from that measured with a DME. The $E_{1/2}$ shift can exist for nitro and nitroso compounds, oximes, carbonyl compounds and other compounds that have complicated reduction mechanisms. There is no shift for metals or quinones[2].

A stirred mercury pool electrode has an even greater sensitivity. To be effective, a propeller should be located 5–10 mm above the mercury surface; should rotate at a constant rate, forcing the test solution down toward the mercury; and should produce no vortex or stirrer vibration. As opposed to diffusion control, which relates the current to scan rate, the established convection control primarily relates the current to stirring efficiency[2]. The mercury pool electrode is illustrated and discussed by Meites[2], pp. 445–455 and Sawyer and Roberts[3], pp. 83 and 85.

Stationary Mercury Drop Electrode

Reservoir Type

For this electrode, contact with the solution is via a mercury drop extruded from a capillary attached to a mercury reservoir. A change in micrometer screw setting or a piston shift determines the drop size. Air bubbles hamper drop size reproducibility and cause easy drop dislodgment. A break in the mercury column causes a gap in the electrical circuit. A hydrophobic coating on the capillary removes some causes of breaks in the mercury column or of drop dislodgment. The capillary is positioned vertically so the drop either hangs from it (HDME) or sits on top of it (SMDE). The SMDE has the greater stability[1].

Diffusion of the depolarizer into the capillary (back diffusion) contaminates the next bit of mercury to be extruded and, when the HDME is used in ASV, the diffusion decreases the signal during stripping. This contaminated mercury is discarded before metering out the next drop to be used. A capillary of a smaller inside diameter, such as with SMDE, and a shorter electrolysis time during ASV cut down on back diffusion. Long electrolysis times during ASV require a different electrode type. An HMDE has been simulated by applying very slow drop times of 18 minutes to a DME[47].

The HMDE presents reproducible data in neutral or acid solutions at potentials from -0.1 to -1.2 V. In alkaline solutions, the drop easily dislodges and is poorly reproducible[1].

Inert-Support Type

Soft glass tubing or (sometimes) plastic tubing encases an inert wire support for the mercury drop. The exposed portion of the inert wire is coated by amalgamation of the wire metal with mercury. A measured mercury drop on this coated portion is the contact with the solution. The wire composition is usually platinum, but sometimes is silver or gold. Silver and gold require less stringent cleaning than platinum but can produce distorted curves caused by formation of intermetallic compounds. One design procedure etches the platinum to about 0.1 mm deep into the encasing material and holds the mercury drop in the indentation with the mercury touching the wire. This design is stable enough to be rotated. Another design describes a wire hook that protrudes from the encasing material, while a third design describes a small spoon shape at the end of the wire. Along with creating a clean surface, special pretreatment procedures improve platinum amalgamation.

When perfectly covered with mercury, these electrodes have polarographic responses similar to HMDE and are mechanically stable, and a single drop is good for several analyses[1]. Stationary mercury drop electrodes (including HMDE and SMDE) are illustrated and discussed by Vydra et al.[1], pp. 134–144, Meites[2], pp. 454–459 and Sawyer and Roberts[3], pp. 83 and 85.

Mercury Film Electrode (MFE)

An MFE consists of an inert-support electrode coated with a mercury film. The film has a large surface area and only a 1- to 100-μm thickness, preferably 1–3 μm. The inert support is platinum, nickel, silver, graphite or glassy carbon. The shape is either cylindrical and mounted on a wire, or is planar and the active surface of the electrode is a rotating disk[1,28].

The metallic-support electrodes form surface oxide films, intermetallic compounds with mercury and mercury films that reproduce poorly. Carbon electrodes are inert and sturdy, having good conductance and high hydrogen overpotential[28].

Mercury is deposited on the specially cleaned inert surface by dipping in mercury or by electrolysis. The mercury electrolysis is either: (1) in a separate solution before the depolarizer-plating process[1]; (2) in the test solution by plating out added mercury ions at -0.20 V before depositing the depolarizer at an appropriately more negative voltage[48]; or (3) in the test solution by plating out added mercury ions simultaneously with the depolarizer[49]. Lund and Salberg[50] found that for Cu and Cd, the i_f is increased as the potential for metal deposition is made more negative.

The Florence electrode applies the third mercury plating method and produces a much thinner film of 0.001- to 0.01-μm thickness on a glassy carbon electrode. This very thin film wipes off easily with damp filter paper, leaving the support electrode ready to repolish for the next analysis[50].

Compared with the HMDE, the MFE has a much larger surface:volume ratio, more rapidly distributes the depolarizer within the mercury and, during stripping, diffuses the depolarizer back to the mercury surface faster. The stirring can be continued during the stripping with no rest period and can be at a faster rate. The total result is better resolution, higher sensitivity and faster analysis[28].

The MFE does not work satisfactorily in highly acidic media[1]. The MFE is less reproducible and less stable than the HMDE and can have a memory effect. Controlling pH with acetate buffering at pH 5.8 decreases the memory effect and, by keeping the hydrogen overvoltage well negative of the peak potential for zinc, permits the analysis of zinc along with copper, lead and cadmium. An extreme stripping poential (e.g., -0.1 V for copper or -0.05 V for zinc), applied long enough to remove the depolarizer, removes the memory effect. Lengthy holds at these potentials also strip some of the mercury film, along with the depolarizer metal. Acidification can shorten the hold time necessary to strip the depolarizer. The MFE capability of highly concentrating the depolarizers in the mercury can cause intermetallic compounds to form in the film. Thicker mercury films reduce the saturation by depolarizers such as Cu, Ga, Pb and Zn[15]. When the desired deposition potential is on the rising portion of the polarographic curve, a thicker film application can shift the curve toward a more positive voltage and the desired deposition voltage to the plateau[32]. MFE are discussed by Vydra et al.[1], pp. 145–146, among others[15,40,50].

Solid Electrode

Refined polarographic techniques also use electrodes with a nonmercury surface, having constant properties and constant active-surface areas. The electrode material is sealed, cemented or pressed into an insulator and the current is passed from the electrode material, through a wire inside the insulator and on to the external electronics. The contact between the insulator and the electrode must be tight enough to prevent creep of the test solution between the two materials. Electrodes are checked after they are prepared, and only electrodes capable of producing well-defined polarographs are used. The electrodes are disks with circular, active areas, small-diameter rods and large-diameter cylinders. They are stationary or rotating. The insulators are glass or chemically resistant, thermally stable polymers like Teflon.

The electrode composition is impregnated graphite or carbon, pyrolytic carbon, glassy carbon, noble metal, chromium or tungsten carbide. The impregnated electrodes are either pastes or solids. The pastes are spectrographically pure graphite or carbon homogenized with a filler as binding material. The filler is a noncontaminating, nonelectroactive, water-insoluble material such as Nujol or Ceresin wax. A smooth surface lends better resolution and reproducibility. A filler, such as paraffin, impregnates the solid nonmetallic electrodes, fills the pores and reduces the i_c. The glassy carbon electrode is gas-impermeable and hard enough to acquire a high polish without wax impregnation. Pyrolytic graphite or glassy-carbon electrodes are preferred.

Extrusion of the paste and cutting it off flush with the insulator renews the paste-electrode surfaces. Cleaning, polishing and pretreating renews the surfaces of the other solid-electrode types. Reproducibility requires great care in either case.

Most stripping analyses use graphite or carbon (especially glassy carbon) stationary-cylindrical or stationary-disk electrodes. Stationary and rotating, disk and ring-disk electrodes are also popular[1]. Most measurements using rotating-solid working electrodes require a three-electrode system.

Poor peak resolution during stripping and surface contamination, as well as poor reproducibility limit the value of solid electrodes. Solid electrodes are of value when conditions prohibit use of mercury electrodes. Some of these conditions are when

1. mercury itself is analyzed;
2. the depolarizer oxidation is anodic of the oxidation potential of mercury;
3. the depolarizer does not form an amalgam;
4. the depolarizer does not react with mercury;
5. the depolarizer reacts irreversibly with mercury; and
6. another solution component reacts with the mercury, forming a film.

Solid electrodes are not recommended for depolarizer mixtures[1].

Opekar and Beran[51] review rotating-disk electrodes, and Bruckenstein[52] reviews rotating electrodes and their application to the study of reactions. Solid electrodes are discussed by Vydra et al.[1], pp. 66, 146–156, 178, Sawyer and Roberts[3], pp. 66–74, 86–111, Ellis[28] and Covington and Lacoste[53].

CURRENT-VOLTAGE EQUATIONS
FOR POPULAR ANALYSIS METHODS

The current and the voltage measured are related to such electrode parameters as size and shape, and also to the mode of measurement. For the DME, Parry and

Osteryoung[54] evaluate analytical-pulse polarography, and Birke[55] presents current-potential-time relationships for DPP when applied to reversible, quasi-reversible and irreversible electrode processes. Equations 23–29 describe controls on current and voltage for systems analyzed by one of the more sensitive measuring modes, ASV, using the popular mercury electrodes, the MFE, the Florence RDE and the HMDE[28,50,56]. Adjustment of parameters mentioned in these equations can maximize the results desired.

The validity of Equations 23–29 requires a constant current during deposition and, for the MFE, requires that there be no concentration gradient in the film. If there is neither a measurable change in the bulk solution concentration, nor in the stirring rate, a constant current is possible. Establishment of a condition of no concentration gradient within the film requires films <10 μm thick and voltage scan rates of <1 V/min.

Equations 23–25 describe the controls on the MFE.

$$E_p = E^{o'} + \frac{2.3\,R\,T}{n\,F} \log \frac{\delta\,\ell\,v\,n\,F}{D\,R\,T} \tag{23}$$

where E_p = peak potential

$E^{o'}$ = formal electrode potential
δ = diffusion-layer thickness
ℓ = mercury-film thickness
v = scan rate
D = diffusion coefficient

Stirring rate, cell geometry and electrode design influence the term δ:

$$i_p = \frac{n^2\,F^2}{e\,R\,T}\,A\,\ell\,v\,C_a = \frac{n^2\,F^2}{e\,R\,T}\,D\,m\,A\,C_b\,v\,t \tag{24}$$

where i_p = peak stripping current

e = base of Naperian logarithms
A = electrode area
C_a = metal concentration in amalgam
m = mass transport coefficient
C_b = metal concentration in the bulk solution
t = duration of deposition

$$H = \frac{R\,T}{n\,F\,v\,t} \tag{25}$$

Approximations require H in Equation 25 to be 10^{-2} for a 3.9% error in i_p, as defined in Equation 24 and H to be less for greater i_p accuracy.

Equation 26 describes the i_p for the Florence RDE when considered as a rotating-disk MFE:

$$i_p = 0.88 \frac{\pi^{1/2} n^2 F^2 D^{2/3}}{e R T p^{1/6}} N^{1/2} A v t C_b = K N^{1/2} A v t C_b \tag{26}$$

where p = kinematic viscosity
N = number of electrode revolutions per second
K = a numerical constant

Equations 27–29 describe the controls on HMDE.

$$E_p = E_{1/2} - \frac{1.1 R T}{n F} \tag{27}$$

where $E_{1/2}$ is half-wave potential.

$$i_p = k m n^{3/2} D^{3/2} r v^{1/2} C_b t \tag{28}$$

where k is a numerical constant and r is mercury drop radius. Reinmuth[58] describes an extra term for Equation 28 when the sweep rate is slow and the radius is small:

$$i_c = A v \left(\frac{d q}{d E} \right) \tag{29}$$

where i_c is charging current and dq/dE is differential double-layer capacity. Since the i_f and the i_r comprise the total current, and a component of i_r is the i_c, a low i_c aids in measurement of a distinguishable i_f. For a thick-film MFE and for the HMDE, i_p is proportional to $rv^{1/2}$ and, for the HMDE, i_c is proportional to r^2v.[28]

For DCASV with a plane mercury film electrode, De Vries[57] discusses the dependence both of peak width at half height, $W\frac{1}{2}$, and of peak current on the rate of potential change and on the mercury film thickness. He concludes that, for high sensitivity and good resolution of closely adjacent half-wave potentials of metals, the mercury film should be as thin as possible. Parry and Osteryoung[54] discuss $W\frac{1}{2}$ relation to pulse amplitude. Figure 5 illustrates $W\frac{1}{2}$. Peak width at half height, $W\frac{1}{2}$, is discussed in the literature[14,28,46,54,57]. Current-voltage control equations are given by various authors[28,54-59].

CELL MODIFICATION

Polarographic cells are designed with volumes from as large as practical to work with down to tens of microliters. With tiny cells, the apparatus is simple, deaeration is fast and effective, and subnanogram metal determination is possible. A drop on a porous glass disk, or even thin layers of solution, can be analyzed. Separation methods, like solvent extractions or ion exchange carried out in conjunction with the polarographic analysis, improve selectivity.

Regardless of cell design, three conditions of measurement must all be reproducible: the electrode condition, the stirring conditions and the deaeration intensity. For precise work, thermostatic control also is necessary. A synchronous, motor-driven stirrer produces better reproducibility of test solution movement than electromagnetic stirrers. A stream of inert gas or ultrasonic waves are two other stirring modes.

A cell with the electrodes and gas inlet tube tightly secured in the top of the stationary cell and with a removable cell bottom is a convenient design.

Isolation of the depolarizer from interfering substances in the test solution is possible by depositing the depolarizer on the working electrode and by stripping it off into a fresh, pure, support-electrolyte solution. Keeping the electrode continually immersed, as in the flow-through cell, ensures no loss of depolarizer for this solution exchange method. In the flow-through cell, the test solution drains as the noninterfering solution is added, and the volume of the cell contents remains constant. The total fresh support-electrolyte solution used is about four times that of the test solution. A flow-through cell with a mercury pool permits replacement of test solution and rinsing of cell components between analyses with a minimum of manipulation[1].

Clem[40] describes an ElectRoCell that facilitates rapid analysis. As the cell is rotated in one direction, the test solution is spread in a thin film on the cell wall and is sparged. During the depolarizer deposition, periodic reversal of cell rotation stirs the test solution. There is no cell motion during the stripping process. The probe locations are near the cell wall, and the 1/8-inch graphite-rod working electrode faces the cell wall in a horizontal position. The mercury and depolarizer are codeposited on the exposed end of the working electrode. A centrally located lucite rod controls the vortex formation. Other cells are mentioned in the book by Vydra et al.[1] and in the references cited by them. Cells are illustrated and described by Vydra et al.[1], pp. 159–163, Sawyer and Roberts[3], pp. 142–146 and Clem[40].

INTERFERENCES AND THEIR ELIMINATION

Interferences can hinder the production of a simple, interpretable current-voltage response. Both peak overlapping and a solution component that reacts with the electrode material can cause interferences.

In ASV, if two depolarizers reduce at the same potential (e.g., Pb and Sn), they can influence each other during plating or stripping; or one depolarizer can be soluble in solution, interfere with the electrode response of the amalgamating depolarizer and distort the polarogram. During ASV, some concentrated, reduced depolarizers in the mercury can form intermetallic compounds that alter peak heights. Often, if one or more of the reduced depolarizers is a noble metal, these intermetallic compounds form. Some examples of these intermetallic compounds are Zn-Au, Cd-Au, Zn-Cu, Sn-Cu, Co-Zn and Zn-Ni[1]. When plated on graphite, the couples Cd-Cu, Pb-Sn, Cd-Sb, Sn-Ag and Fe-Cu form intermetallic compounds that display a decreased current for the more electronegative component, even when a small amount of electropositive component is present. Also, when metals are deposited simultaneously, the metal combinations Pb-Cu, Cu-Ag and Co-Cu can form solid solutions. These solid solutions display an extra peak when the more electronegative metal is oxidized in a more electropositive metal; however, the converse is not true[60].

Naturally occurring organics, organic pollutants and surfactants (including maxima suppressors) can sorb on the working electrode. The sorption decreases the current and the reversibility, and shifts the $E_{1/2}$ in a positive direction. The sorption effect is not as severe for the dropping mercury electrode as for constant surface area electrodes because of its constant surface renewal. The sorption is sometimes increased by pH decrease.

A depolarizer may not be detected or may be only partially detected because some or all of it is in a nonelectroactive complex. Sometimes sorption and complexation are hard to distinguish, and some organics, such as proteins and humic and fulvic acids, can do both processes, presenting compounded interpretation problems[14]. Sorption is discussed in greater detail by Brezonik et al.[14].

Judicious working-electrode and analysis-method selection can improve peak selectivity. For instance, one of the depolarizers can be complexed and the complex can prevent deposition or shift the peak of that depolarizer to a different potential. One depolarizer can be preconcentrated on a solid electrode and can leave the interfering metal in solution. The depolarizer can be plated from one solution and stripped in another[1]. A depolarizer can be plated out at a constant potential and stripped for several seconds by a potential shift; and then, before the oxidized product can diffuse away, a dc polarographic scan can be applied. This double-plating method gives the same sensitivity as ASV but suppresses interferences and increases selectivity of the desired depolarizer[61]. Substituting the HMDE for the MFE or decreasing the total depolarizer concentration in the mercury film reduces the formation of intermetallic compounds[28]. The addition of gallium to form the stronger Cu-Ga and Ni-Ga intermetallic compounds hinders the formation of Cu-Zn and Ni-Zn intermetallic compounds and permits zinc analysis in the presence of copper and nickel. Complex formation (e.g., nickel-citrate) also permits zinc analysis in the presence of nickel[15].

The low pH required for prevention of metal ion adsorption on plastic storage containers shifts the hydrogen ion wave to the zinc wave region. For MFE, the required pH adjustment with buffers leads to possible unnecessary contamination and, at pH levels greater than 6, adsorption can occur. Provided the hydrogen evolution does not interfere, zinc can be determined directly in acid media by DPASV with a HMDE, if all analysis conditions are exactly the same for the unknown and standards[16].

For a metal of several oxidation states, selection is improved by anodic deposition and cathodic stripping[1]. Application of other separation techniques in conjunction with polarographic analysis can also remove interferences.

Mercury reacts with SCN^- and $S^=$ to form a film that interferes with the CSV analysis of halides. These interfering ions must be removed prior to analysis[61]. Interferences are discussed by Vydra et al.[1], pp. 58–84 and others[15,60-65].

APPENDIX I
DEFINITION OF ABBREVIATIONS

Methods

dc Polarography	Direct current polarography
ac Polarography	Alternating current polarography
DPP	Differential-pulse polarography
NPP	Normal-pulse polarography
ASV	Anodic-stripping polarography
CSV	Cathodic-stripping polarography
DCASV	Direct-current anodic-stripping voltammetry
DPASV	Differential-pulse anodic-stripping voltammetry

Electrodes

SCE	Saturated calomel electrode
NCE	Normal calomel electrode
DME	Dropping mercury electrode
HMDE	Hanging mercury drop electrode
SMDE	Sitting mercury drop electrode
MFE	Mercury film electrode
RDE	Rotating disk electrode
RRDE	Rotating ring disk electrode

Potentials

$E_{1/2}$	Depolarizer half-wave potential
E_p	Electrode potential
E_s^o	Standard potential of the half reaction or of the couple involving the aquocomplex ions
$E^{o\prime}$	Standard potential of the half reaction or of the couple involving the aquocomplex ions under experimental conditions of ionic strength, temperature, etc.

Currents

i_r	Residual current
i_f	Faradaic current
i_d	Diffusion-limiting current

i_d	Average i_f measured at a DME drop
i_b	Background current
i_c	Charging or capacitance current
i_p	Peak-stripping current

Constants

R	Gas content
F	Faraday's constant
I	Diffusion current constant
β	Formation constant
$\beta_p{}'$	Apparent formation constant (under experimental conditions of ionic strength, temperature, etc.)
K	A numerical constant
k	A numerical constant

Names of Measured and Calculated Values

e	Electron activity
T	Temperature
a	Species activity
γ	Activity coefficient
C	Molar concentration
D	Diffusion coefficient
C_b	Metal concentration in the bulk solution
m	Mass of mercury drop at time t
t	Age of mercury drop
z	Charge on the ion
α	Electron transfer coefficient
δ	Diffusion layer thickness
l	Mercury film thickness
v	Voltage sweep rate or scan rate
A	Electrode area
C_a	Metal concentration in the amalgam
e	Base of Naperian Logarithms
m	Mass transport coefficient
t	Deposition time
N	Number of electrode revolutions
p	Kinematic viscosity
r	Mercury drop radius
dq/dE	Differential double-layer capacity
$W_{1/2}$	Peak width at one-half the peak height

Units

Hz	Hertz

μa	Microamp
cm	Centimeter
mm	Millimeter
μm	Micrometer
V	Volt
mV	Millivolt
sec	Second
msec	Millisecond
M	Molar

BIBLIOGRAPHY OF POLAROGRAPHIC PUBLICATIONS

Instrumentation Techniques and Theory

Anderson, J. E., D. E. Tallman, D. J. Chesney and J. L. Anderson. "Fabrication and Characterization of a Kel-F-Graphite Composition Electrode for General Voltammetric Applications," *Anal. Chem.* 50: 1051–1056 (1978).

Anfält, T., and M. Strandberg. "A Micro-Computer System for Potentiometric Stripping Analysis," *Anal. Chim. Acta* 103: 379–388 (1978).

Barnes, D., B. G. Cooksey, B. Metters and C. Metters. "Interferences in the Electroanalytical Determination of Copper and Lead in Water and Wastewater," *Water Treatment Exam.* 24: 318–328 (1975).

Bond, A. M., and B. S. Grabaric. "Simple Approach to the Problem of Overlapping Waves Using a Microprocessor Controlled Polarograph," *Anal. Chem.* 48: 1624–1628 (1976).

Bond, A. M., and B. S. Grabaric. "Differential Pulse Polarography and Voltammetry with a Microprocessor-Controlled Polarograph and a Pressurized Mercury Electrode," *Anal. Chim. Acta* 88: 227–236 (1977).

Bond, A. M., and B. S. Grabaric. "Fast Sweep Differential Pulse Voltammetry at a Dropping Mercury Electrode with Computerized Instrumentation," *Anal. Chem.* 51: 126–128 (1979).

Bond, A. M., and B. S. Grabaric. "Correction for Background Current in Differential Pulse, Alternating Current, and Related Polarographic Techniques in the Determination of Low Concentrations with Computerized Instrumentation," *Anal. Chem.* 51: 337–341 (1979).

Bond, A. M., and R. J. O'Halloran. "Use of Pulsed Direct Current Potential to Minimize Charging Current in Alternating Current Polarography," *Anal. Chem.* 47: 1906–1909 (1975).

Bond, A. M., R. J. O'Halloran, I. Ruzic and D. E. Smith. "Fundamental and Second Harmonic Alternating Current Cyclic Voltammetric Theory and Experimental Results for Simple Electrode Reactions Involving Amalgam Formation," *Anal. Chem.* 50: 216–223 (1978).

Bos, M. "Computerized Kalousek Polarography," *Anal. Chim. Acta* 103: 367–378 (1978).

Bos, M., and G. Jasink. "The Learning Machine in Quantitative Chemical Analysis Part 1. Anodic Stripping Voltammetry of Cadmium, Lead, and Thallium," *Anal. Chim Acta* 103: 151–165 (1978).

Brown, A. P., and F. C. Anson. "Cyclic and Differential Pulse Voltammetric Behavior of Reactants Confined to the Electrode Surface," *Anal. Chem.* 49: 1589–1595 (1977).

Bruckenstein, S., and P. R. Gifford. "Micromolar Voltammetric Analysis by Ring Electrode Shielding at a Rotating Ring-Disk Electrode," *Anal. Chem.* 51: 250–255 (1979).

Burrows, K. C., M. P. Brindle and M. C. Hughes. "Modification of Pulse Polargraph for Rapid Scanning and its Use with Stationary Electrodes," *Anal. Chem.* 49: 1459–1461 (1977).

Christie, J. H., J. A. Turner and R. A. Osteryoung. "Square Wave Voltammetry at the Dropping Mercury Electrode: Theory," *Anal. Chem.* 49: 1899–1903 (1977).

Cieslinski, R., and N. R. Armstrong. "Metallized-Plastic Optically Transparent Electrodes," *Anal. Chem.* 51: 565–568 (1979).

Cummings, T. E., and P. J. Elving. "Verification of a Diffusion Current Equation Accounting for Convection and Capillary Shielding at the Dropping Mercury Electrode," *Anal. Chem.* 50: 1980–1988 (1978).

Dalrymple-Alford, P., M. Goto and K. B. Oldham. "Peak Shapes in Semidifferential Electroanalysis," *Anal. Chem.* 49: 1390–1394 (1977).

DeAngelis, T. P., R. E. Bond, E. E. Brooks and W. R. Heineman. "Thin-layer Differential Pulse Voltammetry," *Anal. Chem.* 49: 1792–1796 (1977).

DeAngelis, T. P., R. W. Hurst, A. M. Yacynych, H. B. Mark and W. R. Heineman. "Carbon and Mercury-Carbon Optically Transparent Electrodes," *Anal. Chem.* 49: 1395–1398 (1977).

DeAngelis, T. P., and W. R. Heineman. "Differential Pulse Anodic Stripping Voltammetry in a Thin-Layer Electrochemical Cell," *Anal. Chem.* 48: 2262–2263 (1976).

Dillard, J. W., and K. W. Hanck. "Digital Simulation of Differential Pulse Polarography," *Anal. Chem.* 48: 218–222 (1976).

Dillard, J. W., J. J. O'Dea and R. A. Osteryoung. "Analytical Implications of Differential Pulse Polarography of Irreversible Reactions from Digital Simulation," *Anal. Chem.* 51: 115–119 (1979).

Dillard, J. W., J. A. Turner and R. A. Osteryoung. "Digital Simulation of Differential Pulse Polarography with Incremental Time Change," *Anal. Chem.* 49: 1246–1250 (1977).

EG&G Princeton Applied Research. "Applications Bibliography and Applications Index," *EG&G P.A.R. Tech. Note* 110 (1979).

Eggli, R. "Anodic Stripping Coulometry at a Thin-Film Mercury Electrode," *Anal. Chim. Acta* 91:129—138 (1977).

Eggli, R. "The Detection Limit of Anodic Stripping Coulometry at Mercury-Film Glassy Carbon Electrodes," *Anal. Chim. Acta* 97: 195–198 (1978).

Goto, M., K. Ikenoya and D. Ishii. "Anodic Stripping Semidifferential Electroanalysis with Thin Mercury Film Electrode Formed in Situ," *Anal. Chem.* 51: 110–15 (1979).

Goto, M., K. Ikenoya, M. Kajihara and D. Ishii. ''Application of Semidifferential Electroanalysis to Anodic Stripping Voltammetry," *Anal. Chim. Acta* 101: 131–138 (1978).

Goto, M., and K. B. Oldham. "Semiintegral Electroanalysis: The shape of Irreversible Neopolarograms," *Anal. Chem.* 48: 1671–1676 (1976).

212 ANALYTICAL MEASUREMENTS

Gough, D. A., and J. K. Leypoldt. "Membrane-Covered, Rotated Disc Electrode," *Anal. Chem.* 51: 439–444 (1979).

Hanafey, M. K., R. L. Scott, T. H. Ridgway and C. N. Reilley. "Analysis of Electrochemical Mechanisms by Finite Difference Simulation and Simplex Fitting of Double Potential Step Current, Charge, and Absorbance Responses," *Anal. Chem.* 50: 116–137 (1978).

Heijne, G. J. M., and W. E. Van Der Linden. "The Current-Potential Relationship in Differential Pulse Polarography: A Revision," *Anal. Chim. Acta* 99: 183–187 (1978).

Jagner, D., and K. Årén. "Derivative Potentiometric Stripping Analysis with a Thin Film of Mercury on a Glassy Carbon Electrode," *Anal. Chim. Acta* 100: 375–388 (1978).

Jagner, D. "Potentiometric Stripping Analysis in Non-Deaerated Samples," *Anal. Chem.* 51: 342–345 (1979).

Kolthoff, I. M., and S. Kihara. "Reduction Currents of Films Formed during Reductions at the Hanging Mercury Drop Electrode," *Anal. Chem.* 50: 1003–1005 (1978).

Lowry, J. H., and R. B. Smart. "A Multiple Cell System for Differential Pulse Anodic Stripping Voltammetry at a Hanging Mercury Drop Electrode," *Am. Lab. Dec.* 47–49 (1976).

Menard, H., and F. LeBlond-Routhier. "Dropping Mercury Electrode for Polarography in Glass-Corroding Media," *Anal. Chem.* 59: 687–688 (1978).

Michel, L., and A. Zatka. "An Electrochemical Detector with Dropping Mercury Electrode for High-Performance Liquid Chromatography," *Anal. Chim. Acta* 105: 109–117 (1979).

Mohammed, M. "Application of the Line Shape of Linear Scan and Cyclic Voltammograms for the Determination of the Rate Constant and Reversible $E_{1/2}$ for a Chemical Reaction Preceded by a Reversible Charge Transfer," *Anal. Chem.* 49: 60–61 (1977).

Mooring, C. I., and H. L. Kies. "Characteristics of A. C. Polarograms at High Sweep Rates," *Anal. Chim. Acta* 94: 135–147 (1977).

Morris, J. L., Jr., and L. R. Faulkner. "Normal Pulse Voltammetry in Electrochemically Poised Systems," *Anal. Chem.* 49: 489–494 (1977).

Norvall, V. E., and G. Mamantov. "Optically Transparent Vitreous Carbon Electrode," *Anal. Chem.* 49: 1470–1472 (1977).

O'Halloran, R. J., J. C. Schaar and D. E. Smith. "Rapid Drop Time On-Line Fast Fourier Transform Faradaic Admittance Measurements," *Anal. Chem.* 50: 1073–1079 (1978).

Peerce, P. J., and F. C. Anson. "Digital Device for Precise Determination of Drop Times at Dropping Mercury Electrodes," *Anal. Chem.* 49: 1270–1272 (1977).

Princeton Applied Research. "Why Deaeration and How," *P.A.R. Application Note* 108 (1974).

Princeton Applied Research. "Stripping Voltammetry Some Helpful Techniques," *P.A.R. Tech. Note* 109A (1974).

Princeton Applied Research. "Bibliography of Applications of Electrochemical Instruments," *P.A.R. Tech. Note* 110 (1975).

Rifkin, S. C., and D. H. Evans. "General Equation for Voltammetry with Step-Functional Potential Changes Applied to Differential Pulse Voltammetry," *Anal. Chem.* 48: 1616–1618 (1976).

Rifkin, S. C., and D. H. Evans. "Analytical Evaluation of Differential Pulse Voltammetry at Stationary Electrodes Using Computer-Based Instrumentation," *Anal. Chem.* 48: 2174–2180 (1976).

Roeleveld, L. F., B. J. C. Wetsema and J. M. Los. "Pulse Polarography Part XI. Some Problems in Solving Kinetic Equations," *J. Electroanal. Chem.* 75: 839–844 (1977).

Rowley, P. G., and J. G. Osteryoung. "Construction of a Rotating Ring-Disc Electrode from Irregular Electrode Materials," *Anal. Chem.* 50: 1015–1016 (1978).

Ruzic, I., and M. Sluyters-Rehbach. "The Current-Potential Relationship for Differential Pulse Polarography," *Anal. Chim. Acta* 99: 177–182 (1978).

Seelig, P. F., and H. N. Blount. "Experimental Evaluation of Recursive Estimation Applied to Linear Sweep Anodic Stripping Voltammetry for Real Time Analysis," *Anal. Chem.* 51: 327–337 (1979).

Steeman, E., E. Temmerman and R. Verbinnen. "Subtractive Anodic Stripping Voltammetry at Twin Mercury Film Electrodes," *Anal. Chim. Acta* 96: 177–181 (1978).

Stolzberg, R. J. "Potential Inaccuracy in Trace Metal Specification Measurements by Differential Pulse Polarography," *Anal. Chim. Acta* 92: 193–196 (1977).

Strohl, A. N., and D. J. Curran. "Reticulated Vitreous Carbon Flow-Through Electrodes," *Anal. Chem.* 51: 353–357 (1979).

Thomas, Q. V., R. A. Depalma and S. P. Perone. "Application of Pattern Recognition Techniques to the Interpretation of Severely Overlapped Voltammetric Data: Studies with Experimental Data," *Anal. Chem.* 49: 1376–1380 (1977).

Thomas, Q. V., and S. P. Perone. "Application of Pattern Recognition Techniques to the Interpretation of Severely Overlapped Voltammetric Data: Theoretical Studies," *Anal. Chem.* 49: 1369–1375 (1977).

Turner, J. A., J. H. Christie, M. Vukovic and R. A. Osteryoung. "Square Wave Voltammetry at the Dropping Mercury Electrode: Experimental," *Anal. Chem.* 49: 1904–1908 (1977).

Turner, J. A., U. Eisner and R. A. Osteryoung. "Pulsed Voltammetric Stripping at the Thin-Film Mercury Electrode," *Anal. Chim. Acta* 90: 25–34 (1977).

Turner, J. A., and R. A. Osteryoung. "Rapid Scan Alternate Drop Pulse Polarographic Methods," *Anal. Chem.* 50: 1496–1500 (1978).

van Bennekom, W. P., and J. B. Schute. "High-Performance Pulse and Differential Pulse Polarography Part 1. Theoretical Considerations," *Anal. Chim. Acta* 89: 71–82 (1977).

Wang, J., and M. Ariel. "The Rotating Disc Electrode in Flowing Systems. Part 1. An Anodic Stripping Monitoring System for Trace Metals in Natural Waters," *Anal. Chim. Acta* 99: 89–98 (1978).

Wang, J., and M. Ariel. "The Rotating Disk Electrode in Flowing Systems. Part 2. A Flow System for Automated Anodic Stripping Voltammetry of Discrete Samples," *Anal. Chim. Acta* 101: 1–8 (1978).

Yarnitzky, C., and E. Ouziel. "Nebulizer for Eliminating Oxygen from Polarographic Flow Cells," *Anal. Chem.* 48: 2024–2025 (1976).

Zirino, A., and S. P. Kounaves. "Anodic Stripping Peak Currents: Electrolysis Potential Relationships for Reversible Systems," *Anal. Chem.* 49: 56–59 (1977).

Inorganic Analysis

Bartscher, W., and B. Giovannone. "Controlled-Potential Coulometric Determination of Platinum in Americium-Platinum Alloys," *Anal. Chim. Acta* 91: 139–142 (1977).

Ben-Bassat, A. H. I., J. M. Blindermann and A. Salomon. "Direct Simultaneous Determination of Trace Amounts (ppb) of Zinc(II), Cadmium(II), Lead(II), and Copper(II) in Ground and Spring Waters Using Anodic Stripping Voltammetry: The Analytical Method," *Anal. Chem.* 47: 534–547 (1975).

Bilinski, H., R. Huston and W. Stumm. "Determination of the Stability Constants of Some Hydroxo and Carbonato Complexes of Pb(II), Cu(II), Cd(II) and Zn(II) in Dilute Solutions by Anodic Stripping Voltammetry and Differential Pulse Polarography," *Anal. Chim. Acta* 84: 157–164 (1976).

Blaedel, W. J., and R. C. Engstrom. "Investigations of the Ferricyanide-Ferrocyanide System by Pulsed Rotation Voltammetry," *Anal. Chem.* 50: 476–479 (1978).

Bodini, M. E., and D. T. Sawyer. "Voltammetric Determination of Nitrate Ion at Parts-per-Billion Levels," *Anal. Chem.* 49: 485–489 (1977).

Boese, S. W., V. S. Archer and J. W. O'Laughlin. "Differential Pulse Polarographic Determination of Nitrate and Nitrite," *Anal. Chem.* 49: 479–484 (1977).

Bontempelli, G., B. Corain and F. Magno. "Cathodic Behavior of trans-Dicyanobis (diethylphenylphosphine) Nickel Complex," *Anal. Chem.* 49: 1005–1008 (1977).

Bosserman, P., D. T. Sawyer and A. L. Page. "Differential Pulse Polarographic Determination of Molybdenum at Parts-per-Billion Levels," *Anal. Chem.* 50: 1300–1303 (1978).

Brajer, A. R., T. E. Farley, J. W. Kauffman, L. K. Young, R. J. Williams and J. W. Rogers. "Polarography of Tin(II) Chloride in Acetonitrile," *Anal. Chim. Acta* 91: 165–173 (1977).

Chang, S. K., R. Kozeniauskas and G. W. Harrington. "Determination of Nitrite Ion Using Differential Pulse Polarography," *Anal. Chem.* 49: 2272–2275 (1977).

Clark, G. C., G. J. Moody and J. D. R. Thomas. "The Polarographic Behavior of Copper Complexes of Pilocarpine and Some Related Imidazoles," *Anal. Chim. Acta* 98: 215–220 (1978).

Davis, P. H., G. R. Dulude, R. M. Griffin, W. R. Matson and E. W. Zink. "Determination of Total Arsenic at the Nanogram Level by High-Speed Anodic Stripping Voltammetry," *Anal. Chem.* 50: 137–143 (1978).

EG&G Princeton Applied Research. "Applications Bibliography and Applications Index," *EG&G P.A.R. Tech. Note* 110 (1979).

Elder, J. F. "Complexation Side Reactions Involving Trace Metals in Natural Water Systems," *Limnol. Oceanog.* 20: 96–102 (1975).

Fawcett, N. C. "Determination of Uranium in Plutonium-238 Metal and Oxide by Differential Pulse Polarography," *Anal. Chem.* 48: 215–218 (1976).

Figura, P., and B. McDuffle. "Use of Chelex Resin for Determination of Labile Trace Metal Fractions in Aqueous Ligand Media and Comparison of the Method with Anodic Stripping Voltammetry," *Anal. Chem.* 51: 120–125 (1979).

Forsberg, G., J. W. O'Laughlin and R. G. Megargle. "Determination of Arsenic by Anodic Stripping Voltammetry and Differential Pulse Anodic Stripping Voltammetry," *Anal. Chem.* 47: 1586–1592 (1975).

Gardiner, J., and M. J. Stiff. "The Determination of Cadmium, Lead, Copper, and Zinc in Ground Water, Estuarine Water, Sewage and Sewage Effluent by Anodic Stripping Voltammetry," *Water Res.* 9: 517–523 (1975).

Glodowski, S., and Z. Kublik. "Cyclic and Stripping Voltammetry of Tin in the Presence of Lead in Pyrogallol Medium at Hanging and Film Mercury Electrodes," *Anal. Chim. Acta* 104: 55–56 (1979).

Goyal, R. N., and S. Tyagi. "Polarographic Investigations of Some Copper Chelates of 3-Arylazopentane-2,4-Diones," *Anal. Chim. Acta* 98: 111–114 (1978).

Hanck, K. W., and J. W. Dillard. "Determination of the Complexing Capacity of Natural Water by Cobalt (III) Complexation," *Anal. Chem.* 49: 404–409 (1977).

Hanck, K. W., and J. W. Dillard. "Evaluation of Micromolar Compleximetric Titrations for the Determination of the Complexing Capacity of Natural Waters," *Anal. Chim. Acta* 89: 329–338 (1977).

Henry, F. T., T. O. Kirch and T. M. Thorpe. "Determination of Trace Level Arsenic (III), Arsenic (V), and Total Inorganic Arsenic by Differential Pulse Polarography," *Anal. Chem.* 51: 215–218 (1979).

Humphrey, R. E., and S. W. Sharp. "Polarographic Determination of Chloride, Cyanide, Fluoride, Sulfate, and Sulfite Ions by an Amplification Procedure Employing Metal Iodates," *Anal. Chem.* 48: 222–223 (1976).

Jagner, D. "Instrumental Approach to Potentiometric Stripping Analysis of Some Heavy Metals," *Anal. Chem.* 50: 1924–1929 (1978).

Kanzaki, N., Y. Kanzaki and S. Bruckenstein. "Phase-Selective Fundamental and Second Harmonic Alternating Current Voltammetry of Bismuth (III) and Lead (II) at a Continuously Mercury-Coated Rotating Platinum Disk Electrode," *Anal. Chem.* 49: 1789–1791 (1977).

Kinard, J. T. "Simultaneous Determination of Trace Metals in Industrial and Domestic Effluents by Differential Pulse Anodic Stripping Voltammetry," *J. Environ. Sci. Health* A12: 531–547 (1977).

Kirowa-Eisner, E., and J. Osteryoung. "Direct and Titrimetric Determination of Hydroxide Using Normal Pulse Polarography at Mercury Electrodes," *Anal. Chem.* 50: 1062–1066 (1978).

Komatsu, M. "Polarographic and Anodic Stripping Polarographic Studies of Aliphatic Polyamine Complexes of Several Metal Ions," *Bull. Chem. Soc. Japan* 47: 1636–1641 (1974).

Koval, C. A., and F. C. Anson. "Electrochemistry of the Ruthenium (3+, 2+) Couple Attached to Graphite Electrodes," *Anal. Chem.* 50: 223–229 (1978).

Laitinen, H. A., and N. H. Watkins. "Cathodic Stripping Coulometry of Lead," *Anal. Chem.* 47: 1352–1358 (1975).

Lal, S., and P. S. Jain. "Polarographic Maxima in the Cobalt-8-Hydroxyquinoline System," *Anal. Chim. Acta* 96: 353–358 (1978).

Lund, W., and D. Onshus. "The Determination of Copper, Lead and Cadmium in Sea Water by Differential Pulse Anodic Stripping Voltammetry," *Anal. Chim. Acta* 86: 109–122 (1976).

Luther III, G. W., and A. L. Meyerson. "Polarographic Analysis of Sulfate Ion in Seawater Samples," *Anal. Chem.* 47: 2058–2059 (1975).

Moore, W. M. "Voltammetric Determination of Iron (II) and Iron (III) in Standard Rocks and Other Materials," *Anal. Chim. Acta* 105: 99–107 (1979).

Moorhead, E. D., and W. H. Doub, Jr. "Digital Microcoulometric Measurements of Cadmium Anodic Stripping at the Micrometer Hanging Drop Electrode," *Anal. Chem.* 49: 199–205 (1977).

Nuor, S. K., and O. Vittori. "Determination du cadmium, de l'indium et du tellure par polarographic impulsionelle sans séparation préalable," *Anal. Chim. Acta* 91: 143–148 (1977).

Odier, M., and V. Plichon. "Le cuivre en solution dans l'eau de mer: forme chimique et dosage," *Anal. Chim. Acta* 55: 209–220 (1971).

Parkinson, B. A., and F. C. Anson. "Adsorption and Polymeric Film Formation at Mercury Electrodes by Solutions of Lead (II) and Chelating Ligands Containing a Thioether Group," *Anal. Chem.* 50: 1886–1891 (1978).

Peter, F., and R. G. Reynolds. "Percholoric Acid-Free Digestion of Blood in Microdetermination of Lead by Anodic Stripping Voltammetry," *Anal. Chem.* 48: 2041–2042 (1976).

Petrie, L. M., and R. W. Baier. "Thin-Mercury-Film Voltammetry of Inorganic Lead (II) Complexes in Seawater," *Anal. Chem.* 50: 351–357 (1978).

Pilkington, E. S., and C. Weeks. "Determination of Trace Elements in Zinc Plant Electrolyte by Differential Pulse Polarography and Anodic Stripping Voltammetry," *Anal. Chem.* 48: 1665–1669 (1976).

Pinchin, M. J., and J. Newham. "The Determination of Lead, Copper and Cadmium by Anodic Stripping Voltammetry at a Mercury Thin-Film Electrode," *Anal. Chim. Acta* 90: 91–102 (1977).

Poldoski, J. E., and G. E. Glass. "Anodic Stripping Voltammetry at a Mercury Film Electrode: Baseline Concentrations of Cadmium, Lead and Copper in Selected Natural Waters," *Anal. Chim. Acta* 101: 79–88 (1978).

Pool, D., L. K. Young, R. J. Williams and J. W. Rogers. "Cyclic Voltammetric Investigation of the Reduction of Tin (II) Chloride in Acetonitrile," *Anal. Chim. Acta* 92: 361–368 (1977).

Princeton Applied Research. "Differential Pulse Stripping Analysis of Tap Water," *P.A.R. Appl. Note* AN-107 (1973).

Princeton Applied Research. "Determination of Arsenic by Differential Pulse Polarography," *P.A.R. Appl. Note* 117 (1974).

Princeton Applied Research. "Differential Pulse Polarography of Chromium," *P.A.R. Appl. Note* 122 (1974).

Princeton Applied Research. "Coulometric Analysis of Copper in Brass," *P.A.R. Appl. Note* 124 (1974).

Princeton Applied Research. "Bibliography of Applications of Electrochemical Instruments," *P.A.R. Tech. Note* 110 (1975).

Propst, R. C. "Cathodic Pulse Stripping Analysis of Iodine at the Parts-per-Billion Level," *Anal. Chem.* 49: 1199–1205 (1977).

Schieffer, G. W., and W. J. Blaedel. "Anodic Stripping Voltammetry with Collection at Tubular Electrodes for the Analysis of Tap Water," *Anal. Chem.* 50: 99–110 (1978).

Shafiqul Alam, A. M., O. Vittori and M. Porthault. "Determination of Selenium (IV) in Acidic Solutions with A. C. Polarography and Differential Pulse Polarography," *Anal. Chim. Acta* 87: 437–444 (1976).

Shuman, M. S. "Team Looks at Metal Speciation," *E.S.E. Notes* 14: 1–5 (1978).

Shuman, M. S., and L. C. Michael. "Reversibility of Copper in Dilute Aqueous Carbonate and its Significance to Anodic Stripping Voltammetry of Copper in Natural Waters," *Anal. Chem.* 50: 2104-2108 (1978).

Vydra, F., and T. V. Nghi. "Application of a Rotating Disc Electrode and a Rotating Cell with Stationary Electrode in Stripping Voltammetry for the Determination of Lead and Zinc," *Anal. Chim. Acta* 91: 335-338 (1977).

Whitnack, G. C. "Single-Sweep Polarographic Techniques Useful in Micropollution Studies of Ground and Surface Waters," *Anal. Chem.* 47: 618-621 (1975).

Young, R. L., J. E. Spell, H. M. Siu, R. H. Philp and E. R. Jones. "Determination of Nitrate in Water Samples Using a Portable Polarographic Instrument," *Environ. Sci. Technol* 9: 1075-1077 (1975).

Zirino, A., S. H. Lieberman and C. Clavell. "Measurement of Cu and Zn in San Diego Bay by Automated Anodic Stripping Voltammetry," *Environ. Sci. Technol.* 12: 73-79 (1978).

Zur, C., and M. Ariel. "The Determination of Cadmium in the Presence of Humic Acid by Anodic Stripping Voltammetry," *Anal. Chim. Acta* 88: 245-251 (1977).

Organic Analysis

Afghan, B. K., A. V. Kulkarni and J. F. Ryan. "Determination of Nanogram Quantities of Carbonyl Compounds Using Twin Cell Potential Sweep Voltammetry," *Anal. Chem.* 47: 488-494 (1975).

Arnac, M., and G. Berboom. "Voltammetric and Chronopotentiometric Studies of the Quinone and Hydroquinone System in Anhydrous Formic Acid," *Anal. Chem.* 49: 806-809 (1977).

Beckett, A. H., N. N. Rahman and W. F. Smyth. "A Polarographic Study of some N-Oxygenated Products of N-Ethyl-β-Methoxy-β-(3'-Triflouromethylphenyl)-ethylamine(SK and F 40625A)," *Anal. Chim. Acta* 92: 353-360 (1977).

Betso, S. R., and J. D. McLean. "Determination of Acrylamide Monomer by Differential Pulse Polarography," *Anal. Chem.* 48: 766-770 (1976).

Brooks, M. A., and M. R. Hackman. "Trace Level Determination of 1,4-Benzodiazepines in Blood by Differential Pulse Polarography," *Anal. Chem.* 47: 2059-2062 (1975).

Brunt, K. "The Polarographic Behavior of the Antidepressant Drug Chlorimipramine," *Anal. Chem. Acta* 98: 93-99 (1978).

Burgard, D. R., and S. P. Perone. "Computerized Pattern Recognition for Classification of Organic Compounds from Voltammetric Data," *Anal. Chem.* 50: 1366-1371 (1978).

Canterford, D. R. "A. C. Polarographic Determination of Polyethylene Glycols — Application to Analysis of Photographic Processing Solutions," *Anal. Chim. Acta* 94: 377-384 (1977).

Canterford, D. R. "Polarographic Determination of Hydroxylamines: Application to Analysis of Photographic Processing Solutions," *Anal. Chim. Acta* 98: 205-214 (1978).

Cardwell, T. J., and G. Svehla. "Cyclic Voltammetry of Some Indophenol Derivatives in Aqueous Buffered Solutions," *Anal. Chim. Acta* 88: 163–169 (1977).

Carruthers, C. "Polarography of Azocarcinogens and Related Azo Compounds," *Anal. Chim. Acta* 86: 273–276 (1976).

Chambers, J. Q., D. C. Green, F. B. Kaufman, E. M. Engler, B. A. Scott and R. R. Schumaker. "Voltammetry and Potentiometry of Tetrathiafulvalene Halides," *Anal. Chem.* 49: 802–806 (1977).

Chan, H. K., and A. G. Fogg. "Determination of Ciclazindol in Biological Fluids by Differential Pulse Polarography," *Anal. Chim. Acta* 98: 101–109 (1978).

Chiu, S-t, and L. Paszner. "Potentiometric Titration Behavior of Sodium Sulfide, Methyl Mercaptan, Dimethyl Sulfide, Dimethyl Disulfide and Polysulfides in Mixed Alkaline Solutions and Sulfate Pulping Black Liquors," *Anal. Chem.* 47: 1910–1916 (1975).

Coetzee, J. F., G. H. Kazi and J. C. Spurgeon. "Determination of Polynuclear Aromatic Hydrocarbons by Anodic Differential Pulse Voltammetry at the Glassy Carbon Electrode in Sulfolane and Acetonitrile as Solvents," *Anal. Chem.* 48: 2170–2174 (1976).

Davidson, I. E., and W. F. Smyth. "Direct Determination of Thioamide Drugs in Biological Fluids by Cathodic Stripping Voltammetry," *Anal. Chem.* 49: 1195–1198 (1977).

Davis, D. G., and R. W. Murray. "Surface Electrochemistry of Iron Porphyrins and Iron on Tin Oxide Electrodes," *Anal. Chem.* 49: 194–198 (1977).

Dryhurst, G., and L. G. Karber. "Differential Pulse Voltammetric Oxidation of Polyriboxanthylic Acid at the Pyrolytic Graphite Electrode and a Method for Detection and Determination of Traces of Xanthine and Xanthosine-5'-Monophosphate in Polyxanthylic Acid," *Anal. Chim. Acta* 100: 289–300 (1978).

EG&G Princeton Applied Research. "Applications Bibliography and Applications Index," *EG&G P.A.R. Tech. Note* 110 (1979).

Elton, R. K., and W. E. Geiger, Jr. "Analytical and Mechanistic Studies of the Electrochemical Reduction of Biologically Active Organoarsenic Acids," *Anal. Chem.* 50: 712–777 (1978).

Fenn, R. J., S. Siggla and D. J. Curran. "Liquid Chromatography Detector Based on Single and Twin Electrode Thin-Layer Electrochemistry: Application to the Determination of Catecholamines in Blood Plasma," *Anal. Chem.* 50: 1067–1073 (1978).

Fike, R. R., and D. J. Curran. "Determination of Catecholamines by Thin-Layer Linear Sweep Voltammetry," *Anal. Chem.* 49: 1205–1210 (1977).

Fogg, A. G., and Y. Z. Ahmed. "Determination of Oxyphenbutazone and Phenylbutazone by Differential Pulse Polarography after Derivatization," *Anal. Chim. Acta* 94: 453–456 (1977).

Fujiwara, S., Y. Umezawa and H. Ishizuka. "A Polarographic Study of Amino Acids in Aqueous Solutions," *Bull. Chem. Soc. Japan* 44: 1984–1986 (1971).

Haring, B. J. A., and W. V. Delft. "Determination of Nitrilotriacetic Acid in Water by Derivative Pulse Polarography at a Hanging Mercury Drop Electrode," *Anal. Chim. Acta* 94: 201–203 (1977).

Howe, L. H. "Differential Pulse Polarographic Determination of Acrolein in Water Samples," *Anal. Chem.* 48: 2167–2169 (1976).

Jacobsen, E., and M. W. Bjørnsen. "Polarographic Determination of Folic Acid in Pharmaceutical Preparations," *Anal. Chim. Acta* 96: 345–351 (1978).

Jacobsen, E., and B. Korvald. "Differential Pulse Polarographic Determination of Hyrdocortisone in Pharmaceutical Preparations," *Anal. Chim. Acta* 99: 255–261 (1978).

Jacobsen, E., J. H. Pederstad and B. Øystese. "Determination of Bacitracin by Differential Pulse Polarography," *Anal. Chim. Acta* 91: 121–128 (1977).

Jemal, M., and A. M. Knevel. "Polarographic Behavior of Benzylpenicillenic Acid," *Anal. Chem.* 50: 1917–1921 (1978).

Kadish, K. M., and V. R. Spiehler. "Differential Pulse Polarographic Determination of Digoxin and Digitoxin," *Anal. Chem.* 47: 1714–1716 (1975).

Kolthoff, I. M., and S. Kihara. "Voltammetric Determination of Ultratraces of Albumin, Cysteine, and Cystine at the Hanging Mercury Drop Electrode," *Anal. Chem.* 49: 2108–2109 (1977).

Lowry, J. H., R. B. Smart and K. H. Mancy. "Differential Pulse Polarography of Phenylarsine Oxide," *Anal. Chem.* 50: 1303–1309 (1978).

Lund, W., and L. Opheim. "Automated Polarographic Analysis Part III. Determination of Chlordiazepoxide and Diazepam in Tablets," *Anal. Chim. Acta* 88: 275–279 (1977).

McLean, J. D., V. A. Stenger, R. E. Reim, M. W. Long and T. A. Hiller. "Determination of Ammonia and Other Nitrogen Compounds by Polarography," *Anal. Chem.* 50: 1309–1314 (1978).

Moiroux, J., and P. J. Elving. "Effects of Absorption, Electrode Material, and Operational Variables on the Oxidation of Dihydronicotinamide Adenine Diculeotide at Carbon Electrodes," *Anal. Chem.* 50: 1056–1062 (1978).

Moiroux, J., and P. J. Elving. "Optimization of the Analytical Oxidation of Dihydronicotinamide Adenine Dinucleotide at Carbon and Platinum Electrodes," *Anal. Chem.* 51: 346–353 (1979).

Moore, W. M., and V. F. Gaylor. "Phase-Selective Cathodic Stripping Voltammetry for Determination of Water-Soluble Mercaptans," *Anal. Chem.* 49: 1386–1388 (1977).

Opheim, L. "Determination of Noretisterone in Tablets by Differential Pulse Polarography," *Anal. Chim. Acta* 89: 225–229 (1977).

Opheim, L. "Determination of Crotonaldehyde in Ethanol by Differential Pulse Polarography," *Anal. Chim. Acta* 91: 331–334 (1977).

O'Shea, T. A., and K. H. Mancy. "Characterization of Trace Metal-Organic Interactions by Anodic Stripping Voltammetry," *Anal. Chem.* 48: 1603–1607 (1976).

Patrick, R. A., T. J. Cardwell and G. Svehla. "Polarographic Study of Analytically Important Indophenol Derivatives," *Anal. Chim. Acta* 88: 155–162 (1977).

Princeton Applied Research. "Analysis of Diols by Differential Pulse Polarography," *P.A.R. Appl. Note* AN-120 (1973).

Princeton Applied Research. "Tetrahydro-2-ethylanthraquinone (T-2EA) and 2-Ethylanthraquinone (2EA)," *P.A.R. Appl. Brief* A-1 (1974).

Princeton Applied Research. "Maleic and Fumaric Acids," *P.A.R. Appl. Brief* M-2 (1974).

Princeton Applied Research. "Ascorbic Acid and Fumaric Acid in Fruit Juices," *P.A.R. Appl. Brief* A-4 (1975).

Rogstad, A., and K. Høgberg. "The Polarographic Determination of Nitrovin," *Anal. Chim. Acta* 94: 461–465 (1977).

Romer, M., L. G. Donaruma and P. Zuman. "Spectrophotometric and Polarographic Analysis for Phenobarbital, N-Methylphenobarbital and N-Methoxymethylphenobarbital in a Mixture," *Anal. Chim. Acta* 88: 261–273 (1977).

Rubel, S., and M. Wojciechowski. "Analytical Applications of Triethylenetetraminehexaacetic Acid. Part I. The Influence of TTHA on the Polarographic Reduction of Some Metal Ions," *Anal. Chim. Acta* 99: 105–113 (1978).

Smyth, M. R., T. S. Beng and W. F. Smyth. "A Spectral and Polarographic Study of the Acid-Base and Complexing Behavior of Bromazepam," *Anal. Chim. Acta* 92: 129–138 (1977).

Smyth, M. R., and J. G. Osteryoung. "A Pulse Polarographic Investigation of Parathion and Some Other Nitro-Containing Pesticides," *Anal. Chim. Acta* 96: 335–344 (1978).

Smyth, M. R., and J. G. Osteryoung. "Polarographic Determination of Some Azomethine-Containing Pesticides," *Anal. Chem.* 50: 1632–1637 (1978).

Smyth, M. R., and W. F. Smyth. "The Application of Differential Pulse Polarography to the Determination of a Pharmacologically Active Benzhydrylpiperazine Derivative and Its Major Electroactive Metabolites in the Plasma and Urine of Animals," *Anal. Chim. Acta* 94: 119–127 (1977).

Stoltzberg, R. J. "Determination of Ethylenediaminetetraacetate and Nitrilotriacetate in Phytoplankton Media by Differential Pulse Polarography," *Anal. Chim. Acta* 92: 139–148 (1977).

Wade, A. L., and F. M. Hawkridge. "Direct Determination of Pentochlorophenol by Differential Pulse Polarography," *Anal. Chim. Acta* 105: 91–97 (1979).

Wienhold, K., and H. Sohr. "Simultaneous Polarographic Determination of Adenosine-5′-Monophosphate and Adenosine-5′-o-Monothiophosphate at the Nanogram Level," *Anal. Chim. Acta* 89: 297–302 (1977).

Youssefi, M., and R. L. Birke. "Determination of Sulfide and Thiols in the Presence of Vitamin B_{12a} by Pulse Polarography," *Anal. Chem.* 49: 1380–1385 (1977).

REFERENCES

1. Vydra, F., K. Štulík and E. Juláková. *Electrochemical Stripping Analysis,* J. Tyson, trans. Ed. (New York: John Wiley & Sons, Inc., 1976).
2. Meites, L. *Polarographic Techniques* (New York: Wiley-Interscience, 1965).
3. Sawyer, D. T., and J. L. Roberts, Jr. *Experimental Electrochemistry for Chemists* (New York: Wiley-Interscience, 1974).
4. Meites, L. In: *Treatise on Analytical Chemistry,* I. M. Kolthoff and P. J. Elving, Eds. (New York: Wiley-Interscience, 1963).
5. Ficker, H. K., H. N. Ostensen, R. H. Schlossel, F. Scott, M. Spritzer and L. Meites. *Anal. Chim. Acta* 98: 163 (1978).
6. Grenier, J. W., and L. Meites. *Anal. Chim. Acta* 14: 482 (1956).

7. Breda, E. J., L. Meites, T. B. Reddy and P. W. West. *Anal. Chem. Acta* 14: 390 (1956).
8. Baumgarten, S., R. E. Cover, H. Hofsass, P. B. Pinches and L. Meites. *Anal. Chim. Acta* 20: 397 (1959).
9. Galus, Z. *Fundamentals of Electrochemical Analysis,* G. F. Reynolds, trans. Ed. (New York: John Wiley & Sons, Inc., 1976).
10. Roe, D. K., and P. Eggiman. *Anal. Chem.* 48: 9R (1976).
11. Roe, D. K. *Anal. Chem.* 50: 9R (1978).
12. Kolthoff, I. M., and J. J. Lingane. *Polarography* (New York: Wiley-Interscience, 1948), pp. 163–164.
13. Franke, J. P., and R. A. de Zeeuw. *Anal. Chem.* 50: 1374 (1978).
14. Brezonik, P. L., P. A. Brauner and W. Stumm. *Water Res.* 10: 605 (1976).
15. Abdullah, M. I., B. R. Bert and R. Klimek. *Anal. Chim. Acta* 84: 307 (1976).
16. Blutstein, H., and A. M. Bond. *Anal. Chem.* 48: 759 (1976).
17. Hem, J. D. "U. S. Geological Survey Water-Supply Paper 1473," U. S. Government Printing Office, Washington, DC (1970).
18. Butler, J. N. *Ionic Equilibrium, A Mathematical Approach* (Reading, MA: Addison-Wesley Publishing Co., Inc., 1964).
19. DeFord, D. D., and D. N. Hume. *J. Am. Chem. Soc.* 73: 5321 (1951).
20. Lind, C. J. *Environ. Sci. Technol.* 12: 1406 (1978).
21. Heineman, W. R., and P. T. Kissinger. *Anal. Chem.* 50: 166R (1978).
22. Jehring, H. *Electrosorptionanalyse mit der Wechselström Polarographie* (Berlin: Akademie Verlag, 1974).
23. Erdey-Gruz, T. *Kinetics of Electrode Processes* (Budapest: Hungarian Academy of Sciences, 1975).
24. Vielstich, W., and W. Schmickler. In: *Electrochemie II: Kinetik Ekektrochemische Systeme, Grundzuge der Physikalischen Chemie, Band VI,* R. Haase, Ed. (Darmstadt: Steinkopf, 1976).
25. Albery, J. "Electrode Kinetics," *Oxford Chem. Ser.* 14 (1975).
26. Conway, B. E. *Chem. Can.* 28: 28 (1976).
27. Conway, B. E., and H. Angerstein-Kozlowska. *Nat. Bur. Stand. (U.S.), Spec. Publ.* 455: 107 (1976).
28. Ellis, W. D. *J. Chem. Ed.* 50: A131 (1973).
29. Flato, J. D. *Anal. Chem.* 44(11): 75A (1972).
30. Meyer, M. L., T. P. DeAngelis and W. R. Heineman. *Anal. Chem.* 49: 602 (1977).
31. Moody, J. R., and R. M. Lindstrom. *Anal. Chem.* 49: 2264 (1977).
32. Branica, M., L. Sipos, S. Bubic and S. Kozar. *Nat. Bur. Stand. (U.S.), Spec. Publ.* 422: 917 (1976).
33. Watling, H. R., and R. J. Watling. *Water SA* 1: 28 (1975).
34. Riley, J. E., Jr. *Anal. Chem.* 50: 541 (1978).
35. Princeton Applied Res. Corp. *Instruction Manual, Polarographic Analyzer Model 174A,* Princeton, NJ (1974).
36. Barker, G. C. *Proc. Cong. Modern Anal. Chem. in Ind.,* St. Andrews, 199 (1957).
37. Barker, G. C., and A. W. Gardner. *Z. Anal. Chem.* 173: 79 (1960).
38. Osteryoung, J., and K. Hasebe. *Rev. Polarog.* 22: 1 (1976).
39. Burge, D. E. *J. Chem. Ed.* 47: A81 (1970).
40. Clem, R. G. *MPI Applications Notes* VIII: 1 (1973).

41. Miaw, L. L., P. A. Boudreau, M. A. Pichler and S. P. Perone. *Anal. Chem.* 50: 1988 (1978).
42. Blutstein, H., and A. M. Bond. *Anal. Chem.* 48: 248 (1976).
43. Perone, S. P., and T. R. Mueller. *Anal. Chem.* 37: 2 (1965).
44. Zirino, A., and M. L. Healy. *Environ. Sci. Technol* 6: 243 (1972).
45. Buffle, J., F. L. Greter, G. Nembrini, J. Paul and W. Haerdi. *Z. Anal. Chem* 282: 339 (1976).
46. Barendrecht, E. In: *Electroanalytical Chemistry,* Vol. 2, A. J. Bard, Ed. (New York: Marcel Dekker, Inc., 1967).
47. Velghe, N., and A. Claeys. *J. Electroanal. Chem.* 35: 229 (1972).
48. Matson, W. R., D. K. Roe and D. E. Carritt. *Anal. Chem.* 37: 1594 (1965).
49. Florence, T. M. *J. Electroanal. Chem.* 26: 293 (1970); 27: 273 (1970).
50. Lund, W., and M. Salberg. *Anal. Chim. Acta* 76: 131 (1975).
51. Opekar, F., and P. Beran. *J. Electroanal. Chem. Interfacial Electrochem.* 69: 1 (1976).
52. Bruckenstein, S., and B. Miller. *Acc. Chem. Res.* 10:54 (1977).
53. Covington, J. R., and R. J. Lacoste. *Anal. Chem.* 37: 420 (1965).
54. Parry, E. P., and R. A. Osteryoung. *Anal. Chem.* 37: 1634 (1965).
55. Birke, R. L. *Anal. Chem.* 50: 1489 (1978).
56. Roe, D. K., and J. E. A. Toni. *Anal. Chem.* 37: 1503 (1965).
57. deVries, W. T. *J. Electroanal. Chem.* 9: 448 (1965).
58. Reinmuth, W. H. *Anal. Chem.* 33: 185 (1961).
59. Brainina, Kh. Z. *Talanta* 18: 513 (1971).
60. Roizenblat, E. M., and Kh. Z. Brainina. *Electrokhimia* 5: 396 (1969).
61. Yarnitsky, C., and M. Ariel. *J. Electroanal. Chem.* 10: 110 (1965).
62. Dennis, B. L., G. S. Wilson and J. L. Moyers. *Anal. Chim. Acta* 86: 27 (1976).
63. Jacobsen, E., and H. Lindseth. *Anal. Chim. Acta* 86: 123 (1976).
64. Shuman, M. S., and G. P. Woodward, Jr. *Anal. Chem.* 48: 1979 (1976).
65. Abdullah, M. I., O. A. El-Rayis and J. P. Riley. *Anal. Chim. Acta* 84: 363 (1976).

CHAPTER 8

INSTRUMENTATION FOR DETECTION AND MEASUREMENT OF IONIZING RADIATION

W. L. Beck, Jr.

Oak Ridge Associated Universities

C. D. Berger

Industrial Safety and Applied Health Physics Division
Oak Ridge National Laboratory

J. D. Berger

Oak Ridge Associated Universities
Oak Ridge, Tennessee

R. E. Goans

Waldorf, Maryland

INTRODUCTION

Fundamental principles of radiation detection have been known for many years. During approximately the first 40 years of the twentieth century, there were limited applications of these principles in areas of medicine (X-rays and radium), geophysics (cosmic and naturally occurring terrestrial radiation) and basic physics research. Then, in the early 1940s, the nuclear age was born and the operational control, safety and health protection problems associated with this energy source required new and improved radiation monitoring technologies. As a result, during the past 30 to 40 years the field of radiation monitoring instrumentation has experienced tremendous growth and development and, today, there are few physical agents that can be detected and quantified with such sensitivity and accuracy as is possible with ionizing radiation.

Radiation cannot be detected with the human senses. Therefore, one must use some device that can respond to the radiation, coupled with a system to measure the magnitude or intensity of that response. Many types of devices or instruments have been used, both to detect, as well as to measure, radiation. The majority of these make use of the ionization produced in them, for example, gas-filled detectors and semiconductor materials. Others, such as scintillation crystals, chemical dosimeters and solid-state materials, depend on excitation or changes in molecular structure that affect certain physical properties of the material.

Devices may incorporate electronic circuitry to provide continuous and immediate information, or they may be of a passive nature, integrating some chemical or physical effect, to be analyzed later by separate instrumentation. The physical characteristics and performance criteria of any given device are dictated by its intended application, e.g., primary standard, secondary standard, process control, laboratory analysis, field monitoring, etc.[1,2].

This chapter discusses the basic principles of detection instruments and devices, and provides representative examples of a few of the many commercially available items. Equipment illustrated here is of a very limited selection. For a more complete review the reader is referred to the technical catalogs and other information available from the manufacturers and vendors of radiation monitors[3]. The display or description of a specific product is not to be construed as an endorsement of that product or its manufacturer by the authors or their employers.

A brief glossary of radiation monitoring terms and units used in the text has been included as an appendix to this chapter. It should be emphasized that, as with any sophisticated monitoring devices, interpretation and application of radiation measurements requires special knowledge and should not be attempted by those untrained in this profession.

GAS-FILLED DETECTORS

Theory

The gas-filled detector is one of the oldest and most widely used devices for detection and measuring ionizing radiation. This type of detector consists basically of two electrodes at different electric potentials, separated by an appropriate counting gas. Interaction of radiation with detector components or the gas itself may result in the formation of one or more ion pairs within the gas. These ions are collected at the electrodes, producing currents or pulses that are related to the amount or intensity of the incident radiation.

A variety of configurations of gas-filled detectors is possible, depending on the particular application. The general design to be used in the following simplified discussion on gas detector theory, is a positively charged small-diameter, center electrode (anode) within a negatively charged cylindrical electrode (cathode) (Figure 1).

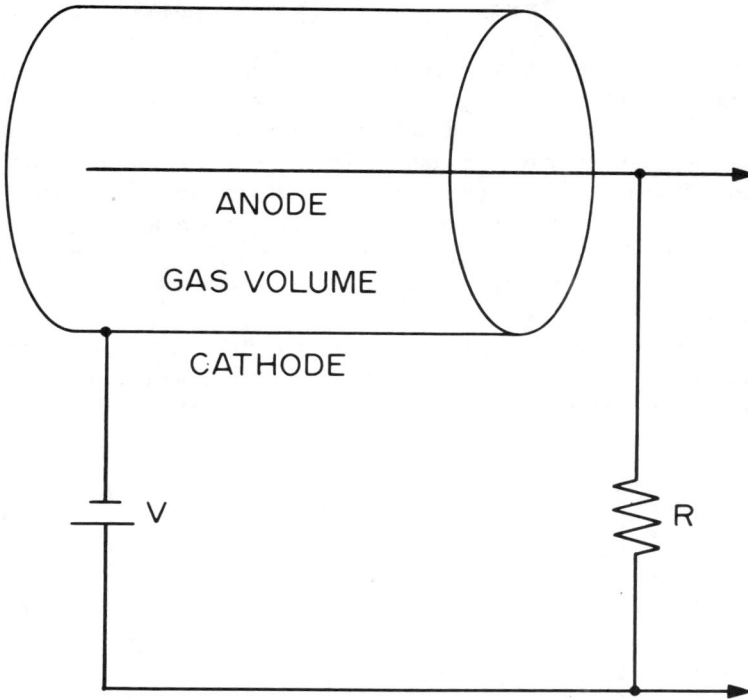

Figure 1.   Diagram of a simplified gas-filled detector.

Gas Amplification

When a potential difference is applied to the detector electrodes, an electrical field is produced within the gas volume. The intensity, E, of this field at a distance, r, from the anode is directly related to the potential, V, and the radii of the anode, r_a, and the cathode, r_c, by the following equation:

$$E = \frac{V}{r \log \dfrac{r_c}{r_a}}$$

Electrons produced in the ionization process migrate toward the positive anode due to the electrical field. As the electron accelerates, it attains a velocity that is a function of the gas characteristics and the strength of the electrical field. Since the electrical field intensity increases nearer the anode, the electron will continue to accelerate, developing a kinetic energy,

$$E_k = 1/2\,m_o\,v^2$$

where m_o is rest mass of the electron and v is its velocity.

During their migration, electrons will collide with gas molecules. If the electrode potential is low, the velocity attained by electrons is small and they may combine with positive ions, other molecules or atoms before reaching the anode, a process known as recombination. As the potential is increased, the migration time decreases and a greater fraction of the initially produced ions reach the electrodes. Eventually, one electron is collected for each initially produced ion pair. The voltage range throughout which this occurs is called the ionization region (Figure 2).

At voltages above the ionization region, the electrical field intensity continues to increase. The kinetic energy acquired by the electrons, particularly in the immediate vicinity of the anode, is great enough to cause the accelerated electrons to

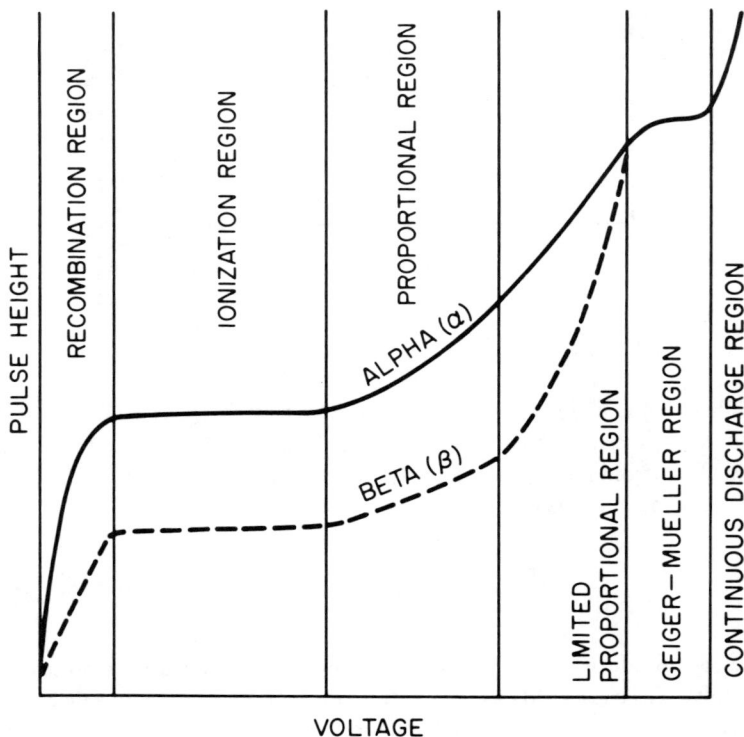

Figure 2.   Pulse height versus electrode voltage, illustrating various regions of operation for gas-filled detectors.

ionize the filling gas molecules. This results in a chain reaction or avalanche effect near the anode, producing further amplification in the number of electrons collected. The number of electrons collected is proportional to the initial number of ionizations in the detector gas. This voltage range is called the proportional region. In this region, the amplification factor, i.e., the ratio of the number of electrons collected to the number of initial ionizing events, may be as high as 10^5–10^6.

At voltages above the proportional region there is a voltage range called the limited proportionality region. This is a transition region and is not useful for detection operation.

Above the limited proportionality region, collisions near the anode produce secondary ionizations, and the avalanche spreads along the entire length of this electrode. A saturation current is reached as all of the ions along the anode are collected. The system is now operating in the Geiger-Müeller (GM) region, where gas amplifications of approximately 10^8 are attained. In this region, the initial ion pair acts only as a triggering mechanism to initiate the avalanche; the total electron current is a function of the counter characteristics (voltage, dimensions, counting gas, etc.) and is independent of the number of initial ion pairs formed.

Above the GM region, the electrode potential is sufficient to cause spurious or continuous discharges in the gas, thus operation of gas-filled detectors above the GM region is impractical. A more detailed discussion of gas detector theory is available in texts dealing with radiation instrumentation[4-7].

Pulse Behavior

A collection of electrons at the anode produces a voltage change or pulse that is a function of the total charge collected, as well as the capacitance of the anode and associated circuitry (Figure 3). This pulse is characterized by a very rapid voltage increase as the electrons are collected, followed by a longer exponential decay due to circuit impedances and the migration of the heavier positive ions to the cathode.

While the positive ion cloud is in the vicinity of the anode, immediately following an avalanche, electrons that are produced by further ionizing events are not able to reach the anode. Consequently, these events will not be recorded. The period of time during which this condition exists is the dead time of the detector (the time necessary for the migrating positive ion sheath to travel to what is termed the critical radius of the chamber). As the positive ions pass beyond this critical radius, the sensitivity of the detector to further ionizations increases until the detector is capable of registering another event. This period between the dead time and the time of returning sensitivity is the recovery time. The sum of the dead time and the recovery time is the shortest time that can elapse before consecutive events can be detected and recorded. This period is known as the resolving time.

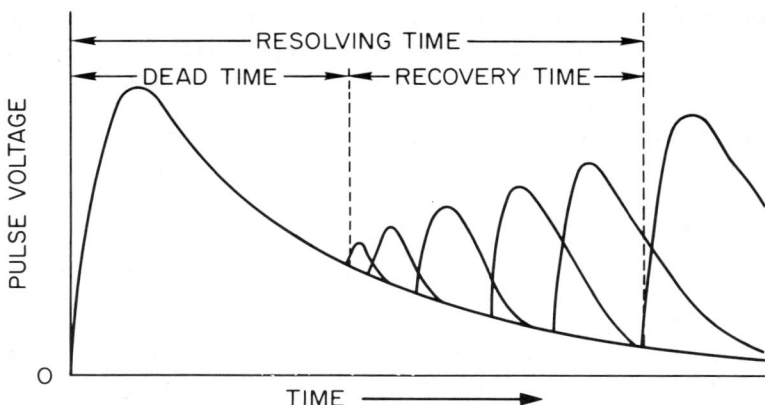

Figure 3. Graphic representation of a typical gas detector output pulse showing recovery, dead and resolving times.

For a specific detector, an increase in electrode voltage causes production of a pulse of sufficient size that it can be recognized by an electronic counting system. This is referred to as the threshold voltage. The counting rate increases rapidly with voltage above the threshold until the voltage is such that any primary ionizations will result in an output pulse of sufficient triggering size. The detector is now operating in its plateau region (Figure 4). Small increases in voltage will increase the pulse size but will produce practically no increase in the count rate. The length and slope of the plateau are determined by the nature of the detector design and filling gas. Typical values of these characteristics would be a length of a few hundred volts, with a slope of 5% per 100 volts. Detectors normally are operated at approximately the mid-point of the plateau, where a slight change or drift in electrode voltage will produce a minimum change in the counting rate. Above the plateau region, the count rate rises rapidly with increased voltage until the detector goes into a continuous discharge.

Quenching

With the arrival of a positive ion at the cathode, an electron from the cathode surface is attracted to the ion to produce a neutralized atom. Usually this electron enters the atom at an elevated energy level, and an energy transition to the ground state occurs with the resulting emission of a low-energy photon. This photon may be of sufficient energy to eject a photoelectron from the cathode material. If this photoelectron enters the sensitive gas volume, a discharge will be reinitiated, and the detector will cease to be functional as a radiation counter. This phenomenon

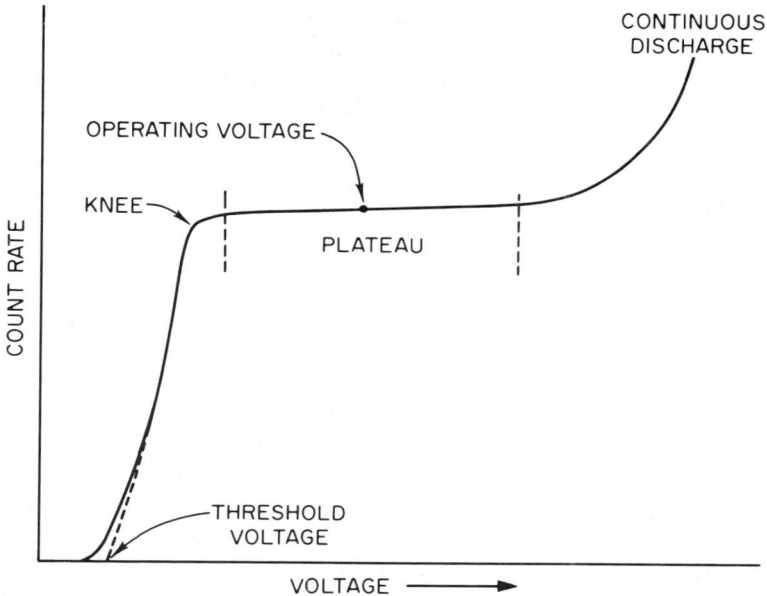

Figure 4. Representative gas-detector voltage plateau curve.

occurs primarily in detectors operated in the GM region. Elimination of this self-perpetuating discharge, permitting return of the detector to a stable state, is termed quenching.

Quenching may be accomplished by electronic circuits, which decrease the detector voltage after each discharge to a value such that gas amplification cannot occur. The more common method, however, is to add small amounts of polyatomic or halogen gases to the counting gas. These quenching agents have low ionization potentials, broad and intense absorption bands in the low-energy photon region and, when excited, dissociate rather than emit radiation. The positive ions of the main counting gas are then neutralized by electrons obtained from the quenching agent. Because of the large number of collisions that occur during travel to the cathode, the positive ion sheath reaching this electrode is composed almost entirely of quenching gas ions. At the cathode, neutralization of these positive quenching agent ions occurs. The excited molecules dissociate and the detector discharge is terminated. There is a possibility that some ions of the counting gas may reach the cathode and produce a repetitive discharge. As the electrode voltage increases, this possibility also increases, and above the GM region it becomes almost a certainty. This is a major source of continuous discharge.

Since some of the quenching agent dissociates with each counting event, detectors using a polyatomic gas as the quenching agent have a limited useful life. Halogen quenching agents recombine after dissociation; therefore, loss is negligible and the detector life is theoretically infinite.

Applications

Typical operating characteristics of gas-filled detectors are summarized in Table I, with specific applications provided as follows.

Ionization Instruments

Ionization chambers are versatile in the detection and measurement of all types of ionizing radiation. The usual electrode configuration is either cylindrical or parallel-plate. The latter, operated in the pulse mode, is used for pulse height analysis of densely ionizing radiation, such as alpha particles. Use in this mode is limited at low particle energies due to inherent electronic noise. Parallel-plate configurations are also employed as primary standard instruments.

The cylindrical ionization chamber, the more common geometry, is used in the current mode for measuring photon radiation levels. With special thin walls, the ion chamber may also be used to detect alpha or beta particles. In this mode, the electron current is a direct measure of the ionization occurring in the gas volume. The counting gas is frequently dry air, although other gases may be chosen for particular measurement applications. Gas fillings are usually maintained at atmospheric pressure, but may be considerably higher to provide greater detection efficiency. Figures 5 through 8 show several examples of ionization instruments.

Proportional Instruments

Proportional instruments find application primarily in charged particle (alpha and beta) counting and spectrometry. The large difference in the specific ionization produced by alpha and beta particles permits pulse size discrimination between these radiations. The detector shape may be rectangular, cylindrical or hemispherical. Sizes vary widely, depending on the intended application.

Proportional counters must have thin windows to permit passage of particulate radiations. These windows are usually aluminized Mylar,® with density thickness of 1–1.5 mg/cm². Some detectors are of the windowless type, where the sample is placed inside the sensitive gas volume. Applications are illustrated in Figures 9 through 12.

Table I.   Summary of Typical Gas-Filled Detector Characteristics

Type Detector	Operating Voltages	Amplification Factor	Principal Radiation Detected	Counting Gas	Approximate Dead Time (μsec)	Applications
Ionization	50–300	1	beta, gamma, X-ray	Air	N.A.	Portable survey instruments, standardization and calibration of source and field intensities, spectrometry
Proportional	1000–2500 (high stability required)	10^4–10^6	alpha, beta	Argon/methane (90%/10%) air, propane	1–2	Portable survey instruments, lab sample counters, spectrometry, source standardizations
Geiger-Mueller	400–1600	10^7–10^9	beta, gamma, X-ray	Noble gases (He, Ne, A) with halogen or polyatomic quenching agents	100–300	Portable survey instruments, lab sample counters, area monitors

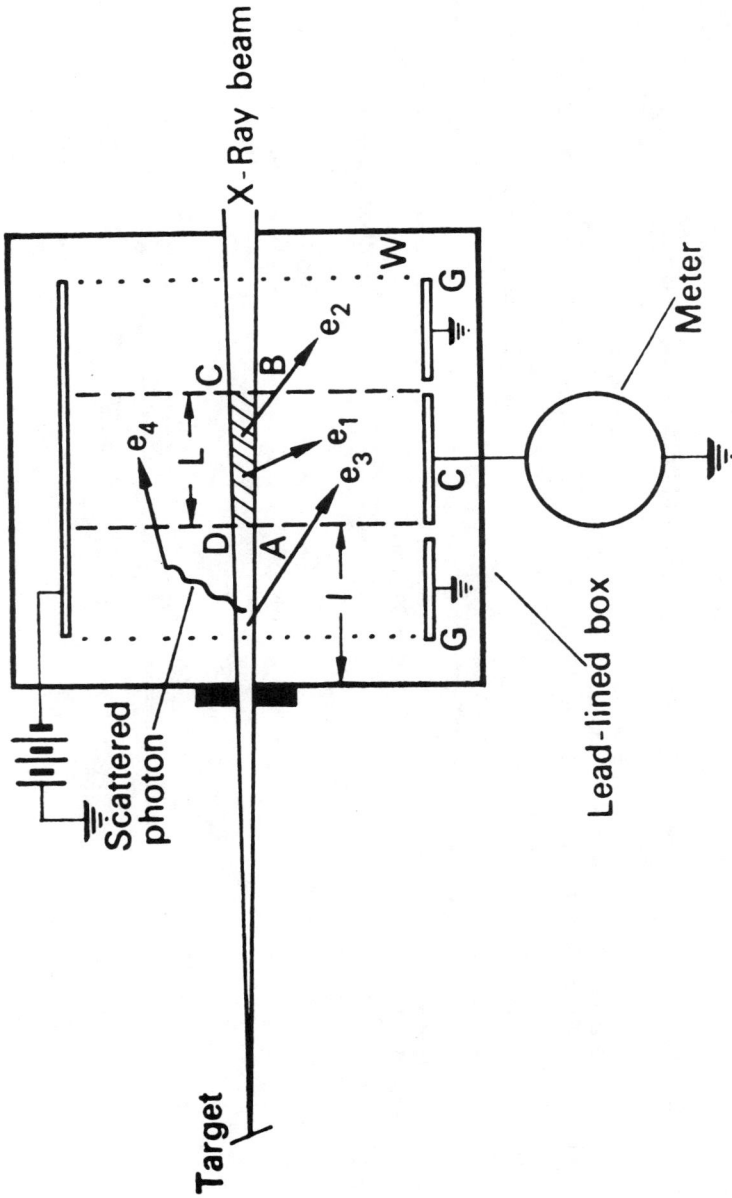

Figure 5. Free air chamber used as a primary standard for measurements of photon exposure. To obtain electronic equilibrium that is representative of photon interactions with air, the sensitive gas volume is bounded by an air "wall," thus the instrument is also known as an air-wall chamber. The counting gas is air at atmospheric pressure.[8]

Figure 6. Medical dose calibrator. Radionuclides intended for medical diagnosis or therapy are placed inside the ion chamber well to verify their activity prior to patient use. Modifying factors are selected to convert the chamber current to digitally displayed units of activity.

Geiger-Mueller Instruments

Geiger-Mueller counters are sensitive radiation detectors for both laboratory and survey meter applications. Since pulse sizes are independent of the initial ionizations produced, GM instruments cannot be used for spectrometry. The energy dependence at lower energies prevents accurate intensity measurements of photon radiation of mixed or unknown energy. This type of instrument, however, is widely used for the detection of beta particles and photons and for determining particle emission rates under controlled or known conditions of geometry and energy. The GM detector (or tube) is often cylindrical and relatively small compared to other gas-filled detectors. Tube walls are typically 30 mg/cm² thick,

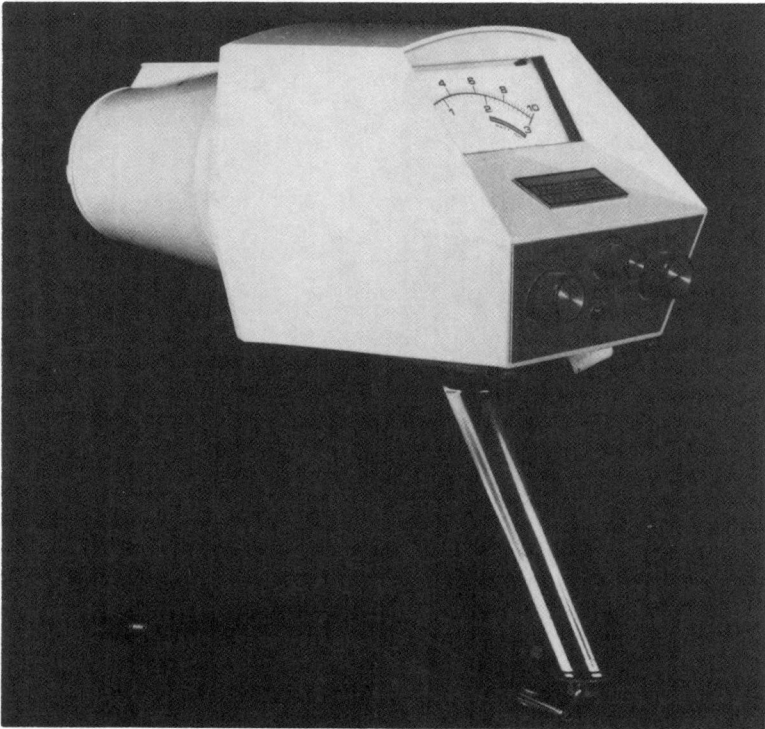

Figure 7.   Portable survey meter, used primarily to measure photon exposure rates. The chamber has a removable shield to permit detection of particlate radiation (beta and alpha); the meter indication for these radiations, however, is not quantitatively correct. This style of meter, known as a "cutie-pie" (abbreviated $QT\pi$) uses an air chamber at atmospheric pressure. The maximum range of this meter is 1000 R/hr.

thus preventing passage of alpha particles and low-energy (<200 keV) beta particles into the sensitive gas volume. This tube is encased in a heavier metal shield with a rotating or sliding opening. Use of this shield permits further discrimination of beta particle energies and allows limited differentiation between the penetrating and nonpenetrating radiations.

Thin mica or mylar coverings of 1–2 mg/cm^2 are used on some GM tubes to permit detection of alpha and beta particles of energy down to approximately 30 keV. These detectors are called end-window tubes and, in addition to being directionally dependent, are relatively fragile.

Several GM instruments are shown in Figures 13 through 17.

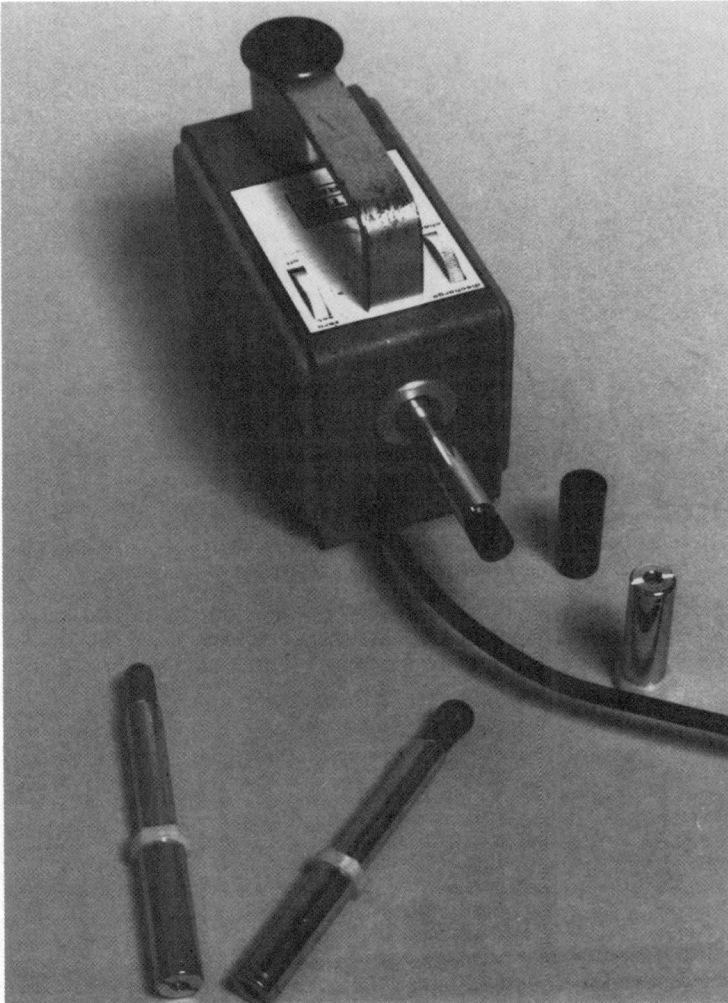

Figure 8.   Condensor-ionization chamber. This device is used to integrate radiation exposure over a predetermined time for the purposes of calibrating radiation-emitting equipment (e.g., teletherapy or radiography machines). The chamber is charged to several hundred volts. Because of its good insulating features, it acts as a capacitor to maintain that charge. As the device is exposed to radiation, neutralization of the ions produced inside the air chamber reduces the electrode charge. The change in voltage is measured by a reader-charger instrument, which converts this charge to an accumulated exposure. A variety of chamber sizes is available, depending on the energy and intensity of the radiation to be measured.

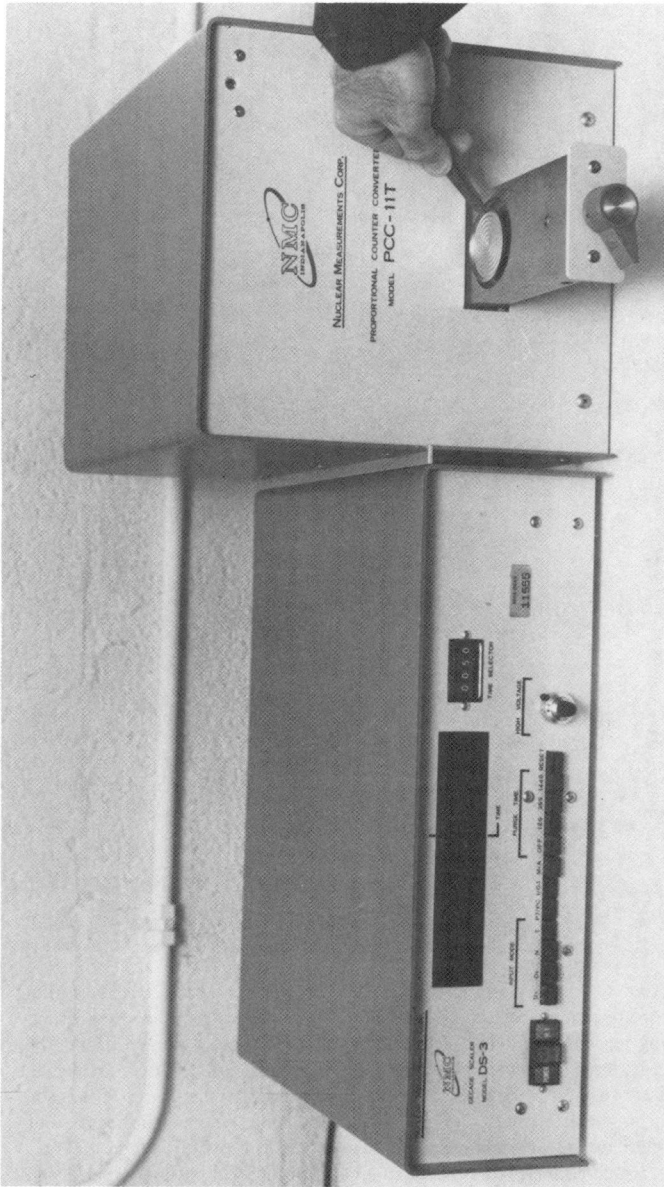

Figure 9.   Laboratory sample counter. This counter utilizes an argon-methane mixture (P-10 gas), which flows through the chamber at a pressure just slightly above atmospheric. The chamber itself is a hemisphere containing a small wire loop as the anode. Primary applications of this instrument are determinations of alpha and/or beta activity in environmental samples. Special sample preparation is required in most cases for accurate results. The main advantage of this type instrument is its ability to distinguish between alpha and beta radiation, based on relative pulse sizes.

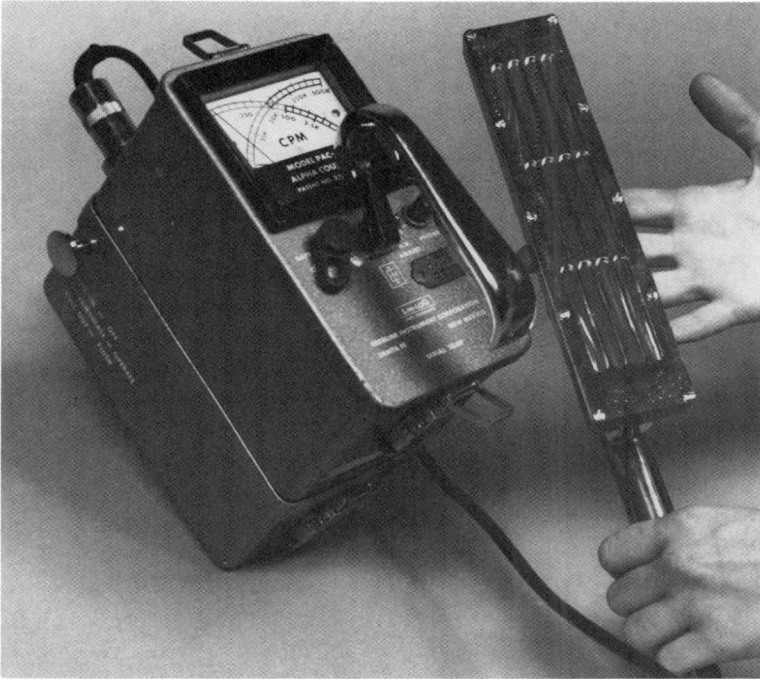

Figure 10.   Portable alpha survey meter. Propane is the counting gas used for this alpha survey meter. A small cylinder of propane is housed in the lower section of the instrument case. The instrument is operated as a gas-flow counter. A thin, aluminized Mylar probe window permits passage of alpha particles of commonly encountered energies. The meter has a logarithmic scale with a maximum range of 5×10^5 cpm. An audible output may be obtained by use of a speaker unit or headphones. There is an alternate probe available for monitoring beta radiation with this instrument.

SCINTILLATION DETECTORS

Theory

A scintillation phosphor is a material able to convert energy deposited in it by ionizing radiation into pulses of light. In most scintillation counting applications, the ionizing radiation of interest is in the form of gamma rays, alpha particles or beta particles, ranging in energy from a few thousand to a few million electron volts (keV and MeV, respectively). There are four general types of scintillators used for photon and charged particle detection: organic crystals, inorganic crys-

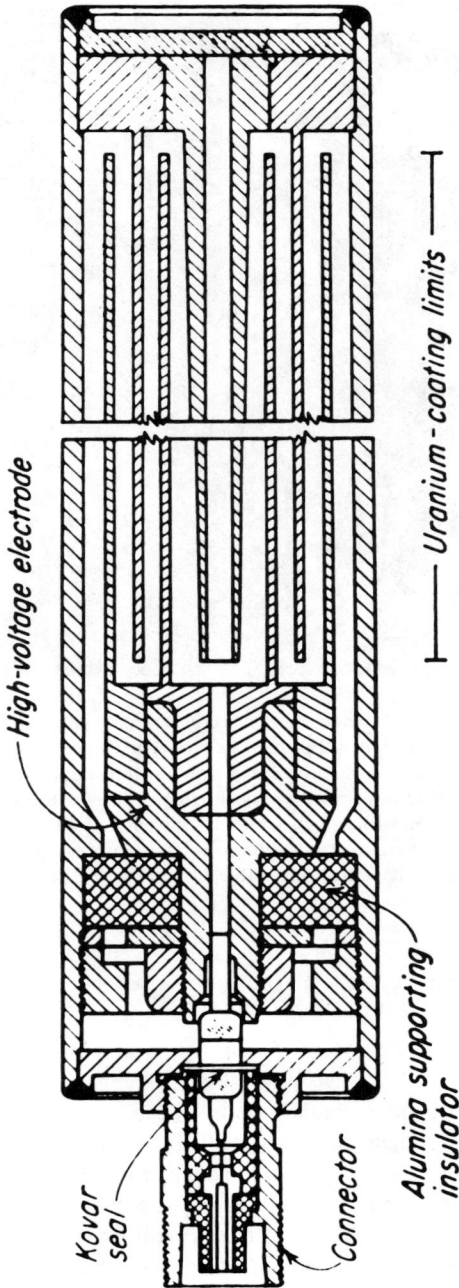

Figure 11. Fission chamber. Fissionable materials, incorporated into the chamber in the form of a foil, interact with thermal neutrons to cause fission. The fission fragments produce ionization of the chamber gas. The large amount of energy released per reaction makes it possible to detect neutrons in mixed neutron-gamma field. This detector utilizes U-235 to monitor thermal neutrons. Pu-239 and U-233 may also be used for thermal neutron measurement and U-238 and Np-237 for fast neutrons. The efficiency of these detectors is related to the quantity of fissionable material used in the chamber. This detector type has a useful range from approximately 10 to 10^5 neutrons/cm^2/sec and is primarily used for continuous monitoring of nuclear reaction levels[9].

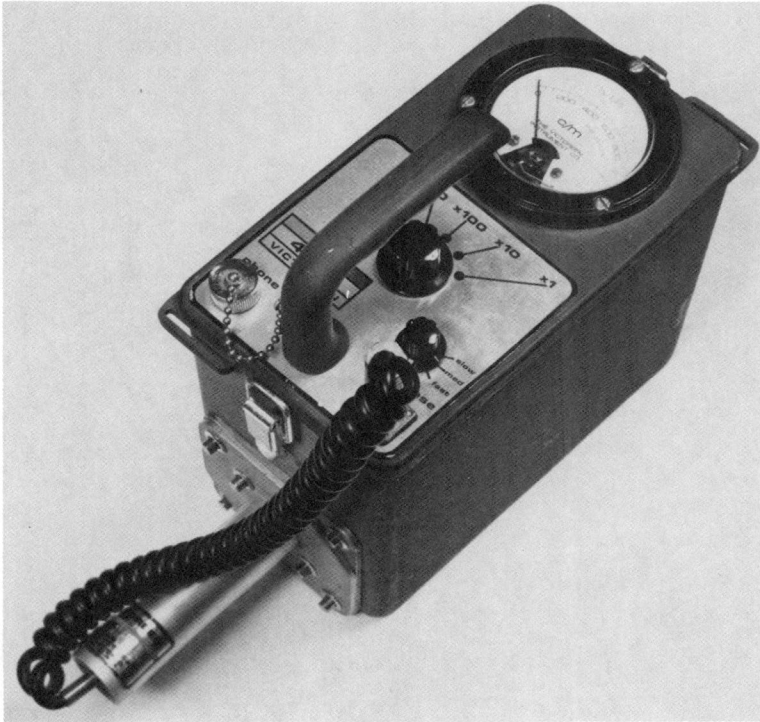

Figure 12. BF_3 neutron survey instrument. The detector in this instrument contains boron trifluoride (BF_3) gas, which has a high cross section for interaction with thermal neutrons. This interaction causes the boron to disintegrate into an alpha particle and a lithium-6 ion, both of which are detected through the particle detection processes described previously. The BF_3 tube can be placed inside a hydrogenous moderator to permit monitoring of higher energy neutrons. This instrument readout is generally neutrons/cm^2/sec.

tals, liquids and plastics. Desirable characteristics of scintillators include high density for greater stopping power, large pulse height for detection of low-energy interactions, short light decay time for fast counting applications, and mechanical durability for fabrication purposes. Table II gives selected properties for the common inorganic crystals and typical values for plastic and liquid scintillators.

Small quantities of impurity atoms are added to scintillator material to create luminescence centers. These impurities are known as activators. The interactions of radiation within the scintillator produce electron vacancies (holes) and free electrons, which move through the scintillator until they are captured at activator sites. The impurity atoms are thereby raised to excited states. If a transition

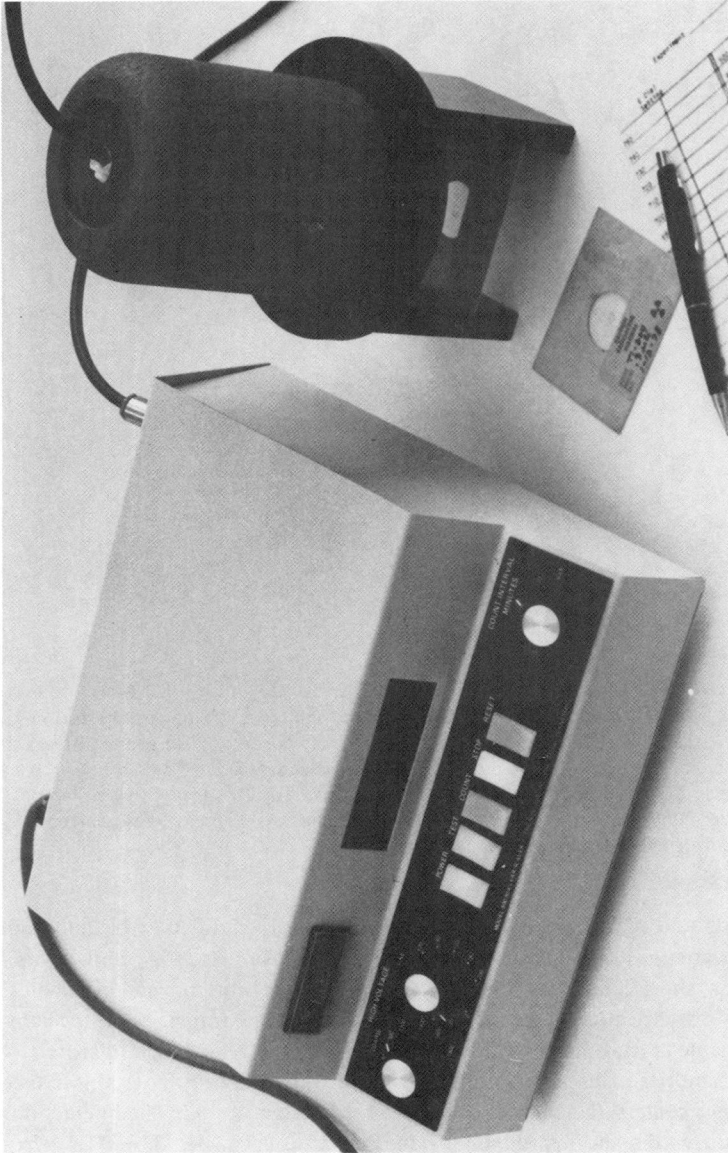

Figure 13. Laboratory sample counter, used for determining beta-gamma activity in samples collected from air, water, surfaces, etc. Samples may require special preparation for optimum counting yield. This counter has an end-window GM tube, which is placed inside a lead shield to reduce background interference. A variable-position sample holder is provided.

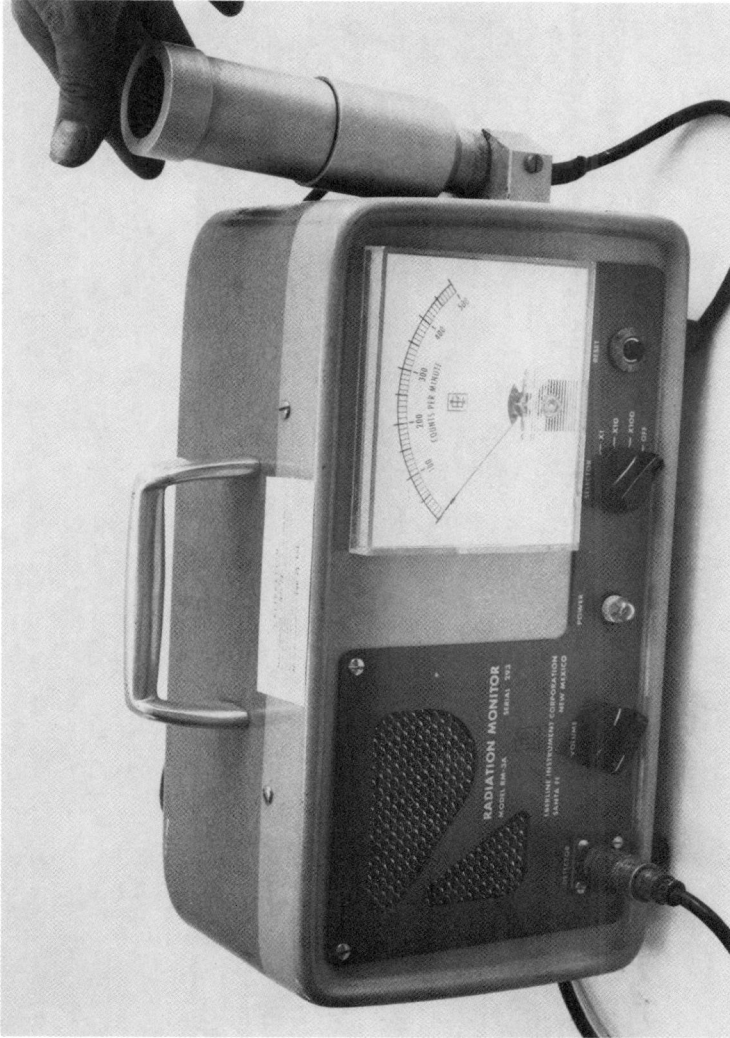

Figure 14. Area monitor, which consists of an end-window GM detector connected to a count rate meter. The instrument has an audible output for easier detection of changes in count rates and a preset alarm. Area monitors are used to provide a continuous indication of general area radiation levels or to monitor personnel or equipment for surface contamination or radioactive material content.

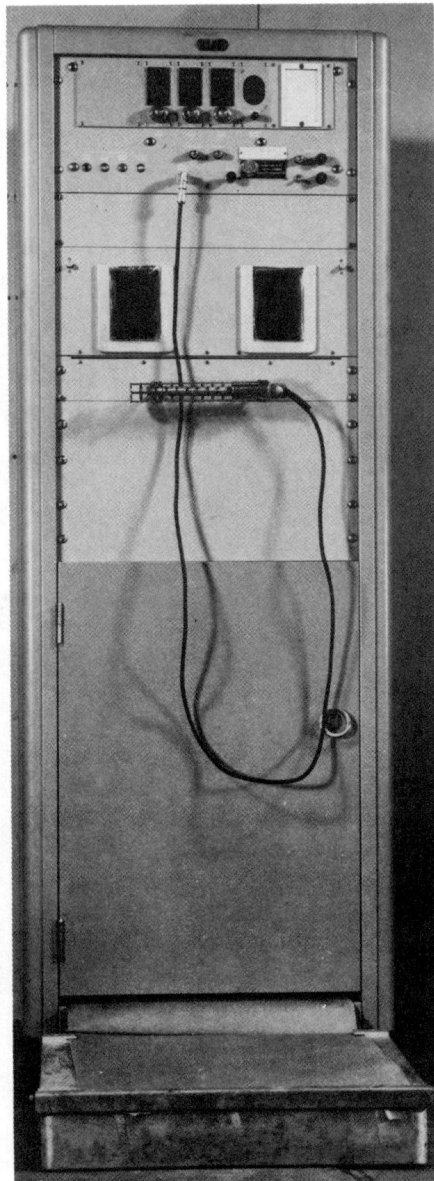

Figure 15.   Hand and foot monitor. This instrument is designed for monitoring hands and feet for contamination. It is an extension of the area monitor concept, employing separate GM tubes and readouts for monitoring various body regions.

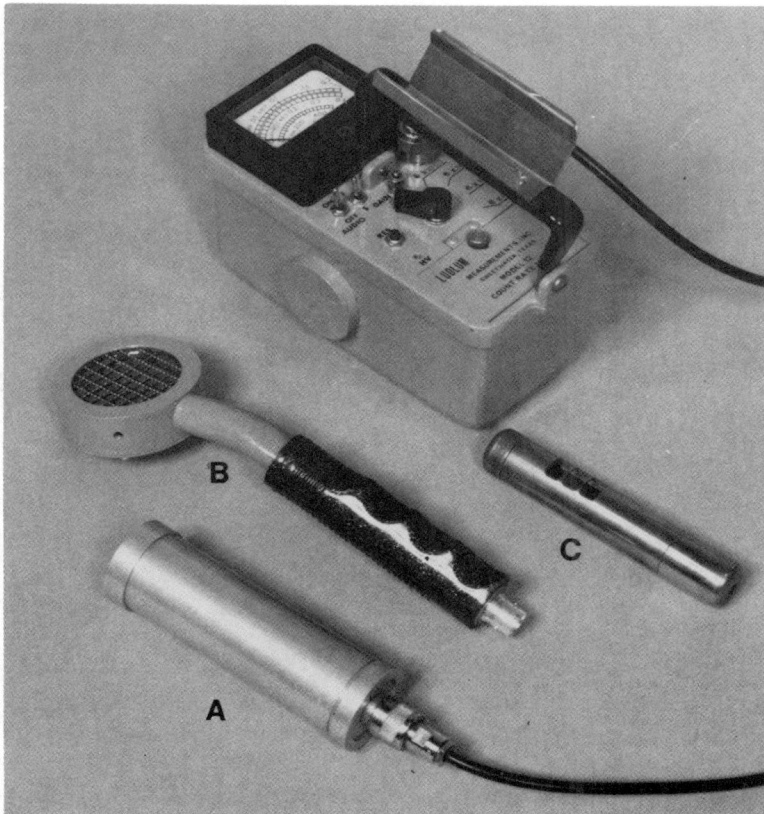

Figure 16. Portable GM survey meter, used for the detection of beta and gamma radiation and, if calibrated for the energies being monitored, to measure photon exposure levels. Several types of GM tubes can be used with this meter. Meter units are in both mR/hr and cpm. A typical upper range is 50 mR or 500,000 cpm. An audible output may be provided for prompt indication of count rate variations. Three types of detector probes are shown: A—Thin end window; B—Pancake end window; and C—Side window.

occurs between an excited state and the ground state, a photon is emitted. Thus, a portion of the original energy of the radiation that passed through the scintillator is converted to light.

The light pulse emitted from the scintillator is converted to an electron pulse by use of a photomultiplier (PM) tube (Figure 18). This tube has a photocathode that is sensitive to light in the energy region emitted by the scintillator. It produces a number of electrons per light pulse that is proportional to the intensity of the

Figure 17.   Extension probe survey meter, which has a small GM tube for monitoring levels up to 1000 R/hr. The telescoping probe enables the user to maintain a distance of approximately 10 feet from the detector during monitoring.

Table II.   Properties of Common Scintillators

Material	Wavelength of Maximum Emission (nm)	Decay Constant (μsec)	Density (g/cm^3)
NaI(Tl)	410	0.23	3.67
CaF$_2$(Eu)	435	0.9	3.19
CsI(Na)	420	0.63	4.51
CsI(Tl)	565	1.0	4.51
^{6}LiI(Eu)	470	0.94	3.49
Plastics	350–450	0.002–0.020	1.06
Liquids	350–450	0.002–0.008	0.86
ZnS	450	0.04–0.4	4.1
Anthracene	440	0.03	1.25

light. The photocathode is followed by a series of dynodes, with a positive potential gradient. Each electron, emitted by the photocathode and striking the first dynode, will cause additional electrons to be emitted and accelerated toward the second dynode. There, each accelerated electron will, in turn, produce several more electrons. This process is continued through several stages. The electrons are finally collected at the anode of the photomultiplier tube as a voltage pulse. Detailed discussions of scintillation theory and applications are available elsewhere[4, 11-13].

Applications

Typical scintillation instrument characteristics are summarized in Table III. Descriptions of specific instrument applications are provided in Figures 19 through 24.

SEMICONDUCTOR DETECTORS

Theory

The semiconductor radiation detector behaves analogously to the gas ionization chamber, except that the charge is carried by electrons and electron vacancies (holes), instead of by electrons and positive ions. The mean energy required to produce an ionizing event in a semiconductor material is approximately one-tenth

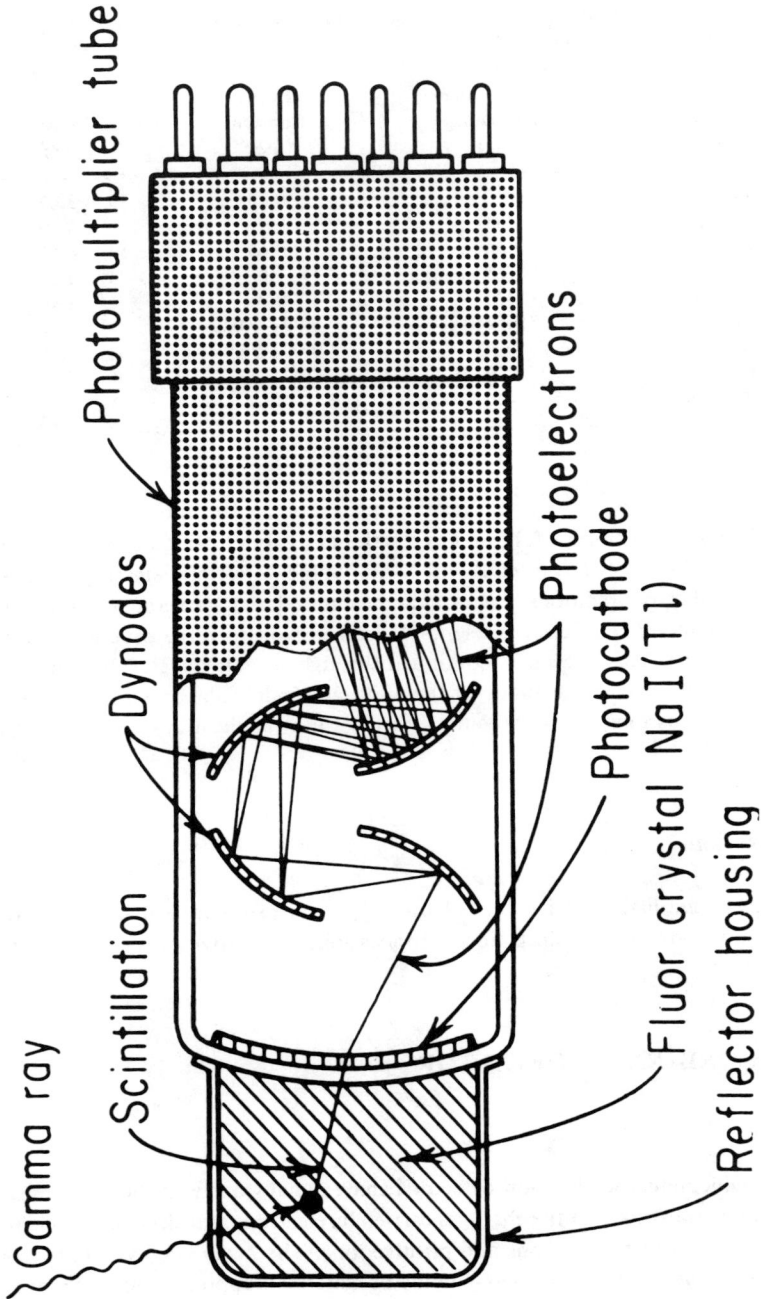

Figure 18.   Diagram of a scintillation photomultiplier tube assembly[10].

Table III.   Summary of Typical Scintillation Instrument Characteristics

Scintillator	Principal Radiations Detected	Application	Remarks
NaI	Gamma, X-ray	Gamma spectroscopy, whole-body counting, medical imaging	Efficiency dependent on photon energy and crystal size
ZnS	Alpha	Portable survey meter, lab sample counter	Low transparency but high efficiency of conversion, low background
Anthracene	Beta	Lab instrument	Slow response
NE213	Neutron, gamma	Portable survey meter, neutron spectrometry	Good gamma discrimination
Liquid Scintillators	Beta, low-energy gamma	Lab instrument	Poor energy resolution, uniform geometry

that required in most counting gases. This property makes the use of a solid as a detector very attractive because the sensitive layer can be very thin, yet possess a high stopping power.

Semiconductor materials used as radiation detectors are crystals in which the atoms are joined together by covalent bonding of their valence electrons. Ionization results in the disruption of these bonds, producing electrons and holes that are free to migrate through the crystal lattice. These electrons and holes can be collected by placing a potential difference across the crystal; however, the available electron current from pure semiconductor material is small. It must be increased to make the semiconductor practical as a radiation detector. This increase is accomplished by adding impurities to the lattice structure. These impurities are referred to either as donors or acceptors if they have an excess of electrons or holes, respectively. Semiconductor material containing donor impurities is called n-type. It if contains acceptor impurities it is called p-type.

If p-type material is in contact with n-type material, the electrons from the n-type material migrate across the junction and fill the holes in the p-type material to create a depleted region known as an n-p junction. This depletion region acts as the sensitive volume of the detector.

Two general types of semiconductor detectors are used, depending on the application. The primary differences between them are the physical techniques used to form the n-p junction and the depleted region. One of these is the diffused-junction detector. It has a large depletion region where the number of donors equals the number of acceptors. This is generally accomplished by diffusing

Figure 19.   Gamma well-detector, a commonly used laboratory instrument for measuring the activity of samples containing low levels of gamma-emitting radionuclides. It is basically a NaI crystal and PM tube assembly with a hole in the crystal for insertion of a sample. It has high detection efficiency (especially for lower photon energies) and good energy resolution.

lithium into the semiconductor lattice. The larger depletion region is necessary to provide sufficient energy absorption of x and gamma radiation, as well as the more penetrating and energetic charged particles.

The second type of semiconductor radiation detector is the surface barrier detector. Surface barrier detectors are similar in basic construction to the diffused junction detectors except that a p-type layer is produced in a slab of n-type silicon by chemically etching the surface of the detector and allowing spontaneous oxidation to take place at the surface. This spontaneous oxidation induces a high density of holes, forming a p-type layer. More information on semiconductor detectors can be found in the literature[4,6,14,15].

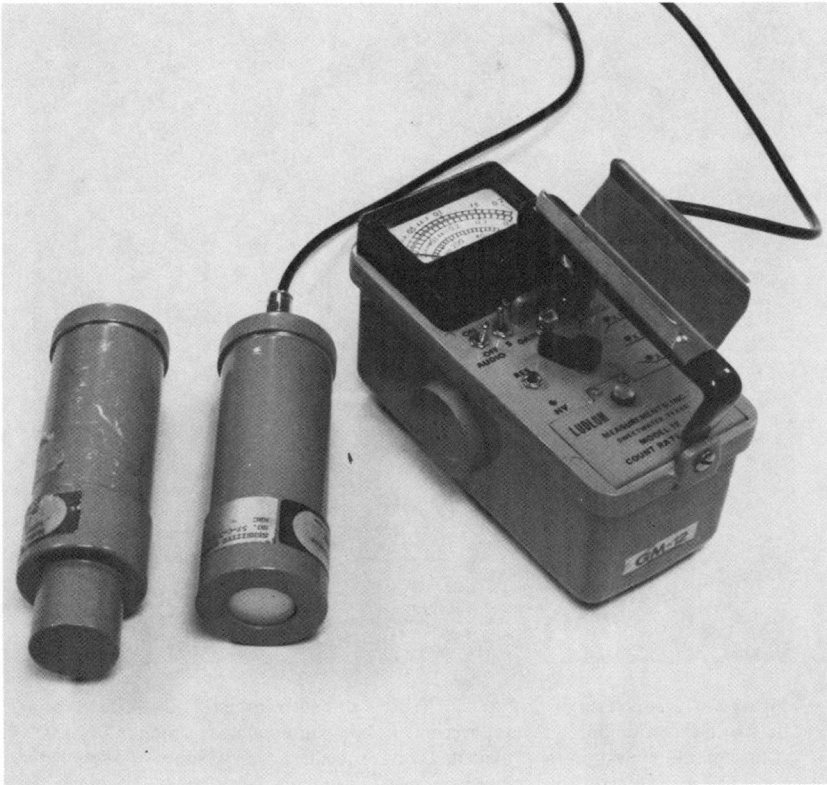

Figure 20.   Portable gamma survey instrument. This instrument uses a NaI crystal for gamma-ray detection. Because the detecting medium is a solid, the detection efficiency is relatively high. The upper range of this instrument is 500,000 cpm. An audible output is provided.

Applications

There are two types of diffused junction detectors: the lithium-drifted silicon detector, Si(Li), and the lithium-drifted germanium detector, Ge(Li). Si(Li) detectors are the choice for X-ray and energetic charged particle detection. The energy resolution of Si(Li) detectors for electrons is ~1–3 keV for electron ener-

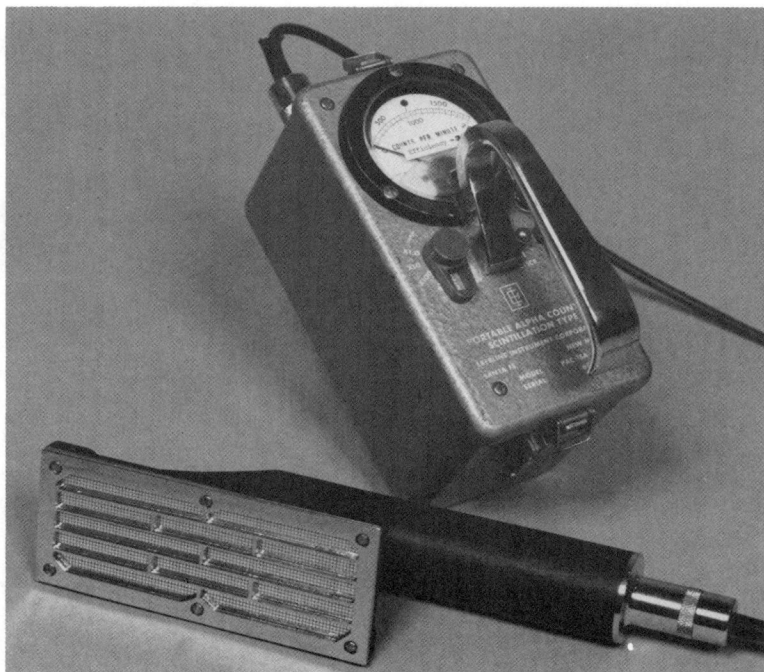

Figure 21.   Alpha survey meter, which uses a zinc sulfide scintillator shielded from light by a thin aluminized mylar covering. The maximum meter reading is 2,000,000 cpm. An audible output is provided. Damage to the mylar window or other light leaks can lead to erroneous meter response.

gies up to 1000 keV. Germanium is used instead of silicon for some semiconductor detectors because the average energy needed to create an electron-hole pair in germanium is lower and its stopping power greater, since the atomic number of germanium (32) is much higher than that of silicon (14). The results are better energy resolution and increased probability of photon interaction. Consequently, germanium is preferred over silicon for gamma ray detection. The most spectacular feature of the Ge(Li) detector for measuring gamma radiation is its superior energy resolution. Hyperpure germanium (HPGe) detectors are being used increasingly for low-energy (< 100 keV) spectroscopy. However, room temperature thermal excitation leads to the formation of electron-hole pairs in germanium; therefore, these detectors must be operated at liquid nitrogen temperature to prevent this phenomenon from occurring. Various properties of the Ge(Li), HPGe and Si(Li) detectors are presented in Table IV.

Figure 22. Liquid scintillation counter, a laboratory instrument used primarily because of its high efficiency for low-energy photon and beta radiations. The sample is mixed with the scintillation media. Energy resolution is poor, limiting applications for spectrometry.

Surface-barrier detectors are used primarily for alpha particle or low-energy beta particle measurement. (Refer to Table V for typical characteristics of this type detector.)

The sensitivity and energy independence of semiconductor detectors contributes to their excellent performance in spectroscopy. This is illustrated in Figure 25, which compares gamma spectra obtained using NaI(Tl) scintillation and Ge(Li) semiconductor detectors. Due to their fragile nature, special electronic requirements and other physical characteristics, applications of these detectors are generally limited to laboratory analysis. Examples of several semiconductor detectors and detector systems are shown in Figures 26 to 28.

Figure 23.   This instrument is a gamma camera, used for visualization of tumors, organs or other anatomical structures by detection of the photons emitted by radionuclides concentrated in these structures. It employs a single, large-area NaI crystal coupled to many photomultiplier tubes. Resolution is limited by the combination of the resolutions of the collimator and the detector imaging system.

MISCELLANEOUS DETECTION METHODS

Chemical Dosimeters

Excitation and ionization, produced when radiation interacts with matter, may result in chemical changes. In some systems these changes are related to the total radiation absorbed. The material may therefore function as a radiation dosimeter. Although the exact mechanisms of some of these chemical changes are not thoroughly understood, it is known that the yield is strongly influenced by the specific ionization of the radiation and that equal energies dissipated in material will provide the same qualitative effect, regardless of the type of radiation.

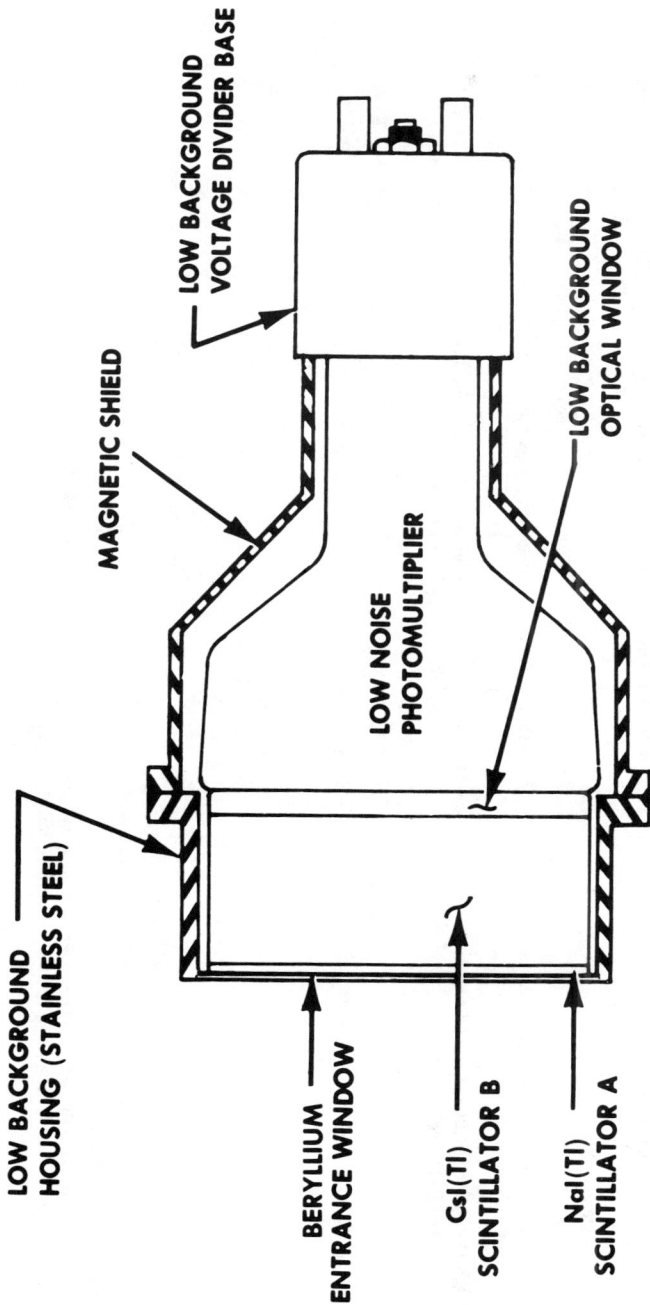

Figure 24.   Phoswich detector, designed for high-efficiency measurement of low-energy photon radiation in the presence of an ambient background. This is achieved through pulse shape analysis and discrimination of the different light outputs from two optically coupled scintillating materials (e.g., NaI and CsI). Photo courtesy of Harshaw Chemical Company.

Table IV.   Diffused Junction Semiconductor Detector Characteristics

Detector Type	Typical Dimensions	Use	Typical Resolution (FWHM)[a]
Ge(Li) Coaxial Detector	10–75%[b,c]	High-energy gamma spectroscopy 100 keV $\leq E < 5$ MeV	$\leq$1.9 keV @ Co-60 1.332 MeV gamma
High-Purity Germanium Coaxial Detector	8–15%[b]	High-energy gamma spectroscopy 100 keV $\leq E < 5$ MeV	$\leq$1.9 keV @ Co-60 1.332 MeV gamma
High-Purity Germanium Coaxial Detector with Ultrathin Entrance Window	$\geq$60 cm^2 active photon area for low-energy photons	Simultaneous low- and high-energy gamma spectroscopy 10 keV $\leq E < 5$ MeV	$\leq$2.0 keV @ Co-60 1.332 MeV gamma 0.95 keV @ Cd-109 88 keV gamma
Planar High-Purity Germanium Detector	Active area 3–20 cm^2, active depth, 0.6–4 cm, multiplexed arrays to 100 cm^2 or more[c]	Low-energy photon spectroscopy 3 keV $\leq E \leq 300$ keV	$\leq$800 keV @ Co-57 122 keV gamma
Si(Li) X-ray Detectors	Active area 3 cm^2	X-ray spectroscopy to 60 keV	<300 eV @ Fe-55 5.9 keV X-ray

[a] FWHM = full width at half maximum.
[b] Expressed in terms of relative photopeak efficiency referenced to a standard 3 $\times$ 3 inch NaI detector.
[c] Larger sizes may be available in custom units or in multiplexed arrays.

Table V.   Surface-Barrier Detector Characteristics

Type	Typical Dimensions	Use
Partially Depleted Si	Active area: 700–900 mm^2 Active thickness: 100–5000 mm	High-resolution charged particle spectroscopy FWHM[a]: 10–30 keV for alpha 6–20 keV for beta
Totally Depleted Si	Active area: 25–340 mm^2 Active thickness: 150–3000 mm	Particle identification FWHM: 15–25 keV for alpha 6–20 keV beta
Planar Totally Depleted Si	Active area: 10–300 mm^2 Active thickness: 15–100 mm	Heavy-ion time-of-flight research
Heavy-Ion Partially Depleted Si	Active area: 100–900 mm^2 Active thickness: 600 mm	Heavy-ion spectroscopy
Position-Sensitive, Ion-Implanted Si	Active area: 80–350 mm^2 Active thickness: 100–1000 mm	Simultaneous measurement of charged-particle energy and position

[a] FWHM = full width at half maximum.

Chemical dosimeters can be of various forms and physical states (solid and liquid). In general, they are divided into two classes: (1) those using water as a solvent, and (2) all others. The many aqueous systems include the Fricke dosimeter (the most widely used chemical dosimeter), which determines the absorbed dose to the solution by measuring the increase in ferric ion concentration in a ferrous sulfate solution. An example of the second class is acid production in chlorinated hydrocarbons, such as chloroform or trichloroethylene.

The chemical dosimeter is relatively simple to prepare and calibrate. The response is fairly linear with energy and the radiation absorption properties closely approximate those of biological systems. These materials can also be immersed or implanted. On the negative side, they require high radiation exposures for measurable reactions, have limited "shelf-life," are both dose rate and temperature dependent, and the chemical change is not always stable. Their principal use is for the measurement of high-intensity radiation fields (primarily photons), although particulate radiations also can be measured. Their usefulness for neutron dosimetry is limited due to activation of the chemicals and by their inability to distinguish between fast and thermal neutron energies. Extensive reviews of chemical dosimetry can be found in the literature.[5,16].

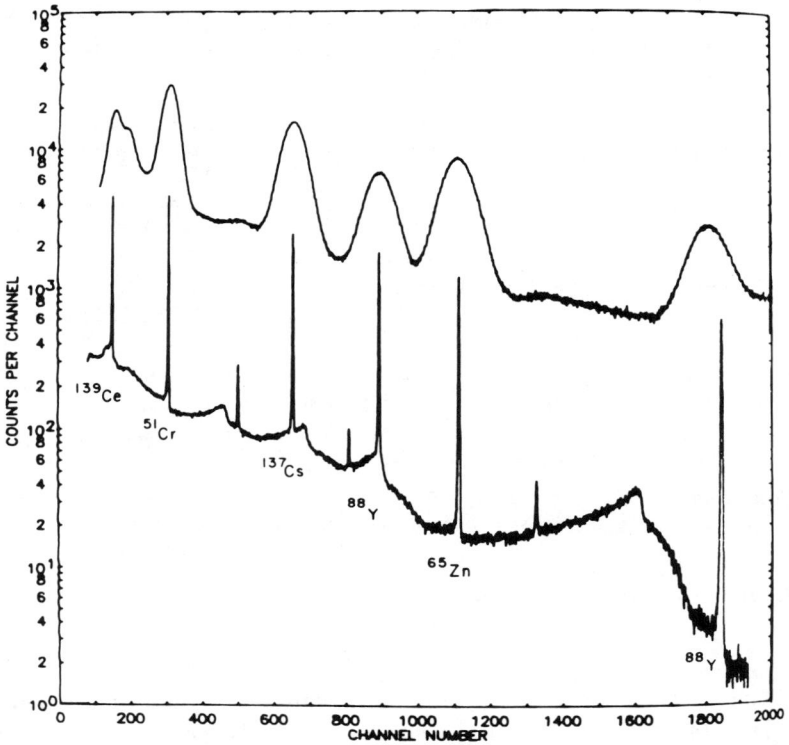

Figure 25.   Gamma ray spectra of a mixed radionuclide source. The upper curve was obtained with a 5 inch NaI(Tl) well crystal; the lower curve was obtained with a 60-cm^3 Ge(Li) detector[14].

Cloud Chamber

A cloud chamber is one of a family of nuclear particle detectors that produce visible tracks. The operation of a cloud chamber depends on the existence of a supersaturated region within a gas containing a condensable vapor. Dust and other particles can serve as centers for condensation within the supersaturated region. Also, ions—positive or negative—serve as condensation nuclei because their presence reduces the saturation vapor pressure in their immediate vicinity. It is this phenomenon that causes a "track" of charged particles to appear as vapor trails in the cloud chamber. There are two general forms of cloud chambers—the expansion type and the diffusion type. The difference between the two is the method employed to obtain a supersaturated region.

Figure 26. Silicon surface barrier detectors for alpha and beta spectroscopy.

In the expansion cloud chamber, a gas containing a condensable vapor is allowed to come to equilibrium with enough liquid present to produce saturation. Adiabatic expansion then takes place, causing supersaturation. After expansion, heat flows slowly into the chamber, raising the temperature and eliminating the supersaturated condition. The sensitive time during which conditions are optimum for track formation is short (from a few milliseconds to about three seconds, depending on the chamber design). A diagram of a typical expansion cloud chamber is shown in Figure 29.

The diffusion cloud chamber operates by allowing diffusion of a condensable vapor from a warm region (where no saturation occurs) to a colder region (where supersaturation takes place). Generally, methyl and ethyl alcohol vapors are best suited for this chamber type (Figure 30).

Both types of cloud chambers employ an electrical field (on the order of 20–50 V/cm) to remove "unwanted" ions. These unwanted ions come from any ambient ionizing radiation, such as cosmic rays. If not removed, the ion density may be enough to remove the supersaturated condition so that "tracks" cannot be produced. Both chambers generally operate at atmospheric pressure, but some

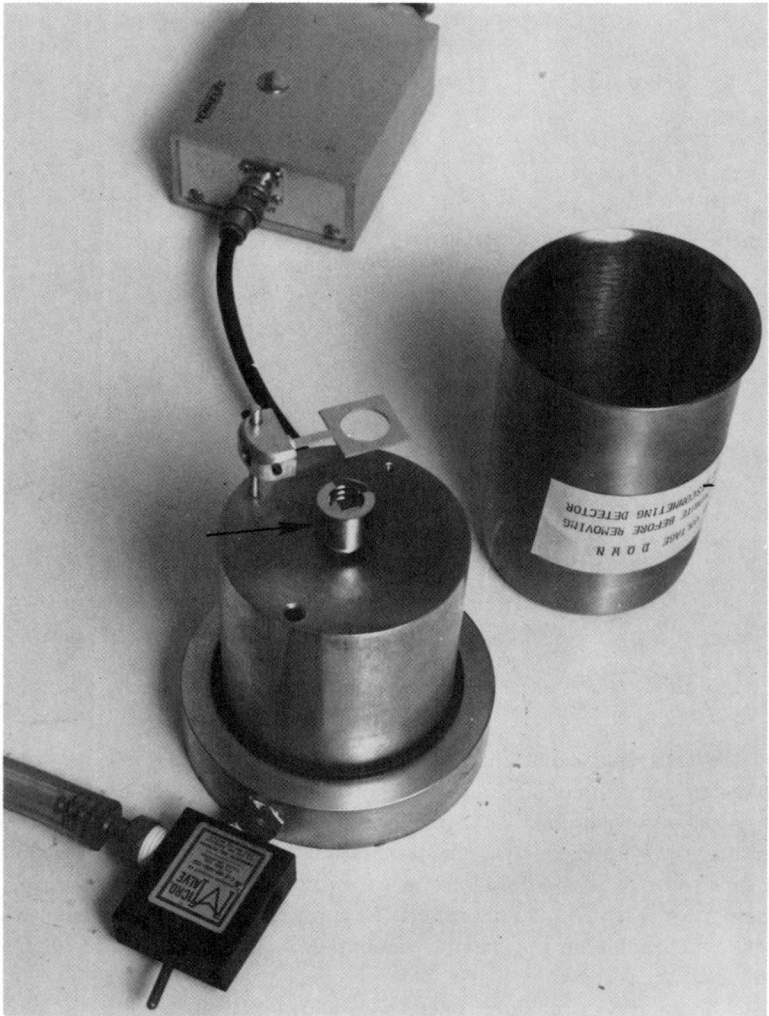

Figure 27. Silicon surface barrier alpha spectroscopy system. The vacuum chamber is used to prevent loss of alpha particle energy due to air absorption. The arrow indicates the surface barrier detector.

types can be operated at very low (10 mm Hg) or very high (200 atm) pressures.

Application of cloud chambers is limited to visualizing and photographing tracks of nuclear particles. The reader is referred to other sources[17,18] for more information concerning these devices and their applications.

Figure 28.   Ge(Li) gamma ray spectroscopy system. The dewar contains liquid nitrogen for cooling the detector unit.

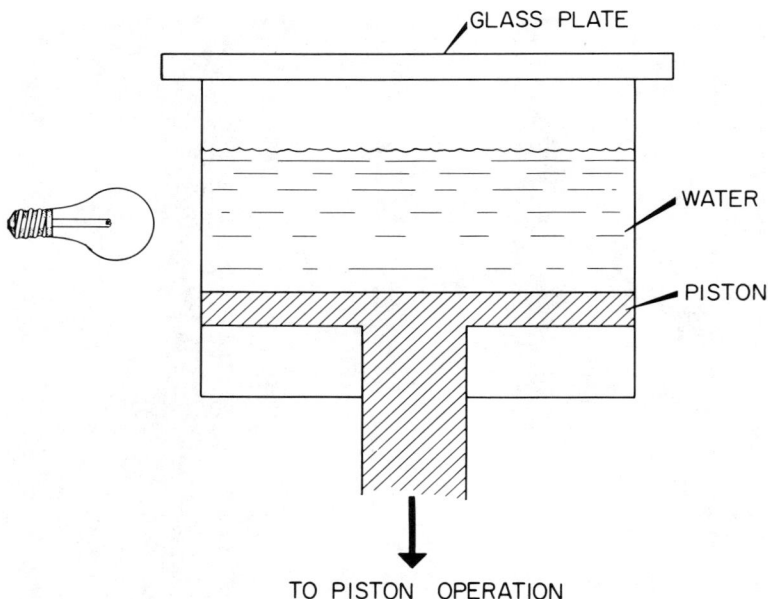

Figure 29.   Diagram of an expansion cloud chamber.

Calorimeters

Since a fraction of the radiation energy absorbed in matter will eventually dissipate as heat, it is possible to relate temperature increases in exposed material to the intensity of the radiation field. This can be achieved by using a device known as a calorimeter. Calorimeters for radiation monitoring commonly use thermocouples to measure the temperature rise in a liquid or metal during exposure. Other types measure vaporization rate in a small volume of liquid nitrogen. The calorimeter is relatively insensitive and is only applicable for high radiation levels. For further discussion of calorimetry the reader is referred to the literature[6].

Many additional devices, instruments and methods have been used for radiation detection and measurement. However, due to their relatively limited use, they are not discussed here. Therefore, the reader is referred to any complete text on radiation monitoring equipment[4-6,14].

PERSONNEL MONITORING DEVICES

Introduction

The purpose of personnel monitoring is to determine the radiation dose received by the radiation worker. Personnel monitoring is required when a worker

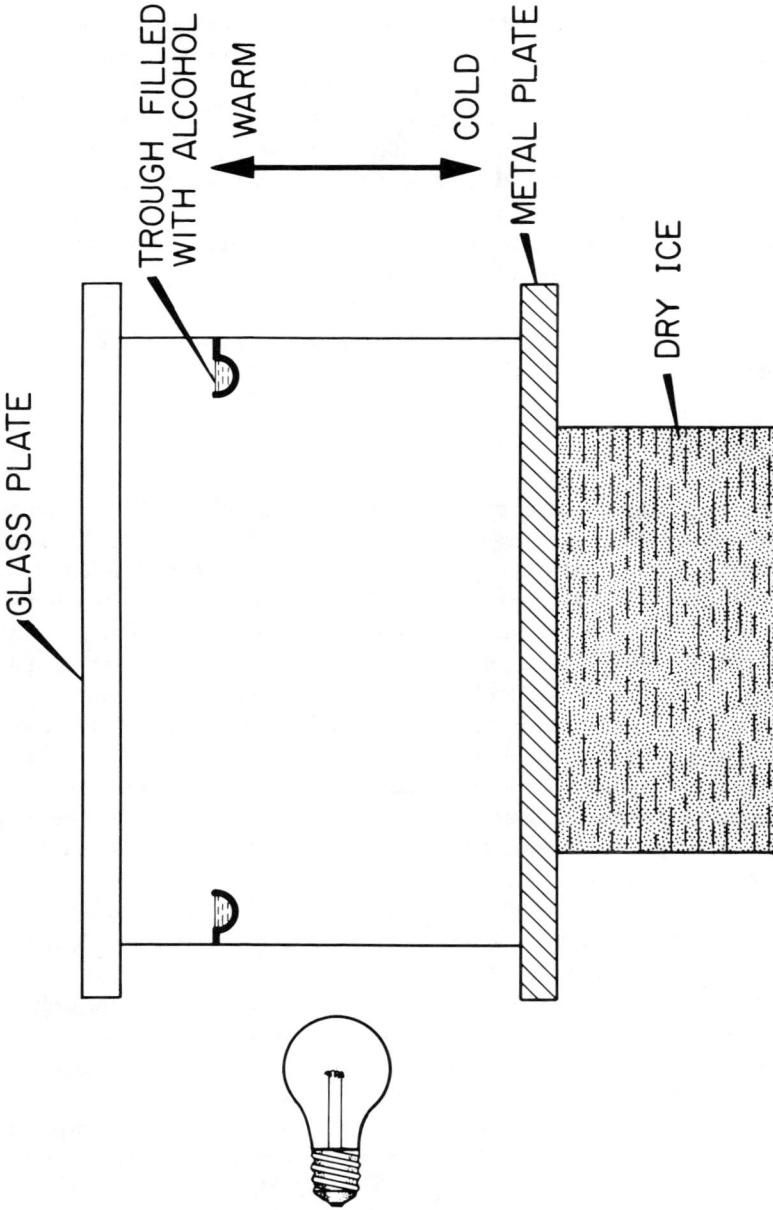

Figure 30.   Diagram of a diffusion cloud chamber.

might receive as much as 25% of the radiation protection standards as determined in statutory regulations.

A number of difficulties may be encountered in personnel monitoring. Even for external radiation monitoring, where the objective is simply to measure the radiation dose from external sources, one must remember that the dosimeter reading is only an estimte of the dose at one particular location on the worker's body. Thus, even if the dosimeter is absolutely accurate in its measurement, the dose at this one point may not reflect the actual radiation insult to a specific body organ. The dose to a particular organ is the best estimate of the radiation insult that may result in biological effect. The methods and accuracy of external radiation monitoring have been reviewed in recent reports[19-23].

External Radiation Monitoring

Film Dosimeters

Photographic film has been routinely used as a personnel radiation dosimeter since the mid-1940s. Film for radiation dosimetry is not significantly different from normal photographic film. It consists of a silver bromide emulsion on a cellulose backing and is wrapped in a light-tight paper covering. Radiation interacts with the film and causes chemical changes that free elemental silver, just as light does in the normal photographic process. The film is then developed, and the relative darkening of the film is an indication of the radiation exposure.

Although the film is the measuring element in a film dosimeter, the film itself is almost always contained within a badge. The badge serves two major purposes. First, it protects the film from puncture and, to some degree, from other environmental insults that would be detrimental to the film. The second and more important reason is to provide filters or attenuators that allow the energy of the incident radiation to be estimated. Filter systems may be simple, consisting of a single metal filter such as copper, aluminum or lead, or complex, with four or more different filters incorporated into the front and back of the badge (Figure 31). These filters, because of their different absorption properties for different radiation energies, produce patterns on the piece of developed film that allow the experienced dosimetrist to determine the energy of the radiation and, ultimately, the dose to the worker.

The operating characteristics for film dosimeters and other personnel monitoring devices are summarized in Table VI. In general, film dosimeters are adequate over the exposure range normally required in personnel monitoring. However, recent questions about the effects of low-level radiation make it desirable to measure levels below the minimum detection level for film, even though monitoring below this level is not required.

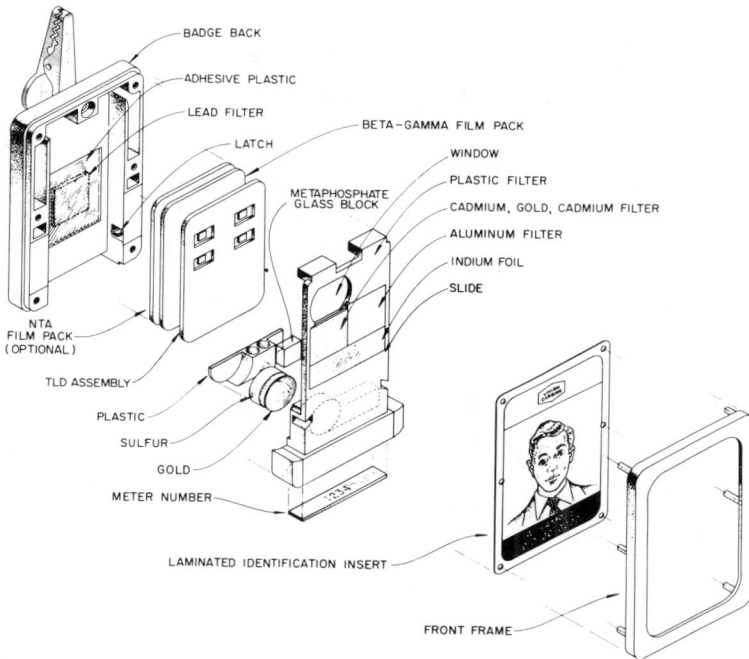

Figure 31. Film badge assembly, exploded view.

The major disadvantage of film dosimeters, however, is their energy dependence. Due to photoelectric interactions in film, low-energy radiations produce a greater darkening for a given exposure than high-energy radiations. Thus, a measured darkening of the film may be the result of either a small amount of low-energy radiation or a larger amount of higher energy radiation. The filter systems mentioned earlier therefore must be used to determine the exposure. Other potential disadvantages include the requirement for wet chemistry processing of film and a readout and data interpretation procedure before the final evaluation is completed. Also, film is relatively sensitive to environmental conditions such as heat and humidity.

There are several advantages of film dosimetry. The film produces a semipermanent record of the exposure. The developed film can be analyzed more than once, and in the hands of the experienced dosimetrist can yield information relative to the direction from which the radiation came, as well as the type of radiation. In some situations, information can be determined as to whether the exposure to the badge was over an extended period of time or was a single rapid exposure. All this information is useful to the health physicist in determining whether the exposure to the film represents the true dose to the worker.

Table VI. Characteristics of Personnel Monitoring Methods for External Radiation

Method	Detector	Type of Radiation Measured	Typical Range[a]	Application
Film Badge	Silver bromide	Beta, X-ray, gamma, (neutron)[b]	30 mR–1000 R	Routine monitoring
TLD Badge	Thermoluminescent dosimeter crystals	Beta, X-ray, gamma, neutron	1 mR–10^5 R	Routine monitoring and accident dosimetry
Pocket Dosimeter	Ion Chamber	X-Ray, gamma, (beta, neutron)	0–200 mR 0–5 R 0–200 R	Special monitoring—permits direct reading of exposure
Personnel Radiation Monitor or Pocket Alarm	Miniature GM tube	Gamma, (beta)	1 mR/hr up	Warning of high radiation levels. Most Personnel Radiation Monitors (PRM) are not dosimeters but indicate relative radiation intensity by the repetition rate of "chirps." The PRM is used as an aid in minimizing exposure when working in varying radiation fields. Some new models do have measuring capability

[a] In units of exposure: mR (milliroentgen) and R (roentgen).
[b] Only specifically designed devices can detect the radiations in parentheses.

Thermoluminescent Dosimeters

Many crystalline materials are capable of trapping or storing free electrons, which are produced during the ionization process. On heating, these electrons are released and drop to a lower energy state. This transition is accompanied by the emission of energy in the form of visible light. The intensity of the light emitted is proportional to the number of electrons trapped which is, in turn, proportional to the amount of radiation received by the crystalline material. This phenomenon is referred to as thermoluminescence, and devices using such crystals as a detecting element are known as thermoluminescent dosimeters (TLD). The two most commonly used TLD materials are lithium fluoride and calcium fluoride.

Thermoluminescent crystals were first used as radiation dosimeters during the 1960s, but have already become one of the most widely used personnel monitors. In general, the TLD has a greater measurement range capability and is less energy dependent than film and other personnel monitoring devices. The major disadvantage of thermoluminescent dosimeters is that the readout procedure removes the information stored in the crystal, making subsequent analysis impossible. Table VI summarizes characteristics of TLD personnel monitoring devices. Examples of TLD badges are shown in Figure 32.

Pocket Ionization Chambers

Another major method of personnel monitoring is the use of small condensor ionization chambers, referred to as pocket dosimeters. These fountain pen-sized devices (Figure 33) are used primarily for real-time monitoring. They are generally divided into two groups—direct reading and indirect reading dosimeters. Both devices operate on the principle that a central electrode is charged relative to its outer shell. As radiation interacts within the ionization chamber, the ions produced are collected at the electrode, reducing its charge.

In the direct reading dosimeter this results in the movement of a visible hairline across an exposure scale and, as the name implies, the exposure can be evaluated visually by looking through the dosimeter, much like one would a telescope. The indirect reading unit requires an auxiliary reader to measure the voltage change on the chamber electrode and to convert it to a radiation exposure value.

The major disadvantages of pocket dosimeters are that they have a limited range; they may "drift" or leak charge, which gives a false high reading; and they usually have poor energy response characteristics. It should be emphasized that the pocket chamber is almost always used in conjunction with an integrating personnel monitoring device, such as the TLD or film badge.

Miscellaneous Personnel Monitoring Devices

There are a number of other monitoring devices that have been used to determine the radiation dose to a worker. Among these are special glass dosimeters,

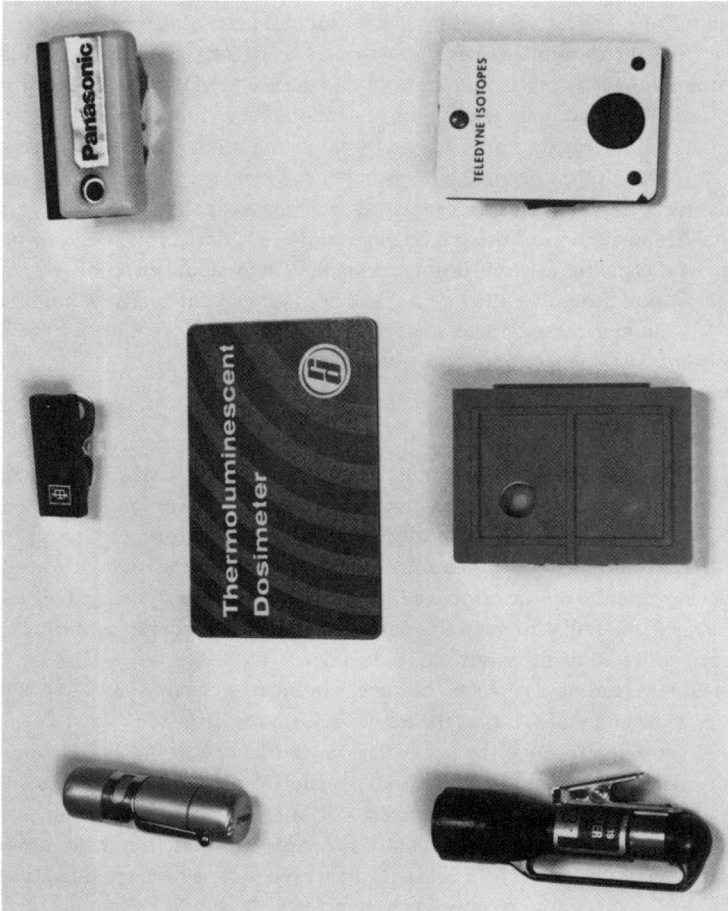

Figure 32.   Examples of thermoluminescent dosimeter monitoring badges.

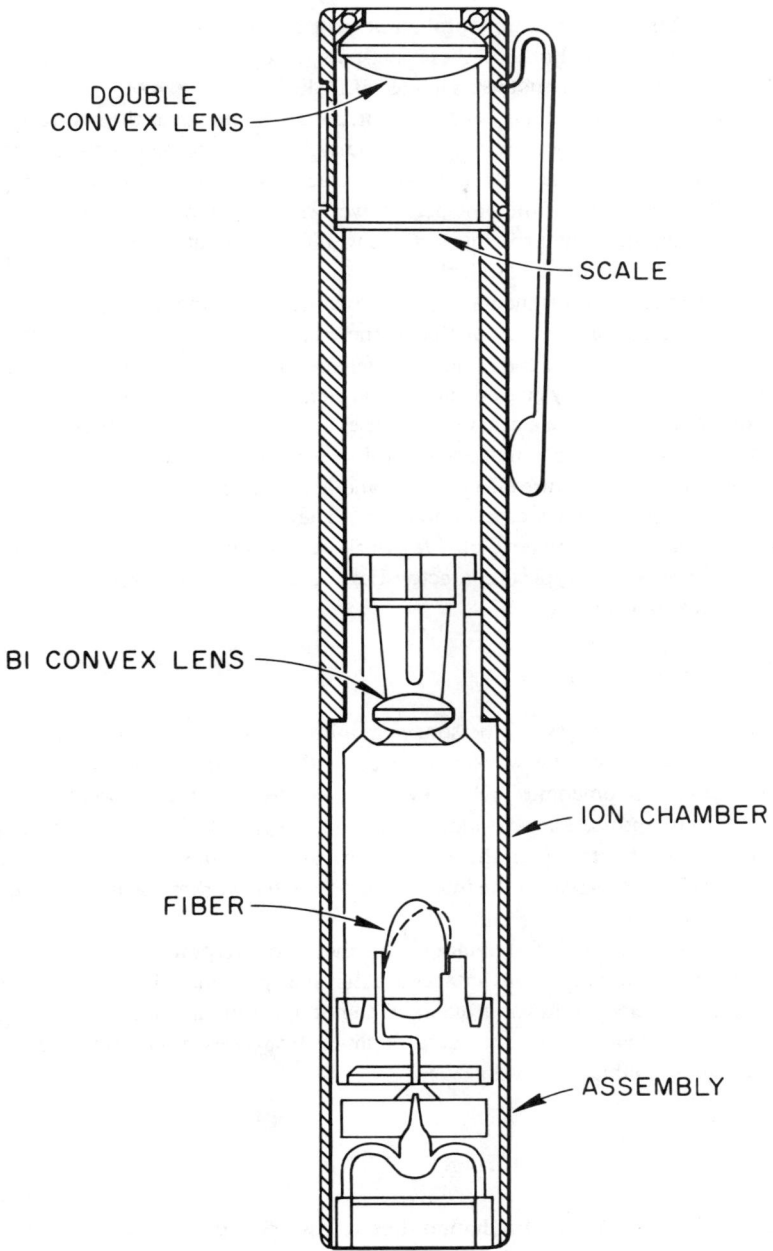

Figure 33.   Diagram of a pocket ionization chamber.

generally referred to as radiophotoluminescent dosimeters, and several types of chemical dosimeters. Both of these methods generally have been replaced in recent years by new techniques, such as TLD dosimeters.

A device that is commonly used for neutron measurements is the neutron activation dosimeter. Many materials, particularly metals, become radioactive when they interact with neutrons. By measuring the activity induced in these dosimeters and with a knowledge of the time period over which the radiation exposure took place, the number of neutron interactions within the activation foil can be determined and the neutron dose calculated.

Another class of instruments considered as personnel monitoring devices are "personnel radiation alarms" or "personnel radiation monitors." These devices generally do not measure the radiation level and most do not integrate the radiation dose; however, they give an audible warning that the worker is being exposed to radiation. The audible alarm rate increases as the radiation field increases, thus the worker is warned of changes in the intensity of the radiation field. These devices contain a miniature GM detector and are relatively small—approximately the size of a cigarette pack or a fountain pen. They generally respond only to X or gamma radiation and are not suited for work with radionuclides emitting particulate radiations only. Typical characteristics for the personal alarm monitor are summarized in Table VI.

Area Monitors

The above techniques for personnel monitoring involve the wearing of the monitoring device on the worker to measure the radiation exposure. Another technique that is sometimes used to evaluate the dose to radiation workers is the use of area monitors. The area monitor is usually a gas-filled detector that monitors the radiation intensity in the general work area. The dose to the worker then must be estimated based on the intensity and time the worker spent in the radiation environment.

The major advantage of area monitors is their rapid response to changes in the area radiation level; the disadvantages are that they provide radiation level information for preselected sites and generally do not integrate the radiation dose. The accuracy of individual exposure data obtained from area monitoring therefore may be questionable.

Internal Radiation Monitoring

Determining the internal radiation dose to a worker is an extremely complex task. This can be accomplished by several different techniques. It can be calculated based on the concentrations of radionuclide contamination in the worker's environment, which may be taken internally. (Air and water sample analyses are the primary sources of this information.) Another technique is by analysis of biological samples such as urine, blood, feces or sweat. Radionuclides are excreted through normal biological processes, as are stable elements. The rate of excretion is determined primarily by the chemical form of the radionuclide. By measuring the concentration of radioactive material being excreted, the total body content of the material can be calculated. These monitoring techniques provide an indirect measurement of the radioactivity residing in the worker's body. Their accuracy is affected by necessary assumptions and estimates of the body's intake and elimination rates.

A third method for determining the internal dose to a worker is through the use of a device called a whole-body counter. As the name implies, these devices are designed to measure the radioactivity deposited in the worker's body. They are capable of measuring only gamma or X radiation emitted from internally deposited nuclides. The detectors for these instruments may be sodium iodide scintillation crystals, solid-state semiconductor detectors or proportional counters. For medium- to high-energy photon emitters, whole-body counters can determine the amount of radionuclide in the body with an accuracy approaching 90%. Additional information on whole-body counters is available in the literature[24].

ACKNOWLEDGMENTS

The authors would like to acknowledge the patience and skill of Ms. Glenda Fritts of the Oak Ridge Associated Universities for typing and compiling this report.

GLOSSARY

This glossary contains simplified definitions of some of the frequently used terms in radiation physics. For additional or more complete definitions, the reader is referred to the Radiological Health Handbook[25].

Absorbed Dose: The amount of energy given up to matter by ionizing radiation per unit mass of irradiated material. The "rad" and the "gray" are units of absorbed dose.

Activity: The number of nuclear disintegrations occurring in a given quantity of material per unit time. The "curie" and the "becquerel" are the units of activity.

Alpha Particle: A densely ionizing particle emitted from the nucleus during radioactive decay, having a mass and charge equal in magnitude to a helium nucleus.

Background Radiation: Ionizing radiation arising from radioactive material other than that directly under consideration.

Beta Particle: Charged particle emitted from the nucleus during radioactive decay, having a mass and charge equal to that of an electron.

Count: The external indication of a device designed to enumerate ionizing events. This term is often used erroneously to designate a disintegration, ionizing event or voltage pulse.

Decay, Radioactive: Disintegration of the nucleus of an unstable nuclide by the spontaneous emission of charged particles and/or photons.

Efficiency: A measure of the probability that a count will be recorded when radiation is incident on a detector.

Exposure: A measure of the ionization produced in air by X or gamma radiation. The "roentgen" is a unit of exposure.

Gamma Ray: Very penetrating electromagnetic radiation of nuclear origin. Except for the origin, it is identical to an X-ray.

Ionization: The process by which a neutral atom or molecule acquires either a positive or negative charge.

Ionizing Radiation: Any electromagnetic or particulate radiation capable of producing ions (either directly or indirectly) in its passage through matter.

Monitoring, Radiological: Periodic or continuous determination of the amount of ionizing radiation or radioactive contamination present in an occupied region as a safety measure for purposes of health protection (e.g., area monitoring, personnel monitoring, etc.).

Neutron: Elementary particle with a mass approximately the same as that of a hydrogen atom and electrically neutral.

Resolution: The ability of a detector to register the arrival of an ionizing particle photon as a separate event from the arrival of another ionizing particle or photon a short time before. Also the ability of a detector to distinguish between different photon or particle energies (e.g., energy resolution), an important specification for spectrometric devices.

Stopping Power: The relative ability of a material to absorb radiation energy and thereby "stop" the radiation.

X-Rays: Penetrating electromagnetic radiations having wavelengths shorter than those of visible light, usually produced by bombardment of a metallic target with fast electrons in a high vacuum.

REFERENCES

1.   Unruh, C. M. *Methods of Measurement and Interpretation of Results in Radiation Monitoring and Dosimetry, Proc. IAEA,* Radiation Protection Monitoring, Vienna (1969).

2. National Council on Radiation Protection and Measurements. *Instrumentation and Monitoring Methods for Radiation Protection,* NCRP 57 (1978).
3. "Buyers Guide," *Nucl. News,* 22(4) (1979).
4. Knoll, F. *Radiation Detection and Measurement* (New York: John Wiley & Sons, Inc., 1979).
5. Attix, F. H., and W. C. Roesch. *Radiation Dosimetry,* Vol. II, (New York: Academic Press, Inc., 1966).
6. Price, W. J. *Nuclear Radiation Detection* (New York: McGraw-Hill Book Co., 1964).
7. Snell, A. H. *Nuclear Instruments and Their Uses* (New York: John Wiley & Sons, Inc., 1962).
8. *Design of Free Air Ionization Chamber,* NBS Handbook 64 (1957).
9. Baer, W., and R. T. Bayard. *Rev. Sci. Instr.* 24: 138 (153).
10. O'Kelley, G. D. "Detection and Measurement of Nuclear Radiation," National Academy of Sciences Publication NAS-NS-3105 (1962).
11. Heath, R. L. *Scintillation Spectrometry* (Elmsford, NY: Pergamon Press, 1964).
12. Crouthamel, C. E. *Applied Gamma Ray Spectrometry* (Elmsford, NY: Pergamon Press, 1960).
13. Hine, G. J., and G. L. Brownell. *Radiation Dosimetry* (New York: Academic Press, Inc., 1956).
14. National Council on Radiation Protection and Measurements. *A Handbook of Radioactivity Measurements Procedures,* NCRP 58 (1978).
15. Wang, C. H., D. L. Willis and W. D. Loveland. *Radiotracer Methodology in the Biological, Environmental, and Physical Sciences* (Englewood Cliffs, NJ: Prentice-Hall, Inc., 1975).
16. Harmer, D. E. *Nucleonics* 17: 72 (October 1959).
17. Lapp, R. E., and H. L. Andrews. *Nuclear Radiation Physics* (Englewood Cliffs, NJ: Prentice-Hall, Inc., 1964).
18. Das Gupta, N. W., and S. L. Ghosh. *Rev. Mod. Phys.* 18: 227 (1946).
19. Griffith, R. V., D. E. Hankins, R. B. Gammage and L. Tommasino. "Recent Developments in Personnel Neutron Dosimeters—A Review," *Health Phys.* 36 (1979).
20. Chabot, E., Jr., M. A. Jimenez and K. E. Skrable. "Personnel Dosimetry in the U.S.A.," *Health Phys.* 34 (1978).
21. Plato, P. "Testing and Evaluating Personal Dosimetry Services in 1976," *Health Phys.* 34 (1978).
22. Wadman, W. W. III. *Proc. Health Physics Society Eleventh Midyear Topical Symp. on Radiation Instrumentation* (1978).
23. Francois, H., E. D. Gupton, R. Maushart, E. Piesch, S. Somasundaram and Z. Spurny. *Personnel Dosimetry Systems for External Radiation Exposures,* Technical Reports Series No. 109, International Atomic Energy Agency, Vienna (1970).
24. Wang, Y. *Handbook of Radioactive Nuclides* (Cleveland: CRC Press, Inc., 1969).
25. *Radiological Health Handbook* (Washington, DC: U.S. Department of Health, Education, and Welfare, 1970).

CHAPTER 9

THE TEA ANALYZER:
CHEMILUMINESCENT DETECTOR—APPLICATION
TO THE DETERMINATION OF N-NITROSAMINES

Tsai-Yi Fan

James Ford Bell Technical Center
General Mills, Inc.
Minneapolis, Minnesota

David H. Fine

New England Institute for Life Sciences
Waltham, Massachusetts

Arthur L. Lafleur

Thermo Electron Corporation
Analytical Instruments Division
Waltham, Massachusetts

INTRODUCTION

The Thermal Energy Analyzer (TEA) is a chemiluminescent (CL) detector developed specifically for the detection and determination of N-nitrosamines, whose general formula is shown below:

$$\begin{array}{c} R_1 \\ \diagdown \\ \diagup \\ R_2 \end{array} \!\! N\!-\!NO$$

where R_1 and R_2 can be virtually any organic group. The chemical structures of several N-nitrosamines, which will be discussed frequently in this chapter, are shown in Figure 1.

N-nitrosamines are a well-established class of animal carcinogens[1-3]. Following

273

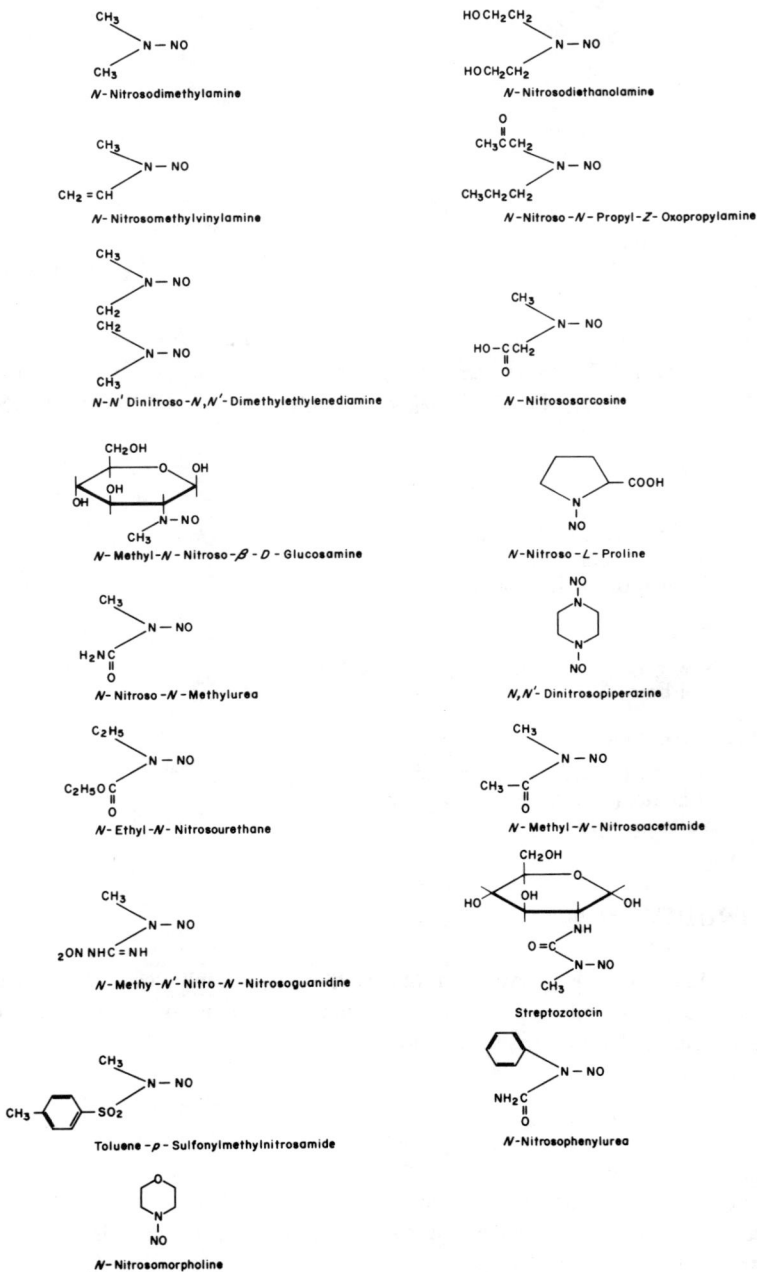

Figure 1. Structures of N-nitrosamines.

an industrial accident, Barnes and Magee[4] became interested in the toxicology of N-nitrosodimethylamine (NDMA). They reported severe liver damage in rats after administration of NDMA. Two years later, Magee and Barnes[5] reported the induction of liver tumors in rats by feeding NDMA. However, the possible health hazard of N-nitrosamines to man did not elicit concern until they were shown to occur in the environment. Ender et al.[6] reported the isolation and identification of NDMA in the fish meal, which had caused liver disorders in mink and several ruminants. An outbreak of acute liver toxicity in sheep in Norway was also traced to the occurrence of NDMA in nitrite-preserved fish meal[7]. The presence of NDMA in fish meal probably originated from the reaction between nitrite, which was used as the preservative, and methylamines, which occur naturally in fish. These findings stimulated active research on N-nitrosamines, especially the occurrence of N-nitrosamines in fish and meat products treated with nitrite.

The carcinogenicity of N-nitrosamines has been studied extensively with laboratory animals. More than 100 of the 130 different N-nitrosamines tested in animals have been shown to be carcinogenic[1-3]. Certain N-nitrosamines have been tested on a wide variety of animal species. For example, N-nitrosodiethylamine (NDEA) was studied for carcinogenic activity in at least twelve species of animals, which included monkeys, mice, rats, rabbits, dogs and sheep. NDEA was shown to be a carcinogen in all these species. Many N-nitrosamines are extremely potent carcinogens. Dose response studies at low dose levels showed that the apparent "no effect levels" were seen only when as low as 1 mg/kg NDMA and NDEA and 5 mg/kg N-nitrosopyrrolidine (NPYR) in the diet were fed to rats[8-10].

Until recently, it was assumed that man's exposure to N-nitrosamines was very limited. Within the past few years this has been shown to be an incorrect assumption. It becomes increasingly apparent that N-nitrosamines are some of the most potent carcinogenic compounds that people may be exposed to on a daily basis. Formation of N-nitrosamines requires the participation of a nitrosating agent and a nitrosatable amine. Secondary and tertiary amines are well established as nitrosatable amines[11-15]. The formation of N-nitrosamines from the nitrosation of primary and quartenary amines also have been reported at extremely low yields[16-20]. All classes of amines are ubiquitous in the environment[16,17,21,22]. They are intermediate products in protein metabolism and also are important biochemical constituents in animals and plants. Furthermore, they have been used widely in pesticide and pharmaceutical formulations[12,23-27] and in consumer products such as anticorrosion and emulsifying agents[28,29].

A variety of nitrosating agents have been reported in the literature. Among them, nitrite is most well known. It reacts with secondary amines to form N-nitrosamines according to the following equations[13,15]:

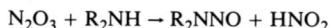

$$2HNO_2 \rightleftharpoons N_2O_3 + H_2O$$

$$N_2O_3 + R_2NH \rightarrow R_2NNO + HNO_2$$

The kinetics of the reactions have been studied extensively[30-34]. It is a third-order reaction: first order with respect to amine and second order with respect to nitrous acid. The reaction rate is pH dependent, with maximum rates at pH 3.4 (pKa of nitrous acid) for the nitrosation of secondary amines[30] and at approximately pH 2.5 for the nitrosation of secondary amino acids[34]. Because the nitrosation reaction is favored in weakly acidic conditions, formation of N-nitrosamines in the stomach has been an area of active research[35-37].

Nitrosation is catalyzed by the presence of thiocyanate and halide ions[30,38], formaldehyde[39] and freezing conditions[31]. The formation of N-nitrosamine is inhibited by the presence of ascorbic acid[40,41], α-tocopherol[42] and by many naturally occurring compounds[40,43,44]. Nitrite is widely distributed in the environment[45]. The use of sodium nitrite in meat products is a food preservation method practiced for a thousand years. Nitrite can accumulate in vegetables under abused storage conditions[46]. Furthermore, nitrite is generated in the human body in large quantities. It is formed in the oral cavity by the bacterial reduction of nitrate[47] and synthesized in the intestinal tract from ammonium compounds[18,19]. The availability of nitrite in the environment is a prime factor in the widespread occurrence of N-nitrosamines[48,49].

The nitrosation of secondary amines by labile N-nitrosamines has been reported[50,51]. The transnitrosation reaction has also been demonstrated with S-nitro[52], C-nitro[53,54], C-nitroso, O-nitroso (organic nitrite) and O-nitro (organic nitrate)[55] compounds as nitrosating agents. Nitrogen oxides such as N_2O_3 and N_2O_4 are also very powerful nitrosating agents[56]. Some nitrosating agents, in contrast to nitrite, are very active in neutral and alkaline conditions. Among all nitrosating agents, however, C-nitro compounds probably pose the most serious threat of contributing to environmental contamination by N-nitrosamines (with the exception of nitrite) because of their use in many consumer products[53].

Fan and Tannenbaum[57] studied the thermal stability of N-nitrosodialkylamines and cyclic N-nitrosamines under various pH conditions. Little loss of N-nitrosamines was observed when they were heated at 110°C for several hours. Thus, once formed, many N-nitrosamines are quite stable if exposure to light is avoided[58,59].

In the past decade, the knowledge about the occurrence of N-nitrosamines in the environment has accumulated rapidly. Numerous reviews have appeared recently. Fine[60] provided a general background on the formation, analysis and human exposure aspects of N-nitrosamines. Scanlan[61], Crosby[21] and Crosby and Sawyer[62] prepared comprehensive reviews concerning the occurrence of N-nitrosamines in foods. The more recent literature about N-nitrosamines in meat and cheese products was reviewed by Gray and Randall[63] and Gray et al.[64]. Krull et al.[65] discussed N-nitrosamines in consumer products and in the workplace. Human exposure to N-nitrosamines was reviewed by Fine[49] and Fine et al.[66-71].

DETERMINATION OF N-NITROSAMINES

N-nitrosamines are often present in trace quantities in extremely complex matrices. A series of sample workup steps is usually required before final analysis can be made. The sample workup generally includes:

1. isolation, to remove N-nitrosamines from the original matrix by distillation or solvent extraction;
2. cleanup, to remove potentially interfering compounds by chemical or physical manipulations;
3. concentration, to reduce the sample volume in order to increase the detection sensitivity;
4. separation, to separate a mixture of N-nitrosamines into individual, identifiable N-nitrosamines; and
5. detection, to detect and quantify N-nitrosamines in the sample.

The analysis of N-nitrosamines in food has been reviewed by Crosby[21], Eisenbrand[72] and Scanlan[61]. The isolation, cleanup and concentration procedures are outside the scope of this chapter and will not be discussed. The TEA analyzer can be used as a detector for gas chromatography (GC) or high-performance liquid chromatography (HPLC). Discussion is therefore limited to the separation and detection of N-nitrosamines by GC and HPLC.

Gas Chromatography

The literature on GC analysis of N-nitrosamines is abundant. In early work, the flame ionization detector (FID) was used as the GC detector. The FID is a universal detector for organic compounds. In a complex matrix, hundreds of organic compounds are present. Even after elaborate cleanup procedures, the sample still may contain sufficient interfering compounds to make the chromatogram unidentifiable[73-75]. For this reason, the FID is used for the determination of N-nitrosamines only in extremely simple systems[16,17,76-78].

The introduction of nitrogen-selective detectors (NSD) was a major advance in the analysis of N-nitrosamines. Two types of NSD have been used. The first type is an alkali flame ionization detector (AFID) which is a normal FID with a small salt pellet placed on the burner jet[79]. The addition of the salt pellet enhances the detector's response to nitrogen compounds. The AFID has been used in the detection of N-nitrosamines in food materials by Crosby et al.[80], Fazio et al.[81], Fiddler et al.[82] and Howard et al.[83]. The second type is an electrolytic conductivity detector (CECD) described by Coulson[84], in which N-nitrosamines are thermally degraded to ammonia, which is measured conductimetrically. CECD has been used for the analysis of N-nitrosamines in cigarette smoke condensates[85] and in various foods[73,86,87]. Palframen et al.[75] evaluated the performance of both types of nitrogen-selective detectors and found greatly enhanced selectivity for N-nitrosamines over the FID.

The electron capture detector (ECD) provides extremely high sensitivity toward electron-absorbing compounds but is insensitive to N-nitrosamines. However, N-nitrosamines can be derivatized to electron-absorbing compounds. Several derivative methods have been described: (1) N-nitrosamines are oxidized to N-nitramines[88-91]; (2) N-nitrosamines are denitrosated to form secondary amines, which are then converted to heptafluoro butyryl derivatives[92-94] or trifluoroacetyl derivatives[95]; and (3) N-nitrosamines are reduced to hydrazines, which condense with 3, 5-dinitrobenzaldehyde[96].

Despite the increased selectivity obtained by the use of nitrogen-specific detectors and by the preparation of derivatives responding to the electron capture detector, it is generally agreed by N-nitrosamine researchers that the identification of N-nitrosamines based solely on the coincidence of retention times in GC is of doubtful validity. Extracts of biological origin contain many nitrogen compounds, even after extensive cleanup. Coeluting nitrogen compounds will lead to false identification of N-nitrosamines using nitrogen-specific detectors. Mass spectrometry (MS) is required to obtain an unequivocal confirmation of the identity of a compound. In a study conducted by Goodhead and Gough[97] of 154 food samples that had been shown to contain NDMA by a Coulson electrolytic conductivity detector, only half the findings were confirmed by MS.

MS techniques used for N-nitrosamine identification can be conveniently divided into those based on low or high resolution. Low-resolution MS has been used as a confirmatory method. Essigmann and Issenberg[73] and Fazio et al.[74] screened food extracts with a nitrogen-specific detector. When peaks at the retention times of N-nitrosamines such as NDMA and N-nitrosopyrrolidine (NPYR) were observed, the identity of the particular N-nitrosamines was confirmed by a complete low-resolution mass spectrum.

High-resolution MS was used for the determination of N-nitrosamines based on a few characteristic ion fragments. With this technique, identity confirmation and quantification can be obtained simultaneously. MS data from a number of N-nitrosamines were given by Pensabene et al.[98]. In general, a molecular ion (M^+) of moderate intensity was observed along with a weak m/e 30 peak due to the NO^+ ion. Osborne[99] and Telling et al.[100] monitored NO^+ with a high-resolution MS for the analysis of N-nitrosamines in food. Telling et al.[100] discussed the possible interferences from ion fragments of all combinations of C, H, O and N at mass 30. At a resolving power of 15,000, $C^{18}O$ is the only ion fragment that can cause interference. Parent ion monitoring requires more skill but improves the detection sensitivity by fivefold over NO^+ monitoring[100-102]. Use of high-resolution MS to monitor parent ions has been a favored technique for the identification of N-nitrosamines.

It is desirable to set the resolution of MS sufficiently high to obtain unambiguous identification. Dooley et al.[103] and Gough and Webb[104] reported that trimethylsilane (HSiMe$_3$) had a similar retention time and MS characteristics to NDMA. It was indistinguishable from NDMA, even at a resolution of 7000. Trimethylsilane is often used as an antifoam agent during sample workup or to deactivate GC columns. Stephany et al.[105] monitored the parent ion at a moderate resolution of 4000 for the analysis of N-nitrosamines in food; however, they proposed to use two capillary GC columns of vastly different polarity. The presence of N-nitrosamines was considered unequivocally confirmed only when the quantification by MS was identical with both GC columns.

Even using high-resolution MS, undesirable pitfalls can be experienced. High-resolution MS can be carried out with the precise ion monitoring technique or the peak matching technique. Peak matching is regarded as the most reliable high-resolution MS method, which has been used by Bryce and Telling[101], Gough and Webb[102] and Pensabene et al.[106]. In this technique, the mass regions of selected ions of a reference compound and the N-nitrosamine are alternatively scanned every few seconds, in contrast to the precise ion monitoring technique, in which the precise mass of an ion is monitored constantly. Gough et al.[107,108] compared both high-resolution MS techniques with low-resolution MS monitoring of selected ions. Extracts of a variety of foods and biological materials were analyzed for the presence of N-nitrosamines. Wide discrepancies in both qualitative and quantitative determinations using low-resolution MS techniques and precise ion monitoring techniques were found in comparison to the results obtained by peak matching. This study demonstrated the danger of uncritical acceptance of MS identification without looking into the detailed methodology.

The analysis of N-nitrosamines by GC-TEA will be discussed in the section entitled Foods, p. 286. The introduction of the TEA analyzer is perhaps the most important development in the analysis of N-nitrosamines. The instrument offers high sensitivity and selectivity toward N-nitrosamines unmatched by other detectors. In the work of Gough et al.[107,108] just mentioned above, TEA detection was used as the fourth technique to compare the performance against three MS techniques. The analytical results obtained from the TEA analyzer agreed with those from peak-matching high-resolution MS, both qualitatively and quantitatively. A similar observation was reported by Havery et al.[109].

High-Performance Liquid Chromatography (HPLC)

Because of the limited availability of the instrument until recently, the published works on N-nitrosamines using HPLC are very scant. Cox[110] described an HPLC procedure that allowed the separation of six volatile N-nitrosamines. Iwaoka et al.[111] reported the separation of synisomers and antiisomers of

N-nitrosamino acids by HPCL using a UV detector at 254 nm. However, the selectivity of UV detectors is such that it is inadequate for the analysis of N-nitrosamines in food extracts[71]. Iwaoka and Tannenbaum[112] described a selective detector for HPLC based on the automatic analytical procedure of Fan and Tannenbaum[113]. N-nitrosamines are denitrosated with UV irradiation and the resulting nitrite is analyzed by a modified Griess reagent. The detection system has been used for the detection of N-nitrosoproline (NPRO) in raw bacon[114]. Singer et al.[115] described a similar colorimetric detection system for N-nitrosamides after thermal denitrosation. Thermal denitrosation is applicable only to N-nitrosamides, and photolytic denitrosation is subject to interferences[113]. Furthermore, the colorimetric determination of nitrite must be carried out in an aqueous medium. All these limitations prevent the widespread adoption of these techniques for the analysis of N-nitrosamines. The TEA analyzer can be interfaced with HPLC (see the section entitled Water, p. 291). The sensitivity and selectivity of the TEA analyzer coupled with the ability of HPLC to handle both volatile and nonvolatile N-nitrosamines provides a very valuable tool for N-nitrosamine research.

THE TEA ANALYZER

During the last decade, many analytical procedures have been developed with a sensitivity at low g/kg levels in complex matrices such as foods[21,61]. However, extensive cleanup is often required to remove interfering substances. False-positive identifications are frequently encountered even with a nitrogen-specific detector and gas chromatography[97]. Concentration of the extract as much as 1000-fold is sometimes necessary to achieve high sensitivity. These analytical procedures are complicated and time consuming. Because of its sensitivity and selectivity, the use of the TEA Analyzer as the detector for either GC or HPLC can eliminate these problems. The principle of operation of the TEA Analyzer has been described by Fine and Rufeh[116] and Fine et al.[117,118].

Principle of Operation

The TEA Analyzer is a chemiluminescent detector. Figure 2 shows the schematic arrangement of the detector. When an N-nitrosamine is introduced into the TEA, it enters first into a pyrolysis chamber, where the N—NO bond in the molecule is cleaved. Nitrosyl radicals produced from the bond cleavage are swept by a carrier gas into cold traps. Interfering substances are removed by the cold traps and a Tenax cartridge. Nitrosyl radicals escape both types of traps and enter the reaction chamber where they react with ozone to produce NO_2 in an excited state.

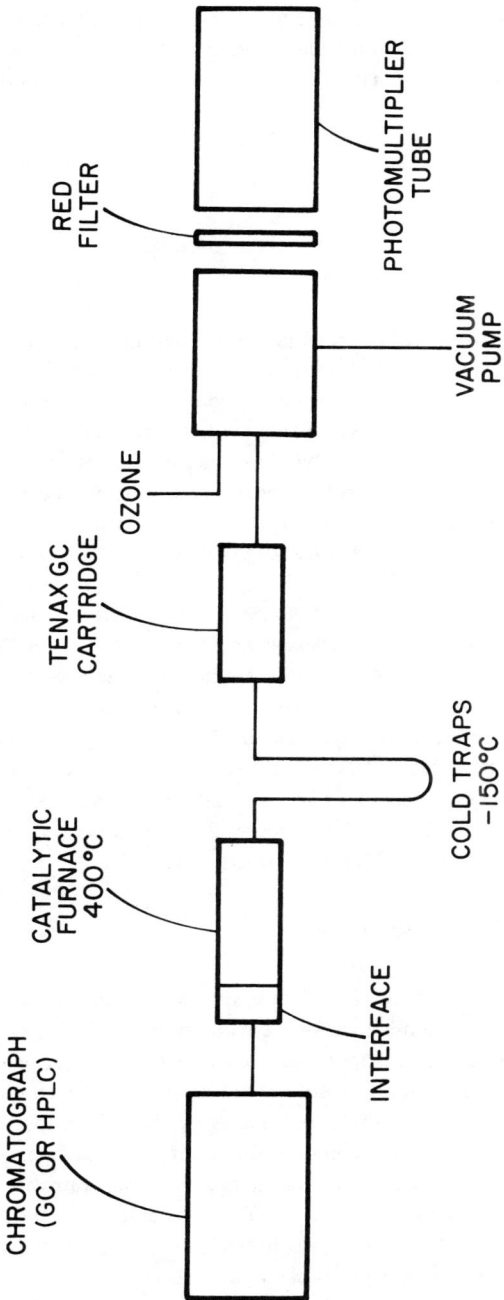

Figure 2. Scheme of the TEA Analyzer.

The excited NO_2 decays to the ground state, resulting in light emission. A photomultiplier tube receives the light emission and amplifies it electronically to give a response. A vacuum pump is used to maintain the instrument under moderate vacuum.

Description of TEA

Pyrolysis Chamber

Fine and Rufeh[116] described the calculation of bond dissociation energy for the N—NO bond in a molecule. The bond strength of N—NO bond was estimated to be in the range of 8 to 40 kcal/mol, while the bond strengths of C—N, C—C and C—H in most organic compounds typically are in the range of 70 to 90 kcal/mol. It is likely that in most cases the N—NO bond is the weakest in a N-nitrosamine molecule. Flash heating in a pyrolysis chamber results in the rupture of the N—NO bond and release of a nitrosyl radical, which is a stable gas at normal temperature and pressure[117]. The other organic fragments decompose or rearrange to more stable species.

The temperature of the pyrolysis chamber is maintained at constant temperature between 300 and 600°C[119]. Operating the instrument with the pyrolysis chamber at optimal temperature is a key factor in obtaining a satisfactory performance from the TEA. If the temperature is too low, the pyrolysis of more stable N-nitrosamines will be incomplete. On the other hand, if the temperature is too high, excessive decomposition of N-nitrosamines will introduce more interfering substances into the reaction chamber and increase the "noise" level of the detector. The optimal temperatures of the pyrolysis chamber are approximately 450°C for GC operation and 500–550°C for LC operation.

Cold Traps and Tenax Cartridges

Both are designed to remove interfering substances. The cold trap for GC operation is a U-shaped stainless steel tube with an external diameter of ¼ inch (6.4 mm). The cold traps for LC operation are two glass impingers with a capacity of approximately 150 ml each. Because the TEA is operated under moderate vacuum (1–2 torr), the seemingly large volume of the cold traps will not contribute significantly to the "dead volume" for either GC or LC operation. Sharp symmetrical peaks are easily obtained (e.g., the chromatograms in the work of Fine and Rounbehler[120] and Fine et al.[117]. The temperature of the cold traps is maintained constant by immersion in a cold bath, which can be provided by mixing liquid nitrogen or dry ice with suitable organic solvents. Desired temperatures are obtained by selecting the proper solvent; for example, a − 150°C cold trap is provided by mixing liquid nitrogen in isobutane; a $-120\,°C$ cold trap by liquid nitrogen in ethanol; and a $-78\,°C$ cold trap by dry ice in ethanol or acetone. The colder the temperature of the cold traps, the fewer the interfering substances and the more selective will be the detector. However, the temperature of the cold trap should not be so cold that it freezes out the solvents coming from an HPLC. For GC operation, the temperature is usually maintained around $-150\,°C$. For HPLC operation, it depends on the elution solvent used. Baker and Ma[121] used an aqueous elution solvent system for HPLC. The cold traps were maintained near $0\,°C$ with ice and water. Because of the high cold trap temperature, they experienced the reduction of sensitivity and selectivity and unsatisfactory instrument performance. In our laboratories, normally two cold traps are used for HPLC operation, with the first trap kept at a temperature sufficient to trap the elution solvent, and the second trap at $-120\,°C$ to $-150\,°C$ to remove highly volatile interfering substances. On rare occasions, three traps at three different temperatures have been used for certain elution solvents, such as those containing large amounts of water.

A stainless steel tube that has an i.d. of ¼ inch (6.4 mm) and a length of 4 inches (10 cm) is filled with Tenax GC packing material and installed after the cold traps as the second line of defense against extremely volatile interfering substances. At room temperature, the nitrosyl radical will pass through the Tenax cartridge without hindrance, while other substances will be retarded. The compounds trapped in the cartridge will be eluted slowly through the column and emerge as a flat and undetectable peak.

Reaction Chamber

In the reaction chamber, nitrosyl radicals react with ozone according to the following equations[117]:

$$NO + O_3 \rightarrow NO_2{}^* + O_2$$

Excited NO_2 ($NO_2{}^*$) is formed. $NO_2{}^*$ will decay to its ground state with concurrent light emission, which is detected to produce a TEA response. However, competing reactions also may occur. $NO_2{}^*$ may collide with other molecular species and not undergo light emission. The collision involves two or more molecules and is more likely to occur at high pressure. Figure 3 shows the effect of pressure on the fraction of nitrosyl radicals that emit light after reacting with ozone. The TEA Analyzer is operated under very low pressure to reduce the side reactions and increase the decay of $NO_2{}^*$.

Photomultiplier Tube

The photomultiplier tube is used to receive the light emission from the reaction chamber and produce a signal after electronic amplification. Light emission

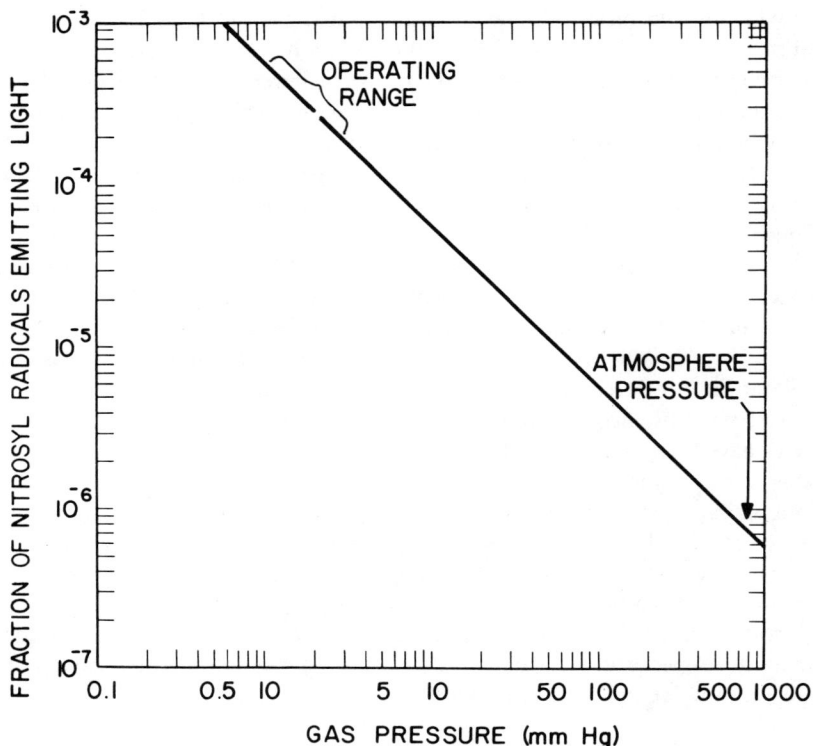

Figure 3.   Effect of pressure on the fraction of nitrosyl radicals that emits light.

resulting from the decay of NO_2^* is in the near infrared (IR) region of the spectrum between 0.6μ and 3.0μ[122]. A red filter is installed between the reaction chamber and the photomultiplier tube to eliminate light with wavelengths shorter than 0.6μ. Compounds such as carbon monoxide and ethylene can react with ozone to produce the light emission with wavelengths shorter than 0.6μ[117]. Figure 4 shows the spectral response characteristics of the photomultiplier tube, the wavelength cutoff of the red filter and the IR wavelength of light emission. Photomultiplier tubes manufactured at present can only detect a small portion of the light emission. Due to the combination of a red cutoff filter and the poor photomultiplier tube characteristics, only the narrow region between 0.6 and 0.8μ is monitored.

Selectivity

For a compound to produce a response from the TEA Analyzer, a series of obstacles have to be overcome:

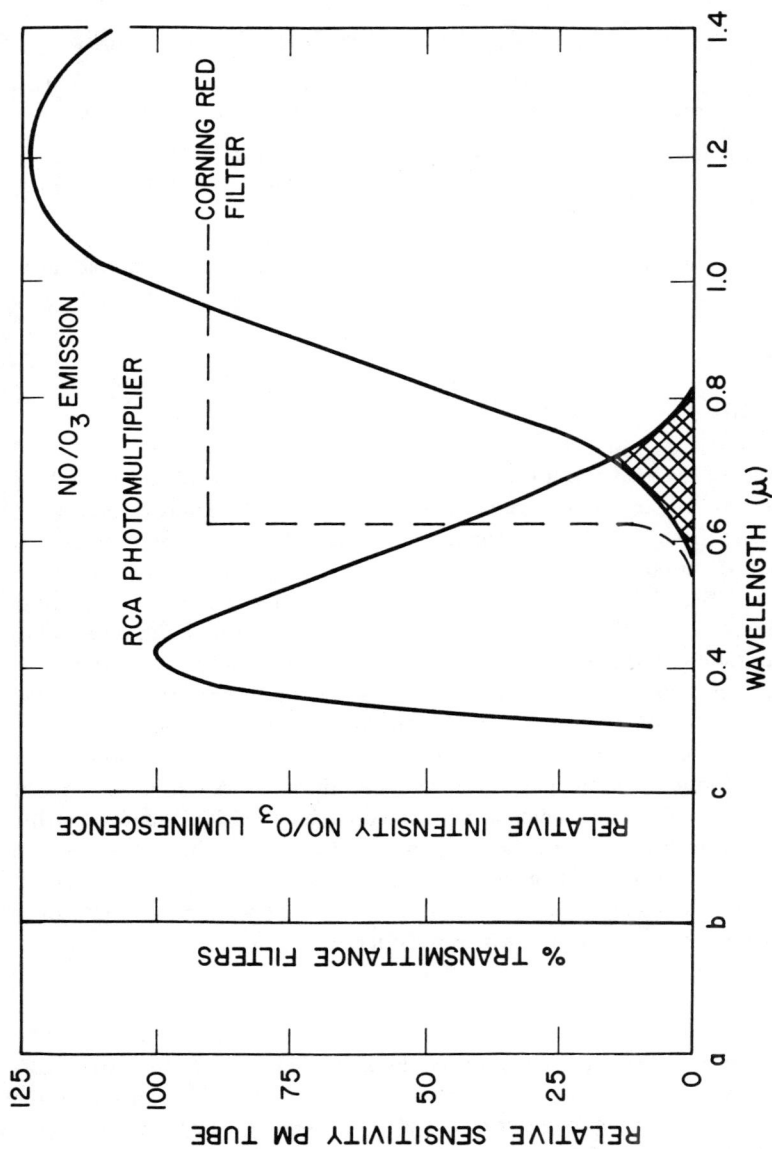

Figure 4.   Distribution of chemiluminescent emission in the IR showing filter cut-off point and responsivity of photomultiplier.

1. The compound must be sufficiently labile that it decomposes within a few seconds on a catalytic surface at temperatures between 400 and 550°C.
2. The pyrolyzed products must survive the cold traps at a temperature as low as −150°C.
3. The pyrolyzed products must be able to pass through the Tenax cartridge without any resistance.
4. The pyrolyzed products must be able to react with ozone.
5. The reaction with ozone must emit light in the wavelength range between 0.6 and 0.8μ.
6. The chemiluminescent reaction must be sufficiently rapid (probably less than one second) that the reaction occurs before the reactants are pumped out of the reaction chamber by the vacuum pump.

For the above reasons, the TEA Analyzer is extremely selective for N-nitrosamines.

Operation

Direct Mode

The use of the TEA for direct analysis of a sample can be performed simply by injecting the sample into the heated pyrolysis chamber[123]. One nitrosyl radical will be released from one N—NO group in a molecule[118]. As a result, equimolar amounts of various mono N-nitroso compounds will produce the same response from the TEA (Table I). For this reason, the direct mode is very useful as an N-nitrosamine-specific group detector[123,124]. Fine et al.[124] tested the method in complex food systems. NDMA (a volatile N-nitrosamine) and N-nitrosodiphenyl-amine (a nonvolatile N-nitrosamine) were spiked in fresh beef or fresh herring. The spiked samples were extracted with dichloromethane. Without any cleanup and concentration steps, extracts corresponding to 1 ng of N-nitrosamines were injected into the TEA Analyzer. Recovery of N-nitrosamines was in the range of 70% to 100%. The experiments demonstrated the high sensitivity and selectivity of the instrument. However, the direct mode of operation can only serve as a group detector; it does not show the identity of N-nitrosamines being detected. With the development of GC and HPLC interfaces, the direct mode operation is seldom used except for the quick screening purpose.

GC Mode

A GC-TEA system has been described by Fine et al.[125]. The interface of GC-TEA is constructed with a narrow-bore stainless steel tube, which introduces the effluent of GC into the pyrolysis chamber of the TEA. Fine and Rounbehler[126] reported the analysis of N-nitrosamine mixtures by GC-TEA. The relationship between TEA response and N-nitrosamine concentration is linear for at least five orders of magnitude. The maximum linear range has not been fully explored.

Table I. TEA Response Factors for Different N-Nitroso Compounds

N-nitroso Compound	Mol wt	Conc. (μg/ml)	Measured Response (integrated units)	Response per Nitrosyl Group	Relative Response Nitrosyl Mole Basis
N-nitrosodimethylamine	74	0.964	235	18.1	1.00
N-nitrosodiethylamine	102	1.07	204	19.4	1.07
N-nitrosodipropylamine	130	0.84	128	19.8	1.09
N-nitrosodiphenylamine	198	1.86	189	20.1	1.11
N-nitroso-N-ethylanaline	150	1.17	145	18.6	1.03
9-Nitrosocarbazole	196	1.99	167	17.6	1.03
N-nitroso-N-methyl urethane	132	0.52	74	18.8	1.04
N-nitroso-N-phenylbenzylamine	212	2.10	166	16.7	0.92
Ethyl-N-nitrososarcosinate	146	2.15	258	17.5	0.97
N-methyl-N-nitroso-N-nitroguanidine	147	4.71	590	18.4	1.02
N-nitrosopiperidine	114	0.93	172	21.2	1.17
N,N'-Dinitrosopiperazine	144	1.99	434	15.7	0.87

Figure 5 shows the GC-TEA chromatogram of 14 volatile N-nitrosamines, with each at the 5 ng level.

The analysis of N-nitrosamines in foods was demonstrated by Fine and Rounbehler[127] and Fine et al.[128]. The method involved only three steps: (1) vacuum distillation, (2) solvent extraction and (3) concentration. Total time for analysis was less than 90 minutes, which compared favorably with other conventional analytical procedures (e.g., those by Fazio et al.[81]). Figure 6 illustrates the selectivity and

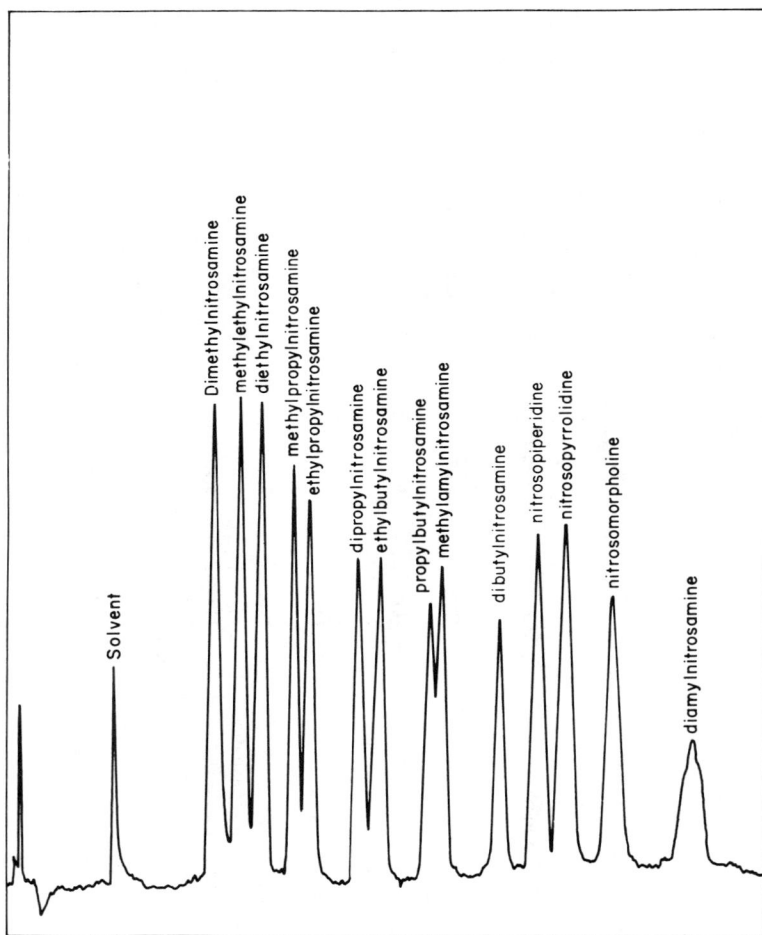

Figure 5.   GC-TEA chromatogram for a standard mixture of 14 volatile nitrosamines, each at a concentration of 0.5 μg/ml. GC column is 15% FFAP on Chromosorb W. Temperature programmed at 5°C/min from 140–210°C.

sensitivity of GC-TEA. Four N-nitrosamines spiked in canned beef at 2 μg/kg level were determined. A clean chromatogram with easily measurable peaks was obtained.

Table II presents the comparison of various GC detectors for sensitivity, linear-

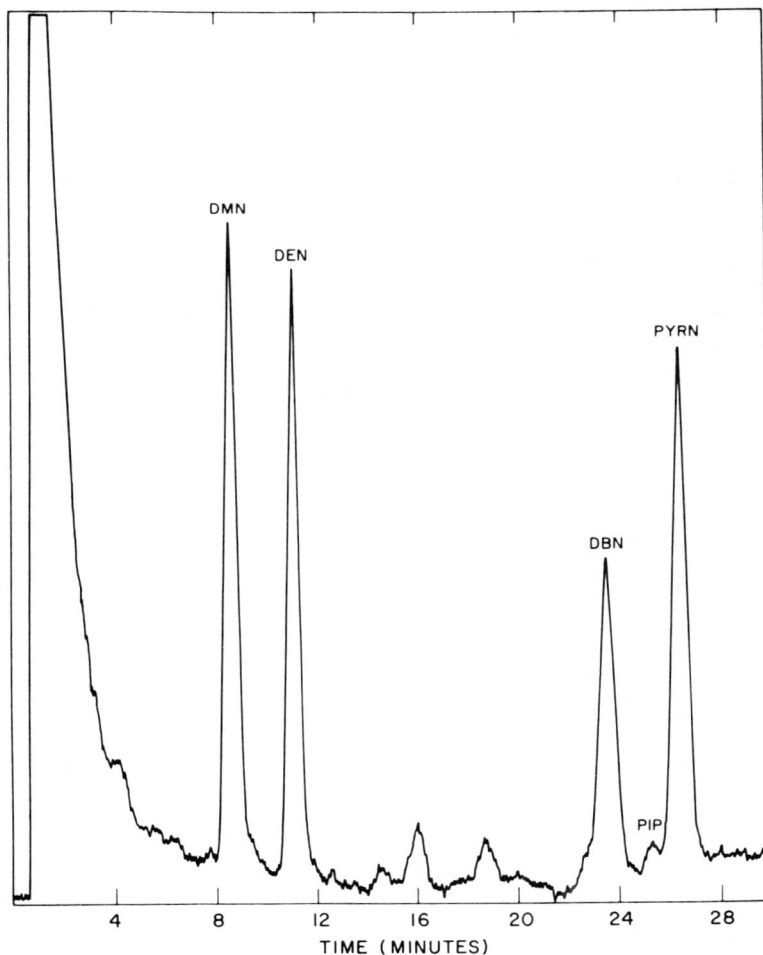

Figure 6.   GC-TEA chromatogram from canned beef extract spiked with a mixture of N-nitrosodimethylamine (DMN), N-nitrosodiethylamine (DEN), N-nitrosodibutylamine (DBN) and N-nitrosopyrazine (PYRN), each at 2 μg/kg. GC column is 15% FFAP on Chromosorb W. Temperature programmed at 5°C/min from 140–210°C.

Table II. Comparison of Performance Parameters of GC Detectors

Detector	Sensitivity (ng)[a]	Reference	Linear Range	Reference	Selectivity
Flame Ionization Detector	10	73	10^7	3	Universal detector for organic compounds
Alkali Flame Ionization Detector	10	74	10^3	3	Enhanced response to nitrogen-containing compounds
Coulson Electrolytic Conductivity Detector	5	73			Selective for nitrogen-containing compounds
Electron Capture Detector			5×10^2	3	Responds to electron-absorbing compounds, such as nitrate, halogens and conjugated carbonyls
(a) Nitramines	0.02	54			
(b) Heptafluorobutyl derivatives	4	129			
Thermal Energy Analyzer	0.5	126	better than 10^6	126	Selective for N-nitrosamines

[a] Sensitivity is calculated based on the response of N-nitrosodimethylamine at a signal:noise ration of 5:1.

ity range and selectivity. The electron capture detector is the most sensitive detector; however, it has a short linearity range and poor selectivity in the presence of other nitro compounds, halogen compounds and conjugated carbonyls. Furthermore, preparation of electron-absorbing derivatives is required. Both nitrogen-specific detectors are 10–20 times less sensitive than the TEA Analyzer and not specific to N-nitrosamines. Overall, the TEA is clearly the superior detector for GC analysis of N-nitrosamines[130].

LC Mode

An HPLC-TEA system was described by Fine and Rounbehler[120]. Because of the solvents delivered at high pressures by the HPLC pump and the high vacuum required by the TEA Analyzer, the design of an interface between the HPLC and the TEA needed special consideration. The present interface employs a 0.1-mm-diameter solvent inlet concentric with a 3-mm atomizer nozzle[131]. The nozzle geometry combined with a large pressure drop serves to atomize the liquid into small droplets. Argon carrier gas, introduced through the nozzle at the beginning of the pyrolysis chamber, also sweeps the atomized droplets through the length of the chamber.

HPLC is applicable to both volatile and nonvolatile N-nitrosamines. Figure 7 shows a typical HPLC-TEA chromatogram for a mixture of volatile and non-volatile N-nitrosamines[132]. The application of HPLC-TEA to the analysis of N-nitrosamines in biological samples was reviewed by Krull et al.[133]. the application of HPLC-TEA to foods was reported by Fine et al.[134]. N-nitrosamines were spiked at the 10 μg/kg level in liquor (vodka), fish, cooked bacon, luncheon meat and dried beef. The food samples were simply extracted with organic solvent and after the extracts were concentrated, they were injected into an HPLC-TEA. No cleanup was necessary.

The performance characteristics of HPCL-TEA were reported by Fine et al.[71]. A linearity range of at least four orders or magnitude was observed. This is shown in Figure 8. N-nitrosocarbazole was detected at the 0.4 ng level with a signal:noise ratio of 5. The benefit of the TEA analyzer as an HPLC detector was clearly seen when it was compared against a UV detector[134]. Large quantities of UV-absorbing substances in the cooked bacon extract caused the UV detector to be off-scale for most of the chromatogram. By comparison, the TEA showed only two small peaks following an initial solvent front.

Artifacts

Since the analysis of N-nitrosamines is done routinely at μg/kg levels, there is concern about possible false negatives and false positives. Since most N-nitrosa-

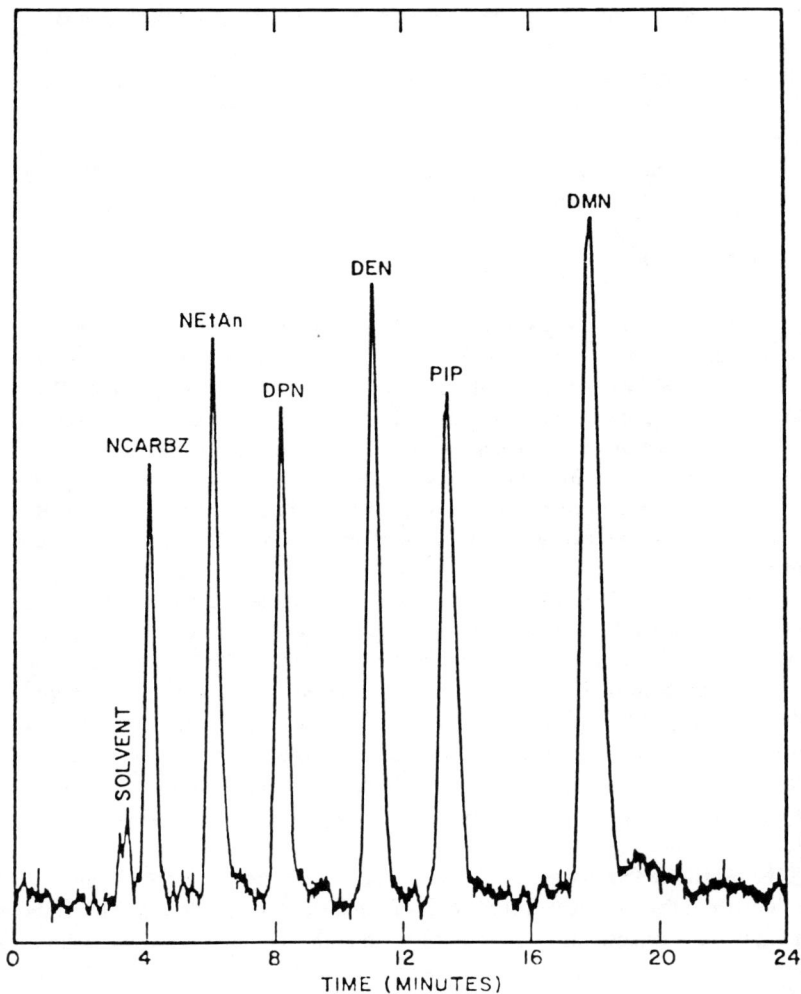

Figure 7. HPLC-TEA chromatogram of a 10-$\mu\ell$ injection of a solution containing 20 μg/ml each of N-nitrosocarbazole (NCARBZ), nitroso-N-ethylaniline (NEtAn), N-nitrosodipropylamine (NDPA), N-nitrosodiethylamine (NDEA), N-nitrosopiperidine (NPIP) and N-nitrosodimethylamine (NDMA). Column used was μ-Bondapack NH$_2$; solvent was 85% isocotane/15% chloroform; flowrate was 1.0 ml/min.

mines are carcinogens, it is also important to correctly determine the presence or absence of N-nitrosamines in any given sample. Krull et al.[135] have discussed in detail the cause and prevention of artifacts during the analysis of N-nitrosamines.

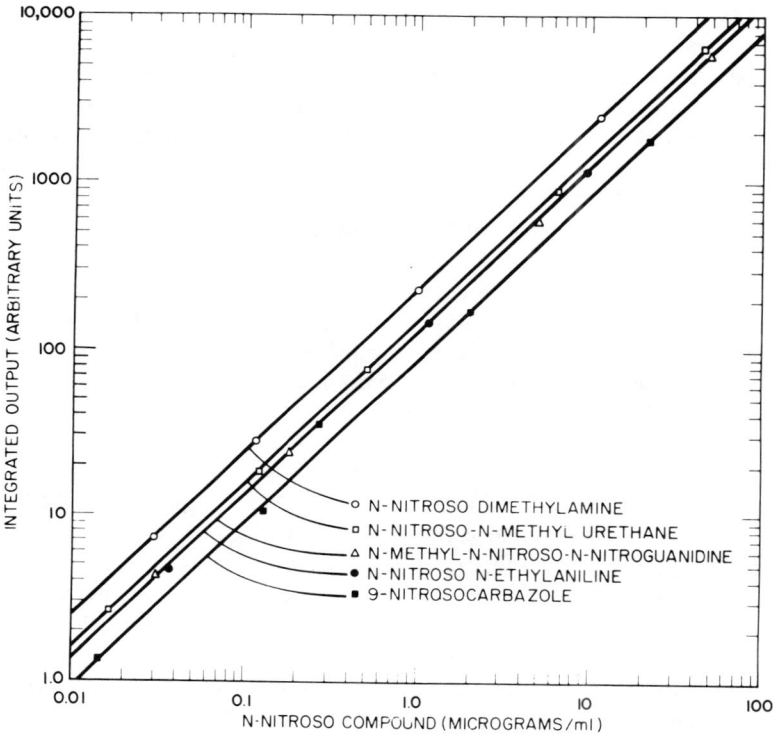

Figure 8.   HPLC-TEA calibration for five nitrosamines.

False Negatives

A false negative is failure to detect a substance that is present in a sample. Most N-nitrosamines examined gave near-molar response on the TEA Analyzer. False negative response has never been observed in the TEA analysis of N-nitrosamines. However, some N-nitrosamines may be sensitive to light, pH or other conditions. Accidental destruction of these compounds may occur during the sample workup. Selection of inappropriate solvent may result in poor efficiency in the solvent extraction. False negative problems can also arise when drastic conditions are used to concentrate an extract, causing the loss of volatile N-nitrosamines through evaporation. The best way to safeguard against the occurrence of false negatives is to carry out recovery experiments. The N-nitrosamine under study or an N-nitrosamine with similar properties should be spiked into the sample. Exactly the same analytical procedure is followed for the analysis of the sample. The recovered amount of the spiked N-nitrosamine will show quickly whether there is a false negative problem.

False Positives

A false positive is the misidentification of an instrument response resulting from chemical compounds other than N-nitrosamines. As is true for all selective detectors, occasional interferences do occur on the TEA. Although many steps are incorporated in the instrument to eliminate interferences, it is still impossible to control the problem completely.

There are three major types of false positives:

Unsaturated olefins. Some compounds may yield unsubstituted olefins of low molecular weight in the pyrolysis chamber, which can chemiluminesce with ozone[118]. This type of interference can be eliminated if a small Tenax cartridge is installed between the cold traps and the reaction chamber (see the section entitled Cold Traps and Tenax Cartridges, p. 282).

Nitrosyl-radical releasing compounds. This type of interface is more of a problem. Fine et al.[118] tested a wide array of chemical compounds on the TEA Analyzer for possible interference. Several compounds, including inorganic and organic nitrite and polynitrobenzenes, responded on the TEA Analyzer. Since then, several published works reported interferences from compounds that could release nitrosyl radical in the pyrolysis chamber and thus are impossible to eliminate as interferences[53,136–140]. Table III lists the interfering compounds. All organic nitrites and organic nitrates can respond on the TEA Analyzer. It is likely that most nitramines could respond on the TEA as well. Aliphatic C-nitro and C-nitroso compounds are more labile than their aromatic counterparts. Many aromatic C-nitroso compounds did not show any response to the TEA, while aliphatic C-nitroso compounds respond on the TEA[140]. Electron withdrawing groups on an aromatic ring tend to promote loss of nitrosyl radical on pyrolysis (e.g., pentachloronitrobenzene). Nitro groups can withdraw electrons, which may explain the fact that polynitrobenzenes respond on the TEA.

Formation of N-nitrosamines in GC injection port. When an N-nitrosamine is detected in a sample, it is likely that the precursor is also present. If the concentrations of the precursors are sufficiently high, formation of N-nitrosamines could occur in the hot GC injection port. This was indeed demonstrated by Fan and Fine[142]. A herbicide formulation was found to contain 1200 μg/ml of NDMA when the sample was injected into GC-TEA without any treatment whatsoever. However, when the herbicide was diluted more than ten times before analysis, the NDMA level in the herbicide was found to be only 360 μg/ml. Use of the HPLC-TEA, which eliminated the hot injection port, resulted in an estimated 360 μg/ml of NDMA in the herbicide. NDMA found in the herbicide decreased from 1200 to 370 μg/ml as the injection port temperature was decreased from 230 to 150°C. All evidence pointed to the formation of NDMA in

Table III.   TEA Responsive Compounds that are not N-nitrosamines

Compound	Molar Response Ratio[a]	Reference
Sodium Saccharin	1×10^{-4}	134
C-nitro Compound		
Tetranitromethane	3.1	53
2,2',4,4',6,6'-Hexanitrodiphenylamine	1.4	118
2,3-Dimethyl-2,3-dinitrobutane	1.1	53
2,2-Dinitropropanol	9×10^{-1}	53
1,5-Difluoro-2,4-dinitrobenzene	1×10^{-1}	53
2,4,6-Trinitro-3,5-dimethyl-t-butylbenzene	1×10^{-1}	53
2-Nitro-2-bromo-1,3-propanediol	8×10^{-2}	53
2-Nitro-1-propanol	2×10^{-2}	53
2,6-Dinitrochlorobenzene	2×10^{-2}	53
α,α,α-Trifluoro-2,6-dinitro-N,N-dipropyl-p-toluidine(trifluralin)	1×10^{-2}	53
Pentachloronitrobenzene	4×10^{-3}	53
C-nitroso Compound		
2-Nitroso-2-acetoxymethylpropane	b	141
2-Nitroso-2-methylpropane	b	141
2-Nitroso-2-chloromethylpropane	b	141
O-nitro Compound		
Ethylene glycoldinitrate	2×10^{-1}	137
N-propylnitrate	7×10^{-1}	137
Isopropylnitrate	7×10^{-1}	137
N-butylnitrate	8×10^{-1}	137
Isopentylnitrate	9×10^{-1}	137
O-nitroso Compound		
n-Pentylnitrite	1.0	118
Isopentylnitrite	1.0	118
Sodium nitrite	1.0	118
Nitric acid	1.0	118

[a] Molar response ratio $= \dfrac{\text{TEA response per mole of compound}}{\text{TEA response per mole of NDMA}}$

[b] Due to a faulty pyrolysis chamber, the response ratios were estimated incorrectly. The response ratios should be less than 1.0 because each compound contains only one nitroso group.

the GC injection port. In fact, Freed and Mujsce[143] generated N-nitrosamines in a GC injection port by injecting secondary amine solution onto a precolumn packed with KNO_2 to avoid the synthesis of carcinogenic N-nitrosamines to be used for standards.

Special Techniques

Mineral Oil Distillation Procedure

Fine et al.[128] developed a mineral oil distillation technique for the analysis of volatile N-nitrosamines in foods. The technique was later simplified by Fine and Rounbehler[127] and adopted by Eisenbrand et al.[144], Hansen et al.[114], Spiegelhalder et al.[141] and Stephany and Schuller[145]. Havery et al.[109] compared the multidetection method developed by Fazio et al.[81] and the mineral oil distillation method for the analysis of N-nitrosamines in 106 cured meat samples. The results from the survey showed good agreement between the two analytical methods.

The detailed description of the mineral oil distillation procedure can be seen in the work of Fine[49] and Havery et al.[109]. Briefly, the sample is minced and weighed (approximately 20 g) into a 500-ml flask containing 20–30 ml of mineral oil. The contents are heated slowly under vacuum. The distillate is collected in a cold trap immersed in a liquid nitrogen bath. Vacuum distillation is continued for about 40 minutes, by which time the oil temperature has reached about 100°C. The distillate is extracted with dichloromethane. The extract is then concentrated and analyzed by GC-TEA. It should be noted that because of the high selectivity of TEA, the extract can be concentrated immediately without the need for cleanup. The simple distillation technique and the elimination of cleanup reduce the analytical time remarkably. The recoveries of 14 volatile N-nitrosamines added at 10 μg/kg level averaged 92% by the mineral oil distillation procedure.

Internal TEA Confirmation Technique

The separation mechanism of GC is based on the difference in the vapor pressure and solubility of the compounds in the liquid phase. In HPLC, compounds elute according to either their size or their polarity. Because of the fundamental difference in the separation mechanism, the elution order by GC and HPLC is vastly different[132]. Using a selective detector, an identification can be made with confidence if both retention times of GC and HPLC coincide with an authentic N-nitrosamine standard.

Parallel GC-TEA, HPLC-TEA procedures have proved invaluable in avoiding incorrect identification of N-nitrosamines[136]. In the screening of pesticides, a peak with a retention time identical to N-nitrosodiethanolamine (NDEA) was detected by HPLC-TEA. However, when the same sample was analyzed by GC-TEA, no trace of NDEA appeared in the chromatogram (Figure 9). Therefore, NDEA was ruled out as the HPLC peak. The internal TEA confirmation technique has also been used successfully to identify NDMA in air[146] and NDMA and NDPA (N-nitrosodipropylamine) in herbicides[147]. In both cases, GC-HRMS

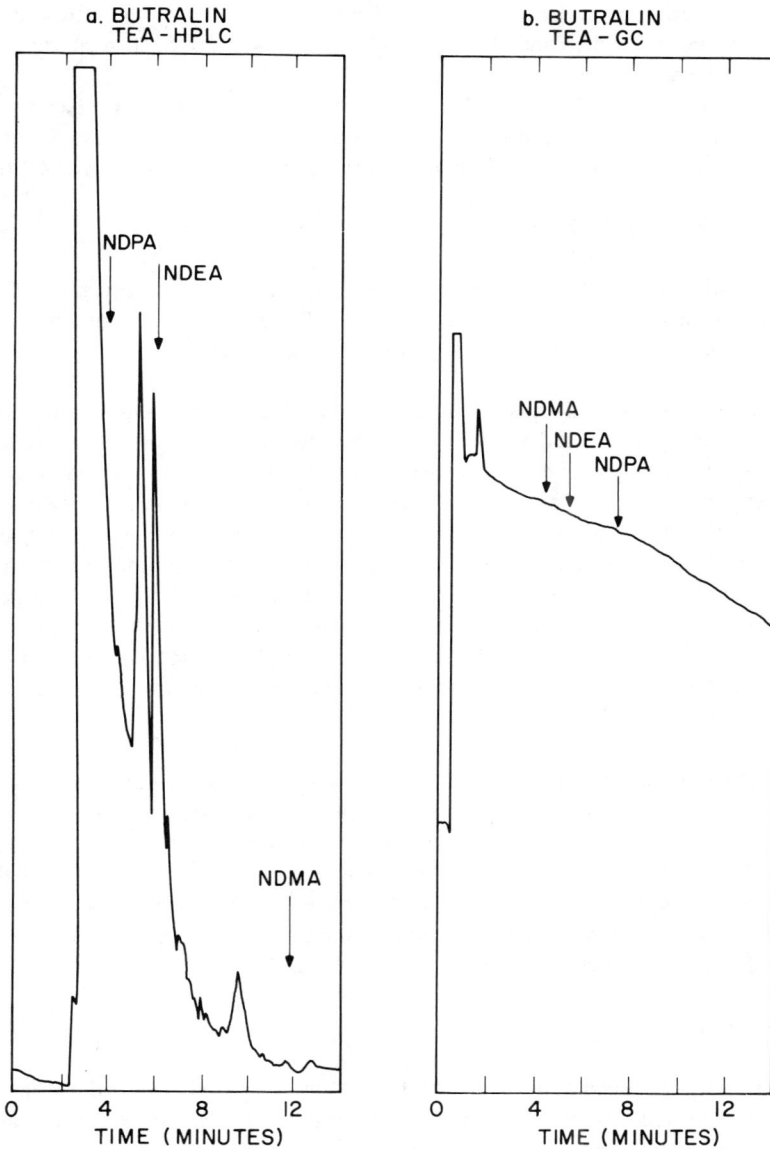

Figure 9. TEA results from butralin herbicide: (a) HPLC condition: μ-Porasil column, 95% hexane/5% acetone solvent, 2 ml/min flowrate; (b) GC conditions: Porapak P. column, 230°C isothermal temperature conditions.

(high-resolution mass spectrometry) confirmed the identity of the nitrosamine. The presence of NDMA and NDEA in human blood was also confirmed using the internal TEA confirmation technique[66].

This confirmatory technique is particularly valuable when the availability of the sample is limited. The technique requires only 1 ng of N-nitrosamine, which is considerably less than the quantity needed for mass spectrometric confirmation.

Comprehensive Analytical Procedures

Analytical schemes to analyze all classes of N-nitrosamines using the TEA Analyzer have been proposed by Fan et al.[136]. Depending on the complexity of the sample, two different analytical approaches can be used:

Simple Sample. Examples of a simple sample are pesticide and pharmaceutical formulation. A liquid sample can be introduced using either GC or HPLC without any pretreatment. A solid sample can be simply dissolved in a suitable solvent and analyzed. The strategy of the comprehensive analysis is to begin with an HPLC elution solvent that is sufficiently polar to elute all compounds at t_o (retention time of nonretained species). A peak observed at t_o indicates that the presence of N-nitrosamines is suspected. Less polar solvent mixtures are then used to elute the compounds, and their retention times are compared with authentic standards. The analytical scheme is presented in Figure 10. This method has been applied to the analysis of N-nitrosamines in agricultural and home-use pesticides[23,136,137,148].

Complex Sample. Examples of complex samples are foods and blood. This type of sample requires more exhaustive treatment. The strategy here is to classify all N-nitrosamines into four categories (Figure 11). Various methods are used to isolate and separate different classes of N-nitrosamines (Figure 12). The volatile N-nitrosamines (Class I) are isolated by the mineral oil distillation procedure and analyzed by GC-TEA. Solvent extraction is used to isolate three other classes of N-nitrosamines, which are analyzed by HPLC-TEA.

Frozen Animal Procedure (FAP)

FAP was developed by Rounbehler et al.[149] for the analysis of volatile N-nitrosamines in small laboratory animals. The animal is frozen in liquid nitrogen and homogenized to a fine powder in a blender. N-nitrosamines in the powder are isolated by mineral oil distillation and analyzed by GC-TEA. FAP was used by Rounbehler et al.[149] to study the in vivo formation of N-nitrosamines in animals after feeding the precursors. It was also used for the analysis of N-nitrosamines in human blood[66].

Sample

Elution solvent: acetone
Column: μ Porasil

No peak; no detectable N-nitroso compound in the sample	Peak beyond nonretention volume	Peak at nonretention volume

Sample contains extremely polar TEA-responsive substances

Elution solvent: 45% acetone, 55% hexane

No peak; sample contains highly polar TEA-responsive substances	Peak at nonretention volume	Peak beyond nonretention volume

Compare retention time to know polar N-nitroso compounds (e.g., NDE1A)

Elution solvent: 5% acetone, 95% hexane

No peak; sample contains polar TEA-responsive substances	Peak at nonretention volume	Peak beyond nonretention volume

Sample contains extremely low polar TEA-responsive substances

Compare retention time to known intermediate and low polar N-nitroso compounds (e.g., NDMA, NDEA, etc.)

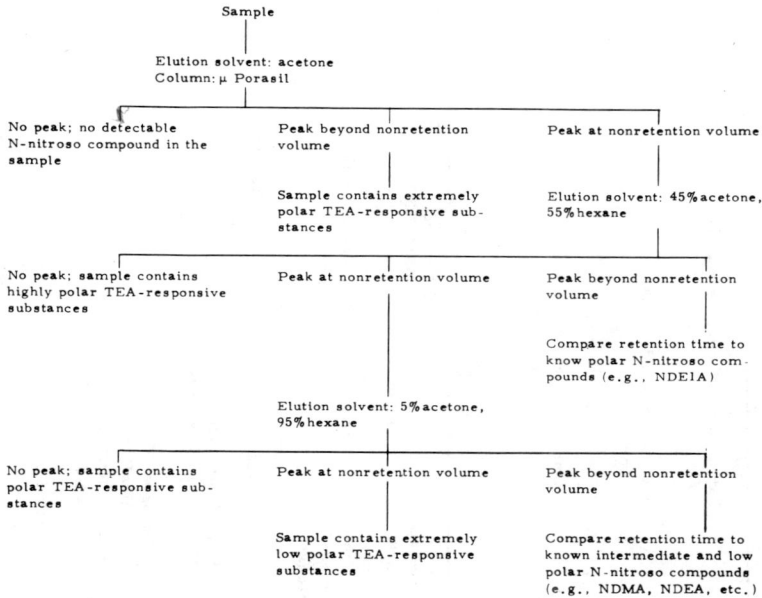

Figure 10. Analytical scheme for the screening of pesticide samples for N-nitroso compounds using the TEA Analyzer.

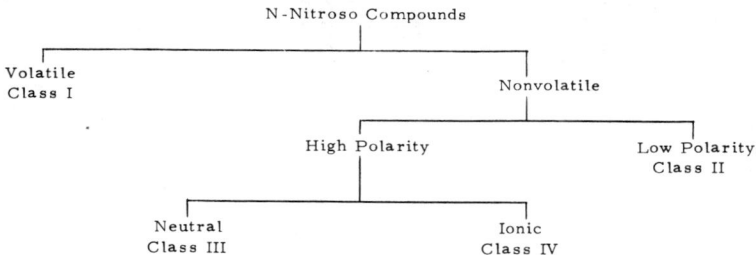

N-Nitroso Compounds

Volatile Class I		Nonvolatile

	High Polarity	Low Polarity Class II

Neutral Class III	Ionic Class IV

Figure 11. Classification of N-nitroso compounds into four broad, overlapping classes.

APPLICATIONS OF THE TEA ANALYZER

A wide variety of materials have been investigated for N-nitrosamine contamination using GC-TEA or HPLC-TEA. The literature compilation is shown in Table IV.

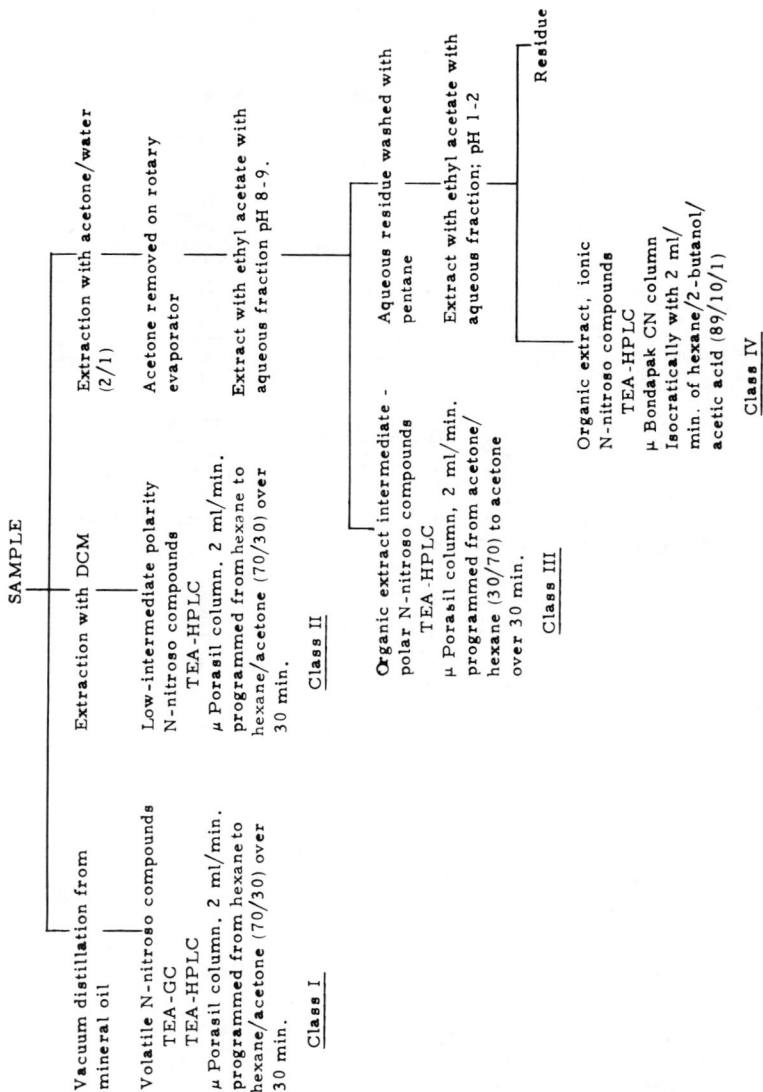

Figure 12.   Comprehensive analytical scheme for the analysis of N-nitroso compounds in biological and environmental samples.

Table IV.   Application of TEA for the Analysis of N-nitrosamines in Various Materials

Source	N-Nitrosamine Detected[a]	Reference
I. Air		
Rocket Fuel Factory	NDMA	69, 70, 146, 150, 151
Amine Factories	NDMA	146, 150, 152
Tire Chemical Plant	NDMA, NMOR, NDPHA	153, 154
Aircraft Tire Plant	NMOR	153, 154
Tannery	NDMA	153, 155
Indoor Air	NDMA	156
Urban Air	NDMA	66, 146, 150
II. Foods		
Bacon	NPYR, NDMA	42, 114, 130, 157
	NPRO, NHPYR	109, 158, 159
Bacon, Cook-out Fat	NPYR, NDMA	42, 160
Cured Meat Products	NDMA, NPYR, NHPYR, NSAR, NPRO, NHPRO	144, 161, 162
Meats	NPRO	121
Prepared Meals	NDMA, NPYR	130, 145, 157
Cooked Fish	NDMA, NPYR	130, 157
Canned Luncheon Meats	NDMA, NPIP	130, 157
Cheese	NDMA	130, 157
Beer	NDMA	141
Animal Feed	NDMA, NPYR	163, 164
III. Water		
Deionized Water	NDMA, NDEA	82, 108, 152
Drinking Water	Unidentified TEA peaks	132, 137, 165
IV. Industrial, Agricultural and Consumer Products		
Herbicides and Pesticides	NDMA, NDPA, NDBA	23, 35, 138, 147, 148, 166
Pharmaceuticals		
(a) Aminopyrine	NDMA	144
(b) Prescription and over-the-counter drugs	Unidentified TEA peaks	167
Cosmetics, Lotions, Shampoos	NDE1A	28
Cutting Fluids	NDE1A	29
	NMOR	168
V. Biological Fluids		
Blood	NDMA, NDEA	67, 169
Gastric Juices	NDMA, NDEA, NDPA	170

[a] See Figure 1 for chemical nomenclature and structure.

Air

Trapping Technique

Three trapping techniques have been used for the removal of N-nitrosamines from air.

1. Cryogenic trap[145]. Air is pumped through two successive cold traps, each containing 1 ml of 1 N KOH at a flowrate of 1-3 ℓ/min. The temperature of the cold trap is maintained by a dry ice-acetone bath.
2. Ambient alkaline trap[59]. The air was sampled through two successive impinger traps, each containing 100 ml of 1 N KOH at a flowrate of 1-4 ℓ/min.
3. Sorbent trap[171]. Air was drawn through a preconditioned Tenax (35/60 mesh) cartridge at a flowrate of approximately 1 ℓ/min.

After trapping, N-nitrosamines are extracted into an organic solvent (e.g., dichloromethane), which is concentrated and analyzed by GC-TEA. N-nitrosamines in a sorbent trap can also be analyzed by GC after thermal desorption in a GC injection port.

The possibility that artifact formation of N-nitrosamines may occur in the trap during air sampling has been tested by Fine et al.[69] and Fisher et al.[59]. Under ordinary conditions, no artifact formation was observed in any type of trap. However, formation of N-nitrosamines from dimethylamine and nitrogen oxides occurred in the cryogenic trap when high concentrations of ozone were present. Ambient alkaline traps were found to be free of artifact formation even in the presence of ozone.

N-nitrosamines in the Air of Factories

N-nitrosamines have been detected in several factories. Fine et al.[146,150] reported the presence of NDMA as an air pollutant in Baltimore, Maryland and Belle, West Virginia. The source of NDMA in Baltimore was traced to a factory manufacturing unsymmetrical dimethylhydrazine, which was synthesized from NDMA[69,151].

NDMA in Belle was found outside a factory that manufactured and used dimethylamine[146,151]. The possibility that NDMA may be formed in the air from the reaction between dimethylamine and nitrogen oxides was investigated by Cohen and Bachman[152]. They surveyed several amine-producing factories. NDMA was found only in the vicinity of the amine production facilities. NDMA was not detected off the plant properties, even though a high concentration of dimethylamine was detected. The results suggested that NDMA was probably produced as an unwanted by-product. Atmospheric formation of N-nitrosamine did not seem to occur.

N-nitrosomorpholine   (NMOR),   NDMA   and   N-nitrosodiphenylamine (NDPHA) were reported in the air of a tire chemical plant[153,154]. NDPHA was made in the plant as an antioxidant for the manufacture of tires. Morpholine was used as a raw material for the synthesis of tire chemicals. Transnitrosation of morpholine by NDPHA is likely to be the source of NMOR[50]. A storage tank for dimethylamine was located outside the plant, which may account for the presence of NDMA in the air.

N-nitrosamines were also found in the air of an aircraft tire plant and a tannery[153-155]. The sources of N-nitrosamines were not known.

Indoor Pollution

NDMA was reported in both mainstream and sidestream smoke[156]. This finding prompted a study of the contribution of tobacco smoke to N-nitrosamines pollution in indoor atmospheres. Low levels of NDMA were detected in the air containing heavy tobacco smoke.

Urban Air

Studies in several urban areas have detected little N-nitrosamines in the air[66,150]. NDMA was found at very low levels in only 3 out of 40 sites. This suggests that airborne N-nitrosamines is not a daily widespread air pollutant, but rather a localized problem associated with certain industries.

Foods

For the analysis of volatile N-nitrosamines in foods, the mineral oil distillation process has been proved to be a rapid and reliable technique and is gaining in popularity. The distillate is extracted by an organic solvent (dichloromethane was most frequently used). The extract is often concentrated in a Kuderna-Danish tube with a Snyder column. A small quantity of isooctane (0.2–1.0 ml) can be added as a keeper to prevent the complete evaporation of the extract. The concentrate is analyzed by GC-TEA.

In recent years, nonvolatile N-nitrosamines in foods have begun to attract attention. These N-nitrosamines were normally isolated from foods by solvent extraction. The extract was concentrated and could be analyzed by either of two analytical approaches. The first approach is to convert N-nitrosamines to volatile derivatives and analyze by GC[114,158,161,172,173]. The other approach is to analyze it directly with HPLC[114,121,134,174].

Both volatile and nonvolatile N-nitrosamines have been found in foods (Table

IV). They are probably formed from the reaction between nitrite and amines. Amines occur naturally in foods, and nitrite is added to foods as a preservative or may be present as a natural constituent of foods.

Water

Volatile and nonpolar, nonvolatile N-nitrosamines can be extracted from water by organic solvents such as dichloromethane. Polar compounds are more difficult to extract from water and can be freeze-dried to remove the water and dissolved in a polar solvent such as acetone or methanol. The extract is then concentrated and analyzed by GC-TEA or HPLC-TEA.

NDMA and NDEA were reported in the deionized water by several workers[82, 107, 108, 152]. Deionized water that was treated with an ion exchange resin was found more likely to contain N-nitrosamines. Angeles et al.[175] have presented a mechanism for the formation of N-nitrosamines in an ion exchange resin column.

Drinking water supplies in many cities have been screened by using GC-TEA and HPLC-TEA. Volatile N-nitrosamines were not detected in the drinking water[132, 137, 165]; however, nonvolatile compounds were detected by HPLC-TEA but were not identified. Industrial wastewaters were also frequently contaminated with N-nitrosamines[69, 152]. N-nitrosamines may be formed in various industrial processes.

Industrial, Agricultural and Consumer Products

Many of these products are in liquid form and can be injected directly on a GC-TEA or HPLC-TEA. Solid materials can be dissolved in organic solvent and filtered to remove any insoluble residue. The solution is then concentrated and analyzed. A mineral oil distillation may be employed if interference problems occur.

Herbicides and Pesticides

Many herbicides and pesticides are secondary or tertiary amines, and are nitrosatable to form N-nitrosamines[26, 27]. Ross et al.[147] were the first to report the presence of N-nitrosamines in commercial herbicide formulations. NDPA was found in trifluralin (a dinitroaniline herbicide); NDMA was found in herbicides formulated as dimethylamine salts. The presence of NDPA in a trifluralin herbicide was confirmed by Hotchkiss et al.[138]

Dinitroaniline herbicides are generally applied by soil incorporation. Ross et al.[166] measured NDPA in a tomato field during the application of trifluralin.

NDPA was not detected in the air, soil or runoff water. Apparently, NDPA in the herbicide was diluted beyond detection, since only a small amount of herbicide was applied onto a large field. NDPA was also not detected in crops treated with various dinitroaniline herbicides[176].

The U.S. Environmental Protection Agenty (EPA) conducted a comprehensive survey of N-nitrosamine contamination in various commercial herbicide and pesticide formulations[23,148]. High levels of N-nitrosamines were found primarily in the substituted amine, dinitroaniline and amine salt formulations, whereas triazine herbicides were generally free from N-nitrosamine contamination.

Pharmaceuticals

Like pesticides, many pharmaceuticals are formulated with secondary and tertiary amines. Several drugs have been shown to be readily nitrosated by nitrite[77,177]. Eisenbrand et al.[144] reported the presence of NDMA in 68 samples of aminopyrine, which is a tertiary amine containing a dimethylamine moiety. It was believe that NDMA was formed via the nitrosation of aminopyrine by nitrogen oxides in the air.

Krull et al.[65] examined 73 prescription and over-the-counter drugs. Three drugs responding to TEA corresponded to the presence of 40 to 81 μg/kg of N-nitrosamines. The identity of the N-nitrosamines was not established.

Cosmetics, Lotions and Shampoos

Fan et al.[28] reported the contamination of NDE1A in a variety of cosmetics, lotions and shampoos. Most of the samples contained diethanolamine derivatives or triethanolamine as ingredients. The source of NDE1A in these products is not completely known. However, bronopol (2-nitro-2-bromo-1,3-propanediol) has been shown to be a nitrosating agent[53,54]. This may explain the presence of NDE1A in some cosmetic products using bromopol as a bactericide.

Cutting Fluids

Fan et al.[29] reported the presence of NDE1A in commercial cutting fluids at concentrations as high as 3%. Cutting fluids are used in machine shops as lubricants, coolants and anticorrosion additives. Synthetic cutting fluids may contain up to 40% triethanolamine and 18% sodium nitrite. Such high concentrations of N-nitrosamine precursors present in the cutting fluid result in the formation of massive amounts of NDE1A.

Biological Fluids

Recently, Stephany et al.[168] reported the presence of trace amounts of NDMA and NDEA in the blood of a volunteer human subject. Lakritz et al.[170] reported the presence of NDMA, NDEA and NPYR in patients with disorders of the gastrointestinal tract. Webb and Gough[169] detected NDMA in blood and urine samples from unidentified sources. N-nitrosamines can be analyzed in the blood by frozen animal procedure (see the section entitled Frozen Animal Procedures, p. 298). N-nitrosamines in urine and gastric juices can be extracted with organic solvent and analyzed by GC-TEA or HPLC-TEA.

CONCLUSIONS

The use of the TEA Analyzer in many studies has demonstrated that N-nitrosamines are widely present in the environment. In recent years, the TEA Analyzer has become a popular instrument for the analysis of N-nitrosamines. The activity in N-nitrosamine research is expected to accelerate. The following areas will benefit greatly from the application of the TEA Analyzer.

Screening

The simplicity, speed and sensitivity of the TEA Analyzer in N-nitrosamine analysis means that more large-scale screening of N-nitrosamines in various foodstuffs, consumer products and other environmental samples, such as those conducted by Bontoyan et al.[23], Eisenbrand et al.[144], Gough et al.[130] and Spiegelhalder et al.[141] will take place.

In Vivo Formation

The human body is exposed daily to nitrite from a variety of sources. The question of whether nitrite circulated in the body will lead to in vivo formation has fascinated many researchers. However, N-nitrosamines are metabolized rapidly in the body. This has made the study of in vivo formation extremely difficult. Rounbehler et al.[149] have investigated the formation of N-nitrosamines in mice from the precursors at concentrations comparable to environmental levels. Fine et al.[67] were able to detect extremely small quantities of N-nitrosamines in human blood with the sensitive TEA detector. Much more in this area remains to be explored.

Worker Exposure to N-nitrosamines

N-nitrosamines have been found in the air of a number of factories. These findings may be only the beginning of research efforts in this direction. More research in this area will help to identify the population groups that are exposed to high levels of N-nitrosamines. This knowledge will not only reduce the worker's exposure to these potent carcinogens, but may also contribute to epidemiological studies to determine whether N-nitrosamines are human carcinogens.

REFERENCES

1. Druckrey, H., R. Preussmann, S. Ivankovic and D. Schmahl. *Z. Krebsforsch.* 69: 103 (1967).
2. Magee, P. N., and J. M. Barnes. *Adv. Cancer Res.* 10: 163 (1967).
3. Magee, P. N., R. Montesano and R. Preussmann. In: *Chemical Carcinogens,* C. E. Searle, Ed., American Chemical Society Monograph No. 173, Washington, DC (1976), Chapter 11.
4. Barnes, J. M., and P. N. Magee. *Brit. J. Ind. Med.* 11: 167 (1954).
5. Magee, P. N., and J. M. Barnes. *Brit. J. Cancer* 10: 114 (1956).
6. Ender, F., G. N. Harve, A. Helgebostad, N. Koppang, R. Madsen and L. Ceh. *Naturwissenschaften* 51: 637 (1964).
7. Sakshaug, J., E. Sognen, M. A. Hansen and N. Koppang. *Nature* 206: 1261 (1965).
8. Druckrey, H., A. Shildbach, D. Schmahl, R. Preussmann and S. Ivankovic. *Arzneimittelfors* 13: 841 (1963).
9. Preussmann, R., D. Schmahl, G. Eisenbrand and R. Port. In: *Proc. 2nd Int. Symp. on Nitrite in Meat Products* (Wageningen, Netherlands: Center for Agricultural Publishing and Documentation, 1977).
10. Terracine, B., P. M. Magee and J. M. Barenes. *Brit. J. Cancer* 21: 559 (1967).
11. Kawabata, T., H. Ohshima and M. Ino. In: *IARC Scientific Publ. No. 19* (1978).
12. Mirvish, S. S. *Toxicol. Appl. Pharmacol.* 31: 325 (1975).
13. Ridd, J. H. *Quart Rev. Chem. Sci.* 15: 418 (1961).
14. Smith, P. A. S., and R. N. Leoppky. *J. Am. Chem. Soc.* 89: 1147 (1967).
15. Turney, T. A., and G. A. Wright. *Chem. Rev.* 59: 497 (1959).
16. Fiddler, W., R. Doerr, J. Ertel and A. E. Wasserman. *Nature* 236: 307 (1972).
17. Fiddler, W., J. W. Pensabene, R. C. Doerr and A. E. Wasserman. *Nature* 236: 307 (1972).
18. Tannenbaum, S. R., D. Fett, V. R. Young, D. D. Land and W. R. Bruce. *Science* 200: 1487 (1978).
19. Tannenbaum, S. R., J. S. Wishnok, J. S. Hovis and W. W. Bishop. In: *IARC Scientific Publ. No. 19* 155–159 (1978).
20. Wartheson, J. J., R. A. Scanlan, D. P. Bills and L. M. Libbey. *J. Agric. Food Chem.* 23: 898 (1975).
21. Crosby, N. T. *Residue Rev.* 64: 77 (1976).
22. Lijinsky, W., and S. S. Epstein. *Nature* 225: 21 (1970).
23. Bontoyan, W. R., M. W. Law and D. P. Wright. *J. Agric. Food Chem.* 27: 631 (1979).
24. Eisenbrand, G., O. Ungerer and R. Preussmann. In: *IARC Scientific Publ. No. 9* 71 (1974).
25. Lijinsky, W., E. Conrad and R. van de Bogart. *Nature* 239: 165 (1972).

26. Lijinsky, W., L. Keefer, E. Conrad and R. van de Bogart. *J. Nat. Cancer Inst.* 49: 1239 (1972).
27. Sen, N. P., B. A. Donaldson and C. Charbonneak. In: *IARC Scientific Publ. No. 9* 75 (1974).
28. Fan, T. Y., U. Goff, L. Song, D. H. Fine, G. P. Arsenault and K. Biemann. *Food Cosmet. Toxicol.* 15: 423 (1977).
29. Fan, T. Y., J. Morrison, D. P. Rounbehler, R. Ross, D. H. Fine, W. Miles and N. P. Sen. *Science* 196: 70 (1977).
30. Fan, T. Y., and S. R. Tannenbaum. *J. Agric. Food Chem.* 21: 237 (1973).
31. Fan, T. Y., and S. R. Tannenbaum. *J. Agric. Food Chem.* 21: 967 (1973).
32. Mirvish, S. S. *J. Nat. Cancer Inst.* 44: 633 (1970).
33. Mirvish, S. S. In: *IARC Scientific Publ. No. 3* (1972).
34. Mirvish, S. S., J. Sams, T. Y. Fan and S. R. Tannenbaum. *J. Nat. Cancer Inst.* 51: 1833 (1973).
35. Alam, B. S., I. B. Saporoschetz and S. S. Epstein. *Nature* 232: 116 (1971).
36. Mysliwy, T. A., E. L. Wicks, M. C. Archer, R. C. Shank and P. M. Newberne. *Brit. J. Cancer* 30: 279 (1974).
37. Sen, N. P., D. C. Smith and L. Schwinghamer. *Food Cosmet. Toxicol.* 7: 301 (1969).
38. Boyland, E., E. Nice and K. Williams. *Food Cosmet. Toxicol.* 9: 639 (1971).
39. Keefer, L. K., and P. P. Roller. *Science* 181: 1245 (1973).
40. Fan, T. Y., and S. R. Tannenbaum. *J. Food Sci.* 38: 1067 (1973).
41. Mirvish, S. S., L. Wallcave, M. Eagen and P. Schubik. *Science* 172: 65 (1972).
42. Fiddler, W., J. W. Pensabene, E. G. Piotrowski, J. G. Phillips, J. Keating, W. J. Mergens and H. L. Newmark. *J. Agric. Food Chem.* 26: 653 (1978).
43. Bogovski, P., R. Preussmann and E. A. Walker. In: *IARC Scientific Publ. No. 3* (1972).
44. Bogovski, P., M. Castegnaro, B. Pignatelli and E. A. Walker. In: *IARC Scientific Publ. No. 3* 127 (1972).
45. Wolff, I. A., and A. E. Wasserman. *Science* 177: 15 (1972).
46. Heisler, E. G., J. Siciliano, S. Krulick, J. Feinberg and J. H. Schwartz. *J. Agric. Food Chem.* 22: 1029 (1974).
47. Tannenbaum, S. R., A. J. Sinskey, M. Weisman and W. Bishop. *J. Nat. Cancer Inst.* 53: 79 (1974).
48. Fine, D. H. In: *IARC Scientific Publ. No. 18* 133–140 (1978).
49. Fine, D. H. In: *IARC Scientific Publ. No. 19* 267 (1978).
50. Buglass, A. J., B. C. Challis and M. R. Osborne. In: *IARC Scientific Publ. No. 9* 94 (1974).
51. Singer, S. S., W. Lijinsky and G. M. Singer, In: *IARC Scientific Publ. No. 19* 175–181 (1978).
52. Davies, R., M. J. Dennis, R. C. Massey and D. J. McWeeney. In: *IARC Scientific Publ. No. 19* 183 (1978).
53. Fan, T. Y., R. Vita and D. H. Fine. *Toxicol. Lett.* 2: 5 (1978).
54. Schmeltz, I., and A. Wenger. *Food Cosmet. Toxicol.* 17: 105 (1979).
55. Fan, T. Y., and D. H. Fine. Unpublished results (1978).
56. Challis, B. C., A. Edwards, R. R. Hunma, S. A. Kyrtopoulos and J. R. Outram. In: *IARC Scientific Publ. No. 19* 127 (1978).

57. Fan, T. Y., and S. R. Tannenbaum. *J. Food Sci.* 37: 274 (1972).
58. Chow, Y. L. *Can. J. Chem.* 45: 53 (1967).
59. Fisher, R. L., R. W. Reiser and B. A. Lasoski. *Anal. Chem.* 49: 1821 (1977).
60. Fine, D. H. *Advan. Environ. Sci. Technol.* 9: (In press).
61. Scanlan, R. A. *CRC Crit. Rev. in Food Technol.* 5: 357 (1975).
62. Crosby, N. T., and R. Sawyer. *Adv. Food Res.* 22: 1 (1975).
63. Gray, J. I., and C. J. Randall. *J. Food Protection* 42: 168 (1979).
64. Gray, J. I., D. M. Irvine and Y. Kakuda. *J. Food Protection* 42: 263 (1979).
65. Krull, I. S., G. Edwards, M. H. Wolf and D. H. Fine. In: *N-nitrosamines,* J.-P. Anselme, Ed., American Chemical Society Symposium Series No. 101, Washington, DC (1979), p. 175.
66. Fine, D. H., J. Morrison, D. P. Rounbehler, A. Silvergleid and L. Song. In: *Toxic Substances in the Air Environment,* J. D. Spengler, ed., Air Pollution Control Association, PA (1977), p. 168.
67. Fine, D. H., R. Ross, D. P. Rounbehler, A. Silvergleid and L. Song. *Nature* 265: 753 (1977).
68. Fine, D. H., D. P. Rounbehler, T. Fan and R. Ross. In: Cold Spring Harbor Conferences on Human Cell Proliferation, Vol. 4, H. H. Hiatt, J. D. Watson and J. A. Winsten, Eds., Cold Spring Harbor, NY (1977), p. 293.
69. Fine, D. H., D. P. Rounbehler, A. Rounbehler, A. Silvergleid, E. Sawicki, K. Krost and G. A. DeMarrais. *Environ. Sci. Technol.* 11: 581 (1977).
70. Fine, D. H., D. P. Rounbehler, E. Sawicki and K. Krost. *Environ. Sci. Technol.* 11: 577 (1977).
71. Fine, D. H., D. P. Rounbehler, A. Silvergleid and R. Ross. In: *Proc. 2nd Int. Symp. on Nitrite in Meat Products* (1977).
72. Eisenbrand, G. *IARC Scientific Publ. No. 9* 6 (1974).
73. Essigmann, J. M., and P. Issenberg. *J. Food Sci.* 37: 684 (1972).
74. Fazio, T., J. N. DaMico, J. W. Howard, R. H. White and J. O. Watts. *J. Agric. Food Chem.* 19: 250 (1971).
75. Palframan, J. F., J. Macnab and N. T. Crosby. *J. Chromatog.* 76: 307 (1973).
76. Foreman, J. K., J. F. Palframan and E. A. Walker. *Nature* 225: 544 (1970).
77. Roper, H., and K. Heyns. In: *IARC Scientific Publ. No. 19* 219 (1978).
78. Scanlan, R. A., and L. M. Libbey. *J. Agric. Food Chem.* 19: 570 (1971).
79. McNair, H. M., and E. J. Bonelli. "Basic Gas Chromatography," Varion Instrument Division, Palo Alto, CA (1969), p. 81.
80. Crosby, N. T., J. K. Foreman, J. F. Palframan and R. Sawyer. *Nature* 238: 342 (1972).
81. Fazio, T., J. W. Howard and R. H. White. In: *IARC Scientific Publ. 3* 16 (1972).
82. Fiddler, W., J. W. Pensabene, R. C. Doerr and C. J. Dooley. *Food Cosmet. Toxicol.* 15: 441 (1977).
83. Howard, J. W., T. Fazio and S. O. Watts. *J. Assoc. Off. Anal. Chem.* 53: 269 (1970).
84. Coulson, D. M. *J. Gas Chromatog.* 3: 134 (1965).
85. Rhodes, J. W., and D. E. Johnson. *J. Chromatog. Sci.* 8: 616 (1970).

86. Crosby, N. T., J. K. Foreman, J. F. Palframan and R. Sawyer. In: *IARC Scientific Publ. No. 3* 38 (1972).
87. Sen, N. P. In: *IARC Scientific Publ. No. 3* 25 (1972).
88. Althorpe, J., D. A. Goddard, D. J. Sissons and G. M. Telling. *J. Chromatog.* 53: 371 (1970).
89. Sen, N. P. *J. Chromatog.* 53: 301 (1970).
90. Telling, G. M. *J. Chromatog.* 73: 79 (1972).
91. Walker, E. A., M. Castegnaro and B. Pignatelli. *Analyst* 100: 817 (1975).
92. Alliston, T. G., B. G. Cos and R. S. Kirk. *Analyst* 97: 915 (1972).
93. Eisenbrand, G. *IARC Scientific Publ. No. 3* 64 (1972).
94. Walker, E. A., and M. Castegnaro. In: *IARC Scientific Publ. No. 14* (1976).
95. Pailer, M., and H. Klus. *Fach. Mitt. Ost. Tabakregie* 203 (1971).
96. Hoffman, D., G. Rathkamp and Y. Y. Liu. In: *IARC Scientific Publ. No. 9* 159 (1974).
97. Goodhead, K., and T. A. Gough. *Food Cosmet. Toxicol.* 13: 307 (1975).
98. Pensabene, J. W., W. Fiddler, C. J. Dooley, R. C. Doerr and A. E. Wasserman. *J. Agric. Food Chem.* 20: 274 (1972).
99. Osborne, D. R. In: *IARC Scientific Publ. No. 3* 43 (1972).
100. Telling, G. M., T. A. Bryce and J. Althorpe. *J. Agric. Food Chem.* 19: 937 (1971).
101. Bryce, T. A., and G. M. Telling. *J. Agric. Food Chem.* 20: 910 (1972).
102. Gough, T. A., and K. S. Webb. *J. Chromatog.* 64: 201 (1972).
103. Dooley, C. J., A. E. Wasserman and S. Osman. *J. Food Sci.* 38: 1096 (1973).
104. Gough, T. A., and K. S. Webb. *J. Chromatog.* 79: 57 (1973).
105. Stephany, R. W., J. Freudenthal, E. Egmond, L. G. Granberg and P. Schuller. *J. Agric. Food Chem.* 24: 536 (1976).
106. Pensabene, J. W., W. Fiddler, R. A. Gates, J. C. Fajan and A. E. Wasserman. *J. Food Sci.* 39: 314 (1974).
107. Gough, T. A., K. S. Webb, M. A. Pringuer and B. J. Wood. *J. Agric. Food Chem.* 25: 663 (1977).
108. Gough, T. A., K. S. Webb and M. F. McPhail. *Food Cosmet. Toxicol.* 15: 437 (1977).
109. Havery, D. C., T. Fazio and J. W. Howard. In: *IARC Scientific Publ. No. 19* 41 (1978).
110. Cox, G. B. *J. Chromatog.* 83: 471 (1973).
111. Iwaoka, W. T., T. Hansen, S. T. Hsieh and M. C. Archer. *J. Chromatog.* 103: 349 (1975).
112. Iwaoka, W. T., and S. R. Tannenbaum. In: *IARC Scientific Publ. No. 14* 51–56 (1976).
113. Fan, T. Y., and S. R. Tannenbaum. *J. Agric. Food Chem.* 19: 1267 (1971).
114. Hansen, T., W. Iwaoka, L. Green and S. R. Tannenbaum. *J. Agric. Food Chem.* 25: 1423 (1977).
115. Singer, G. M., S. S. Singer and D. G. Schmidt. *J. Chromatog.* 133: 59 (1977).
116. Fine, D. H., and F. Rufeh. In: *IARC Scientific Publ. No. 9* 53 (1974).
117. Fine, D. H., D. Lieb and F. Rufeh. *J. Chromatog.* 107: 351 (1975).
118. Fine, D. H., F. Rufeh, D. Lieb and D. P. Rounbehler. *Anal. Chem.* 47: 1188 (1975).

119. Fine, D. H., and D. P. Lieb. U.S. Patent 3,996,009 (1976).
120. Fine, D. H., and D. P. Rounbehler. U.S. Patent 3,996,004 (1976).
121. Baker, J. K., and C.-Y. Ma. *J. Agric. Food Chem.* 26: 1253 (1978).
122. Clough, P. N., and B. A. Thrush. *Trans. Faraday Soc.* 63: 915 (1967).
123. Fine, D. H., F. Rufeh and B. Gunther. *Anal. Lett.* 6: 731 (1973).
124. Fine, D. H., F. Rufeh and D. Lieb. *Nature* 247: 309 (1974).
125. Fine, D. H., D. Lieb and D. P. Rounbehler. U.S. Patent 3,996,008 (1976).
126. Fine, D. H., and D. P. Rounbehler. *J. Chromatog.* 109: 271 (1975).
127. Fine, D. H., and D. P. Rounbehler. In: *IARC Scientific Publ. No. 14* (1976).
128. Fine, D. H., D. P. Rounbehler and P. E. Oettinger. *Anal. Chim. Acta* 78: 383 (1975).
129. Brooks, J. B., C. C. Alley and R. Jones. *Anal. Chem.* 44: 1881 (1972).
130. Gough, T. A., K. S. Webb and R. F. Coleman. *Nature* 272: 161 (1978).
131. Oettinger, P. E., F. Huffman, D. H. Fine and D. Lieb. *Anal. Lett.* 8: 411 (1975).
132. Fine, D. H., F. Huffman, D. P. Rounbehler and N. M. Belcher. In: *IARC Scientific Publ. No. 14* 43 (1976).
133. Krull, I. S., T. Y. Fan, M. Wolf, R. Ross and D. H. Fine. In: *LC Symposium I,* G. Hawk, Ed. (New York: Marcel Dekker, Inc., 1978), p. 443.
134. Fine, D. H., R. Ross, D. P. Rounbehler, A. Silvergleid and L. Song. *J. Agric. Food Chem.* 24: 1069 (1976).
135. Krull, I. S., T. Y. Fan and D. H. Fine. *Anal. Chem.* 50: 698 (1978).
136. Fan, T. Y., I. Krull, R. D. Ross, M. H. Wolf and D. H. Fine. In: *IARC Scientific Publ. No. 19* 3 (1978).
137. Fan, T. Y., R. Ross, D. H. Fine, L. H. Keith and A. W. Garrison. *Environ. Sci. Technol.* 12: 692 (1978).
138. Hotchkiss, J. H., J. F. Barbour, L. M. Libbey and R. A. Scanlan. *J. Agric. Food Chem.* 26: 884 (1978).
139. Krull, I. S., U. Goff, M. Wolf, D. H. Fine, M. Hoes and G. P. Arsenault. *Food Cosmet. Toxicol.* 16: 105 (1978).
140. Stephany, R. W., and P. L. Schuller. *Proc. 2nd Int. Symp. on Nitrite in Meat Products* (Wageningen, Netherlands: Center for Agricultural Publishing and Documentation, 1977), pp. 249–255.
141. Spiegelhalder, B., G. Eisenbrand and R. Preussmann. *Food Cosmet. Toxicol.* 17: 29 (1979).
142. Fan, T. Y., and D. H. Fine. *J. Agric. Food Chem.* 26: 1471 (1978).
143. Freed, D. J., and A. M. Mujsce. *Anal. Chem.* 49: 1544 (1977).
144. Eisenbrand, G., B. Spiegelhalder, C. Janzowski, J. Kann and R. Preussmann. In: *IARC Scientific Publ. No. 19* 311 (1978).
145. Stephany, R. W., and P. L. Schuller. In: *IARC Scientific Publ. No. 19* 249 (1978).
146. Fine, D. H., D. P. Rounbehler, N. M. Belcher and S. S. Epstein. *Science* 192: 1328 (1976).
147. Ross, R. D., J. Morrison, D. P. Rounbehler, T. Y. Fan and D. H. Fine. *J. Agric. Food Chem.* 25: 1416 (1977).
148. Cohen, S. Z., G. Zweig, M. Law, D. Wright and W. R. Bontoyan. In: *IARC Scientific Publ. No. 19* 333–342 (1978).
149. Rounbehler, D. P., R. Ross, D. H. Fine, Z. M. Iqbal and S. S. Epstein. *Science* 197: 917 (1977).

150. Fine, D. H., D. P. Rounbehler and N. M. Belcher. In: *IARC Scientific Publ. No. 14* (1976).
151. Fine, D. H., D. P. Rounbehler, E. D. Pellizzari, J. E. Bunch, R. W. Berkley, J. McRae, J. T. Bursey, E. Sawicki, K. Krost and G. A. DeMarrais. *Bull. Environ. Contam. Toxicol.* 15: 739 (1976).
152. Cohen, J. B., and J. D. Bachman. In: *IARC Scientific Publ. No. 19* (1978).
153. Fajen, J. M., G. A. Carson, T. Y. Fan, D. P. Rounbehler, J. Morrison, I. Krull, G. Edwards, A. Lafleur, W. Herbst, U. Goff, R. Vita, K. Mills, D. H. Fine and V. Reinhold. Paper presented at the Annual Meeting of the Air Pollution Control Association, Houston, TX, 1978.
154. Fajen, J. M., G. A. Carson, D. P. Rounbehler, T. Y. Fan, R. Vita, E. U. Goff, M. H. Wolf, G. S. Edwards, D. H. Fine, V. Reinhold and K. Biemann. *Science* 205: 1262 (1979).
155. Rounbehler, D. P., I. S. Krull, U. E. Goff, K. M. Mills, J. Morrison, G. S. Edwards, D. H. Fine, J. M. Fajen, G. A. Carson and V. Reinhold. *Food Cosmet. Toxicol.* 17: 487 (1979).
156. Brunnemann, K. D., and D. Hoffmann. In: *IARC Scientific Publ. No. 19* (1978).
157. Gough, T. A. In: *IARC Scientific Publ. No. 19* 297 (1978).
158. Lee, J. S., L. M. Libbey, R. A. Scanlan and D. D. Bills. *Bull Environ. Contam. Toxicol.* 19: 511 (1978).
159. Lee, J. S., L. M. Libbey, R. A. Scanlan and J. Barbour. In: *IARC Scientific Publ. No. 19* 325 (1978).
160. Mottram, D. S., R. L. S. Patterson, R. A. Edwards and T. A. Gough. *J. Sci. Food Agric.* 28: 1025 (1977).
161. Janzowski, C., G. Eisenbrand and R. Preussmann. *J. Chromatog.* 150: 216 (1978).
162. Janzowski, C., G. Eisenbrand and R. Preussmann. *Food Cosmet. Toxicol.* 16: 343 (1978).
163. Fine, D. H., I. S. Krull, D. P. Rounbehler, G. S. Edwards and J. B. Fox. Paper presented at the Annual Meeting of American Chemical Society, Miami, FL, 1978.
164. Kann, J., B. Spiegelhalder, G. Eisenbrand and R. Preussmann. *Z. Krebsforsch* 90: 321 (1977).
165. Fine, D. H., D. P. Rounbehler, F. Huffman, A. W. Garrison, N. L. Wolfe and S. S. Epstein. *Bull. Environ. Contam. Toxicol.* 14: 404 (1975).
166. Ross, R., J. Morrison and D. H. Fine. *J. Agric. Food Chem.* 26: 455 (1978).
167. Krull, I. S., U. Goff, A. Silvergleid and D. H. Fine. *Arzneimittelfors.* 29: 870 (1979).
168. Stephany, R. W., J. Freudenthal and P. L. Schuller. *J. Royal Netherlands Chem. Soc.* 97: (in press).
169. Webb, K. S., and T. A. Gough. *Anal. Chem.* 51: 989 (1979).
170. Lakritz, L., A. E. Wasserman, R. Gates and A. M. Spinelli. In: *IARC Scientific Publ. No. 19* 425 (1978).
171. Pellizzari, E. D. M., J. E. Bunch, J. T. Bursey, R. E. Berkley, E. Sawicki and K. Krost. *Anal. Lett.* 9: 579 (1976).
172. Nakamura, M., N. Baba, T. Nakaoka, Y. Wada, T. Ishibashi and T. Kawabata. *J. Food Sci.* 41: 874 (1976).

173. Sen, N. P., W. F. Miles, S. Seaman and J. F. Lawrence. *J. Chromatog.* 128: 169 (1976).
174. Fan, T. Y., W. Herbst and D. H. Fine. Paper presented at the 37th Annual Meeting of the Institute of Food Technologists, paper no. 152, Philadelphia, PA, 1977.
175. Angeles, R. M., L. K. Keefer, P. P. Roller and S. J. Uhm. In: *IARC Scientific Publ. No. 19* 109 (1978).
176. West, S. D., and E. A. Day. Paper presented at the 175th Meeting of the American Chemical Society, Anaheim, CA, 1978.
177. Montesano, R., H. Bartsch and H. Bresil. *J. Nat. Cancer Inst.* 52: 907 (1974).

CHAPTER 10

A METHOD FOR DETERMINATION OF HEAVY METAL DISTRIBUTIONS IN MARINE SEDIMENTS

Leonard J. Warren

CSIRO Division of Mineral Chemistry
Port Melbourne, Australia

INTRODUCTION

Sediments are usually regarded as the ultimate "sink" for heavy metals discharged into the environment, yet relatively little is known about the way that heavy metals are bound to sediments or the ease with which they may be released[1].

Part of the difficulty lies in the complex nature of sediments, whether freshwater or marine. Sediments are normally mixtures of several components, including different mineral species and organic debris. In addition, there are a variety of ways the heavy metals may be bound, e.g., by adsorption, on exchange sites of clay minerals, coprecipitated with iron oxides or held inside biological materials such as seagrass fragments, shells and animal feces. Individual metallic sulfide or oxide particles also may be present.

Most attempts to determine the distribution of heavy metals in such sediments have been based on the selective dissolution of the sample by a series of reagents of increasing reactivity[2-6]. For example, the sediment may be treated successively with distilled water to dissolve soluble and weakly sorbed metal, calcium chloride or ammonium chloride to release exchangeable metal, and dilute nitric acid to remove all specifically adsorbed metal. However, apart from problems of contamination, such chemical treatments are seldom selective enough to give an unambiguous indication of the metal distribution[5-7].

An alternative and, in many cases, complementary approach is to first separate the sediment into it main components, without chemical change, and then to analyze the separated components for their heavy metal content[8,9]. This is the basis of the method reported here, in which the components of a sediment that differ in specific gravity are separated into distinct bands in a density gradient in a suitable heavy liquid. The concentrations of heavy metals such as lead, cadmium and zinc in the various density subfractions can then be determined by appropriate chemical analysis. Since the technique physically separates the sediment components, it is possible to study not only the total metal, but also the easily extractable metal associated with each component.

Similar methods have been developed for use in the mineral processing industry[10] and for characterizing soil clays[11], but their potential for environmental analysis has only recently been realized. Using density gradient mineral segregation in a zonal rotor, Francis and Brinkley[12] detected preferential adsorption of radioactive cesium on the micaceous component of a freshwater sediment.

However, the heavy liquid density gradient technique is not easily adapted to cope with sediments containing a large proportion of discrete ultrafine particles. Thus, freshwater and deep sea sediments may give difficulty, but nearshore marine sediments have been studied routinely[8,9]. The technique requires a skilled operator, and care must be taken in handling the toxic heavy liquids, such as tetrabromoethane, and in chemically analyzing density subfractions, which may contain only submicrogram quantities of heavy metal.

EXPERIMENTAL

Sample Preparation

Sediment samples are collected and stored taking due precaution to prevent contamination and inadvertant chemical changes[13]. The sediments are then dried, either by heating at 100°C in a clean stainless steel oven after preliminary filtration, or by freeze-drying. "Caking" of oven-dried samples is reduced by washing once with a small volume of distilled water to remove the bulk of the sea salt and by drying the damp sediment in thin layers. Oven-drying may lead to irreversible changes in some sediment components.

Separation into Size Fractions

The heavy liquid density gradient technique works best with samples that contain a fairly narrow range of particle sizes, and the sediment should, therefore, be separated into a number of sized fractions. The sizing procedure adopted should not contaminate the sample and should be designed to measure either the "original" or the "ultimate" size distribution of the dried sediment sample.

Many sediments contain not only discrete particles and shell and organic fragments, but also small aggregates composed of much finer particles. Some of these "conglomerate particles" are soft and break apart easily; others are quite hard. Conglomerate particles have been observed in wet sediments[8,9] and are not merely an artifact of oven-drying. Since conglomerates can make up a significant proportion of the total sediment weight, it is important to decide at the outset whether to measure the ultimate size distribution of discrete particles by destroying the conglomerates, or the original distribution of effective "particles," including conglomerates.

The size fractionation procedure described here is designed as far as possible to be nondestructive and to measure the actual or original distribution of particle sizes. Samples of each of the "whole" dried sediments are separated into +1000-(S1) and −1000 μm size fractions by sieving on a polyster screen cloth* mounted in a 200-mm-diameter aluminium holder (Tonindustrie Test Sieves). The S1 fraction is treated with ultrasound to remove adhering fines from the coarser particles; it is then dried and resieved.

Ten-gram samples of the undersized material (-1000 μm) are suspended in 500 ml of AR acetone and dispersed with an ultrasonic probe (3 min, 20 kHz, 150 W). Acetone, rather than water, is chosen as the dispersing medium to avoid the possibility of water removing weakly sorbed materials. The suspension, 50 mm high, is allowed to settle for 2 minutes 45 seconds, and the supernatant liquor is then decanted. This settling time is calculated to allow 10-μm spheres of density, 2.5g/cm^3, to fall 50 mm in acetone at 25°C (see Equation 1 below). More acetone is added to the sediment, which is redispersed, if necessary, to free adhering fines. The sedimentation step is then repeated four or five times until the supernatant liquor is reasonably clear. The sediment should then contain only those particles coarser than about 10 μm, but finer than 1000 μm, and is called the S2 fraction. The combined supernatant suspensions are further sized using the same beaker decantation procedure[14], but settling for 4 hours 35 minutes, to effect a cut at about 1 μm. Settling times for other solvents and different particle sizes can be calculated from Stoke's law:

$$V_T = g\,(\rho - \rho_o)\,d^2 / 18\eta \qquad (1)$$

where V_T = the terminal velocity of a particle of diameter, d, and density ρ falling freely in a liquid viscosity η and density ρ_o
    g = the gravitational acceleration

The Stoke's law sedimentation method of sizing assumes that no flocculation occurs during settling.

---

*Zurich Bolting Cloth Manufacturing Co., Ltd., CH-8803 Ruschlikon, Switzerland.

All acetone used is collected and evaporated to yield an oily residue. The residue and subsamples of each of the size fractions—S1 ($+1000$ μm), S2 ($-1000+10$ μm), S3 ($-10+1$ μm) and S4 (-1 μm)—are then analyzed for their heavy metal content.

Separation into Density Subfractions

The size fractions S2 and S3 are further separated into six density subfractions, D1–D6, each corresponding to a chosen range of particle densities (Figure 1). If S1 comprises a significant proportion of the total sample weight, it can be similarly separated. The size fraction S4 is too fine to separate into density subfractions with a conventional centrifuge and it is analyzed only for its total content of heavy metals. In principle, S4 could be fractionated using an ultracentrifuge[11]. By treating each size fraction separately, as shown in Figure 1, problems caused by the differing behavior of the large and the very fine particles in the density gradient are avoided.

A base heavy liquid is chosen with a specific gravity high enough to suspend most of the minerals commonly found in nearshore sediments. Tetrabromoethane (TBE), with a specific gravity of 2.96 g/cm^3 is convenient. However, the vapor is heavy and toxic, and all operations using TBE should be performed in a well-ventilated fume cupboard. Skin contact should be avoided.

Laboratory-grade TBE is purified by redistillation at reduced pressure, and the colorless distillate (boiling at 106°C at 11.8 mm Hg) is mixed with acetone to give a series of liquids at specific gravity between 2.96 and 2.20. If the required specific gravity of a mixture is ρ_M and the total volume is V_T, then the volume of acetone required is given by

$$V_A = V_T (\rho_{TBE} - \rho_m) / (\rho_{TBE} - \rho_A) \tag{2}$$

where $\rho_{TBE} = 2.96$
        $\rho_A = 0.79$

The specific gravities of the various mixtures are chosen so as to separate most of the known minerals in the particular sediment sample. One selection of density bands that proved suitable for Spencer Gulf sediments[8] is compared in Figure 2 with the specific gravity of the main minerals present.

The liquid mixtures are carefully layered into a double-walled centrifuge tube of the type shown in Figures 3 and 4*, in order of decreasing specific gravity. First, 10 ml of pure TBE is dispensed from a funnel with a tap via plastic tubing

---

*Appropriate tubes and associated apparatus are available from T. W. Wingent Ltd., Cambridge, England.

"Whole" sediment sample

SIZE FRACTIONATION

+1000 μm	-1000 + 10 μm	-10 +1 μm	-1 μm	Acetone
S1	S2	S3	S4	soluble
				organics

DENSITY FRACTIONATION

| <2·4 | 2·4-2·55 | 2·55-2·66 | 2·66-2·75 | 2·75-2·95 | >2·95 |
| D1 | D2 | D3 | D4 | D5 | D6 |

DENSITY FRACTIONATION

| <2·2 | 2·2-2·55 | 2·55-2·66 | 2·66-2·75 | 2·75-2·95 | >2·95 |
| D1 | D2 | D3 | D4 | D5 | D6 |

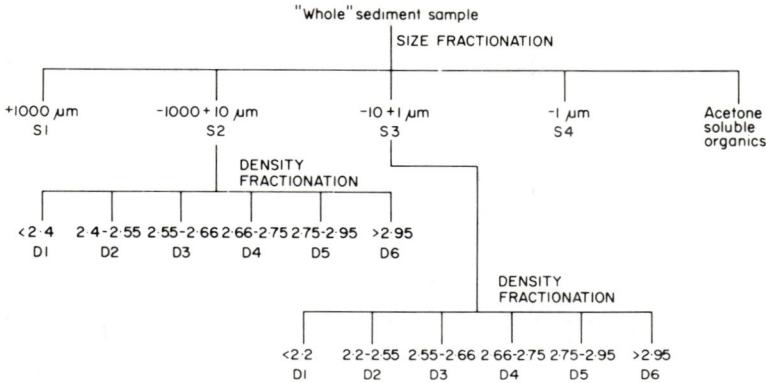

Figure 1.   Sediment fractionation procedure (reproduced courtesy of Ann Arbor Science Publishers, Inc.).

and a Pasteur pipette onto a thin piece of acid-washed cork, which floats up on the liquid in the centrifuge tube. Next, 5 ml of liquid (density 2.75) is dispensed dropwise onto the cork to form a sharp interface between the layers of different density. The procedure is repeated for the layers of density 2.66, 2.55, 2.40 and, if necessary, 2.20.

About 0.4 g of fraction S2 is wet with acetone and carefully introduced on top of the heavy liquid layers. The tube is centrifuged at about 1500 rpm for about 5 minutes. Higher speeds are unnecessary for particles of the S2 fraction and may break the glass. Sample masses greater than 0.5 g should be avoided as some of the denser particles may be entrained in the upper low-density layers. During centrifugation, distinct bands of particles of similar density are formed at the interfaces between the liquid layers (Figure 3).

The various density bands are removed in order by displacement with pure TBE in an apparatus of the type shown in Figures 4 and 5. The funnel A on the aluminium frame G is filled with TBE. The ground glass plate C is rotated over the fixed plate D to form a seal. TBE is introduced into the tubing B and any trapped air removed by squeezing the tube. The plates C and D are then realigned and the seal between the centrifuge tube E and the separator checked. The spring-loaded base plate F should ensure a proper seal. TBE is slowly introduced to the tube E and the layers displaced upwards without mixing. Plate C is rotated to successively cut off each band of particles, which are transferred in turn to the exit tube on the other side of the separator. Each density band is delivered onto a filter paper in a Buchner funnel and washed repeatedly with acetone to remove residual TBE. After drying and weighing, the bands are analyzed for their heavy metal content. Tests on Spencer Gulf sediments showed that TBE did not solubilize lead or cadmium during the density fractionation procedure[8].

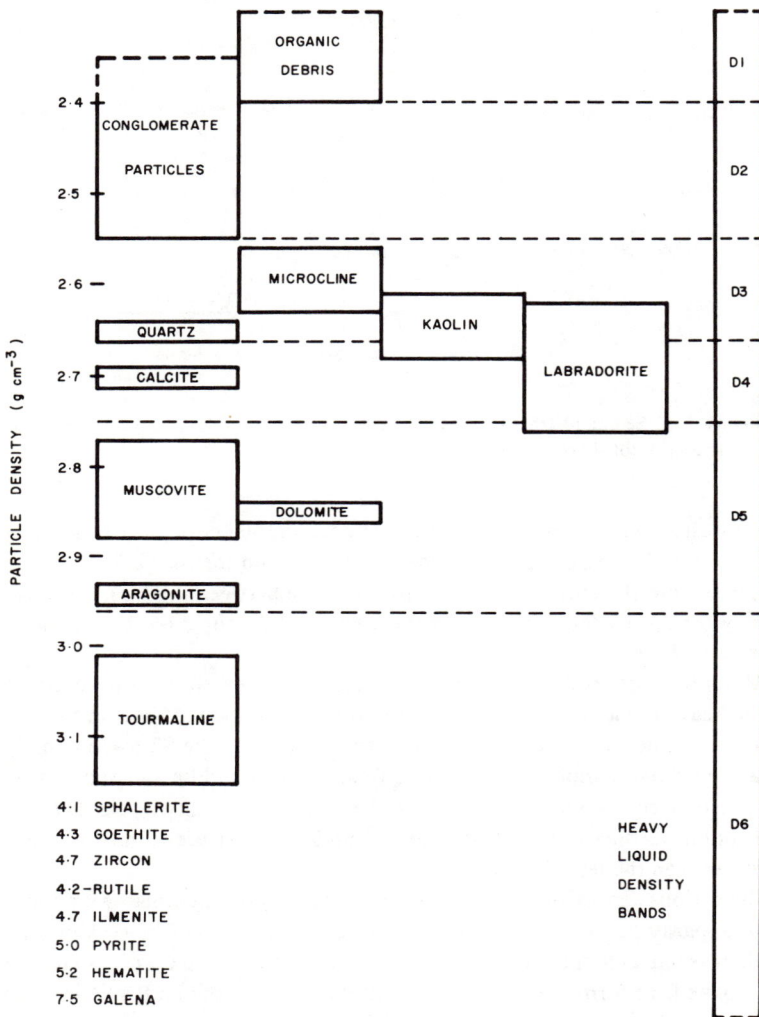

Figure 2. Heavy liquid density bands in relation to the densities of particles present in the sediment.

A slightly different procedure is required for the S3 ($-10+1$ μm) size fraction. The density of the topmost layer of heavy liquid is adjusted to 2.2 instead of 2.4 because the proportion of sample reporting in S3 D1 is considerably greater than in S2 D1. Also, to prevent flocculation of the S3 particles in the TBE/acetone mixtures, polyvinylpyrrolidone (PVP) is added as a dispersant. A 1% solution of

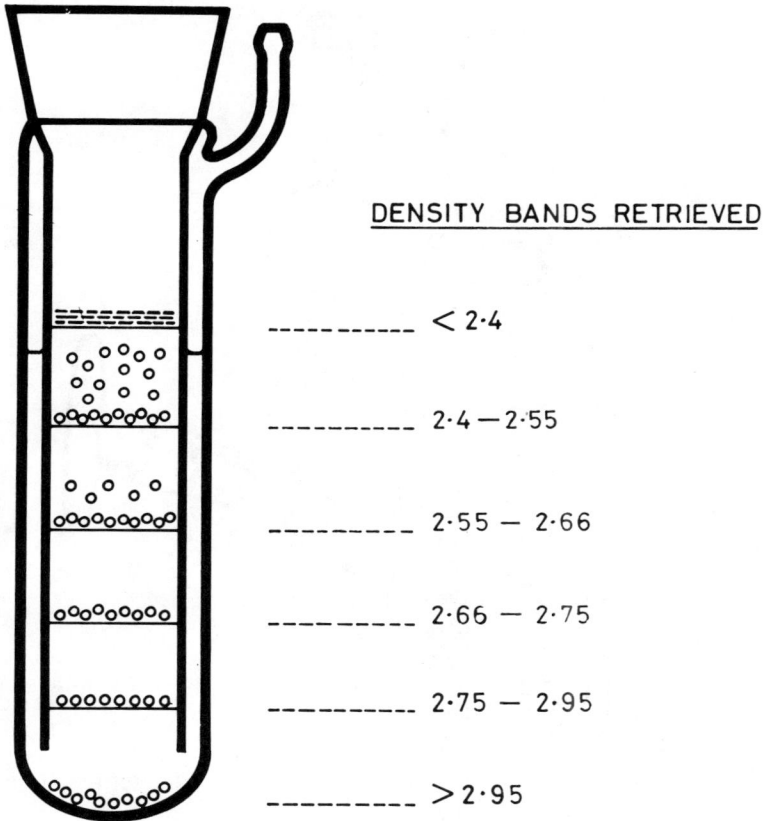

DENSITY BANDS RETRIEVED

---------- < 2·4

---------- 2·4 — 2·55

---------- 2·55 — 2·66

---------- 2·66 — 2·75

---------- 2·75 — 2·95

---------- > 2·95

Figure 3. Mineral bands observed after centrifuging sediment samples in the heavy liquid density gradient (reproduced courtesy of Ann Arbor Science Publishers, Inc.).

PVP in acetone is prepared and used as the diluent in making the heavy liquid layer of density 2.20. Of this 2.20 mixture (with PVP,) 5 ml is added to about 0.3 g of acetone-wet S3 particles and the suspension dispersed with ultrasound. The dispersed suspension is layered on top of the 2.55 mixture in the centrifuge tube by slowly pouring through a small funnel with a short glass rod fused at the end. The density gradient is then centrifuged for up to 60 minutes at 1500 rpm. The system is maintained at a constant temperature during centrifugation to prevent intermixing of the bands.

Successive displacement of the density subfractions with the apparatus in Figure 5 is more difficult with the finer particles, which tend to stick to the cen-

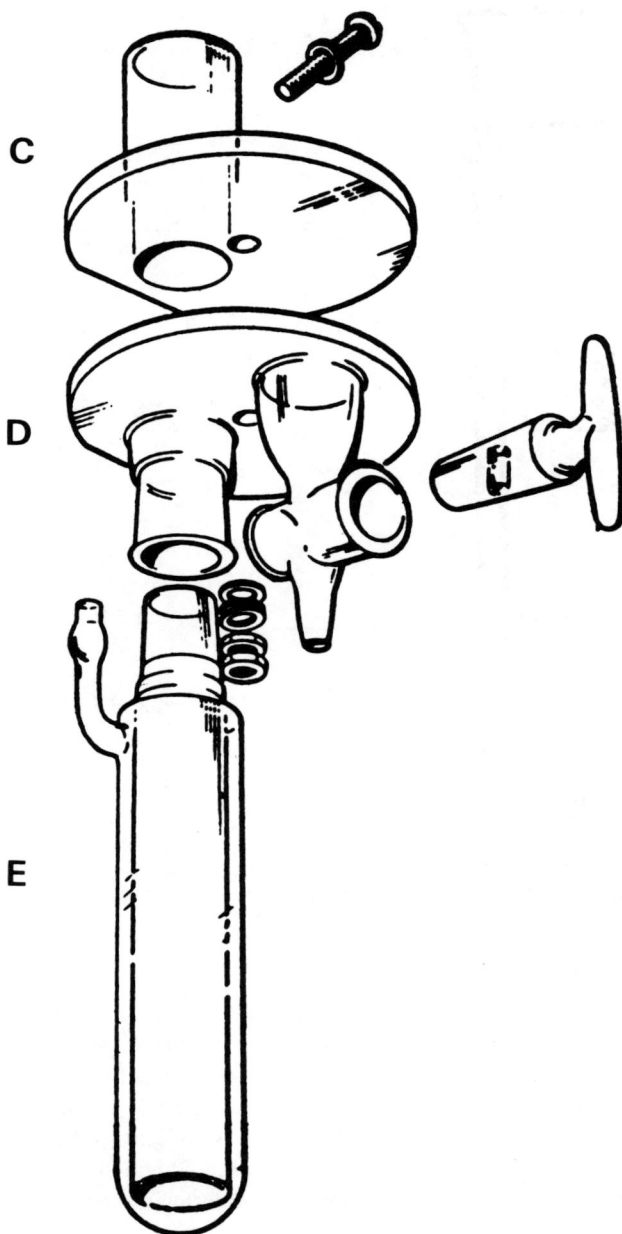

Figure 4. Exploded view of the double-walled centrifuge tube (E) and separator (C and D). Adapted from Muller and Burton[10].

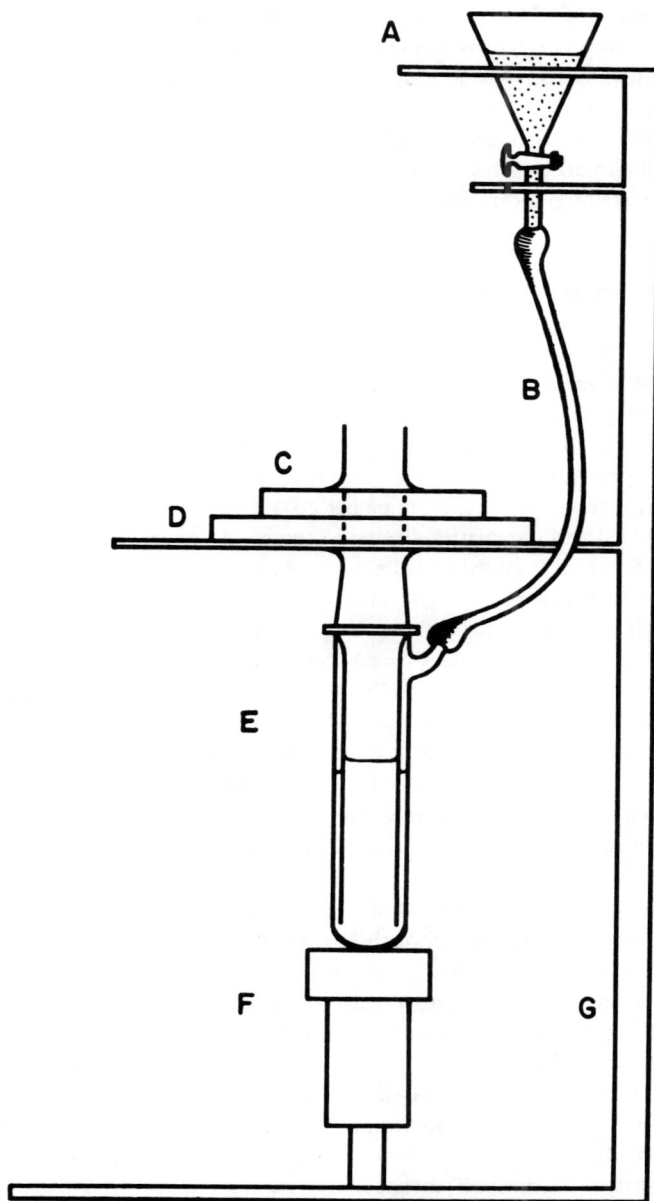

Figure 5.   Apparatus for displacing the mineral bands after centrifuging: A, funnel; B, plastic tubing; C and D, separation plates; E, double-walled centrifuge tube; F, spring-loaded base; G, aluminium frame.

trifuge tube walls. The bands are better removed with a Pasteur pipette, the tip of which is turned at right angles so that suction is applied in a horizontal plane. The viscous floating S3 D1 band also can be removed with a platinum wire loop. The S3 density subfractions are washed with acetone by repeated centrifugation in plastic tubes, and then dried, weighed and analyzed for their heavy metal content.

The finest particles (S4, -1 μm) cannot be separated into density subfractions with a conventional centrifuge.

Analysis for Heavy Metals

The density subfractions range in mass from 1 to 200 mg and may contain less than 1 μg of metal. Careful microanalysis is therefore required. The dissolution method may be chosen to release either the easily extractable metal or the total metal; only the total dissolution procedure is described here.

"Suprapur" nitric acid is added dropwise to the dry sample in a platinum crucible until effervescence ceases (1 ml per 200 mg); Suprapur perchloric acid (2 ml) is then added and the mixture heated to fumes. After cooling, 5 ml of Suprapur hydrofluoric acid is added and the mixture stood overnight, each crucible being covered by a plastic beaker. The mixture is then heated to fumes, but not to dryness. Further hydrofluoric acid may be necessary if there is a high silica content in the sample. Occasional black specks may remain in some digests, but it is assumed that these contain negligible heavy metal. A fine white precipitate may form on cooling the digest, and this has been shown to be potassium perchlorate, formed when potassium levels in the sample are high, due, for example, to the presence of muscovite[8]. There is apparently no coprecipitation of lead, cadmium or zinc with the potassium perchlorate. The digests are made up to 10 ml with double distilled water in acid-washed volumetric flasks; the pH of such solutions is <1.

The concentration of zinc in whole-sediment samples, size fractions and density subfractions is usually high enough to determine by conventional flame atomic absorption spectrometry (AAS) e.g., for Spencer Gulf sediments[8,9] zinc concentrations in the digest solutions ranged from 0.1 to 10 mg/ℓ, while zinc blanks derived from the added Suprapur acids were about 0.05 g/ℓ.

The concentration of lead and cadmium in contaminated sediments is also high enough to measure by flame AAS, provided background correction is made to compensate for the high concentrations of calcium present (up to 5000 mg/ℓ in Spencer Gulf digests[8]). A deuterium source automatic background corrector should be used routinely. Typical concentration ranges for lead and cadmium in digest solutions from contaminated sediments[9] were 0.08–8 and 0.006–0.6 mg/ℓ, respectively, with corresponding blanks of 0.02 and 0.002 mg/ℓ.

Sediment samples taken from uncontaminated areas are rather more difficult to analyze. Levels of lead and cadmium (but not zinc) in the subfractions are generally too low to determine by flame AAS, but may be measured using a graphite furnace attachment, e.g., a Perkin-Elmer Model 300SG atomic absorption spectrometer, an HGA-74 furnace and 2100 furnace controller. A 20-μl aliquot of the digest solution, diluted if necessary, is suitable. Furnace processing conditions for Spencer Gulf sediment digests[8] were: dry, 110°C, 40 sec; ash, 400°C, 20 sec; and atomize, lead, 2300°C, cadmium, 1700°C, 6 sec. Ramping is beneficial on all three cycles, and the use of the lower sensitivity lead line at 283.3 nm in addition to the 217.0-nm line allows a wide concentration range to be covered while maintaining absorbance values between 0.05 and 0.5. Increases in sensitivity are obtained, when required, by use of the gas-stop facility.

Determinations of calcium, iron, aluminium and potassium are also made on the digest solutions to ensure that adequate correction is made for any matrix effects. This approach is preferred to separation of lead and cadmium from the matrix, thus simplifying processing and minimizing the risks of contamination.

Anodic stripping voltammetry is an alternative analytical technique for the determination of trace amounts of lead, cadmium and zinc in the sediment digests[15,16].

Fractionation and digestion of sediment samples with very low metal levels should be carried out in a clean, particle-free atmosphere, e.g., a laminar-flow clean air cabinet designed to deliver "Class 100" air.

Mineralogical Analysis

The full value of the heavy liquid density gradient technique is only realized where the metal concentrations in the various subfractions are related to the mineral composition of those subfractions. Mineralogical analysis is also necessary to assess the cleanness of the cut between different mineral species achieved with a particular density gradient. Mineralogical analysis is carried out on duplicates of the samples prepared for heavy metal analysis. The main minerals and their approximate proportions are obtained by X-ray diffraction (XRD) analysis, aided by microscopic observation of the coarser sized fractions (S1 and S2).

Where there is sufficient sample, a semiquantitative mineralogical analysis is made. The sample is completely dissolved in a mixture of nitric, perchloric and hydrofluoric acids (see the section entitled Analysis for Heavy Metals, p. 324) and the concentrations of the major elements Ca, Mg, Na, K. Fe, Al, Si and Ti determined by direct-flame AAS. The elemental concentrations are assigned to the various minerals identified by XRD, and the proportion by weight of each mineral is calculated. The proportions of calcite, magnesian calcite, kaolin and goethite can be measured independently by differential thermal analysis (DTA), while the density subfractions enable another check to be made on the mineralogical composition.

RESULTS AND DISCUSSION

Sediment Components

In a complex mixture such as a sediment, it is not easy to identify and quantify the main components or to define what can be regarded as a separate "component," e.g., conglomerate particles act as separate kinetic units but are composed of a mixture of different minerals; seagrass fragments may retain epiphytes on their surfaces; metal-containing bacteria adsorbed on sediment particles may be difficult to distinguish from adsorbed free metal.

The semiquantitative mineralogical analysis described in the preceding section is useful in providing a mineral-by-mineral breakdown of the sediment composition, but it does not necessarily characterize the original components of the untreated sediment. For example, destructive semiquantitative analysis can distinguish between sediment samples with quite different mineral compositions, as shown in Table I[8]. Sample ES 26 evidently is composed mainly of shells and shell fragments with a significant proportion of organic debris, whereas sample ES 35 contains less shell grit and organic debris but more quartz, clays and iron oxides.

Table I.   Mineralogical Analysis[a] of Unfractionated Sediments[8]

	wt %[b]	
Mineral	**Sample ES 26**	**Sample ES 35**
Calcite + Magnesian Calcite	55	30
Aragonite	20	5
Quartz	10	30
Kaolinite		15
Mica (muscovite)		10
Goethite	~1	5
Feldspar	5	5
Heavy Minerals (ilmenite, tourmaline, magnetite, etc.)	<0.1	~0.2
Organic Debris	~5	<1
Total	96	100

[a] Analysis by XRD, AAS, DTA and density fractionation.
[b] Values rounded to 5%.

However, by analyzing the density subfractions it is possible to reconstruct a clearer picture of the original sediment components. Results obtained on Spencer Gulf sediments[8,9] (Table II) show that in the S2 size fraction, the mica, kaolin and goethite occurred not as discrete mineral particles, but bound together into conglomerates. Quartz was present both as free particles and in conglomerates. Most of the kaolin occurred in the lightest density subfraction (<2.40), whereas the mica was distributed in both the clay conglomerate (<2.40) and conglomerate (2.40–2.55) subfractions of sample ED 35. Table II[8] also shows that calcite and magnesium calcite were present in whole shells, shell fragments and conglomerates, but that nearly all the aragonite occurred in shell fragments.

Particles in the S3 size fraction are too small to examine routinely with a binocular microscope, and the various components are identified on the basis of their XRD patterns. Their composition could be further elaborated by examination

Table II.   Mineral Compounds of the Density Subfractions of the S2 Size Fraction[8]

Density Subfraction	Sample ES 26 (−1000 + 10 μm)	Sample ES 35 (−1000 + 10 μm)
D1	(9% wt) *Organics:* mostly organic debris; a few soft white conglomerates of magnesian calcite and quartz	(8% wt) *Clay conglomerates:*[a] soft golden brown conglomerates of magnesian calcite, mica, kaolin, quartz and goethite; a few black vitreous particles
D2	(6% wt) *Conglomerates:* soft white conglomerates of magnesian calcite, quartz, a little feldspar	(40% wt) *Conglomerates:* soft brown and off-white conglomerates of magnesian calcite, mica, some kaolin, quartz, traces goethite, feldspar
D3	(46% wt) *Quartz + calcite shells:* quartz and calcitic coralline stems, hexagonal rods, microshells	(42% wt) *Quartz + calcite shells:* mostly quartz; some calcitic shell fragments, a few hexagonal rods
D4	(20% wt) *Magnesian calcite shells:* opaque white shell fragments and whole microshells of magnesian calcite; some coralline stems and tubes	(4% wt) *Magnesian calcite shells:* as for ES 26 but including some black particles of irregular shape
D5	(19% wt) *Aragonite shells:* thin shiny shell fragments of aragonite; some transparent platelets	(6% wt) *Aragonite shells:* as for ES 26
D6	(<0.1% wt) *Heavy minerals:* as for ES 35	(0.2% wt) *Heavy minerals:* translucent brown and green particles of tourmaline, ilmenite; opaque black particles of magnetite; opaque golden rounded particles

[a] The term conglomerate is used here not in the geological sense.

with a scanning electron microscope. Notable differences between the S2 and S3 size fractions of Spencer Gulf sediments[9] were (1) the presence of large amounts of pyrite in S3 D6, but not in S2 D6; (2) the presence of free mica particles in S3 D5, whereas all mica in S2 is tied up in conglomerate particles; and (3) less quartz in the finer size fraction. Table III shows that the cut between different mineral components is not as clean in the S3 subfractions as with S2.

Table III.   Mineral Components of the Density Subfractions of the S3 Size Fraction

Density Subfraction	Mineral Components of Sample ES 23 S3 $(-10+1\ \mu m)$
D1	(15% wt) Mainly magnesium calcite; some quartz; a little mica, kaolin
D2	(73% wt) Mainly magnesian calcite; some quartz and calcite; a little mica, kaolin
D3	(6% wt) Mainly magnesian calcite, calcite and quartz
D4	(3% wt) Mainly magnesian calcite, mica and aragonite; some kaolin
D5	(2% wt) Mainly mica; some aragonite; traces kaolin
D6	(1% wt) Nearly all pyrite (FeS_2); traces of hematite and sphalerite, rutile and zircon

The results in Tables I–III show that the heavy liquid density gradient technique has several advantages for the mineralogical analysis of nearshore marine sediments: free organic debris floats to the top of the gradient and can be separated from inorganic components; aragonitic and calcitic shells are clearly separated; conglomerate particles, because of thin porosity, report in a separate density band.

Reproducibility and Accuracy of the Density Gradient Technique

It is evident from the results of Table IV that the reproducibility of the density fractionation procedure is satisfactory for both S2 and S3 size fractions.

The accuracy of the technique is assessed by comparing the sum of the individual amounts of heavy metal present in the density subfractions with the heavy metal in an equivalent weight of unfractionated material. (This comparison tests *both* the separation technique *and* the analytical method.) It may be seen from Table V that reasonable mass balances can be obtained for zinc, lead and cadmium. Mass balances should be calculated for each density fractionation as a check on the accuracy of the separation and analysis.

Table IV.   Reproducibility of the Density Fractionation Procedure

Density Subfraction	Sample ES 41 S2			Sample ES 41 S3	
	Run 1	Run 2	Run 3	Run 1	Run 2
D1	0.7	0.6	0.6	44.8	51.8
D2	7.6	6.9	8.2		
D3	72.8	73.8	73.6	37.0	35.4
D4	7.7	7.2	7.7	2.9	2.1
D5	10.9	11.3	9.6	14.8	10.4
D6	0.3	0.2	0.3	0.5	0.4

% wt in Each Subfraction

Table V.   Mass Balances for the Density Gradient Technique

Density Subfraction[a]	Mass of Heavy Metal Present (μg)		
	Zn	Pb	Cd
D1	8.92	3.93	0.220
D2	254	110	5.58
D3	37.0	18.4	0.760
D4	28.4	15.7	0.679
D5	12.8	7.53	0.270
D6	39.9	1.12	0.180
Total	381	157	7.69
Unfractionated Sample	403	172	8.12

[a] Nearshore sediment sample ES 36 S2[9].

Metal Concentrations in Size Fractions

The distribution of total sediment weight between the size fractions S1 to S4 in two Spencer Gulf samples[9] is given in Table VI, together with values for the concentrations of zinc, lead and cadmium in each size fraction and in the residue recovered from the acetone. By combining the metal concentrations with the weight distributions, the distribution of heavy metals between the size fractions is also obtained (Table VI).

Results such as those shown in Table VI can be used to study the way in which metal is distributed between the various size fractions, e.g., in the samples in Table VI the S2 size fractions contain the greatest weight and largest amount of metal, but not the highest concentration of heavy metals. Any trend of metal concentration with decreasing particle size can also be evaluated. For example, for

Table VI.   Concentration and Distribution of Metals by Size Fractions[9]

Sample	Size Fraction	Weight Distribution[a] (%)	Metal Concentration ($\mu g/g$)			Metal Distribution (%)		
			Zn	Pb	Cd	Zn	Pb	Cd
ES 26	S2	87	104	54	2.5	66	78	81
	S3	7	362	122	5.4	19	15	14
	S4	3	676	145	4.2	15	7	5
	Acetone Solution		Not measured					
ES 36	S2	87	1110	476	22.4	82	81	81
	S3	8	1760	886	39.1	12	14	13
	S4	4.5	1490	626	30.6	6	5	6
	Acetone Solution	0.1	585	126	56.9	0	0	0

[a] The wt % of size fraction S1 in samples ES 26 and ES 36 was 3 and 0.4, respectively. S1 was not analyzed for heavy metals.

sample ES 26, zinc concentrations in S4 were nearly seven times those in S2, but the same trend is not evident in sample ES 36.

Metal Concentrations in Density Subfractions

The concentrations of zinc, lead and cadmium are unlikely to be the same in the different density subfractions of a particular sample (Figures 6 and 7, Table VII), nor are the metal concentrations in similar subfractions likely to be the same in different samples (Table VII).

Results for many sediment samples in a given area can be averaged to estimate the relative metal levels in different sediment components. For Spencer Gulf sediments[9], it was found that, in general, the heavy metals were concentrated preferentially in certain components, being highest in the organic debris and the conglomerate particles, less in mica particles and magnesian calcite shell fragments, and least in aragonitic shell fragments and quartz and feldspar grains.

Table VII also shows that about 70% of all metal in ES 36 S2 was present in the D2 subfraction. Since D2 consists of conglomerate particles that are themselves composed of much finer particles, it may be expected that the metal concentration versus particle size relationship will be biased toward the larger "particles," thus explaining the results recorded in Table VI for sediment ES 36.

In any given sediment sample, the subfraction that has the highest concentra-

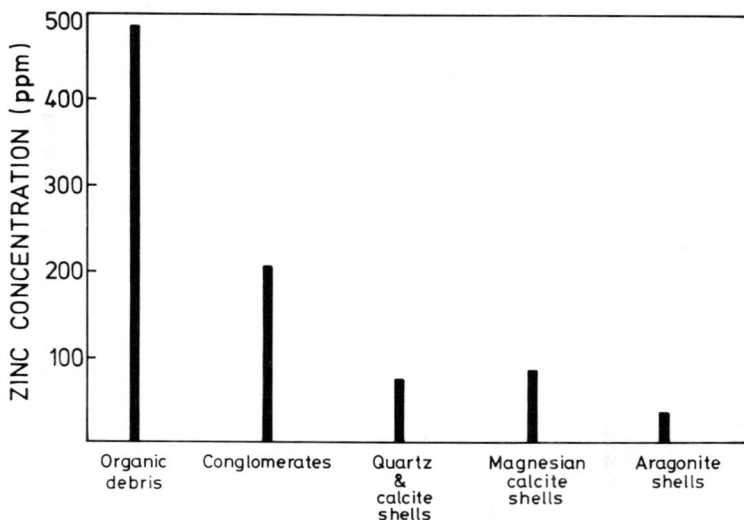

Figure 6.   Concentrations of zinc in the density subfractions separated from
sample ES 26 S2. (See Table II for a description of the subfractions.) The weight
of "heavy minerals" in ES 26 S2 was too small to collect (reproduced courtesy of
Ann Arbor Science Publishers, Inc.).

tion of heavy metal does not necessarily contain the highest proportion of total
metal. Thus, in sediment ES 35 the heavy minerals subfraction S2 D6 contained
792 μg/g of lead, but contributed only 1.5% of the total lead[8] (Figure 7). This
raises the interesting question of whether it is merely the concentration of metal
that is important to the local ecosystem, or whether consideration must also be
given to the distribution of metal between the various sediment components.

On the basis of the lead concentrations given in Figure 7, one might assume
that the heavy minerals were the key source of heavy metal in a food chain that
began in the sediment. However, this argument ignores possible variations in the
ease with which lead may be extracted from the various components and, also, no
account is taken of the fact that nearly 60% of the total lead is present in the conglomerate particles, even though at a much lower concentration than in the heavy
minerals subfraction.

CONCLUSIONS

The heavy liquid density gradient technique is suitable for the separation of the
mineral components and organic debris in nearshore sediments as a basis for the

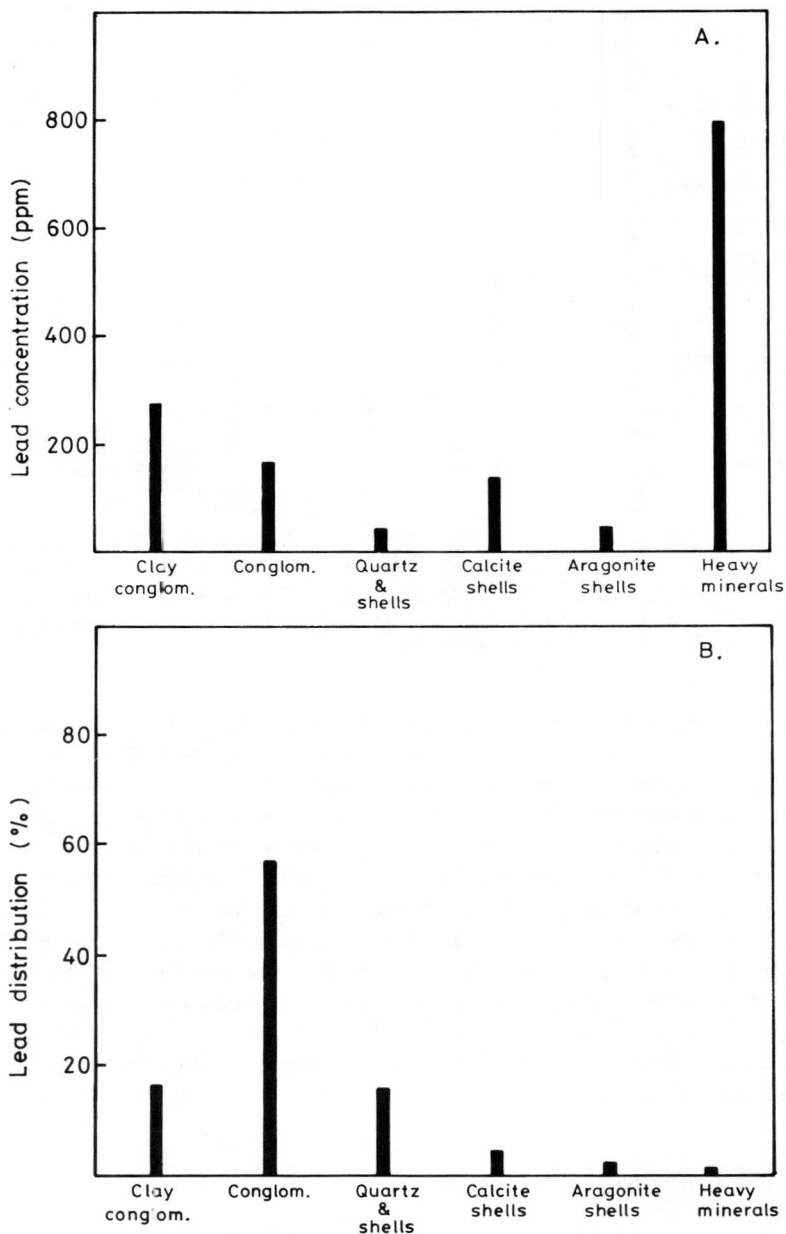

Figure 7. (A) Lead concentrations in the density subfractions of sample ES 35 S2; (B) distribution of lead between the density subfractions of sediment ES 35 S2. (See Table II for a description of the subfractions.) Reproduced courtesy of Ann Arbor Science Publishers, Inc.

Table VII.   Concentration and Distribution of Metals by Density Subfractions[9]

Sample	Density Subfraction	Weight Distribution (%)	Metal Concentration ($\mu g/g$)			Metal Distribution (%)		
			Zn	Pb	Cd	Zn	Pb	Cd
ES 26 S2	D1	8.5	486	436	22	37	43	38
	D2	6.5	204	166	9.4	12	13	13
	D3	45.7	74	45	1.9	30	24	18
	D4	20.3	85	61	5.4	15	15	23
	D5	19.0	35	23	1.9	6	5	8
	D6	<0.1	Not measured					
ES 36 S2	D1	1.0	3300	1460	81.5	2	2	3
	D2	36.6	1900	833	41.9	67	70	73
	D3	35.7	287	143	5.9	10	12	10
	D4	12.4	640	354	15.3	8	10	9
	D5	14.2	247	145	5.2	3	5	4
	D6	0.2	44300	1240	200	10	1	2

assessment of the concentration and distribution of heavy metals. Adequate sensitivity for the determination of zinc, lead and cadmium in the density subfractions is obtained using either direct flame or graphite furnace atomic absorption spectrometry.

The technique avoids many of the problems of nonselective dissoution encountered with chemical methods of determining metal distributions. Its disadvantages are that it requires a skilled operator, uses toxic heavy liquids and is difficult to apply to ultrafine particles. The results obtained are useful for evaluating the environmental significance of contaminated sediments.

ACKNOWLEDGMENTS

The author would like to thank Mr. E. S. Pilkington and Mr. P. Dossis for advice, for performing the heavy metal analyes using the graphite furnace, and for assistance with the density fractionation experiments. The work was supported in part by the International Lead Zinc Research Organization, Inc.

REFERENCES

1.   Jenne, E. A., and S. N. Luoma. "The Forms of Trace Elements in Soils, Sediment and Associated Waters: An Overview of Their Determination and Biological Availability," in *Biological Implications of Metals in the Environment*, R. E. Wildung and H. Drucker, Eds., Conf. 750929, (Springfield, VA: NTIS, 1977), pp. 110–143.

2. Piper, C. S. "Soil and Plant Analysis," University of Adelaide, Adelaide, S. Australia (1950).
3. Agemian, H., and A. S. Y. Chau. "Evaluation of Extraction Techniques for the Determination of Metals in Aquatic Sediments," *Analyst* (*London*) 101: 761–767 (1976).
4. Schmidt, R. L., T. R. Garland and R. E. Wildung. "Copper in Sequim Bay Sediments," in *Battelle Pacific Northwest Laboratory Annual Report,* Part 2 (1975), p. 136.
5. Malo, B. A. "Partial Extraction of Metals from Aquatic Sediments," *Environ. Sci. Technol.* 11: 277–282 (1977).
6. Guy, R. D., C. L. Chakrabarti and D. C. McBain. "An Evaluation of Extraction Techniques for the Fractionation of Copper and Lead in Model Sediment Systems," *Water Res.* 12: 21–24 (1978).
7. Subramanian, V. "Effect of Chemical Treatments in the Mineralogical Studies of Sediments," *Experientia* 31: 12–13 (1975).
8. Pilkington, E. S., and L. J. Warren. "Determination of Heavy-Metal Distribution in Marine Sediments," *Environ. Sci. Technol.* 13: 295–299 (1979).
9. Dossis, P., and L. J. Warren. "Distribution of Heavy Metals Between the Minerals and Organic Debris in a Contaminated Marine Sediment," in *Contaminants and Sediments,* Vol. 1, R. A. Baker, Ed. (Ann Arbor, MI: Ann Arbor Science Publishers, Inc., 1980).
10. Muller, L. D., and C. J. Burton. "The Heavy Liquid Density Gradient and its Applications in Ore Dressing Mineralogy," *Proc. Eighth Commonwealth Mining Metallurg. Cong.,* Melbourne, Australia, Vol. 6 (1965), pp. 1151–1163.
11. Francis, C. W., W. P. Bonner and T. Tamura. "An Evaluation of Zonal Centrifugation as a Research Tool in Soil Science," *Soil Sci. Soc. Am. Proc.* 36(2): 366–376 (1972).
12. Francis, C. W., and F. S. Brinkley. "Preferential Adsorption of ^{137}Cs to Micaceous Minerals in Contaminated Freshwater Sediment," *Nature* (*London*) 260: 511–513 (1976).
13. Batley, G. E., and D. Gardner. "Sampling and Storage of Natural Waters for Trace Metal Analysis," *Water Res.* 11: 745–756 (1978).
14. Pryor, E. J., H. N. Blyth and A. Eldridge. "Purpose in Fine Sizing and Comparison of Methods, Recent Developments in Mineral Processing," *Proc. First Int. Mineral Processing Cong.,* London, England (1953), pp. 11–13.
15. Florence, T. M., and G. E. Batley. "Trace Metals Species in Sea-water," *Talanta* 23: 178–186 (1976).
16. Nurnberg, H. W., "Potentialities and Applications of Advanced Polarographic and Voltammetric Methods in Environmental Research and Surveillance of Toxic Metals," *Electrochim. Acta* 22: 935–949 (1977).

CHAPTER 11

GAS MIXTURES FOR INSTRUMENTAL ANALYSIS

Patrick Carlucci

Matheson Division of Searle Medical Products USA, Inc.
Lyndhurst, New Jersey

INTRODUCTION

Gas mixtures have many diverse uses in industry, research and government. The basic use of a gas mixture, as discussed in this chapter, is as a calibration or reference standard, against which an unknown is compared. The means for this comparing can be just about any analytical instrument one can think of: infrared (IR) detectors, ultraviolet (UV)-photometric detectors, gas chromatographs (GC) using flame ionization, thermal conductivity, flame photometric and electron capture detectors (ECD), mass spectrometry (MS), fluorescence, helium ionization, etc. Most instruments are not self-calibrating, so some known reference is required. The chromatograph, for example, is not a primary chemical procedure, as is titration. A reference is required. The two-component gas mixture is usually the best source for a reference standard since there is a single component contained in a background gas, for example, 1.1% methane in helium. In gas chromatography, for example, the background gas could be the same as the carrier gas and, therefore, only the single component in the mixture would be detected. This is the simplest gas mixture. There are no rules other than the basic chemical/physical principles of partial pressures, materials of construction, and reactivity of components that limit the number of components that a gas mixture can have. Certain gas mixtures used in the petroleum industry routinely have 13 and more components, and mixtures of more than 30 different components have been made in one cylinder. But there are more factors to consider than just the number of components in the gas mixture. The following is a list of key parameters that are addressed in this chapter:

- Mixtures that cannot be made
- Do mixtures separate?
- Calibration and analysis
- Gas mixture accuracy and tolerances
- Sampling the gas mixture for better results
- Other gases
- Impurities
- Storage, handling and safety of compressed gases

The development of compressed gas mixtures in cylinders began in the late 1940s. The first mixtures were of helium-nitrogen, used to provide low dew point atmospheres in altimeters. Such atmospheres provide protection against clouding at low temperatures. They are also useful in leak detection. Later, mixtures such as argon-methane (P-10) and butane-helium ("Q" gas) were prepared for use in various types of nuclear flow counters. These were followed by ethylene oxide mixtures for gas sterilization. In all cases, approximate compositions were considered satisfactory and no analysis was required.

Later applications required gas mixtures with certified compositions. In these early days, analytical methods were limited to wet chemistry, absorption instruments such as the Orsat Analyzer, and gravimetric techniques coupled with combustion and absorption. With the development of more sophisticated instrumentation, such as infrared analyzers and gas chromatographs, it was possible to analyze more gases. These instruments required external calibration standards. Synthetic mixtures with certified compositions were needed for these calibrations. These requirements have continued to grow and become more exacting, with the technology continuing to advance to meet the demands of the scientific community.

Internal calibration procedures have continually been improved to meet more stringent accuracy requirements. In 1963 the gravimetric technique for the preparation of Primary Standards was developed, representing a tremendous gain in gas mixture technology. It provided a new level of accuracy for gas mixtures and, at the same time, increased the number of applications for gas mixtures.

The pattern of progress continues. New applications for gas mixtures arise continually. Semiconductor manufacture, vehicle emission control and air pollution abatement each have done their share to advance gas mixture technology. Now, industrial hygiene mixture requirements are testing the capability and versatility of current mixture technology. These new requirements have advanced gas mixture capabilities. For example, computer-controlled gas mixing is presently being introduced to ensure repeatable precise, multicomponent mixtures.

CAN ANY MIXTURE BE PREPARED?

Certain chemical and physical properties limit the possible mixtures that can be made. Within these limitations, virtually any combination of gases at nearly all concentration levels can be prepared. Insufficient partial pressure, chemical reactions, flammable compositions and cylinder reactions are the limitations that are encountered most commonly. Each of these limitations will be discussed in detail below.

Insufficient Partial Pressure

The most common limitation encountered when considering a gas mixture is low vapor pressure of one or more components. Normally, this occurs when these components are liquids at room temperature or low vapor pressure liquefied gases, such as propane, butane, benzene, hexane or xylene. For such liquefied materials, the component concentration and/or the total pressure is limited by the partial pressure equal to the vapor pressure of the component at ambient temperature. To avoid condensation and thus ensure reasonable stability while providing an economical volume of gas in each cylinder, mixtures are prepared using a component partial pressure less than the saturated vapor pressure at 70°F. This procedure provides a maximum component dew point between 32° and 40° F for most materials.

To illustrate this condition, let us evaluate the preparation of a mixture of ammonia in hydrogen. Ammonia has a vapor pressure at 70° F of 128.7 psia. Normally, a maximum pressure of approximately 90 psia would be used. Using the formula partial pressure = mole fraction × total pressure, we determine that the maximum concentration of ammonia that can be prepared at 2000 psia is 4.5%. Any concentration below this level can be prepared at a total cylinder pressure of 2000 psia, while any concentration above this level must be prepared at a proportionately lower total cylinder pressure, i.e., 9% ammonia can be prepared at a maximum of 1000 psia.

In addition, safety considerations have caused us to impose pressure restrictions on several nonliquefied gases. These materials and their restrictions are shown in Table I.

Table I.   Pressure Restrictions for Safety

Gas	Pressure Restriction
Acetylene	15 psia
Carbon Monoxide	1650 psig
Fluorine	100 psia
Nitric Oxide	<0.75%—2000 psia 0.75–10%—1000 psia over 10%—500 psia

Reactive Components

Mixtures of components that react chemically with one another cannot be introduced into the cylinder. Some of the reactions are obvious. Others are not normally expected to take place except under conditions of extremely high pressure and/or high cylinder temperature; experience taught us that the walls of a compressed gas cylinder can catalyze some of these reactions. Some component combinations that will react are listed below:

- $CO + H_2S$
- $H_2S + SO_2$
- $NH_3 + CO_2$
- $Cl_2 + H_2$
- $HCl + C_2H_4$
- $H_2S + COS$
- $SO_2 + NO_2$

Flammable Mixtures

Mixtures of a fuel and an oxidizer in concentrations at which the mixture is near, at or above the lower flammable limit should not be made. It is possible to make mixtures in which the flammable component is higher than the lower explosive limit if the fuel:oxidizer ratio lies outside the flammable range. It is difficult to make a general statement with regard to this type of mixture since the considerations are extremely complex. Each mixture in this category should be evaluated carefully.

Mixtures containing a flammable component and air generally can be prepared if the concentration of flammable gas is below 50% of the lower flammable limit. The flammable limits of most gases are listed in both the *Matheson Gas Data Book* and the *Unabridged Gas Data Book*.

For gas chromatographic analysis, it is possible to prepare mixtures that provide a reference for both oxygen and a flammable gas in the same cylinder. The technique used is the "oxygen equivalent." Oxygen and argon elute as a single peak, separate from nitrogen, when introduced into a chromatograph employing a thermal conductivity detector with a molecular sieve column and a helium carrier gas at or above ambient temperature. The sensitivity for argon is not the same as that for oxygen, but an argon concentration in a gas mixture can be certified equivalent to a corresponding amount of oxygen. This certified argon can then be used as a calibration standard for oxygen analysis. Suppose there is a requirement for a mixture containing 5% methane, 20% oxygen, and 75% nitrogen. Normally, this mixture would not be made, but by specifying "20% argon-oxygen equivalent," the mixture can be safely be prepared and you will get the calibration standard needed. This substitution is suitable only for gas chromatographs.

Preparation of a mixture containing a flammable gas and an oxidizer requires stringent safety procedures. When feasible, the fuel is diluted with an inert gas to a concentration less than the lower flammable limit before the oxidizer is introduced. If the mixture is being made in air, it is often necessary to synthesize the air, adding nitrogen to the fuel to dilute it before adding the oxygen. The risks involved in the preparation of such mixtures are great, and users are urged not to attempt them. Those wishing to investigate the characteristics of flammable mixtures are referred to the literature[1-3].

Cylinder Reaction

The widespread interest in air pollution and industrial hygiene has created a demand for gas mixtures containing highly reactive components at extremely low concentrations. Research aims at being able to provide these mixtures at concentrations as low as 0.1 ppm. Nonreactive mixtures as low as 0.05 ppm already have been supplied.

Reactive gases such as sulfur dioxide, hydrogen sulfide, hydrogen chloride, ammonia, etc. present different problems. Some of these problems have been solved by the development of special internal treatments of the steel cylinder walls to provide stable concentrations. Processed aluminum cylinders contain mixtures of hydrogen sulfide, sulfur dioxide, oxides of nitrogen, mercaptans and other certain other compounds at low and fractional ppm levels. The stability of these mixtures has been developed to the point that they are routinely supplied in processed aluminum cylinders.

GAS MIXING TECHNIQUES

To prepare a gas mixture of specified composition, it is first necessary to measure the amount of each component to be added to the system. Mass, volume and pressure are the three basic properties of gases used in the preparation of gas mixtures. Depending on the technique involved, one or more of these properties may be employed to achieve the desired result.

Pressure—The Basic Tool of the Gas Mixer

The pressure of a gas in a cylinder can be measured by using a calibrated gauge of the aneroid or bourdon tube type. Assuming that gases behave ideally, the final volume concentration of a gas mixture will be proportional to the partial pressures of the components, according to the following formula:

$$Pp = (MF)\,Pt$$

where   Pp = the absolute partial pressure
         MF = the mole fraction of the component
         Pt = the absolute total pressure of the mixture

Assuming ideal gases A and B to be mixed in the proportion 10% A, 90% B at a total cylinder pressure of 2000 psia, we can use the following formula:

$$Pa = (MFa)\,Pt$$

where   Pa = 0.10 × 2000
         Pa = 200 psia

To prepare this mixture, 200 psi of gas A must be measured into the cylinder, then 1800 psi of gas B. Under ideal conditions, the result is the gas mixture intended.

Since at high pressures very few gases even approach being ideal, the partial pressure mixing technique does not always give accurate results. However, the results can be improved by empirical determination, by the application of compressibility factors to the initial equation and by our own experience factor:

$$Pa = Za\,(MFa)\,Pt$$

Although the partial pressure method is not extremely accurate, it generally provides mixtures of reasonable tolerance at moderate cost. An initial preparation tolerance of between 5 and 20% of the component value, depending on the concentration level, can be expected for mixtures of this type. If, after analysis, the mixture concentrations do not fall within specifications, it is often possible to make adjustments. Adjustments are made by the addition of one or more gas constituents. The mixture is then reanalyzed to ensure that the desired composition has been attained. As the complexity of a mixture increases, corrections of this type become extremely difficult.

A process has been developed to produce multiple-cylinder quantities of the same gas mixture composition. This process permits the use of the same mixture at various locations.

Mass Measurement—The Highest Level of Accuracy

The chemist has long employed gravimetric techniques in many laboratory procedures. With this technique, a class of gas mixtures was introduced to open a new era in the field of calibration gas mixtures. For the first time, calibration gas standards prepared and certified to accuracies exceeding the capability of gas chromatographs and other analytical instruments were available to the analyst.

The gravimetric procedure is a fundamental preparation method. Components are weighed on a high-capacity, high-sensitivity analytical balance. This technique eliminates the errors associated with the partial pressure technique previously described because it is independent of temperature, pressure and compressibility. As long as the cylinder on the balance is tared carefully, any gas* can be weighed accurately into the mixture. A mixture containing any number of components can be prepared by this technique. The preparation tolerances of the gravimetric technique are less than 1% of the component value. General specifications are shown in Table II. The actual accuracy obtainable is a function of the weight of the gases being introduced into the cylinder and of the accuracy of the balance.

Table II.   Gas Mixture Specifications

Category	Range	Preparation Tolerance	Certification Accuracy
Primary Standard	5%–50%	±1% of component	±0.02% absolute or 1% of the component, whichever is smaller[b]
	1%–5%	±2% of component	
	0%–1%	±5% of component	
Certified Standard	10%–50%	±5% of component[a]	2% of component
	50 ppm–10%	±10% of component[a]	2% of component
	10 ppm–50 ppm	±20% of component[a]	5% of component
	3 ppm–10 ppm	±2 ppm	5% of component
Unanalyzed Mixture	This group of mixtures is expected to have essentially the same preparation tolerance as the Certified Standard but cannot be guaranteed as such because these mixtures are not checked by analysis.		

[a] The values presented may vary for components that tend to be unstable or present other blending problems.
[b] Accuracies for hydrogen and helium components may vary from those stated due to their low molecular weight.

Dynamic Blending

Blending gas using a dynamic system can be the answer to certain physical and chemical difficulties encountered with gas mixtures. For example, the following requirements can be met through the use of dynamic blending:

---

* Some reactive gases at low concentrations cannot be made with this technique because of interaction with the cylinder walls, i.e., what is weighed into the cylinder may not come out.

1. experimental systems with constantly changing mixture compositions of the same or similar gas components;
2. high-volume gas mixture applications, where the number of cylinders required to meet the need presents serious material handling problems; and
3. preparation of mixtures of unstable compounds at extremely low concentrations not normally available in cylinders (mixtures must be generated at the use location).

One piece of equipment designed for this procedure is the DYNA-BLENDER®, which can be used onsite.

The DYNA-BLENDER delivers online, premixed gases of different ratios using electronic mass flow controllers. An expansion of the Dyna-Blender concept is an in-house, computer-controlled dynamic blending system, which results in very precise and reproducible gas mixtures in unlimited quantities.

HOMOGENIZING THE MIXTURE

Over the years, a great deal of concern has been expressed regarding the potential for the components of a gas mixture to stratify in the cylinder according to their density as a result of the effect of gravity. Tests have been conducted to determine whether such a phenomenon ever takes place. Results of these tests have shown that once the mixture is homogeneous, it remains so and does not separate.

It is possible to obtain some separation of the components in a gas mixture if one or more of the components are normally liquefied gases. In such a case, the liquefied components may be subject to partial condensation in the cylinder if subjected to sufficiently low temperatures. If the partial pressure of a mixture component approaches the pressure limitation level, condensation of that component will take place when subjected to lower temperatures. The amount of condensation is a function of the temperature and vapor pressure of the material in question.

Consider a simple 40% propane, 60% nitrogen gas mixture in which the partial pressure of propane is 80 psia in a total cylinder pressure of 200 psia. In checking the propane vapor pressure curve, at approximately 42°F the vapor pressure of propane is found to be 80 psia. Therefore, the propane dew point of this mixture is 42°F. As the temperature falls below this level, more and more propane condenses, creating a two-phase system in the cylinder. The liquid phase is 100% propane; the vapor phase contains propane in nitrogen but in a percentage less than the original mixture. At 25°F the vapor pressure of propane is 59 psia. If this were the storage temperature of the cylinder, the vapor phase of the mixture would be approximately 29.5% propane, rather than the original 40%. Before this cylinder can be used it must be warmed to above 42°F, preferably to room temperature (approximately 70°F) and mixed using the following technique. The cylinder is heated at the lower end using a hot water or infrared heat source to produce a connection current within the cylinder. Caution must be taken never to heat the cylinder to a temperature greater than 125°F. If a gas were to be withdrawn while the cylinder were cold, and the cylinder were then warmed and mixed, the propane concentration of the remaining mixture would be higher than the original 40%. For this reason, proper storage and handling of gas mixtures is extremely important.

CALIBRATION AND ANALYSIS

It is not practical to discuss detailed analytical procedures for the wide variety of gas mixtures that are possible. Rather let it suffice to point out that all the tools of modern gas analysis are available—gas chromatographs equipped with a variety of detectors including flame ionization, thermal conductivity, photo ionization, ultrasonic and helium ionization. Atomic absorption, infrared and plasma emission spectroscopy are also employed. This instrumental capability is supplemented and supported by a complete repertoire of wet chemical and gravimetric techniques.

Although the exact analytical technique is not critical to this discussion, the basis of calibration for this collection of expensive instrumentation is nonetheless of utmost importance. Analytical laboratories should be standardized using reference standards from the National Bureau of Standards, whenever such standards are available. Almost all of the remaining calibration standards required for instrumental analysis should be prepared gravimetrically as Primary Standards. These standards are checked either independently by wet chemical analysis, or by direct comparison to a previous standard.

ACCURACY AND TOLERANCES

Gas mixtures can be blended to customer, but to simplify ordering one can standardize on three categories of gas mixtures: Primary Standards, Certified Standards and Unanalyzed Mixtures. The specifications for these standard categories are shown in Table II. In many cases, tolerances and accuracies can be made more stringent if required. Also, it is possible to vary tolerances on the components within a mixture.

Primary Standards

Instrument technology, as it develops, requires more and more stringent accuracy. For maximum instrument accuracy there is only one class of calibration mixtures: Primary Standards. These mixtures are a must for accurate work with chromatographic systems employing electronic digital integration. They can save thousands of dollars when used to calibrate process control instruments.

Primary Standards are prepared gravimetrically using weights certified by the National Bureau of Standards. The accuracies possible with this technique exceed most analytical techniques. Normal tolerances for the Primary Standard are shown in Table II.

Certified Standards

Certified Standards are analyzed calibration mixtures prepared by the most accurate partial pressure and volumetric techniques available. Whenever possible, NBS Standard Reference Materials and/or Primary Standards are used as the reference analytical standard. These mixtures are recommended for applications in which the accuracy of the Primary Standard is not required. General specifications and tolerances for Certified Standards are also shown in Table II.

The example in Table III will help clarify the difference between the Primary Standard and the Certified Standard. A single mixture with four different concentration levels has been selected for examination.

Unanalyzed Mixtures

Mixtures can be prepared without certification; however, these mixtures are not recommended nor can one make any guarantees regarding their accuracy.

Table III.   General Specifications for Tolerances and Standards

| Mixture Requested | Primary Standard | | Certified Standard | |
	Preparation Range	Certification Accuracy	Preparation Range	Certification Accuracy
20 ppm methane	19–21 ppm	±0.2 ppm	16–24 ppm	±1 ppm
1% carbon monoxide	0.98–1.02%	±0.01%	0.9–1.1%	±0.02%
5% carbon dioxide	4.9–5.1%	±0.02%	4.5–5.5%	±0.1%
40% nitrogen	39.6–40.4%	±0.02%	38.0–42.0%	±0.8%
Balance helium				

Summary

The technique used to manufacture a particular gas mixture will depend on the nature of the mixture and the accuracy required. In many cases, combinations of two or more of the techniques previously discussed will be used; in some, modification and innovation will be required.

The cost of a gas mixture is controlled by two factors; material and labor. For any specific gas mixture the material cost is a constant. Thus, the specifications on the preparation and analytical tolerances usually control the cost of the mixture. The more stringent the preparation and analytical requirements, the greater the time required to prepare the mixture. The number of components in a mixture also contributes to its cost. A multicomponent, high-accuracy mixture will necessarily cost considerably more than a two-component certified standard.

SAMPLING FOR BETTER RESULTS

Without a proper transfer system, all the care and work put into the production and analysis of a gas mixture become meaningless. It is important that careful consideration be given to the equipment to be used with a particular gas mixture. There are two common pitfalls:

- the introduction of impurities; and
- adsorption and/or reaction with the transfer system.

To ensure that impurities are not introduced into the mixture, the transfer system should be constructed using metal diaphragm regulators with inert fluorocarbon seat materials such as Kel-F® or Teflon. These regulators should be fitted with outlet valves of the diaphragm packless type. The transfer line should be metal tubing or Teflon with compression fittings or connections. Brass regulators are suitable for gas mixtures containing noncorrosive components. This type of transfer system is a must for mixtures containing very low concentrations of moisture, oxygen, nitrogen or total hydrocarbons, or where these impurities must be excluded from the mixture. The outgassing of hydrocarbons from a rubber diaphragm, or the diffusion of oxygen, nitrogen and water through a rubber diaphragm, or rubber or plastic lines, can greatly influence the accuracy of a calibration.

The proper choice of a regulator depends on the delivery pressure range required, the degree of accuracy of delivery pressure to be maintained and the flowrate required. There are two basic types of pressure regulators: single stage and double, or two stage. The single-stage type will show a slight variation in delivery pressure as the cylinder pressure drops. It will also show a greater drop in delivery pressure than a two-stage regulator as the flowrate is increased. In addition, it will show a higher "lockup" pressure (pressure increase above the delivery set point necessary to stop flow) than the two-stage regulator. In general, the two-stage regulator will deliver a more nearly constant pressure under more stringent operating conditions than will the single-stage regulator.

For gas mixtures containing corrosive compounds, a similar system is recommended except that the materials of construction must be capable of withstanding the corrosive action involved. Even a small amount of corrosive gas must be considered. A minute amount of corrosive gas may not be sufficient to cause significant corrosion, but if the materials of construction are not properly selected, these small concentrations will react with, or adsorb on, the regulator and transfer line. Generally, Teflon or stainless steel transfer lines are recommended. High-purity metal diaphragm regulators in stainless steel are satisfactory for mild corrosives such as ammonia, hydrogen sulfide and sulfur dioxide. Higher concentrations require a model with the necessary corrosion resistance.

Sampling Liquid Phase Mixtures

There are numerous applications that require gas mixtures prepared in a liquid phase. These are commonly used to evaluate petrochemical processes, as aerosol propellants and as calibration standards for process chromatographs analyzing liquified gas streams. Some cylinders shipped with liquid phase gas mixtures are equipped with full-length eductor tubes to ensure consistent liquid composition for nearly the entire contents of the cylinder. For liquid compositions that have very low vapor pressures, a head pressure or "pad" of nitrogen or helium is provided to assist in the withdrawal of the liquid from the cylinder. In some cases, the cylinder is provided with dual valves; one for liquid withdrawal and the other for pressurizing the head space. With dual valve systems a constant pressure can be maintained. This is important with certain liquid phase mixtures. Figure 1 illustrates the two types of cylinders used.

Generally, a pressure regulator cannot be used to sample the contents of a liquid-phase mixture. A manual control should be selected. Before removing material from a liquid cylinder for analysis, it is advisable to clear the eductor tube. This can be accomplished by withdrawing a small amount of material, which, in effect, purges the system.

General Sampling Rule

There is one primary rule in using a gas mixture as a comparison standard. Both the unknown and the calibration standard must be sampled in exactly the

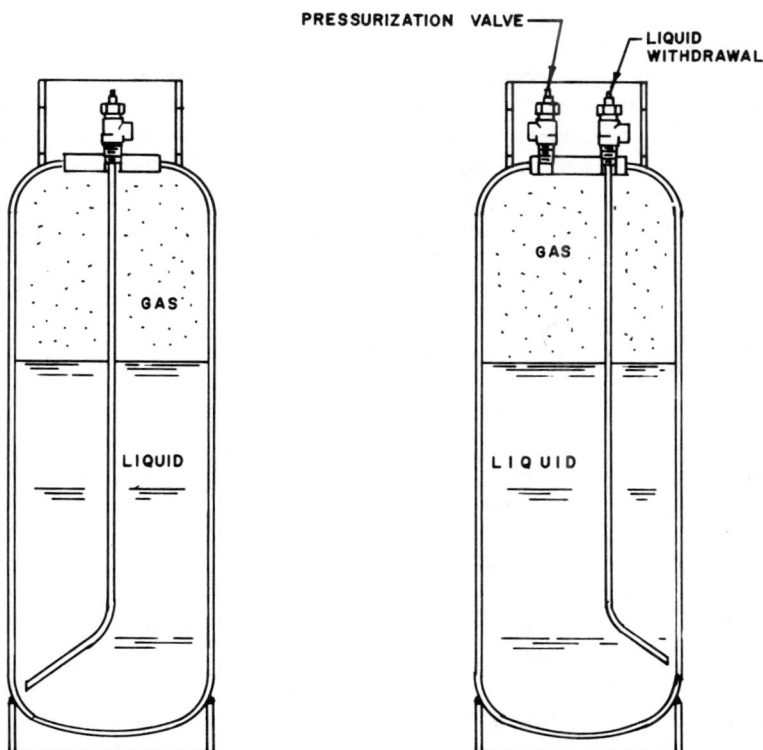

Figure 1.   Typical cylinders for liquid-phase mixtures.

same manner. This means that when a gas sample is injected into a gas chromato-graph, the PV/T relationship for the unknown and the calibration standard must be equal for a one-to-one comparison. For dynamic sampling, similar to that used with infrared analysis, one must ensure that the mass flowrate is equal for both the calibration mixture and the unknown. Unless the unknown and the reference are sampled in the same way, one will not achieve maximum accuracy from his equipment.

One of the prime uses of gas mixtures is the calibration of instruments and ana-lyzers, both in the laboratory and the process plant. These analytical instruments vary greatly among the wide range of chemical, pharmaceutical and petrochemi-cal industries, and within each industry, depending on the application. For example, the Chemical Process Industry (CPI) uses gas mixtures in the monitor-ing of the production of a particular compound, in controlling the quality of the raw materials that are used to form that compound, in determining what, if any, pollution is created in that manufacturing process, and determining how much of that pollution is reaching the atmosphere. In many cases, specific analyzers, i.e., analyzers that respond to a particular chemical speices, are the choice, and these instruments are dedicated to one particular analytical function. However, in many instances the versatility and reliability of the gas chromatograph is utilized. When gas chromatography is the instrument chosen for a particular job, care must be given to the proper selection of not only the calibration standards, but also to the carrier gases along with the fuel gas, and oxidizer gas (if the GC is equipped with a flame ionization detector). Table IV is a chart that can assist in the correct selection of gases, pressure regulation equipment and filters.

IMPURITIES—WHAT TO DO ABOUT THEM

Even the purest gases contain impurities. Mixtures made from these gases must necessarily contain impurities. In most cases, impurities do not affect the system in which a gas mixture is to be used; however, in a few cases the presence of the wrong impurity could be disastrous. It is important to review a system to determine the effect of impurities.

Let us look at some examples of mixture impurities and their effect on systems in which they are employed. A mixture of 50 ppm oxygen in nitrogen could conceivably contain 5 ppm argon. If this mixture were used as a calibration standard for an electrolytic trace oxygen analyzer, the argon would have no effect. If the same mixture were used as a calibration standard on a gas chromatograph, a 10% error in analysis would occur because argon and oxygen combine as a single response under normal analytical conditions using a thermal conductivity detector. It is true that one way to separate argon and oxygen on a gas chromatograph is by using a molecular sieve column immersed in a liquid nitrogen bath. However, this system presents many operating difficulties and is not available in many laboratories. To avoid the problem of argon contamination, both oxygen and argon should be analyzed so that the argon impurity can be taken into consideration.

Hydrocarbon mixtures at low levels present an additional impurity problem. The use of flame ionization detectors (FID) requires special care. Everyone recognizes the requirement for air, hydrogen or fuel gas and carrier gas to be hydrocarbon free, but too often not enough consideration is given to the calibration mixture. When ordering a calibration gas to be used with an FID, it is important to consider the actual use. If the mixture is to be used on a gas chromatograph with an FID, hydrocarbon impurities probably will not be a problem. On the other hand, if this mixture is to be used with an FID that responds only to total hydrocarbons in the gas stream, then the presence of any hydrocarbons other than the ones specified will be detrimental. Consider that, in its simplest form, the FID is a carbon atom counter that considers a methane molecule as one

Table IV.   Guide to Selection of Regulators and Filters

Detector	Gases Required	Pressure Regulators	Filters
Thermal Conductivity	*Carrier gases (1 required)*		Filter 6164T
	Helium, high purity	3104–580	
	Argon, prepurified	3104–580	
	Hydrogen, prepurified	3104–350	
	Calibration gas mixture	Regulator selection is based on mixture components	
Flame Ionization (FID) or Flame Photometric	*Carrier gases (1 required)*		Filter 6164T in all gas lines
	Helium, zero gas	3104–580	
	Nitrogen, zero gas	3104–580	
	Fuel gases (1 required)		
	Hydrogen, zero gas	3104–350	
	40% H_2 / 60% N_2, zero gas	3104–350	
	40% H_2 / 60% He, zero gas	3104–350	
	Oxidizer gas (1 required)		
	Air, zero gas	3104–590	
	Calibration gas mixture	Regulator selection is based on mixture components	
Electron Capture (ECD)	*Carrier gas (1 required)*		a
	Nitrogen, UHP	3104–580	
	10% CH_4 / 90% Ar	3104–350	
	5% CH_4 / 95% Ar	3104–350	
	8.5% H_2 / 91.5% He	3104–350	
	Calibration gas mixture	Regulator selection is based on mixture components	

[a] This stainless-steel heliarc-welded filter has filtration efficiency of 99.999% of particles 0.3μ and larger. Such particles should not get into the instrument.

carbon atom and a propane molecule as three carbon atoms. In such a situation, a mixture containing 1 ppm methane and 1 ppm propane would contain an equivalent methane content of 4 ppm or an equivalent propane content of 1.3 ppm. In either case, the effect is obvious. Therefore, when specifying hydrocarbon mixtures for use with flame ionization detectors that measure total hydrocarbon content, one should requires the analysis in terms of total hydrocarbon content related to the hydrocarbon molecule to be used as a reference.

These examples illustrate the problems that can be encountered when impurities are not considered. In most applications these problems do not arise, but such effects should be considered to ensure maximum effectiveness of a system.

STORAGE, HANDLING AND SAFETY

After cylinders are received, they are usually stored either in a special gas storage area or in the laboratory itself. In general, most gas mixtures find their way directly to the process control shed, laboratory or pilot plant because of their unique nature. Storage areas should be fire resistant, well ventilated, dry and located away from sources of ignition or excessive heat. Where cylinders are stored outdoors they should be protected from the direct rays of the sun, particularly in localities where high ambient temperatures prevail. Only those cylinders actually in use should be allowed in the laboratory or work area. In all cases, storage areas should comply with government regulations.

Ideally, the temperature of a gas mixtures storage area should be about 70°F. If this is not possible, the mixture storage area should be such that the cylinders will not be subjected to temperatures below 50°F. If temperatures fall below 50°F and the mixtures contain a liquefied component, one must be careful that the cylinder has not fallen below the condensation point. If one suspects that a cylinder has been subjected to temperatures below the condensation point, steps should be taken to ensure that all the components are homogeneous. Under these circumstances, the hot water technique previously described will provide the most immediate results. A cylinder should never be heated over 125°F. If the hot water technique is not satisfactory and rolling must be used, it will be necessary to store the cylinder for several days at a temperature between 70°F and 90°F before attempting to homogenize it by rolling. Do not handle or roll the cylinder unless the valve protection cap is in place. Also, when moving cylinders from a storage area into the plant or laboratory, make sure that the valve protection cap is in place. The cylinder should then be transported by means of suitable hand truck provided with a chain or belt to secure the cylinder so that it cannot fall if the hand truck passes over a bump. The hand truck should also be designed with rear wheels and braces to support the weight of the cylinder; this provides ease of movement and precise control.

When the cylinder is brought to its place of use in the laboratory or plant, it should be secured to a wall, bench or some other firm support. A plain chain or a bench clamp and belt should be used. In all cases, the chain or belt should be located high enough on the cylinder body so that the cylinder cannot possibly tumble out of it. If it is not possible to support the cylinder by this means, a cylinder stand should be used that is adaptable to a number of different size cylinders by means of thumb screws, which can be turned until they tighten around the cylinder and hold it in place. Although use of a stand is not as effective as securing a cylinder to a wall or bench, it improves cylinder stability in situations in which other types of support are impractical. Once the cylinder is secure, the cap may be removed, exposing the valve.

GENERAL PRECAUTIONS

Some general precautions for handling, storing and using compressed gas follow.

1. Never drop cylinders or permit them to strike each other violently.

2. Cylinders may be stored in the open, but should be protected from the ground beneath to prevent rusting. Cylinders may be stored in the sun, except in localities where extreme temperatures prevail. In the case of certain gases, the supplier's recommendation for shading should be observed. If ice or snow accumulates on a cylinder, thaw at room temperature not exceeding 125°F.

3. The valve-protection cap should be left on each cylinder until it has been secured against a wall or bench, or placed in a cylinder stand, and is ready to be used.

4. Avoid dragging, rolling or sliding cylinders, even for a short distance. They should be moved by using a suitable hand truck.

5. Never tamper with safety devices in valves or cylinders.

6. Do not store full and empty cylinders together. Serious suckback can occur when an empty cylinder is attached to a pressurized system.

7. No part of a cylinder should be subjected to a temperature higher than 125°F. A flame should never be permitted to come in contact with any part of a compressed gas cylinder.

8. Cylinders should not be subjected to artificially created low temperatures (−20°F or lower) since many types of steel will lose their ductility and impact strength at low temperatures. Special stainless steel cylinders are available for low-temperature use.

9. Do not place cylinders where they may become part of an electric circuit. When electric arc-welding, precautions must be taken to prevent striking an arc against a cylinder.

10. Ground all cylinders, lines and equipment used with flammable compressed gas.

11. Use compressed gases only in a well-ventilated area. Toxic, flammable and corrosive gases should be handled in a hood. Only small cylinders of toxic gases should be used.

12. Cylinders should be used in rotation as received from the supplier. Storage areas should be set up to permit proper inventory rotation.

13. When discharging gas into a liquid, a trap or suitable check valve should be used to prevent liquid from getting back into the cylinder or regulator.

14. When using compressed gases, wear appropriate protective equipment, such as safety goggles or face shield, rubber gloves and safety shoes. Well-ventilated barricades should be used in extremely hazardous operations, such as in the handling of fluorine. Gas masks should be kept available for immediate use when working with toxic gases. These masks should be placed in convenient locations in areas not likely to become contaminated, and should be approved by the U.S. Bureau of Mines for the service intended. Those involved in the handling of compressed gases should become familiar with the proper application and limitations of the various types of masks and respiration aids available.

15. When returning empty cylinders, close the valve before shipment, leaving some positive pressure in the cylinder. Replace any valve outlet and protective caps originally shipped with the cylinder. Mark or label the cylinder "empty" (or utilize standard DOT "empty" labels) and store in a designated area for return to the suppliers.

16. Before using cylinders, read all label information and data sheets associated with the gas being used. Observe all applicable safety practices.

17. Eye baths, safety showers, gas masks, respirators and/or resuscitators should be located nearby but out of the immediate area that is likely to become contaminated in the event of a large release of gas.

18. Fire extinguishers, preferably of the dry chemical type, should be kept close at hand and should be checked periodically to ensure their proper operation.

REFERENCES

1.   Zabetakis, M. G. "Flammability Characteristics of Combustible Gases and Vapors," *Bur. Mines Bull.* 627.
2.   Coward, H. F., and G. W. Jones. "Limits of Flammability of Gases and Vapors," *Bur. Mines Bull.* 603.
3.   Lewis, B., and G. von Elbe. *Combustion, Flames and Explosions of Gases,* 2nd ed. (New York: Academic Press, Inc.,).

CHAPTER 12

PNEUMATIC INSTRUMENTS

Edward J. Farmer

Ed Farmer & Associates
Sacramento, California

INTRODUCTION

General Description

Pneumatic instrument systems are those in which air pressure and flow are used to convey and act on information. Among the oldest forms of process control equipment, they are highly developed and well adapted to most control situations. Electronic instrumentation has made inroads into the once almost exclusively pneumatic process control instrumentation field, but pneumatics continue to retain about 50% of the market.

The pneumatic instrumentation field is far from stagnant. New products incorporating blends of the most modern electronic technology, space-age machining, finishing and manufacturing techniques, and the most modern materials have been introduced at a steady pace.

In current industrial practice, pneumatics are blending with electronics into hybrid systems. The applications of pneumatic instruments are becoming more limited—small systems, process interface for large systems and analog backup for computer control systems—but their enduring place in industry appears assured.

Advantages and Disadvantages

The advantages of pneumatic instruments include a wider ambient temperature tolerance than electronic instruments, total immunity to electromagnetic interference, lower cost than corresponding electronic instruments and absolute safety from causing injury or fire damage. Operators and mechanics also seem to understand pneumatic instruments more easily than they do electronic ones.

Transmission range, speed of response and limited systems capability are the principal detracting aspects.

Pneumatic signals travel through tubes. They are required to expand bellows and are throttled through restrictions. All of these operations require time and some amount of work. The effect is attenuation and phase distortion of the signals.

Figures 1 and 2 comprise a bode diagram illustrating the effect of 6.35-mm instrument tubing on a pneumatic control loop. As length increases, the attenuation of the signal increases rapidly as does the phase shift introduced to the loop. Many processes will tolerate a control loop frequency response as low as 10^{-2} Hz. Pneumatic controls with tubing lengths of a few hundred meters will result in generally acceptable performance for these processes. In more demanding situations, control loop performance will not be adequate. Inadequacies usually manifest themselves as difficult controller tuning problems. These limitations may be mitigated through the use of larger diameter tubing and pneumatic relays, but the cost and complexity of the systems will increase rapidly, thus making the choice of electronic instruments attractive.

It is more difficult and costly to construct a large, complex pneumatic instrument system than an equivalent electronic one. This is due to the limited systems capability of pneumatic instruments. Traditionally, pneumatic instruments have been loop oriented in design—one transmitter, one controller, one valve operator and perhaps one additional receiver instrument per loop. As complex systems requiring additional loads on the signal transmission tubing are added, performance is degraded. The substantial physical space required for pneumatic instruments, auxiliaries and interconnecting tubing also contributes to the cost and difficulty of constructing and maintaining large systems.

The system's capability limitations can be mitigated and the advantages of pneumatics retained by constructing hybrid systems in which pneumatic instruments are used to interface with the process and electronic instruments are used for long distance transmission and central control. This concept is in widespread use in the process industries.

BASIC ELEMENTS

Pneumatic instruments are collections of components arranged to perform the various functions required of instruments. These functions include converting mechanical motion to pressure and vice-versa, amplification of signals, implementation of proportional-integral-derivative control algorithms, and the many others required by control systems engineers and operators.

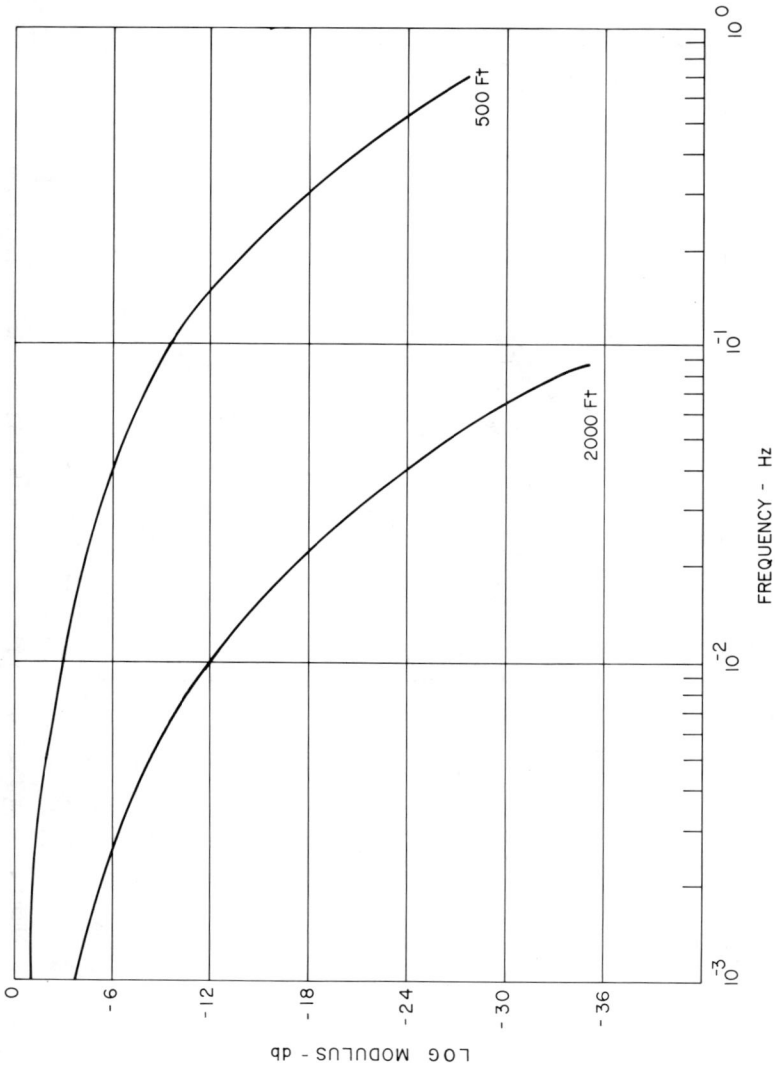

Figure 1.   Amplitude vs frequency plot for 6.35-mm (¼-in.) pneumatic instrument tubing.

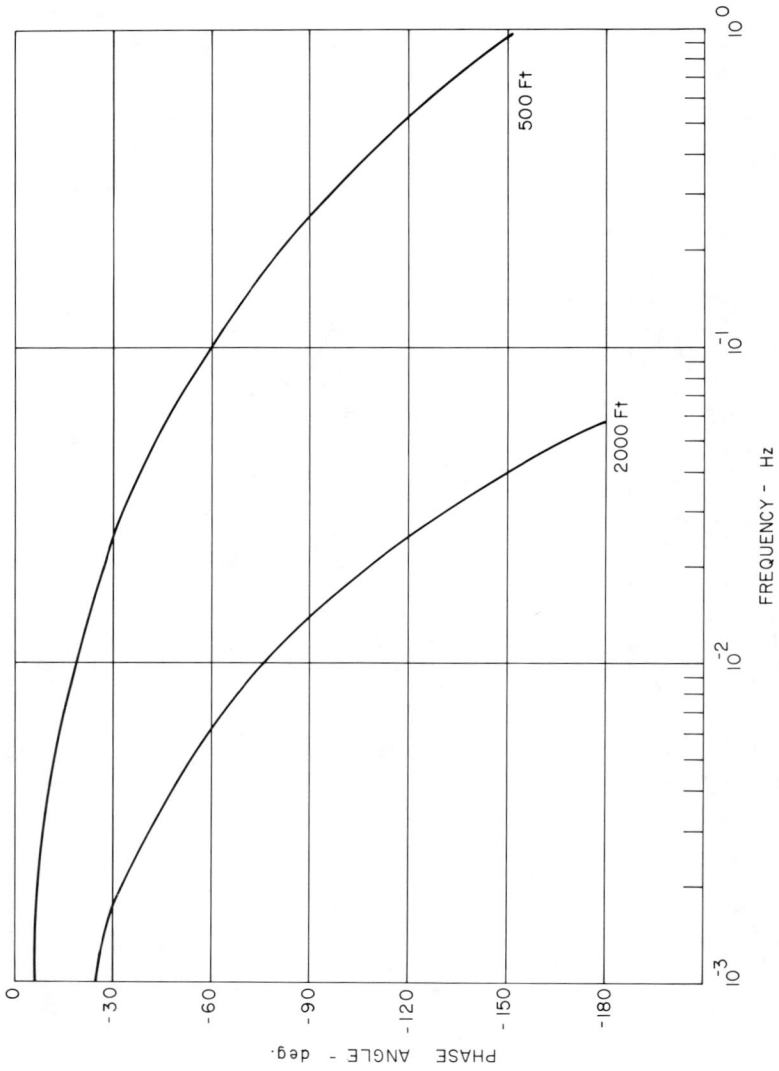

Figure 2.   Phase vs frequency plot for 6.35-mm (¼-in.) pneumatic instrument tubing.

The more significant basic elements used to construct instruments are discussed in this section. Although some illustrations of applications are included, it is important to realize that each instrument manufacturer arranges the basic elements in different ways to produce an instrument with special characteristics and advantages relative to its competition.

Spring

The spring is an important component of pneumatic instruments. It is useful because it provides a force directly proportional to deflection from its equilibrium position (Figure 3). The force is given by

$$F_s = K_s d$$

where F_s = the force exerted by the spring
            d = the deflection
           K_s = the spring constant (Figure 4)

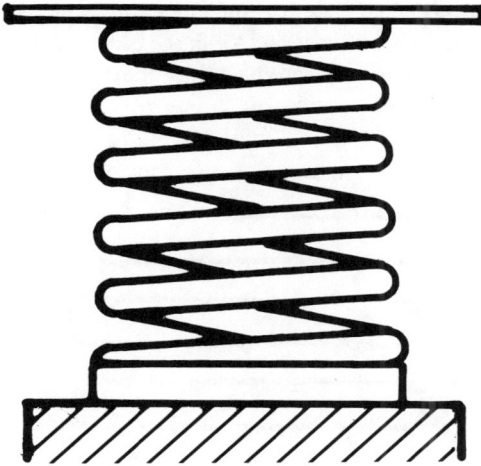

Figure 3.   Typical linear spring.

Springs are used to provide a restoring force for indicator movements, linear opposition for a bellows and other applications in which a force proportional to motion is required.

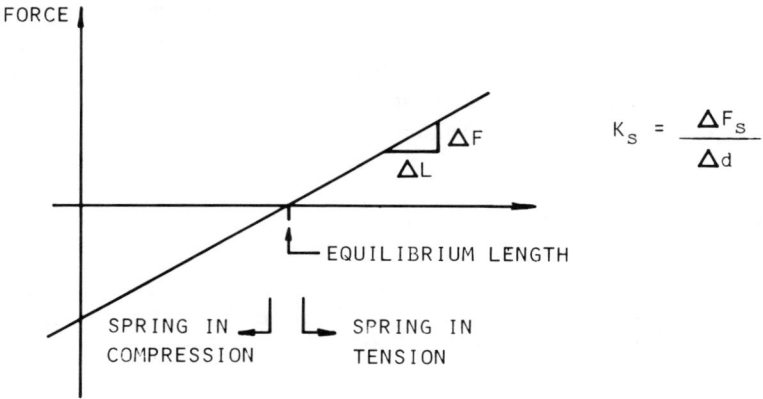

FORCE

ΔF

ΔL

$$K_s = \frac{\Delta F_s}{\Delta d}$$

EQUILIBRIUM LENGTH

SPRING IN COMPRESSION

SPRING IN TENSION

Figure 4.   Force-deflection diagram for typical spring.

Bellows

The bellows is an apparently simple, yet extremely important, component of pneumatic instruments (Figure 5). It is the method by which air pressure is converted to mechanical force and motion. Air in the bellows causes it to expand and extend along its axis. The force it produces is proportional to the pressure inside the bellows and the area of its end plate.

Figure 5.   Bellows.

$$F_B = A_B \times P_B$$

where F_B = the force exerted by the bellows
       A_B = the area of the end of the bellows
       P_B = the air pressure in the bellows

A bellows is a very special device. It must be strong enough to survive the treatment encountered by instruments in process plants. On the other hand, it must be sufficiently flexible and resilient to not exhibit hysteresis. Its expansion must be linear with pressure. Nonlinearities over the range in which it must flex in normal operation are unacceptable. It is also necessary that a bellows undergo millions of expansion-contraction cycles without fatigue failure or changing properties.

Diaphragm

Diaphragms are generally flexible, nonpermeable devices used to couple together pressures or motions in two separate systems without mixing the contents of the two systems. They must be sufficiently strong to withstand the differential pressure between the systems and sufficiently flexible and pliable to linearly transmit the pressure or motion without introducing nonlinearities and offsets.

Within instruments, the greatest application is in the construction of pneumatic relays. The diaphragm is used to provide an impermeable, pressure-sensitive connection between the sensing chamber and the valve assembly.

Flapper Nozzle

The heart of pneumatic instrumentation is the flapper nozzle assembly. It provides the link between mechanical movement or force and air pressure. It consists of a small nozzle through which air is exhausted. The pressure upstream (back pressure) of the nozzle is controlled by manipulating the distance the flapper is from the nozzle. If the flapper is sealing the nozzle, back pressure is essentially supply pressure. If the flapper is well away from the nozzle, pressure drops to a value near zero. In between these extremes, pressure is a function of flapper position. More than a substantial portion of the flapper's travel pressure is approximately linear in flapper position (Figure 6).

By designing the flapper nozzle assembly to operate in this linear region, a linear instrument can be constructed. By combining the flapper nozzle assembly with a pneumatic relay, a sensitive motion to pressure transducer can be constructed. This high gain assembly can be arranged in a feedback configuration so that accuracy and stability are enhanced. The two configurations in common use are the force-balance and motion-balance configurations.

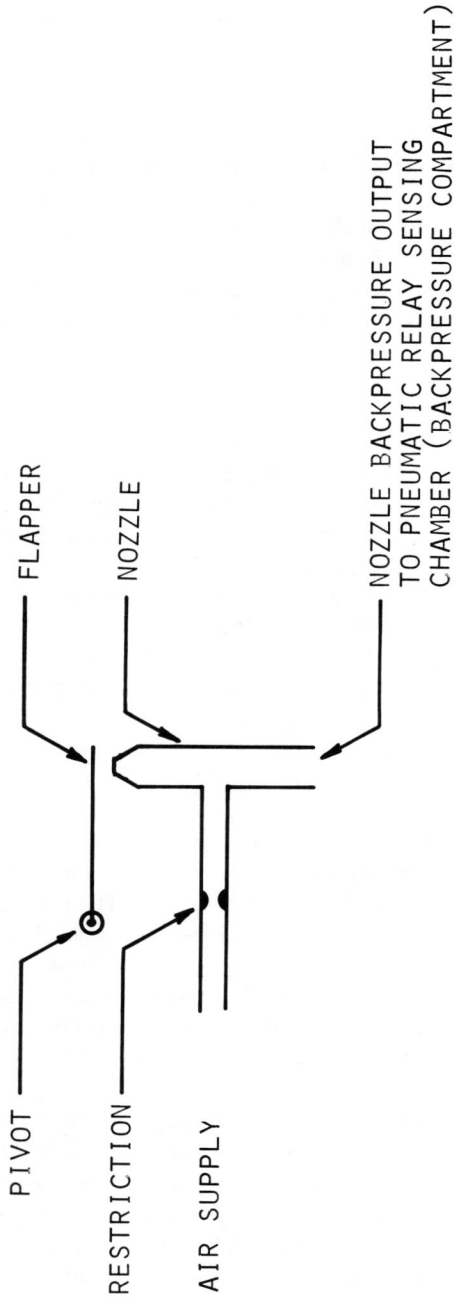

Figure 6.   Flapper-nozzle assembly.

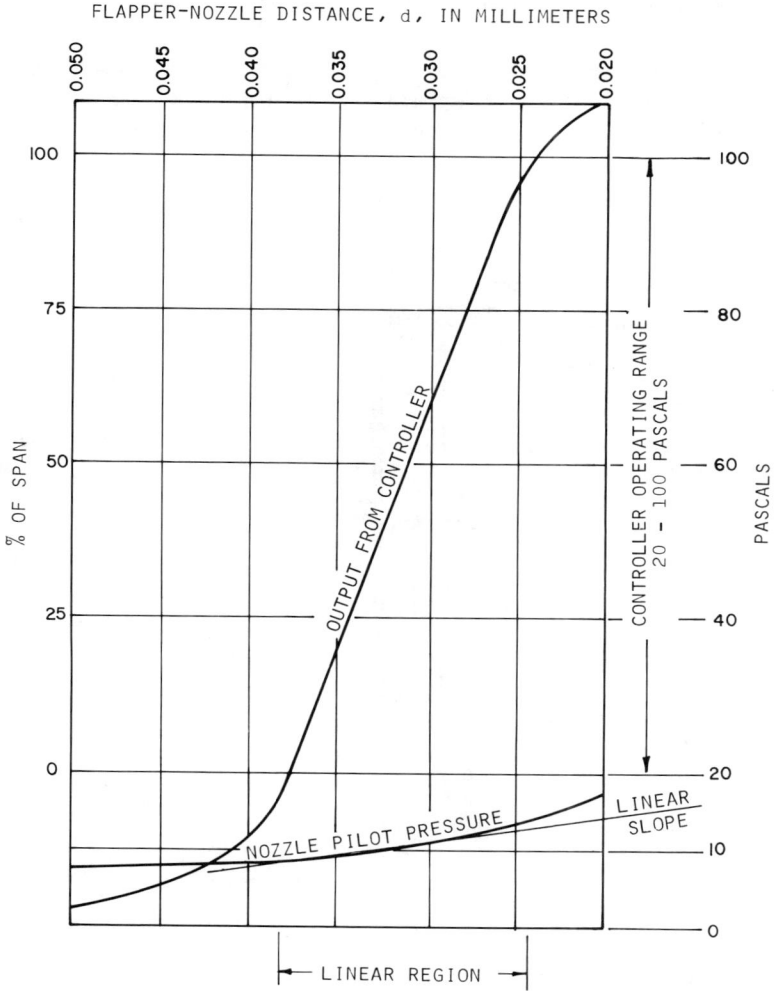

Figure 6a.   Flapper-nozzle relation vs. operating pressure.

Pneumatic Relays

The pneumatic relay is essentially an amplifier. It converts very small changes in pressure—such as nozzle back pressure at a flapper-nozzle assembly—to large changes in pressure as used in pneumatic instruments (Figure 6a).

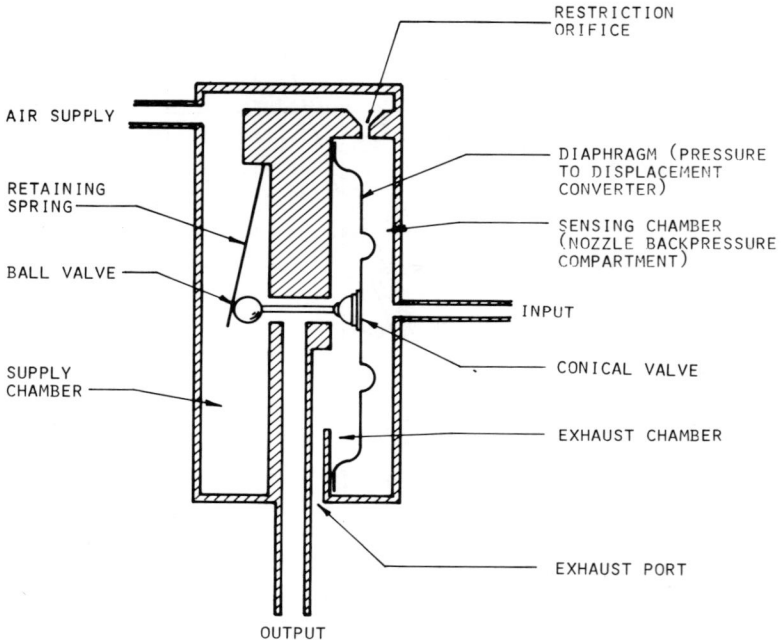

Figure 7.   Foxboro model 40 pneumatic relay.

All pneumatic relays consist of three essential components:

- nozzle back pressure compartment,
- pressure to displacement converter, and
- pilot valve.

The nozzle of a flapper-nozzle assembly is connected to the relay's nozzle back pressure compartment, in which changes in pressure can act on a pressure-to-displacement converter such as a bellows or diaphragm. Motion in the pressure to displacement converter moves a pilot valve, which controls the output pressure by throttling supply air. Amplification takes place through the small nozzle back pressure signal acting on a large diaphragm area and positioning a pilot valve designed to provide sensitive and accurate throttling of supply air.

Pneumatic relays usually are classified as (1) bleed type, in which some supply air is continuously bled past the pilot to exhaust when the relay is operating in its linear region, and (2) nonbleed type, in which no supply air is bled past the pilot. Nonbleed relays consume less air than bleed relays. The most common type in use is manufactured by Honeywell. The relay in most widespread use in industry is probably the Foxboro Model 40 (Figure 7).

The input is connected directly to the sensing chamber, which is connected to the supply chamber by a restriction orifice. Changes in pressure on the input line control the volume of the sensing chamber. The restriction orifice allows a very small change in pressure on the input line to control the pressure in the sensing chamber.

The output line is connected to the supply chamber by the ball valve. It is also connected to the exhaust chamber by the conical valve. A decrease in supply line pressure causes the diaphragm to move to the right. The ball valve moves closer to its seat and increases the amount of throttling. At the same time, the conical valve moves away from its seat, which decreases the amount of throttling between the output line and the exhaust chamber. With supply reduced and exhaust increased, pressure on the exhaust line decreases.

If pressure on the input line increases, the diaphragm and, thus, the valve assembly, moves to the left. Throttling of air from the supply chamber is decreased and throttling of output line air to the exhaust chamber is increased. Pressure in the output line increases.

At the extremes, the output port is either connected to the supply chamber through an open ball valve and closed conical valve, or connected to the exhaust chamber through an open conical valve and a closed ball valve.

The total stroke of the valve from the ball valve being seated to the conical valve being seated is typically about 0.305 mm. A change in pressure of about 5 kpa on the input line can change the output from 20 to 100 kpa. The gain of the relay is thus 80/5 or 16.

Volume Tank and Restrictions

Changes in phase in pneumatic instruments are produced by volume tanks and restriction orifices. Adjustable restrictions are obtained by use of needle valves. An integrating process can be implemented by configuring the elements as shown in Figure 8.

Pressure in the signal line produces flow through the needle valve into or out of the volume tank. This causes pressure in the tank to change toward the pressure in the signal line. The pressure in the tank will always lag behind the pressure in

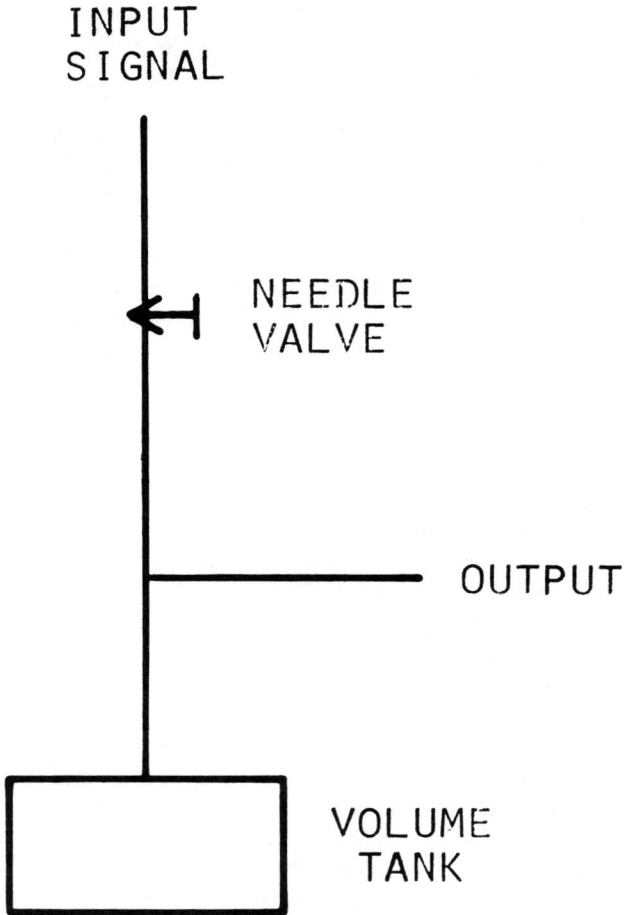

Figure 8.   A simple pneumatic integrator.

the signal line, a characteristic of an integrating process. The time constant of the assembly is determined by the volume of the tank and the settling of the needle valve.

A differentiating process—one in which the output is proportional to the rate of change of the input—can be implemented with a needle valve and a bellows. A simple example is illustrated in Figure 9. A change in the input signal causes the flapper-nozzle-relay assembly to produce a large signal on the output. The large signal will be opposed by the action of the feedback bellows.

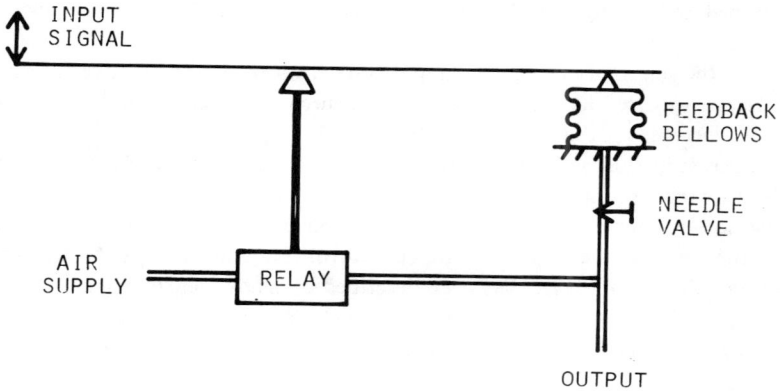

Figure 9.   A simple pneumatic differentiator.

Motion-Balance Mechanism

A motion-balance mechanism is shown schematically in Figure 10. A balance is effected via the flapper-nozzle-relay assembly between motion on the input link and force in the feedback bellows.

At equilibrium, the flapper-nozzle distance, d, is adjusted so that the relay output is mid-range. An upward motion of the input link increases the flapper-nozzle distance, resulting in a decrease in nozzle back pressure and, via the relay, a decrease in output pressure. This allows the spring to compress the bellows and decrease the flapper-nozzle distance. This results in an equilibrium being re-established.

Any change in position of the input link within the linear range of the mechanism will result in a proportional change in output pressure. If the output-input

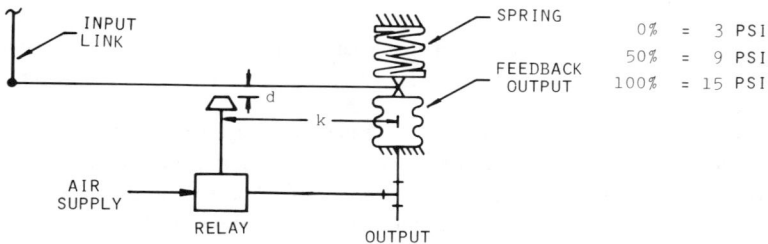

Figure 10.   Motion-balance schematic.

loop is closed through the process, an error-detecting and -correcting device is created.

Since the sensitivity to an error input is determined by the distance change at the flapper-nozzle, the change in output pressure per unit error can be controlled by the placement of the nozzle along the lever. When the nozzle is close to the bellows, it is at its least sensitive position. When at the connection of the input link, it is at its most sensitive position.

Motion-balance mechanisms have been used by many manufacturers. They find their widest use in field-mounted controllers. The Foxboro Model 40, a motion-balance controller, has been described by many control engineers as the best controller ever designed. Even with such enthusiastic praise, the motion-balance mechanism has not been used in a new design in many years.

Force-Balance Mechanism

A force-balance mechanism is shown schematically in Figure 11. At balance, the force exerted by the measurement bellows is equal to the force exerted by the set signal bellows. Tension on the spring has been adjusted to position the flapper nozzle such that a mid-range pressure output is produced, typically 60 kpa.

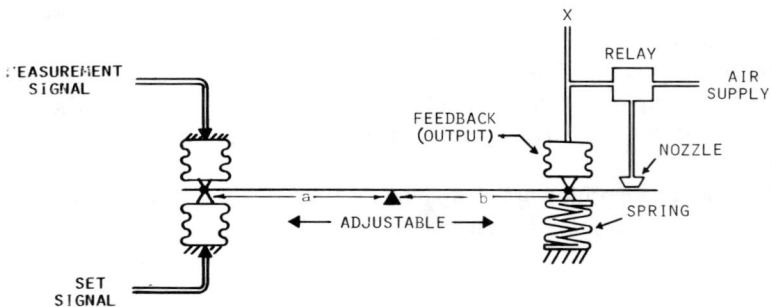

Figure 11.   Force-balance schematic.

If the measurement signal increases, the measurement bellows will expand and cause the lever to move in a counterclockwise direction around the fulcrum. This decreases the flapper-nozzle distance, which causes the relay to increase the output pressure. Pressure increases until an equilibrium is reached.

If the measurement signal decreases in pressure, the measurement bellows will contract and cause the lever to move in a clockwise direction about the fulcrum. This increases the flapper-nozzle distance, which causes the output relay to decrease the output pressure until a new equilibrium is reached.

The position of the fulcrum controls the magnitude of the output pressure change that occurs from a unit change of signal at the set-measurement end of the lever. This is the controller's gain adjustment. Force-balance mechanisms are in widespread use in industry and are the basis of most modern pneumatic controller designs.

INSTRUMENTS

The various components of pneumatic systems can be assembled to form instruments. Operation of a pneumatic force-balance transmitter is quite simple (Figure 12). The net force on the diaphragm is always to the right. The amount of force varies with the differential pressure. The force is transmitted through the force bar (which turns around the fulcrum) to the range rod. The range rod is separated from the supporting structure by a range wheel, which acts as a fulcrum. Force from the force bar pulls the top of the range rod to the left. This movement is detected by the flapper-nozzle assembly, which causes an increase in output pressure of the relay. The output pressurizes the bellows, which attempts to restore the force bar to its original position.

An equilibium is reached when the force exerted by the bellows balances (considering the mechanical advantage of the levers) against the force input from the

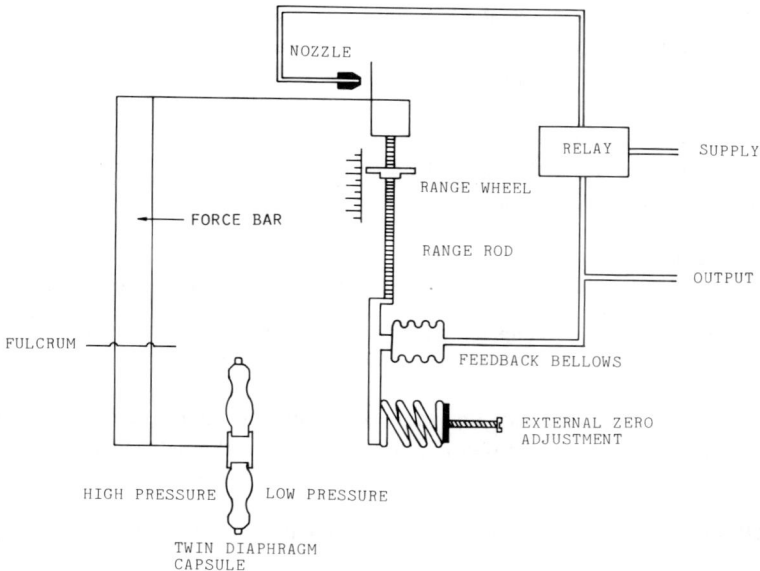

Figure 12.   Force-balance differential pressure transmitter.

diaphragm via the force bar. There is a unique output pressure for every force input. As it turns out, the output pressure is linear in differential pressure at the diaphragm within the limits of linearity of the flapper-nozzle and the other components.

Other instruments can be constructed from the basic elements discussed herein. For example, it is not difficult to envision a receiver instrument consisting of a spring-opposed bellows positioning an indicator-pointer, according to process pressure signals.

Some arrangement drawings and schematics illustrating the operation of several pneumatic instruments are included (Figures 13–15). They are presented without explanation in the interest of brevity. The central point is to illustrate that instruments consist of obvious, inobvious, mundane and brilliant applications of the basic elements.

COMPUTER APPLICATIONS

The benefits of advanced control strategies and the computer systems frequently necessary to implement them are an established part of the operation of many industrial processes. Pneumatic instruments are used frequently to interface the computer system and the process. They also are used frequently for backup in the event of a computer failure.

When computers are added to a plant with an existing pneumatic instrument system, effort is frequently made to productively use the existing pneumatic hardware. This must be approached carefully and with a firm appreciation of the limitations of pneumatic instruments.

Instruments in existing plants probably were not originally selected with the intent of using them with a computer. Consequently, several important features necessary in interfacing the controllers to the computer may not be available. These would include a means of:

1. feeding the controller's set point to the computer,
2. indicating the position of the controller's remote-auto-manual switch to the computer, and
3. preventing reset windup in the controller when the loop is being controlled by the computer.

The successful use of existing instruments depends on the characteristics of the specific instruments and the way the designer resolves the deficiencies.

Pneumatic Instruments and Computer Set Point Control

In computer set point control, a computer is used to generate a set point for an analog controller. The analog controller then manipulates the control loop. The

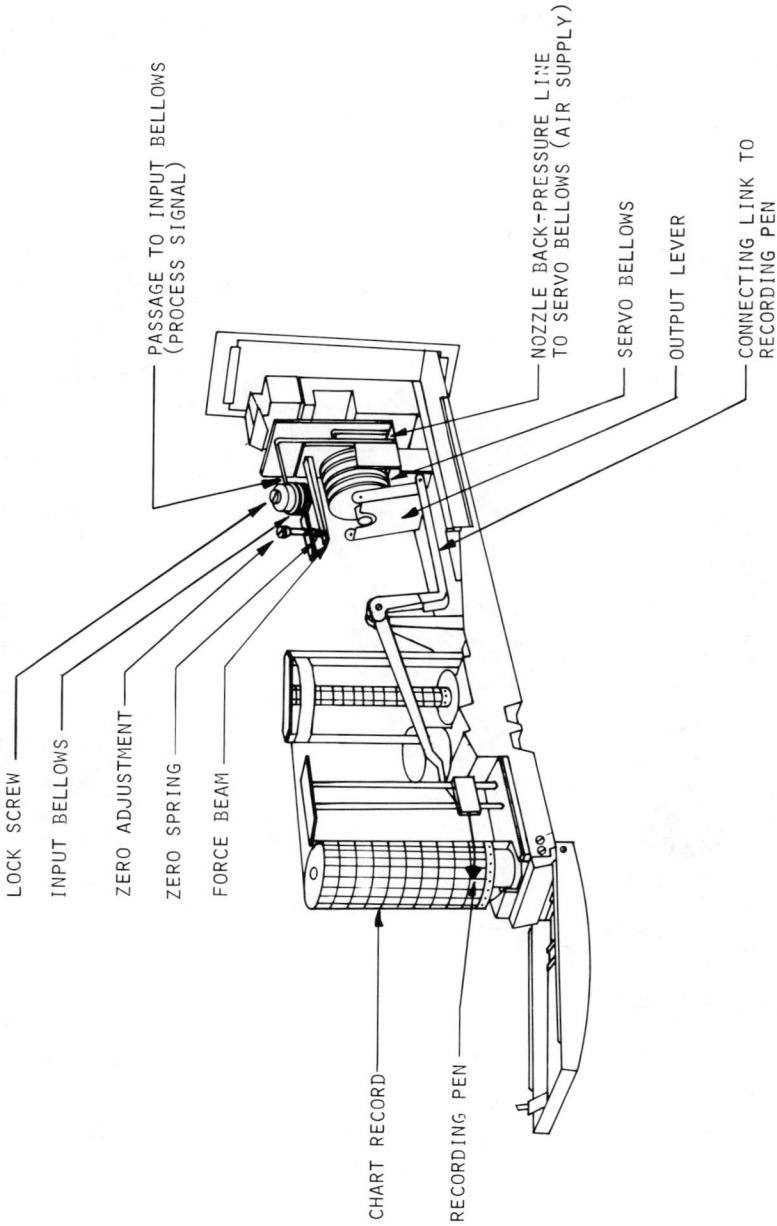

Figure 13.   Typical pneumatic recorder.

PASSAGE TO INPUT BELLOWS (PROCESS SIGNAL)

NOZZLE BACK-PRESSURE LINE TO SERVO BELLOWS (AIR SUPPLY)

SERVO BELLOWS

OUTPUT LEVER

CONNECTING LINK TO RECORDING PEN

LOCK SCREW

INPUT BELLOWS

ZERO ADJUSTMENT

ZERO SPRING

FORCE BEAM

CHART RECORD

RECORDING PEN

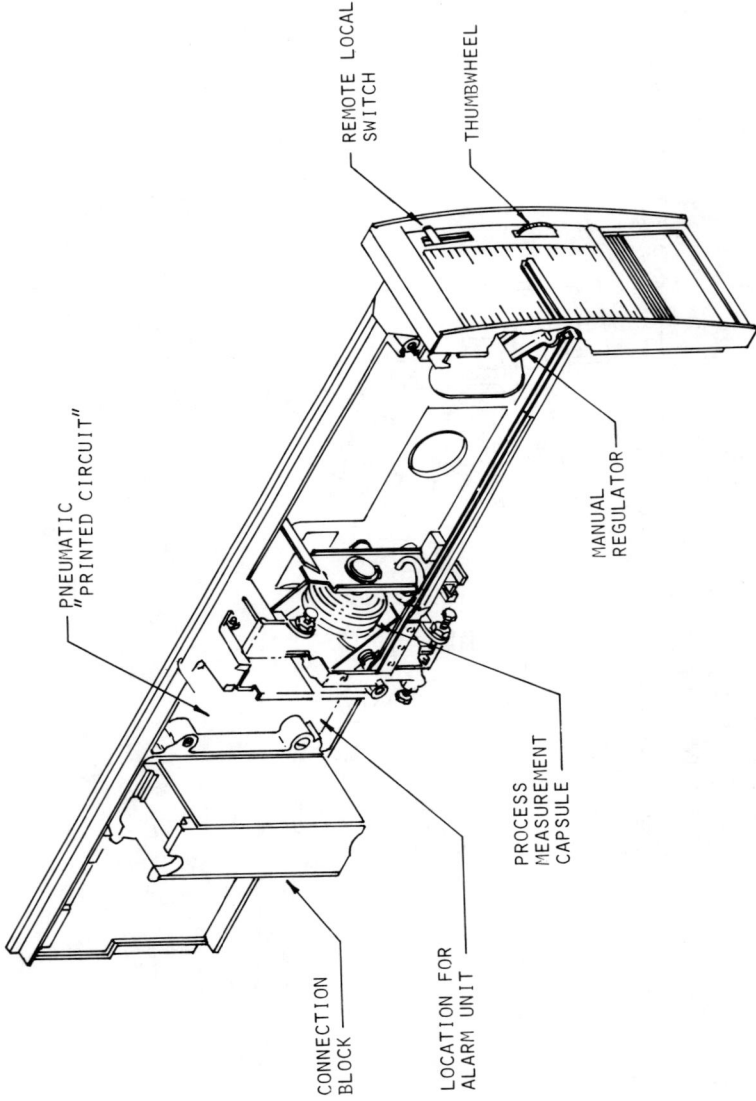

REMOTE LOCAL SWITCH

THUMBWHEEL

PNEUMATIC "PRINTED CIRCUIT"

MANUAL REGULATOR

PROCESS MEASUREMENT CAPSULE

CONNECTION BLOCK

LOCATION FOR ALARM UNIT

Figure 14.   Typical manual loader—indicator.

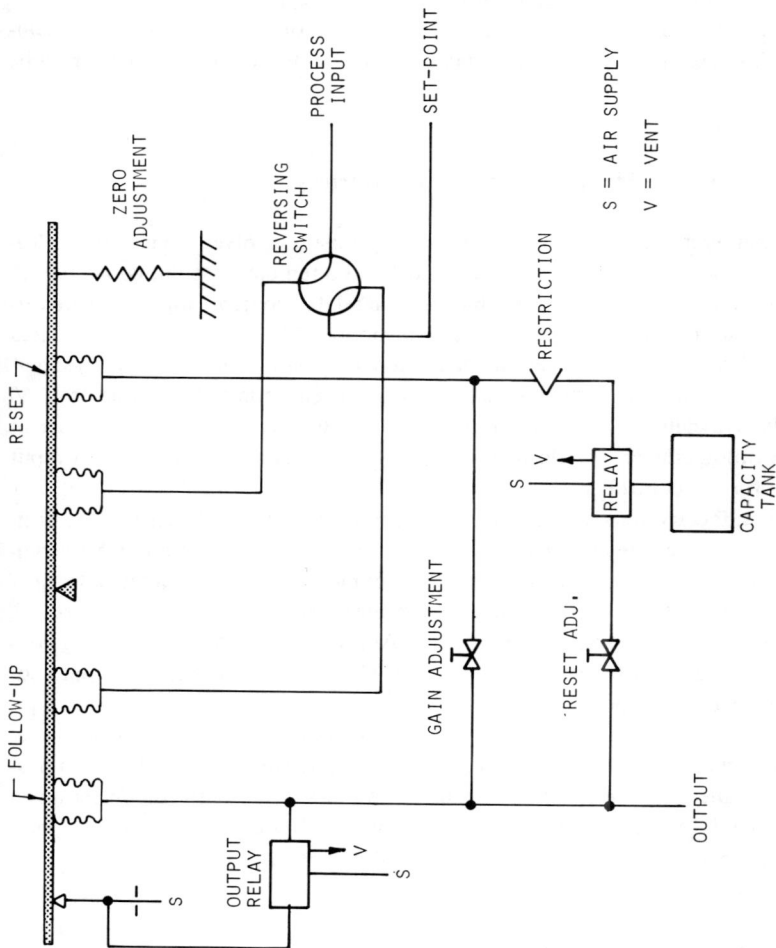

Figure 15.   Simplified schematic of proportional-plus-reset controller (similar to Taylor 1462).

algorithm used to actually control the loop is that designed into the analog controller. Loop tuning is adjusted in a conventional manner and is accomplished by means of the adjustments on the controller. The algorithm used to adjust the set point is contained in a computer and is presented to the controller as an analog signal, or in digital form for conversion by the controller to an analog signal. The algorithm can be based on most any control strategy.

If pneumatic instruments are used as the analog controllers, the obvious problem of interfacing the electronic signals of the computer to the controllers must be addressed.

Interfacing By Current-to-Pressure Converters

The simplest approach to this interfacing problem involves the use of an electronic-to-pneumatic converter of some sort and a remote set point controller. A typical situation might be one in which several related control loops in several different locations are to be supervised by a computer. The computer transmits data from its location to each location where there are analog controllers to be supervised. The transmission medium is typically a current loop (4-20 MA dc or 10-50 MA dc, typically). Current-to-pressure (I/P) converters for each controller at each location convert the electronic signals to pneumatic remote set point input signals for the controllers.

If the I/P converters and the process measurement transmitters used are intrinsically safe, the entire instrument assembly at the controller locations can be completely suitable for hazardous process environments. If the computer fails, or if the electronic transmission circuits between the computer and the controllers fail, the loops can be controlled in the "Local Automatic" mode by the analog controllers. A bump in the control may occur when the I/P converter output suddenly changes in response to a failure.

It is possible to use pneumatic relays to restrict the range over which the set point can move, even if the electronic signal changes substantially. It is also possible to immediately transfer some controllers to local automatic on detecting an unreasonable signal being received. A bump will occur when the set point abruptly changes, but automatic control will continue.

Stepping Motor Interface

Pneumatic controllers especially designed for computer set point control can be obtained with stepping motors attached to the set point dial mechanism. A stepping motor is typically constructed with a winding that, when energized, rotates the motor shaft in one direction and a second winding that, when energized, rotates the shaft in the opposite direction. When a winding is energized, movement is always one increment of rotation. The size of the increment depends on the construction of the motor.

The output of the computer then consists of a series of pulses on one circuit to raise the set point and a similar series of pulses on another circuit to lower the set point. These pulses each increment the stepping motor in the appropriate direction. The rotation of the stepping motor shaft moves the controller's set point mechanism in the same manner as the manual adjustment knob (Figure 16).

With this type of controller, failure of the computer or the electronic communications circuits will result in the last set point received by the controller being retained. Assuming the operators are alerted to the failure, control can be continued in a manually supervised, but totally uneventful, manner.

Since the set point will move from minimum to maximum in a fixed and predetermined number of steps, and since the computer can certainly count the number of impulses it transmits, knowing the position of the set point would not seem to be a problem. In fact, this is not the case. The operator may manually move the set point. Unless this movement is reported to the computer, the set point generation algorithm may operate in error.

A second problem involves coordinating the set point generated by the computer with the output of the supervised controller. It is possible that conditions in a particular control loop, or perhaps process conditions at large, will result in a particular controller not being able to adjust its controlled variables so the set point can be achieved. If the computer continues to hold or move the set point to unachievable values, the controller will, if reset is used, go into reset windup. To avoid this, it may be necessary to transmit the controller output value or the value of the measured variable to the computer. To avoid the possibility of "windup" in the computer's set point adjusting algorithm, it is convenient to report the status (computer/local) selector to the computer. Figure 16 indicates the wiring required to install a typical stepping motor type computer set point controller.

The cost of implementing a set point control loop in terms of computer analog and digital inputs can be seen to be potentially quite high. The apparent simplicity of the set point control concept is somewhat clouded by the actual complexity of an implemented system.

Pneumatic Instruments and Direct Digital Control

In direct digital control (DDC), the algorithm used to manipulate a final control element (FCE) (such as a valve) is resident in a computer. The output signal from the computer to the FCE is either proportional to the desired valve position (absolute position) or the desired change in position (incremental position). Of course, this signal must be interfaced to the process.

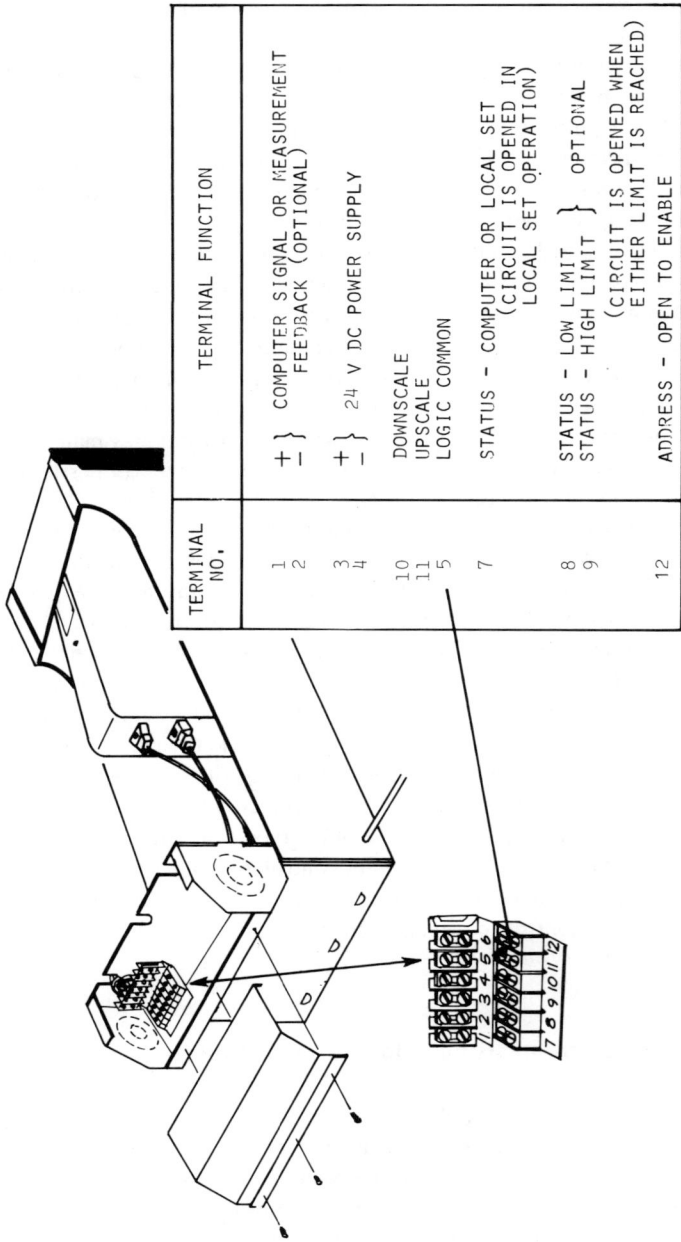

TERMINAL NO.	TERMINAL FUNCTION
1, 2	+ ⎫ COMPUTER SIGNAL OR MEASUREMENT − ⎭ FEEDBACK (OPTIONAL)
3, 4	+ ⎫ 24 V DC POWER SUPPLY − ⎭
10	DOWNSCALE
11	UPSCALE
5	LOGIC COMMON
7	STATUS − COMPUTER OR LOCAL SET (CIRCUIT IS OPENED IN LOCAL SET OPERATION)
8	STATUS − LOW LIMIT ⎫ OPTIONAL
9	STATUS − HIGH LIMIT ⎭ (CIRCUIT IS OPENED WHEN EITHER LIMIT IS REACHED)
12	ADDRESS − OPEN TO ENABLE

WIRING TO REAR OF SHELF

Figure 16.   Stepping motor-type computer set point controller (Foxboro model 130K).

Pneumatic instruments may be used as the interface. The type of computer output signal and the desirability of pneumatic analog backup capability will dictate the design of the pneumatic instrument system.

In the simplest case, an absolute position output from a computer can be converted to a pneumatic signal by an electronic-to-pneumatic converter. The converter could be a separately mounted unit with its output connected to the FCE actuator, or it could be an electronic input pneumatic positioner mounted on the FCE. Either of these arrangements results in the simplest possible approach to the pneumatic portion of the system. All desired control redundancy would be in the electronic portion of the system. Although the approach using an electropneumatic positioner on the FCE is the least complex, simplest to install and least expensive, it may not be completely suitable in all applications. Vibration at the FCE that would not affect a pneumatic positioner may affect an electropneumatic one. Temperatures at the FCE that would not affect a pneumatic positioner may destroy an electropneumatic one. Although intrinsically safe electropneumatic positioners are commonplace, they are only intrinsically safe when used within a certified intrinsically safe system. The cost advantage normally favoring the electropneumatic positioner approach may be lost when special instrument power supplies and intrinsic safety barriers must be included—especially if the electronic-to-pneumatic converters can be located in a safe area, such as a control room or air-purged cabinet and connected to the FCE actuator by pneumatic tubing.

If backup instruments are to be provided and are to be pneumatic, two interfacing problems must be overcome: the electronic signal from the computer must be interfaced to the pneumatic instruments and the pneumatic instruments must be interfaced to the process in such a way that a bump will not occur in transitions between computer control and local automatic control, or vice-versa. These problems are not always simple to overcome because the algorithm and measured variables used in computer control may be significantly different than what can be done with a local automatic controller, capable of only PID algorithms, and only one measured variable input.

CONSTRUCTION

Instrumentation systems, more than anything in engineering, demand proper attention to detail. Excellent conceptual work will not produce the desired results unless all of the details involved in fabricating and installing the equipment are carefully considered and translated into practice.

There are many opinions in practice concerning exactly how various details should be handled. This is a natural enough outgrowth of the age of the field and the many practitioners. Although there are certainly many solutions to the various problems, implementation of some solutions may dictate that other aspects of the installation be handled in specific, and sometimes inobvious, ways.

The approach taken here is to rely on industry standards and common practices in the more technically developed process industries. It is an important duty of the engineer to evaluate local situations in the context of standards and design accordingly.

Pneumatic Instrument System Architecture

The two major constraints on the applicability of pneumatic systems are the somewhat related items of speed of response and transmission distance. It takes time for a pressure change to propagate down a tube connecting the transmitter to a receiver instrument. The propagation time can be on the order of process time constants and can, therefore, become a factor in controlling the process.

Some pneumatic instruments consume air and, thus, require a flow of air in the interconnecting tubing. Almost all instruments require the interchange of at least small volumes of air. The flow of air through a tube will always result in some loss of pressure due to friction in the tubing. The magnitude of this loss is proportional to the flowrate, tubing material and length of the tube. In a specific application in which similar instruments and tubing are used, the length of the tubing determines the performance limits.

Since speed and accuracy objectives are always relative to the specific control task, there is no absolute maximum length of a pneumatic transmission line. When lengths will exceed 100 meters, it is important to analyze the effect on the control system.

The maximum acceptable tubing length will determine the maximum distance between transmitters and receivers and, thus, the location of the instrument board. Since transmitter locations are governed by the process equipment configuration, the location of the instrument board is dependent on the particular physical situation. Traditionally, process plants have been laid out in functional blocks analogous to city blocks. A control house containing the pneumatic instrument board was located in each process block so tubing runs could be kept at acceptable lengths. Coordination between process blocks was accomplished by defining boundary conditions and operating to maintain them. Details were coordinated by telephone.

Central control facilities with the advantages of multiunit overview and coordination were not possible with pneumatic instruments, but the advantages were so compelling that methodology was developed to accommodate pneumatic field instruments and central control facilities.

The approach usually taken involves locating a pneumatic instrument board within range of all transmitters and valves and then connecting it through electro-pneumatic converters and electronic transmission systems to the control equipment at the central facility. With such a system, a process unit can be operated in the traditional fashion from a local panelboard during central system or transmission system failures. When operating normally, the pneumatic instruments are used as the process interface.

The instrument board must be connected to the field instruments in a reliable and economical way. In most plants, the system also must adapt easily to changing requirements. The transmission system also must be made of a material not degraded by conditions existing in the area.

The usual practice is to connect groups of field-mounted devices to junction boxes located in proximity to the devices. One junction box may serve few or many devices. Tubing is used to connect bulkhead unions on the junction box to the field devices. Single-tube runs, such as to valve operators, almost always are made with metallic tubing, frequently plastic coated to reduce corrosion. Two-tube runs, such as to pressure transmitters, usually are made with two-tube jacketed bundles. Tubes may be metallic or plastic, depending on the mechanical strength required.

Multitube bundles are used to connect junction boxes to either the instrument board bulkhead fittings or intermediate junction boxes (Figure 19). Multitube bundles usually incorporate plastic tubing with an overall jacket. They may be run exposed on framing channel, in cable tray, or in conduit, as appropriate for proper protection in the specific environment.

The routing of instrument tubing should consider minimizing the length, but also protection. Tubing should not be run through areas with excessive ambient temperatures or over machinery, such as hydrocarbon pumps, likely to be involved in fires. In areas where damage is likely to occur, the tubing should be sufficiently accessible to allow workmen to quickly repair or replace damaged runs. In some instances, it may be attractive to locate junction boxes at convenient locations so that new bundles can be quickly connected between them when damage occurs on the permanently installed bundle.

The usual difficulty in pneumatic transmission systems is leaking. This usually occurs at valves and fittings; however, in older systems it can be the result of pinholes developing in tubing due to corrosion. ISA R.P. 7 provides testing and troubleshooting information. Instrument systems should be pressure-tested periodically, such as during plant turnarounds, to avoid reliability problems causing forced outages.

Materials are an important concern in the design and construction of pneumatic systems. Brass fittings and PVC-coated copper tubing generally are adequate for most corrosive environments. In some environments, such as atmospheres containing H_2S, stainless steel tubing and fittings are preferred.

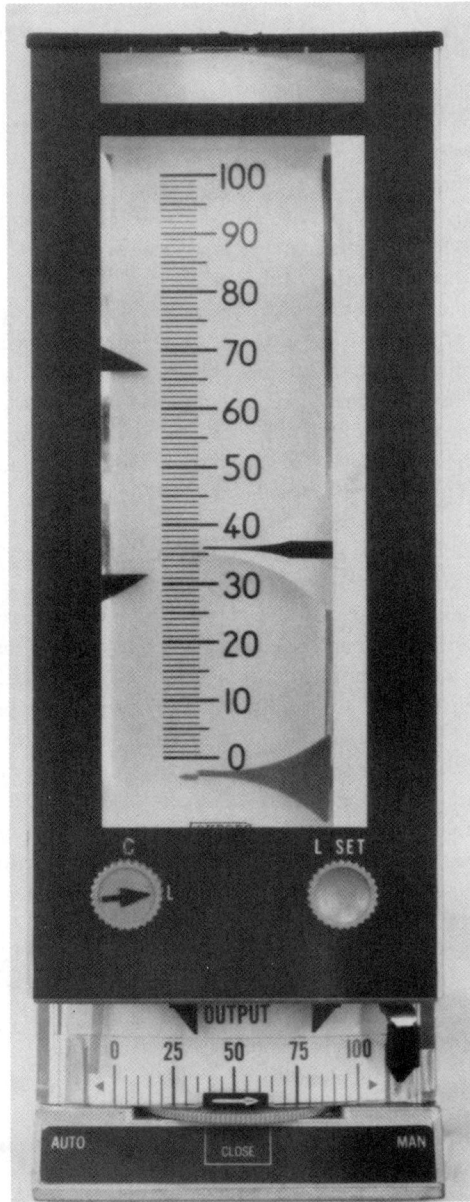

Figure 17.   Stepping motor-type computer set point controller (Foxboro model 130K).

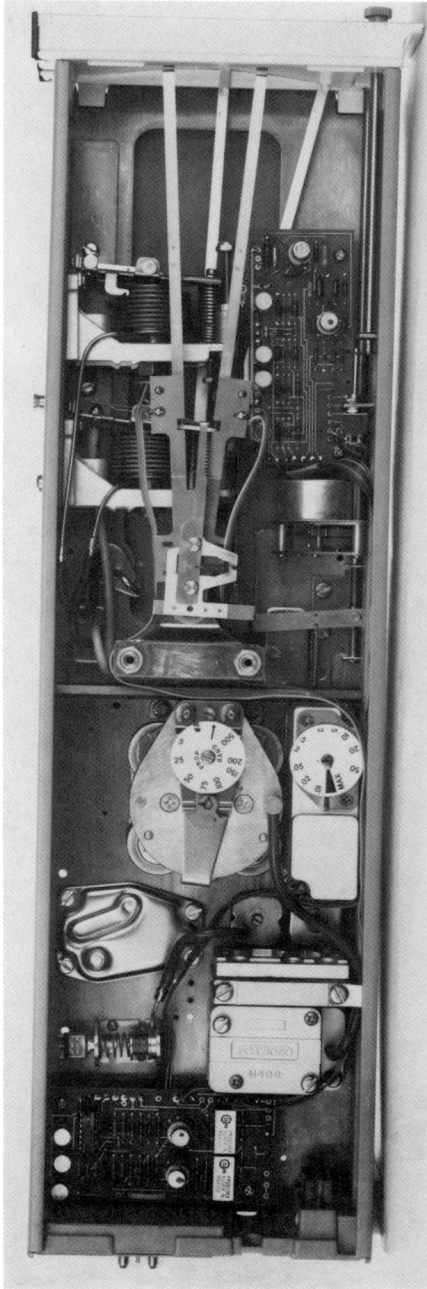

Figure 18.   Stepping motor-type computer set point controller (Foxboro model 130K).

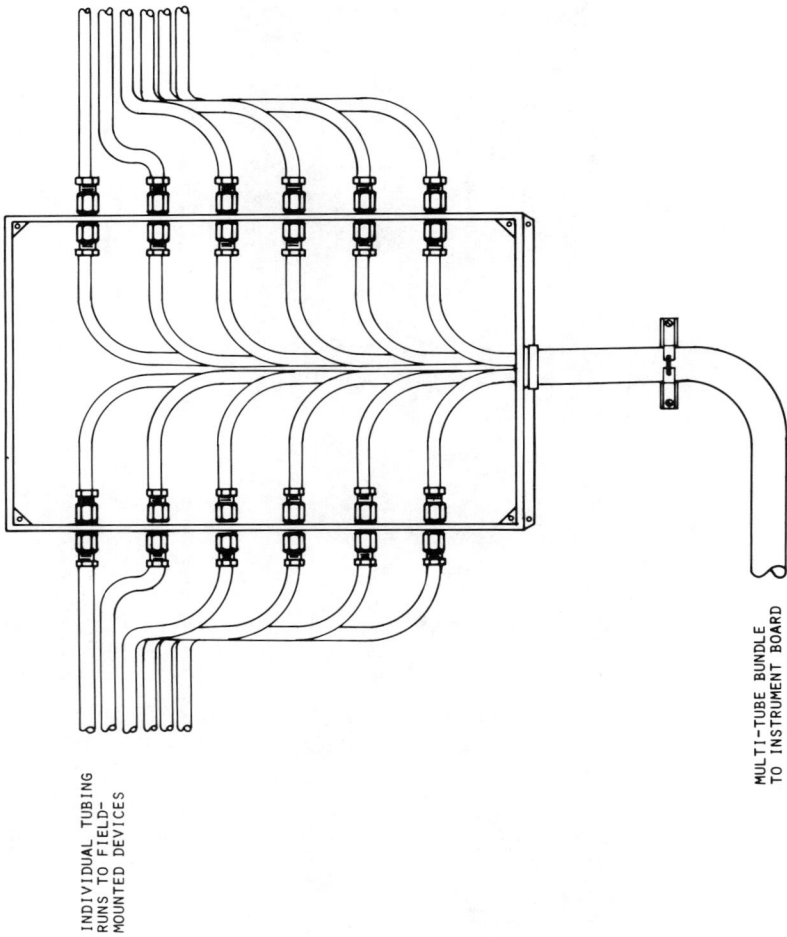

INDIVIDUAL TUBING
RUNS TO FIELD-
MOUNTED DEVICES

MULTI-TUBE BUNDLE
TO INSTRUMENT BOARD

Figure 19.   Pneumatic junction box.

Construction Within Instrument Boards

A well-designed instrument board (Figures 20 and 21) can be completely assembled and tested in a shop and then moved in sections to the control room. There, one should be able to quickly accomplish assembly and connection to field tubing in a logical and orderly manner.

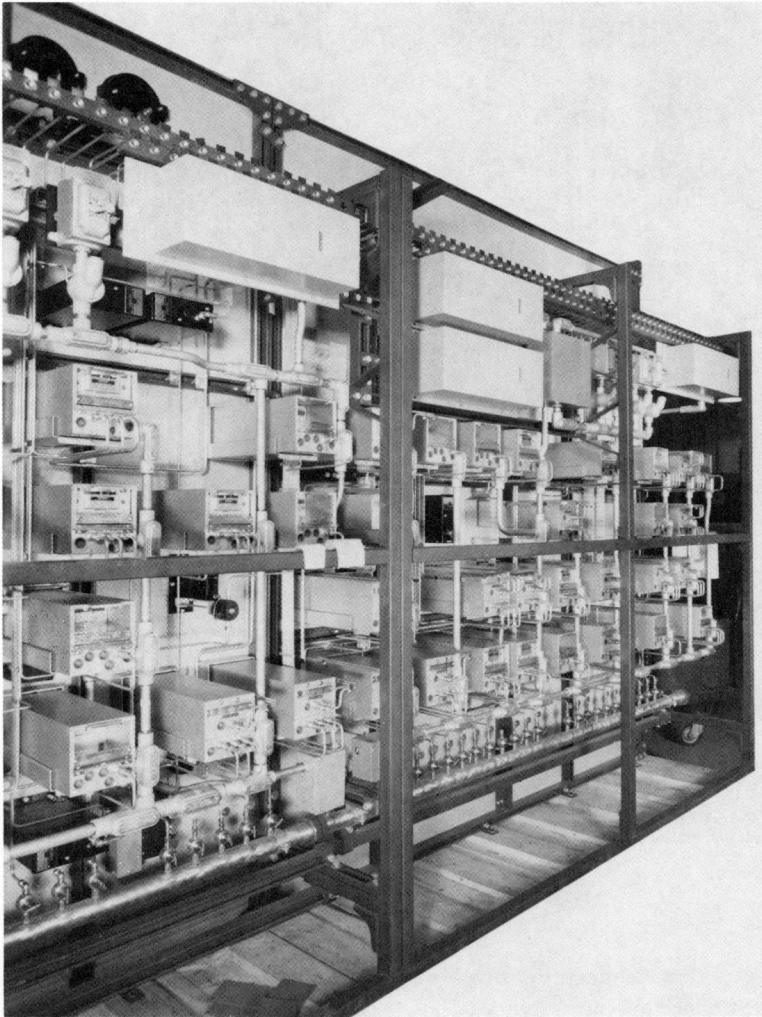

Figure 20. Bulkhead connections.

FIGURE 21
BULKHEAD CONNECTIONS

Figure 21.   Bulkhead connections.

Connections required to assemble sections together should be kept to a minimum. Connections are a source of leaks and, thus, are the cause of major expenditures of maintenance time. Building sections around process units or major items of equipment will reduce the number of connections, as well as usually result in a functional and understandable instrument arrangement. When only a few connections are required between sections, the most satisfactory way to make them in the field is to mechanically assemble the sections and then install the tubing from point to point, as one would do in the shop.

Since it is seldom as easy to do high-quality work in the field as it is in the shop, large numbers of connections are best made by installing bulkhead fittings on special intersection tubing connection plates. Interconnection is then made by running tubing between these plates during installation. Some sort of support, such as framing channel rack, cable tray, or enclosure such as a wireway will be required to support and protect the interconnecting tubing. With plastic tubing, plastic wireway is excellent. An uncluttered and unobstructed route between sections is necessary. Since at least one connection typically will be required from each section to every other section, it is logical to reserve a particular space in each section for the interconnecting tubing and its support system. This space is usually best located at the top or bottom of the board, whichever will *not* be used for making field connections. It should be easily accessible in the environment where the board will be installed, and should be of more than ample size. It is never as easy to *assemble anything in the field!*

One school of thought advocates that all pneumatic boards, no matter how large, should be built as a single unit. The advantages of this approach are completely obvious—fewer connections; no interconnections in the field; maximum assembly in the controlled and usually less costly per connection panel shop environment; and an opportunity for complete factory testing of the entire assembly as a unit in the exact form it will arrive at the job site (providing all goes well in shipment). The disadvantages include the difficulties and high potential for damage in packing, shipping, and unloading; the difficulty for handling at the job site without special and often costly equipment; the substantial additional cost in weight and materials required to construct a large unit; and the difficulty in making onsite modifications and substitutions. Further, the delivery of the instrument board and construction of the control house must be coordinated so the board can be delivered and installed and protected from the weather without severely disrupting the schedule on which the control house is being built. If the assembly is to be placed in an existing control house, removal and reconstruction of a wall or roof may be necessary.

Connections to field tubing are best accomplished through bulkhead fittings located on the top or bottom of the instrument board, depending on whether field tubing will leave the control room under the floor or through the walls or roof. Generally, single or double rows of fittings distributed along the panel will work best. Dense clusters of tubing should be avoided. Each fitting should be clearly marked, and prints should reflect the marking. Since field tubing will generally be in multitube bundles, allow space to connect the tube to the fitting without kinking or straining. A method that works extremely well has field tubing running in ladder-type cable tray above the board. Tubes are then fished between rungs and arced down to the appropriate fitting. Allow sufficient room for easy connection. Making field connections in a poorly designed environment will cost dearly in excess labor and delays.

Connections within the panel may be made with metallic or plastic tubing. Although almost everyone will agree that soldered copper tubing is the best approach within a panel, hardly anyone is willing to pay what it now costs. If craftsmanlike, experienced workmen are not available, the end result will be more satisfactory if fittings are used anyway. Plastic tubing inside panels was once thought of with horror, but ease of assembly and low cost make its consideration mandatory. If plastic wireway, such as Panduit, is used to contain the tubing, as is commonly done with control wiring, a neat-looking job can be done.

One should not economize on fittings. If poor-quality fittings are used, many will break during assembly and many more will leak. Consider the number of steps involved in installing competing fittings. A savings in the unit cost of the fitting is quickly lost when expensive labor is required to accomplish additional steps or make poorly manufactured parts fit without leaking.

Whether plastic or metallic tubing is used in an instrument board, color coding of the tubing should be considered mandatory. When plastic tubing is used in wireway and the interconnections are complex, tagging according to some easily understood scheme is also desirable.

Instrument air must be distributed within an instrument board. This is commonly accomplished by an air header consisting of a large pipe, usually 50.8 mm (2 in.) with shutoff valves located every few centimeters along its top. Generally, the header is installed horizontally in the bottom of the instrument board. It should be arranged to slope slightly away from its source of supply so condensation will be aided by the natural flow of air in the pipe in reaching the header's lowest point. There, a drain valve should be installed on the bottom of the header.

Table I.   Pneumatic Tubing Color Coding Per ISA RP-7.2

Service	Color[a]
Air supply	Red
Transmitted measurement to receiver element	Orange
Controller output to valve, slave, etc.	Yellow
Branch transmitted measurement to alarm element	Green
Branch transmitted measurement to readout element	Blue
Seal (to remote mounted controller)	Purple
Set (to remote mounted controller)	Black
All others	Natural[b]

[a] Color may be applied by use of manufactured colored tubing, paint or colored tape.
[b] Natural means uncolored metal or uncolored plastic.

The diameter of the header should be large enough so that very little pressure drop occurs along it. In the case of 20–100 kpa systems, instrument supply pressure should be between 133.3 kpa and 146.6 kpa. Considering the loss in even a short length of 6.35-mm tubing used to supply an instrument with substantial air consumption, such as a controller, it is apparent that pressure loss along the header must be kept to a minimum. In large boards, multiple headers should be used.

The shutoff valves either should be sized so that opening one to atmosphere does not cause a significant drop in header pressure, or restrictors should be used. Losing an entire process unit because someone bumps an air header valve is not desirable. The likelihood of such an occurrence is reduced by using valves requiring several turns on the handle to open, as opposed to lever-operated valves that offer the unneeded convenience of being able to completely open or close the valve by moving a lever 90°.

Current practice dictates the use of galvanized steel pipe or, better still, galvanized rigid steel electrical conduit for the air header. A cheaper lightweight substitute would be Schedule 40 PVC conduit or tubing. The heavyweight plastic has excellent resistance to mechanical damage, is somewhat flexible and, therefore, less susceptible to vibration damage, and will not usually suffer damage from exposure to ultraviolet light when inside a control panel. It will become brittle and fail, however, when exposed to some hydrocarbon vapors for long periods of time.

Although compressed air is not usually conditioned into instrument air inside a panel, it is usual to provide regulators, filters and, in some instances, dryers. The regulators drop the distribution pressure down to the pressure used by the instruments and are essential. In a well-designed instrument air system, the other components are insurance. Two passes, isolatable with ball valves, should be provided. While changing a filter or replacing a regulator, air pressure at the point of disconnection will be lost. The presence of another regulator, filter, etc. with sufficient isolating valves will allow operation to continue normally. One should not provide a bypass around an air conditioning system. The damage that would result if it were opened is substantial. Figure 22 shows a well-designed air conditioning system for inclusion in an instrument board.

In some instrument boards it may not be practical to locate the header horizontally in the bottom of the board. A vertical header, with the valves mounted on the side, will work quite well. Air should enter the header at least 50 mm (2 in.) above the bottom, and a moisture drain should be installed in the bottom cap. It is wise to consider increasing the diameter of a vertical header to further reduce air velocity in it because airborne moisture could be carried through the header to the instruments instead of being condensed out and trapped in the bottom as intended.

When reliability is extremely important, two air headers can be used to supply

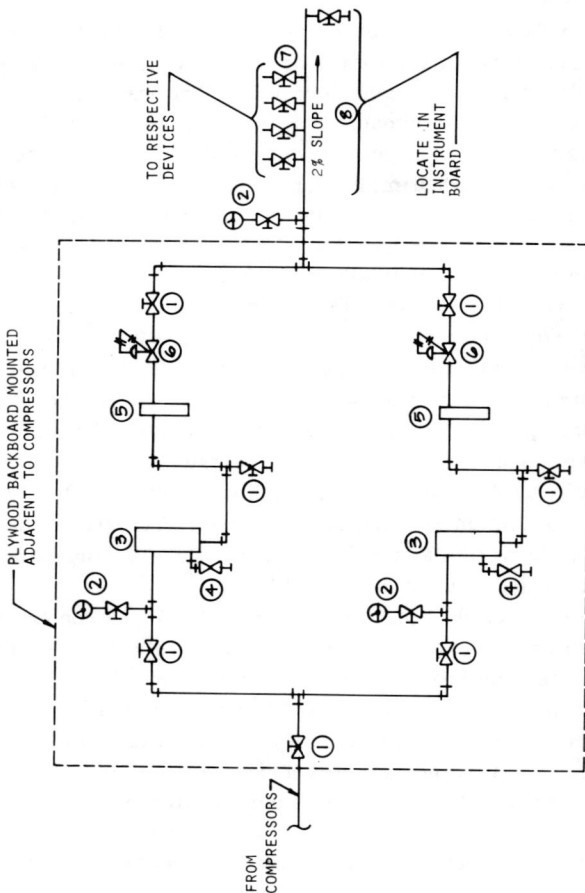

NO.	DEVICE
1	BALL VALVE, BRASS BODY WITH S.S. BALL
2	1–100 PSIG PRESSURE GUAGE
3	AIR DRYER
4	AIR TRAP/WATER DRAIN
5	FILTER
6	REGULATOR
7	SUPPLY VALVE, BRASS BODY 1/4" NPT
8	2" INSTRUMENT AIR HEADER ASSEMBLY WITH DRAIN VALVE

Figure 22.   Air conditioning system.

each instrument with check valves in the supply lines to isolate them from each other if one becomes depressurized. Although a strong argument can be made for dual regulators, only a tenuous one can be made for dual headers. Severe mechanical damage to the instrument board usually would be necessary to damage one and, if the board sustained that amount of damage, a cracked header would be of only incidental concern.

Another header that can be of major convenience in some instrument boards is a test header. This is used in testing the operation of alarm and shutdown devices and can be used to simulate measured variables for control loop testing. Since it is usually connected to nonair-consuming ports on instruments, it need not be as large as the air distribution header; 25.4 mm (1 in.) diameter usually will be sufficient. It should be installed in a manner similar to an air distribution header but should be supplied from the distribution header through a high-quality manual loader so precise pressures can be produced for testing.

Instruments generally are mounted through the front panel and supported at the rear by framing channel. The most acceptable method is to mount multiunit, high-concentration shelves into which the instruments can easily be installed and removed from the front. Tubing connections should be made to the shelves, which then connect to the instruments by means of pneumatic tubes with connectors similar to the familiar "cord set" used with electronic instruments. Auxiliary devices such as pneumatic relays typically are mounted behind the board on framing channel or specially designed brackets. Nothing should be supported by tubing, even metallic tubing. Vibration present in many control rooms can cause resonances that can result in cracks in even well-designed metallic tubing installations. Supporting such heavy loads as relays from tubing will compound the problem, as well as invite damage due to stresses applied in shipping.

AIR SYSTEMS

The most important factor in the success of a pneumatic instrumentation system is the instrument air system. Air must be available continuously, clean and dry (Figure 23).

Two operating pressure ranges are in common use. The first incorporates an 80-kpa span, in which the pneumatic intelligence is transmitted as a pressure between 20 and 100 kpa. The supply pressure for instruments of this range should be at least 133.3 kpa, but not greater than 146.6 kpa.

The second incorporates a 160-kpa span in which the pneumatic intelligence is transmitted as a pressure between 20 and 180 kpa. The supply pressure for instruments of this range should be at least 200 kpa, but not more than 233.3 kpa.

Many pneumatic instruments consume air. The operation of the flapper-nozzle and pneumatic relay is inherently air consuming; consequently, a flow of air to

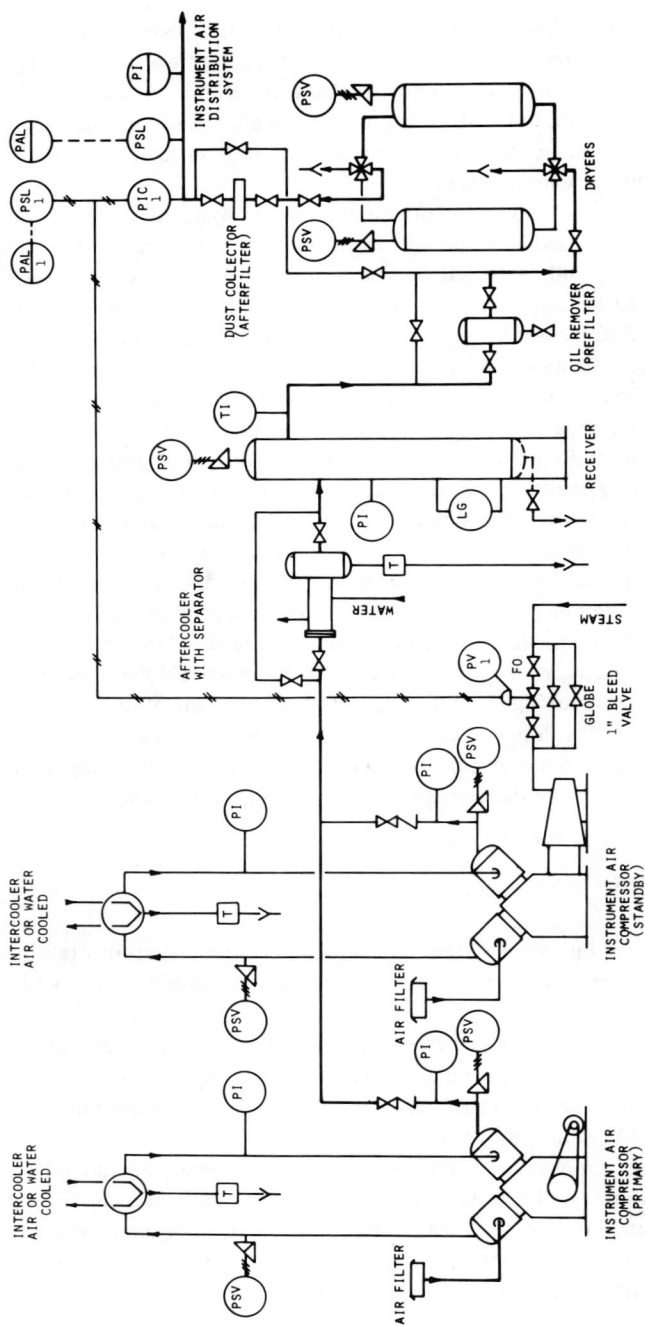

Figure 23.   Instrument air supply system.

instruments is required. If the instruments are not near an air supply, some loss will be encountered in the transmission system and this will have to be accounted for in the design.

Production of Instrument Air

Many types of air compressors are available. The selection of the appropriate type involves consideration of the flowrate required, the desired operating pressure, temperature limitations, method of sealing, method of lubricating and the type of prime mover available. Figure 24 indicates the compressor choices available for various situations.

Reciprocating Compressors

Reciprocating compressors are by far the most common type used in the production of instrument air. Single-stage units can produce pressures in excess of 670 kpa in small sizes and 400 kpa in large sizes. Two stage units can produce pressures in excess of 670 kpa up to about 4.25 SCMM.

These compressors usually consist of a crosshead, in which the rotary motion of a shaft is converted to the back-and-forth motion required by the piston. The rotary shaft is coupled to the piston pushrod through a connecting rod, which is constrained so that the motion of the pushrod is always in a straight line, simplifying packing. The piston may be single or double acting, but double acting is more common in larger sizes.

Cooling may be by air or water, with water more common in larger sizes and higher pressures. The disadvantage of water cooling is that a supply of appropriately treated water must be available or must be created. An advantage of water cooling is that considerable energy can be recovered easily from the compressors. In some cases, space heating for the compressor building or control building can be completely accomplished by recovering heat from compressor cooling water.

Most reciprocating compressors are oil lubricated. Various methods are employed to prevent oil carryover into the instrument air system. Additionally, equipment usually is provided to remove oil present in the compressor discharge before it reaches instruments. Some manufacturers can furnish "nonlubricated" compressors in which graphite carbon or Teflon is used as piston rings. Plastic packing not requiring an oil lubricant is used in the stuffing box. This greatly reduces oil contamination of discharge air. Nonlubricated compressors offer significant advantages in instrument air systems. They significantly reduce the remedial equipment necessary to condition the air and decrease the maintenance associated with it.

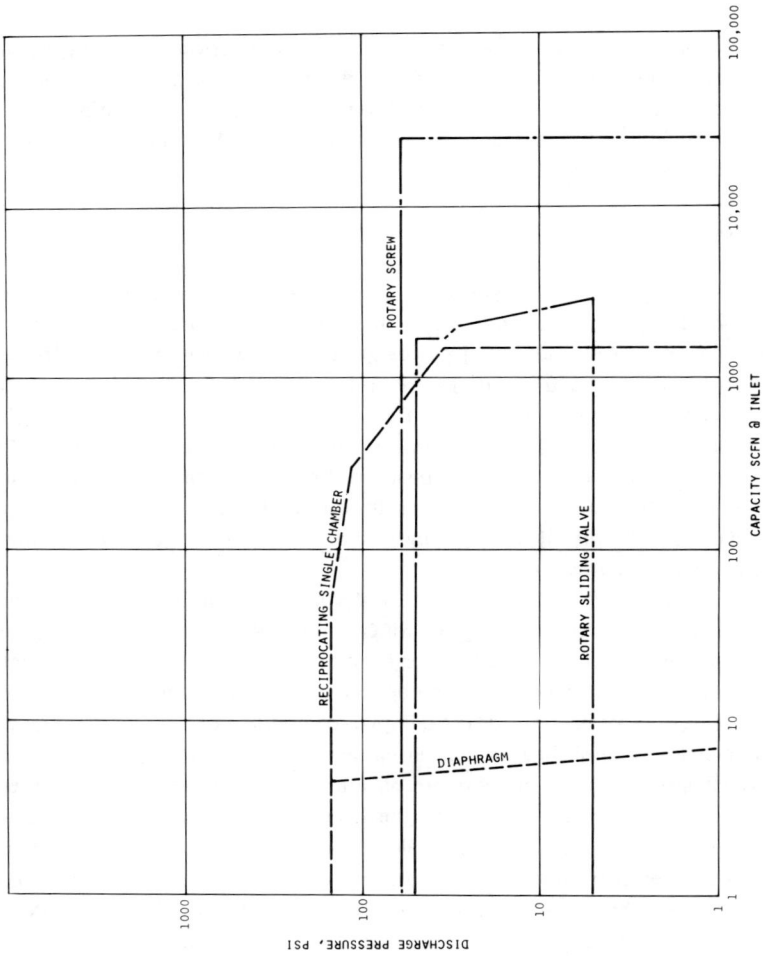

Figure 24. Compressor coverage chart.

Reciprocating compressors are generally designed to be run by electric motors, steam engines, or gas or diesel engines. They frequently are designed with the shaft available at both ends of the unit so two prime movers, usually of different types, can be attached. In this way, availability can be improved significantly since energy source failures such as loss of electricity does not result in the loss of the compressor.

Rotary Compressors

Rotary compressors are basically positive-displacement machines with constant volume-variable pressure discharge. The two types suitable for instrument air service are the screw type and sliding vane type.

The screw type consists of male and female intermeshed screws that, when rotated, cause an axial progression of successive sealed cavities.

The sliding vane type consists of a rotor with a set of sliding vanes running eccentrically in a cavity. Inlet air is trapped by the vanes into pockets that decrease in size as the shaft rotates.

Since it is practical, and desirable, to build these units for high rpm operation, their largest application is in conjunction with steam or fuel turbines.

Control of Compressors

The control of air compressors is not a trivial subject. The appropriate methodology depends on the type, size, prime mover and specific application. When designing a compressor installation, appropriate references outside the scope of this chapter should be consulted.

Receiver Tank

The output of reciprocating compressors pulsates with each piston stroke. It is not desirable to transmit this pulsation to the instrument system. When a compressor fails for some reason, it is not desirable to immediately shut down the plant the instruments are controlling. For these reasons, it is common practice to use the compressors to charge a tank from which the air is drawn. If the tank is sufficiently large so that velocity through it is low, some suspended particulates in the compressor discharge will settle in the tank and not be required to be removed by the air filters.

Instrument Air Conditioning Systems

Compressed air always should be assumed to be wet and dirty. Positive methods should be employed to produce adequately clean and dry air.

Table II summarizes the objectives of an air conditioning system. It is readily apparent that a number of points must be carefully considered in the design. A sometimes overlooked objective of the design also must be the inherent reliability and maintainability of the system. This will require dual dryers, filters and regulators. Since reliability and cost usually are objects of concern, it is important to design toward the simplest system that will accomplish all objectives.

Table II.   Instrument Air Quality Standard

Parameter	Indoor Installations	Outdoor Installations
Contaminants (corrosive, flammable, toxic, or hazardous gases)	Free of all contaminants	Free of all contaminants
Oil Content	Less than 1 ppm (by weight or volume)	Less than 1 ppm (by weight or volume)
Particle Size	3 μm max.	3 μm max.
Moisture (dew point at line pressure)	2°C max.	−7.8°C below minimum ambient

Dryers

Moisture can be removed from air by several methods.

Refrigeration

When air is cooled, moisture suspended in it condenses. The colder it becomes, the greater the percentage of moisture that will be condensed. Since the object is to remove enough moisture so that condensation will not take place at operating temperatures, it is necessary to refrigerate the air to some temperature below the lowest possible operating temperature.

Refrigeration is an extremely effective way to dry instrument air. Of all methods it is the most certain to produce the desired results; however, it requires refrigeration equipment that is complex, large and energy consuming. Typically, electrical energy is required to operate the chiller, thus forcing the instrument system to be dependent on the availability of electricity. It is less reliable and more difficult to maintain than other methods.

Desiccants

Some materials have an affinity for water; consequently, when moist air is passed over them, moisture is removed. These materials are the most effective means of drying air, provided the maximum flowrate for the dryer assembly is not exceeded and the desiccant is regenerated or replaced when necessary. The life of the desiccant depends on the volume of air passed through it and the moisture content of the air. Since these factors sometimes are difficult to determine with precision, establishing a maintenance schedule can be difficult. Once the desiccant is saturated, air will pass through the unit but will not be dried, which could result in instrument problems occurring before the failure of the dryer was detected. Properly used, desiccants are the best method of drying instrument air.

Coalescing Filters

Some filter medias can trap and remove water from a stream of air passing through them. Many claims for the performance of these filters as air dryers have been made, but industry reaction seems mixed. As with a desiccant, the filter life between maintenance events is somewhat variable. If the filter saturates before it is maintained, it will effectively plug and reduce the flow of air available to the instruments. Filters and desiccants share the advantage of having no moving parts.

Centrifugal Dryers

The incoming air is accelerated through a helix or is used to spin a disk, which causes a temperature drop in the air stream, which causes moisture to condense out of the air. Moisture is collected in a trap and removed automatically.

Since construction is simple, operation fully automatic and effectiveness good, these units are an excellent choice for most air drying applications. Since they operate automatically when air is required and have no replaceable or renewable elements, they can be maintained at predetermined intervals with at least some assurance they will operate well in between.

Air Supply Distribution

Economics generally will dictate that an air production system serve as large an area as possible. Since pressure at the instruments must be fairly closely regulated, close attention to the design of the distribution system is essential.

The uses made of air produced by the production system will vary with plant requirements. Potential uses in addition to the instrumentation system include air purging cabinets and operating pneumatic tools. The more uses expected of the air supply and the greater its geographical extent, the less reliable it can be expected to be in supporting its primary task—running the instrument system.

In designing an air distribution system, one should remember that at constant temperature, velocity and, therefore, friction, head loss in the distribution piping can be halved by doubling the pressure. Moving air from the central compressor plant to various utilization locations is best done at as high a pressure as economically feasible. Compressor, tubing and piping code limitations make 670 kpa (100 psig) a good transmission pressure. At utilization locations it is then regulated to the appropriate pressure for the instruments.

BIBLIOGRAPHY

American Petroleum Institute. *Manual on Installation of Refinery Instruments and Control Systems,* 3rd ed., Washington, DC (1974), Part I, Sections 7, 9, 11 and 12.

Anderson, N. A. *Instrumentation for Process Measurement and Control,* 2nd ed., (Radnor, PA: Chilton Book Co., 1972).

Buckley, P. S., and W. L. Luyben. "Designing Long-Line Pneumatic Control Systems," *Instr. Technol.* 61–66.

Callahan, F. J., Jr. *Tube Fitting and Installation Manual,* Crawford Fitting Co., Cleveland, OH.

Caplan, F. "Finding Air Pressure Drop Through Smooth Tubes," *Plant Eng.* 74–75 (January 6, 1977).

Considine, D. M., Ed. *Process Instruments and Controls Handbook* (San Francisco: McGraw-Hill Book Co., 1974), Chapters 3, 15–17.

Farmer, E. J. "Pneumatics in a Digital World," *Instr. and Control Syst.* 31–35 (March 1979).

The Foxboro Company. "Process Control Instrumentation," Foxboro Publication No. 105A (1971).

Gassett, L. D. "Instruments—Pneumatic or Electronic?" *Chem. Eng.* 36 (June 2, 1969).

Howard W. Sams & Co. Instrument Training Course, Volume 1, *Pneumatic Instrumentation* (1978).

Instrument Society of America. "Nomenclature for Instrument Tubing Fittings (Threaded)," ISA RP 42.1 (1965).

Instrument Society of America. "Pneumatic Control Circuit Pressure Test," ISA RP 7.1 (1956).

Instrument Society of America. "Color Code for Panel Tubing," ISA RP 7.2 (1957).

Instrument Society of America. "Quality Standard for Instrument Air," ISA S 7.3 (1976).

Instrument Society of America. "Air Pressure for Pneumatic Controllers and Transmission Systems," ISA S 7.4 (1970).

Magison, E. C. *Electrical Instruments in Hazardous Locations,* 3rd ed. (Philadelphia: Instrument Society of America, 1978).

The Samuel Moore Co., Dekoron Division. "The Effect of Tubing ID on Long Line Response" (April 1978).

Perry, R. H., and C. H. Chilton. *Chemical Engineer's Handbook,* 5th ed. (San Francisco: McGraw-Hill Book Co., 1973), Chapters 6, 22 and 24.

Shinskey, F. G. *Process Control Systems* (San Francisco: McGraw-Hill Book Co., 1967).

Shinskey, G. "Pneumatic Transmission Distances," *The Foxboro Recorder* p. 31 April 1972).

CHAPTER 13

FIBER OPTICS IN INFRARED INSTRUMENTATION AND CONTROL

Riccardo Vanzetti
Vanzetti Infrared & Computer Systems, Inc.
Canton, Massachusetts

INTRODUCTION

The use of optical fibers in conjunction with infrared detectors and signal processing electronics represents the latest advance in the field of noncontact temperature measurement and control. Infrared (IR) detectors have been around for many years, although R&D keeps adding to their numbers and performance characteristics.

Optical fibers are much younger and only recently have become the object of widespread interest, thanks to their ability to carry over long distances optical information signals.

Generally, IR detectors have been used in conjunction with conventional optical elements (lenses, mirrors, prisms). Fiber optics were excluded from consideration because they are made of either glass or plastics, both of which are opaque throughout most of the infrared spectral region. Thus, according to fundamental laws of physics, their marriage to infrared detectors could never work.

Defying all theoretical predictions, coupling fiber optics with infrared detectors resulted in several new families of instrumentation and control systems endowed with superior performance characteristics. How is this possible? Let us view the process step by step. For those not familiar with infrared, we shall start with the fundamentals.

INFRARED RADIATION

IR radiation, sometimes called thermal radiation, is the electromagnetic radiation occupying the spectral area between visible light and the radio microwaves (Figure 1). Infrared radiation is emitted by all physical matter, as a function of temperature and of emissivity, which is a numerical coefficient of proportionality related to the amount of electromagnetic power radiated by a blackbody at the same temperature.

Like visible light, IR radiation is made of photons carrying energy away from its source (the surface of a physical body) and traveling at the speed of light. The power emitted in this process is proportional to the fourth power of the absolute temperature of the emitting surface, as shown by Stefan-Boltzmann's equation (Figure 2):

$$W = \epsilon \, \sigma \, T^4$$

where W = radiant flux emitted per unit area (W/cm^2)
ϵ = emissivity (unity for blackbody source)
σ = Stefan-Boltzmann constant $(5.673 \times 10^{-12} \text{W}/(\text{cm}^2)(°\text{K}^4))$
T = absolute temperature of source (°K)

The power so emitted is of "incoherent" type, which means that the radiation takes place simultaneously at all wavelengths of the infrared spectrum. This is due to the fact that the emitting surface is made of an extremely large number of different "oscillators" (molecules, atoms and subatomic particles), which vibrate, rotate, oscillate and "jump" at their own resonant frequencies.

Intuitively, it is easy to understand why the higher the temperature, the higher the infrared energy emitted by the source. Heat is directly related to the degree of molecular and submolecular motion: the more violently these physical particles move, the higher is the energy content of the photons emitted.

However, the power content of this radiation is not constant throughout the infrared spectrum, but its distribution varies as a function of wavelength and temperature. Figure 3 shows the blackbody radiation curves, that is, the distribution of infrared power emitted by an ideal surface having an emissivity factor $\epsilon = 1$. In the illustration, the slanted line indicates how the wavelength of the peak emission moves towards the visible region of the spectrum as the temperature of the emitting surface increases. This explains why physical matter becomes incandescent at high temperatures. Figure 4 shows the correlation between temperature and the wavelength of the peak emission.

BLACKBODY AND EMISSIVITY

It has been mentioned that the blackbody's emissivity is equal to unity. Blackbody is a term indicating an ideal surface having an infinite number of oscillators,

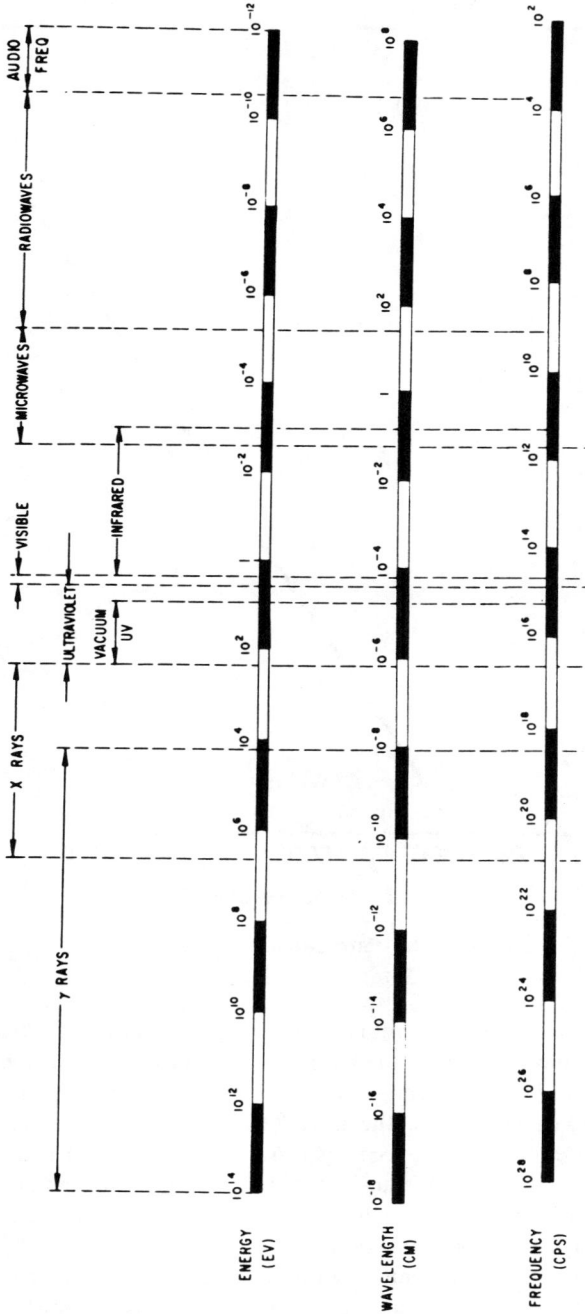

Figure 1.   The electromagnetic spectrum.

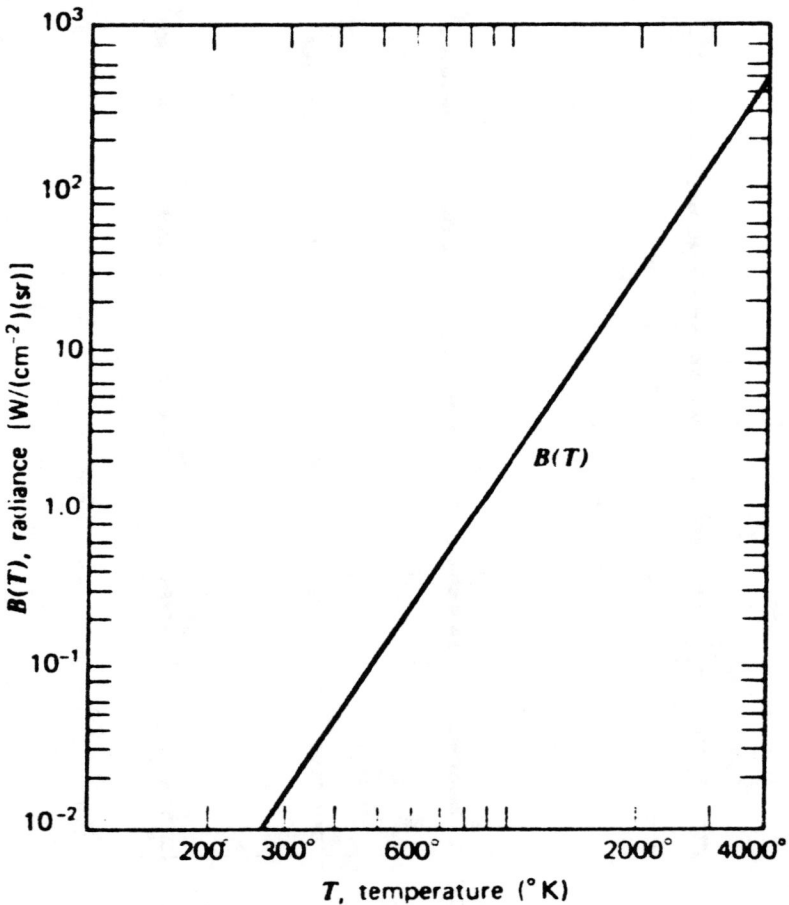

Figure 2.   Stefan-Boltzmann law.

one for each wavelength of the infrared spectrum. All the infrared radiation laws are formulated and are true for the blackbody. In practice, no such ideal body exists because every material emits infrared radiation according to its physical composition and characteristics. The degree to which this emission approximates the blackbody's emission is expressed as a fraction of unity and is called *emissivity*. Its symbol is ϵ. From Stefan-Boltzmann's equation we see that once ϵ is known, it is possible to establish a direct correlation between the power radiated by a physical surface and its temperature.

The concept of emissivity is not very easy to grasp; however, perhaps a comparison with the properties of visible light might help. Consider, for instance, an

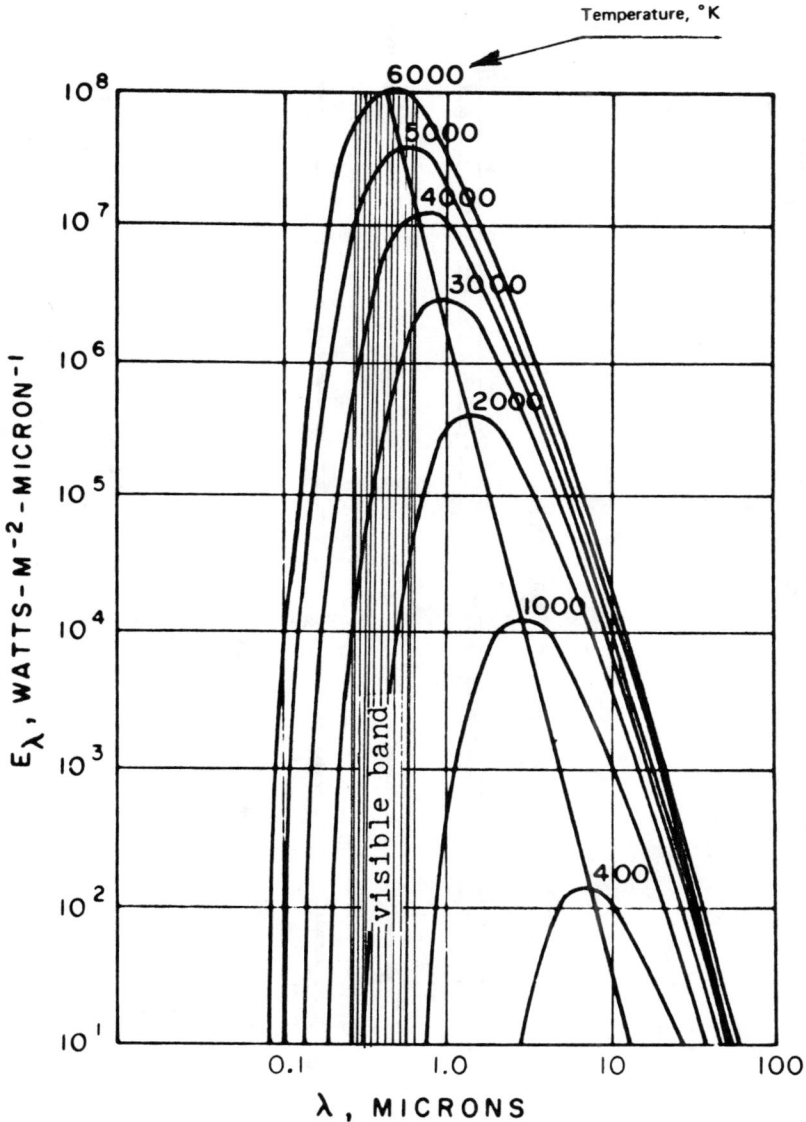

Figure 3.   Blackbody radiation curves.

opaline electric bulb emitting white light. Set on "unity" the amount of light so radiated. If the bulb is now painted in red, blue or any other color, the amount of light radiated shall be a *fraction* of the *unity* emitted when the bulb was white. We could call it "emissivity," and it will always be less than unity.

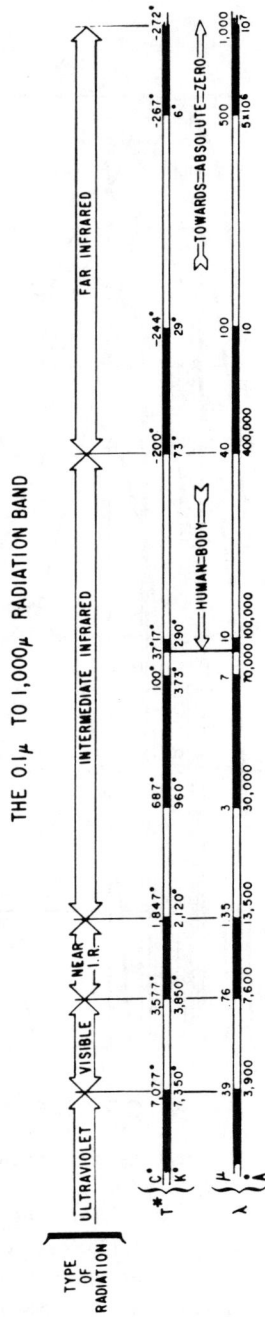

Figure 4.   T° and peak wavelength correlation.

In the infrared domain, the blackbody could be compared to the just-mentioned white bulb, while we could think of any other surface as emitting "infrared colors." And we should also include the color *gray,* with the only difference that the gray radiation covers the whole infrared spectrum (just as the blackbody, only at a lower level) while the other "colors" are only covering a portion of the total infrared spectral area.

Unfortunately, often ϵ is not known, either because it is a variable difficult to control, or simply because it is inconvenient to measure. In such instances, the ratio radiometer offers a practical solution. Its operation is based on the assumption that the incoherent radiation emitted by a nonblackbody surface follows a "graybody" distribution, which is identical to the blackbody curve in shape and wavelength with only the difference of being "dropped" to a lower level in the power scale (Figure 5).

For many surfaces, this assumption can be taken as true in first approximation. Consequently, every temperature of the target is precisely identified by a unique value of the ratio of the power radiated at two different wavelengths, A and B, conveniently chosen and independent from the level at which the corresponding graybody emission curve has been dropped.

Several versions of ratio radiometers have been developed, but until recently their operation was limited to the high-temperature region because of the difficulty of collecting enough radiant energy at the chosen wavelengths when using conventional optics, which necessarily have a small aperture number.

Conversely, the use of fiber optics, which have a remarkably larger aperture number, makes it possible to gather enough energy to allow measurement of much lower temperatures. At the time of this writing, approximately 200°C is the low-end temperature measurable with the most advanced fiber-optics ratio-radiometers. They are quite useful in a host of applications because they measure the temperature of a target independently from the emissivity value of its surface. This is quite a convenient feature in those instances in which the value of surface emissivity is not known, or when it varies during the measurement process due to changes in surface characteristics, such as oxidation and crystallization.

THE INFRARED RADIOMETER

Figure 6 is the basic block diagram of all single-channel infrared radiometric systems. The elements shown in solid lines are always present, while the ones shown in dotted lines are optional, their presence and features being dictated by the system's performance requirements. Ratio detectors and multichannel systems include additional blocks, generally of the same type.

In essence, these systems turn the infrared radiation emitted by the object located in their field of view, into an electrical signal that, after due processing and manipulation, can, in real time (1) be displayed as the temperature of the tar-

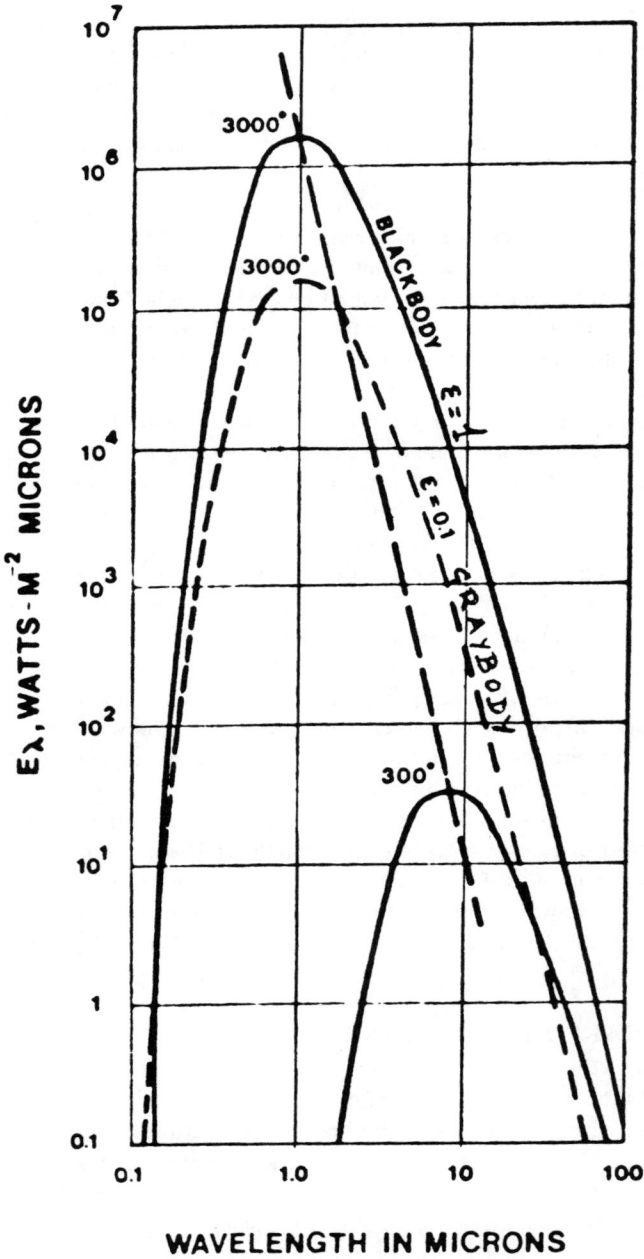

Figure 5.   Blackbody and graybody curves.

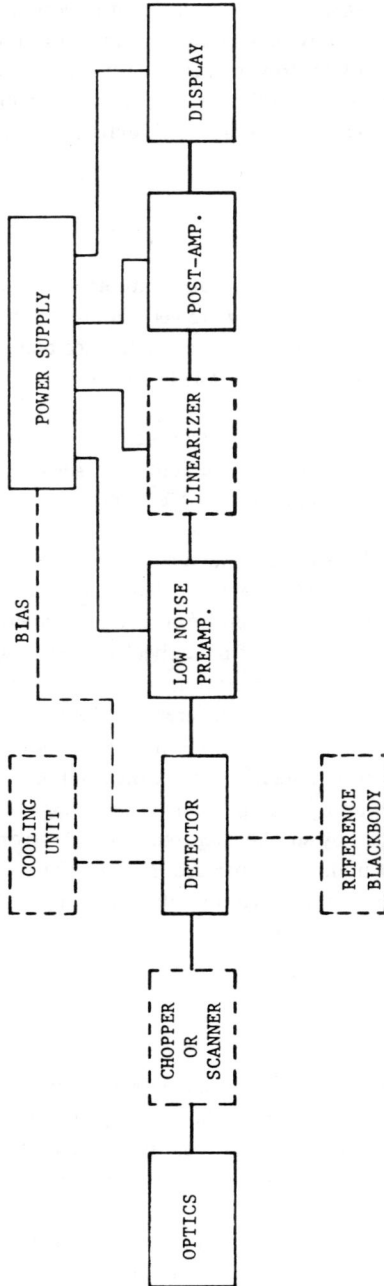

Figure 6.   Basic block diagram of a radiometer.

get, with the desired degree of resolution (both thermal and spatial); (2) be recorded and/or stored in any of several existing ways; and (3) through a feedback loop, control the process to which the target is subjected, for instance heating, cooling or moving, etc. The following sections review in more detail the most important elements shown in the basic block diagram.

THE OPTICS

Until recently, the optics used in infrared radiometry were of the conventional type, such as lenses and prisms. They are very similar to their corresponding elements with the visible radiation, the major difference being the optical materials of which they are made, which must have good transmissivity in the infrared spectral area.

Figure 7 shows the transmission spectral regions of a number of optical materials. These transmission regions are commonly called "windows," and their limits are indicated for a minimum of 10% external transmission and 2 mm of sample thickness.

In addition to refractive elements, such as lenses, reflective surfaces either flat or curved are frequently used as optical elements handling IR radiation. The major advantage is their total freedom from chromatic aberration, which is a particularly serious difficulty in the infrared spectrum because the refractive index varies with wavelength, and the spectral area of interest is generally very broad.

Allowing for the abovementioned differences, the optics until recently used in infrared radiometry were pretty much like the optics used in photographic cameras. The major limitations of such optics are the need for (1) a direct line of sight to the target; (2) a clear optical path, free from signal absorption or interference such as produced by smoke, fog, etc.; and (3) strong radiation from the target due to the small numerical aperture (NA) of these systems. Just recently, these limitations have been eliminated by substituting optical fibers in place of the conventional optical systems used so far.

Optical Fibers

Optical fibers are transparent linear elements inside which radiation propagates by total internal reflection. Figure 8 shows the physical principle on which the fiber operation is based. All the fibers used in infrared instrumentation are made of glasses especially chosen for their ability to transmit the radiation comprised in the chosen spectral region.

The illustration shows that after entering the front surface, all rays that acquire an inclincation smaller than the critical angle are totally reflected inside the fiber

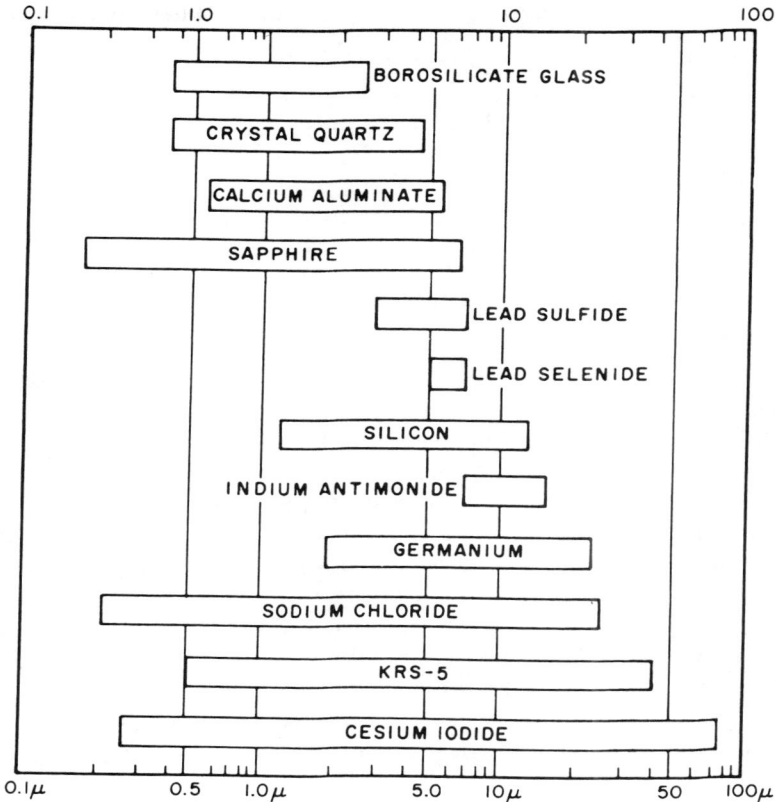

Figure 7.   Transmission "windows" of selected optical materials.

core and keep traveling in this fashion until they reach the opposite end or are totally absorbed, whichever comes first. For a fiber having a critical angle of 67°, the illustration shows an acceptance angle of 70°, which means that all rays incident onto the fiber's front surface at a 35° angle or less with its axis are trapped inside the fiber by total internal reflection.

On the other hand, all incident rays entering the fiber with an inclination larger than the 35° angle will leave the fiber at the first contact with its internal surface. This behavior is commonly called "spilling."

The value of the critical angle is a function of the ratio between the refractive indexes n_1 and n_2 of the glass of which the core is made and of the medium surrounding it. Thus, by controlling the ratio n_1/n_2, we have a means of increasing

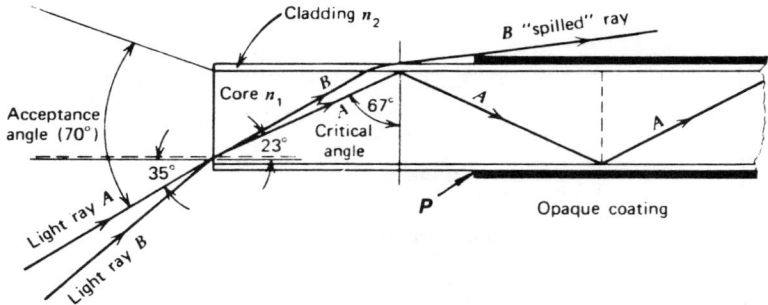

Figure 8. Ray propagation in optical fiber.

or decreasing the acceptance angle of fiber optics, a feature that can be used to obtain special performance characteristics.

Cladding is the name of the technique used to establish a permanent n_1/n_2 ratio. It consists of coating the core with a layer of solid material (usually a different type of glass) having the desired index of refraction. Without this coating, two fibers touching each other would "spill" light into each other since they have the same refractive index. The presence of the cladding prevents this from happening.

Figure 8 shows a fiber core enveloped by a layer of cladding, whose outside surface, to the right of point P, is, in turn, coated with an absorbent layer of material. This will prevent stray rays, such as those surpassing the critical angle (for instance, where the fiber makes a curve) from spilling out: the outer coat will merely absorb them. Also, at a curve, radiation cannot get into the fiber from outside, and every chance of interference with the rays traveling along the fiber is eliminated.

How tight a curve can a fiber make without spilling? As long as the ratio of the bend radius to the fiber diameter is above 40, the losses are negligible. This means that a fiber with a 25-μ core diameter can be wound around a 1-mm mandrel with no significant transmission loss.

Besides their ability to carry radiation around corners and through opaque obstacles, optical fibers provide quite large acceptance angles for incident radiation, thereby making them comparable to "fast" conventional optical systems; that is, systems having a large NA number. NA stands for the numerical aperture number of an optical system and is a measure of its ability to accept incident light rays. This, of course, is a function of the limit angle of acceptance. The larger this is, the larger the cone of radiation entering the optical fiber, transmitted along its length and out of its output end.

When compared with typical radiometers, which have optical f/no generally ranging between 4 and 8, optical fibers can capture remarkably higher radiating power. On the basis of correlation of equivalent f/no, such optical fibers theoretically are more effective by factors ranging from 20 to 88. Figure 9 graphically illustrates a comparison between the cone of radiating energy collected by a conventional lens and an optical fiber.

As mentioned before, the optical fibers used with infrared instrumentation are made of special glass having good transmissivity in that portion of the infrared spectrum covered by the detector used in the system. In other words, fibers and detector must be matched to each other for best system performance. Figure 10 shows the transmission curves of one of the glasses used to make a special family of fibers for low-temperature infrared radiometers. It is interesting that the transmission losses are indicated as a percentage of the radiation signal being trans-

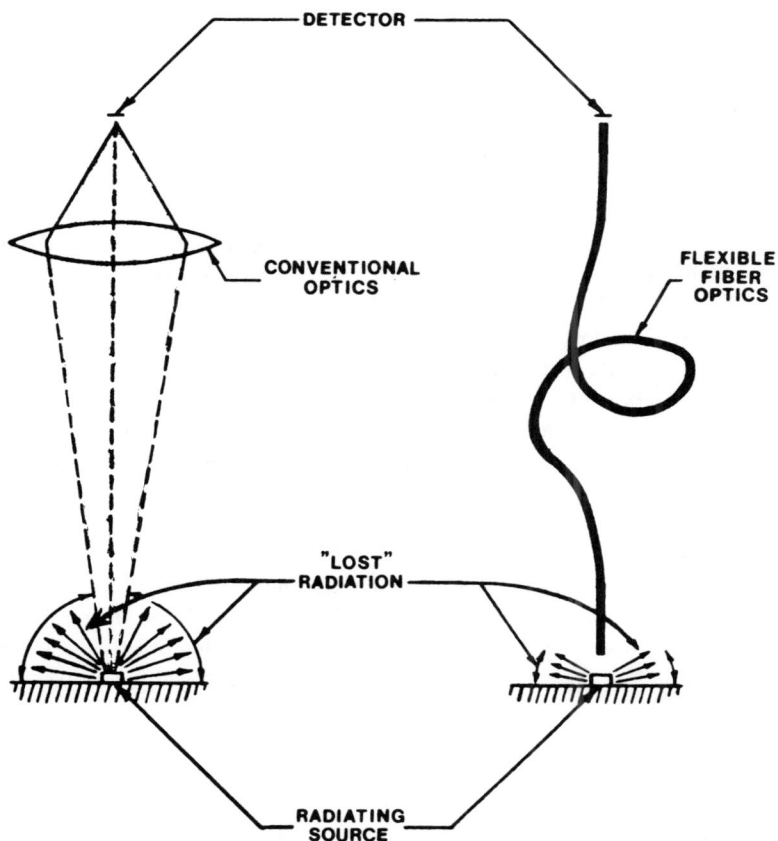

Figure 9.   Capture of radiant energy.

mitted. In other words, the signal magnitude is inversely correlated with the length of the fiber, which means that the signal magnitude will never drop to zero, although it will asymptotically approach the zero level. However, the noise traveling through the fiber also will follow the same laws; therefore, the S:N ratio will remain the same no matter how long the fiber is. Thus, the detector noise and the self-generated noise of the signal processing electronics become the limiting factors in signal readability.

Fiber Bundles

Thus far single fibers have been discussed. For practical purposes, however, fiber bundles are most commonly used. A fiber bundle is an assembly of a number of fibers, anywhere from just a few to many hundreds, lined inside a containing sleeve that can be either rigid or flexible. In Figure 11, fiber bundles of different length and composition are shown. As can be seen from the picture, the ends of these bundles are made rigid to hold firmly in place the terminations of all the fibers. Usually this is achieved with the use of a cementing compound, such as an epoxy resin that hardens to a degree adequate to permit optical finishing of the end surfaces.

Figure 12 shows the geometric representation of an end surface of a very small bundle of fibers. The majority of those used for infrared temperature measurement applications have several hundred fibers to pick up and transmit more signal to the detector. Typically, the outside diameter of a single fiber is 25 μ.

Fiber bundles are divided in two groups: coherent and incoherent. The coherent ones have the same identical geometrical distribution of the fibers at the two ends. In this fashion, the light distribution picked up at the front end is exactly duplicated at the output. In other words, an image is transferred from one end to the other, with only the degradation resulting from (a) the transmission losses, and (b) the resolution allowed by the size of the individual fibers and the spacing in between.

Instead, the incoherent fiber bundles have a random distribution of the fibers at the two ends and no image is transferred from one end to the other. This is the type generally used in infrared fiber optic systems for temperature monitoring and control. Only radiation is available at the output in quantity proportional to the total input, minus the transmission losses.

Most of the optical fiber bundles used with infrared radiometers are of limited length, generally 1 or 2 meters long, occasionally up to 10 meters. If they are not terminated with a lens, they have a field of view of approximately 60°. This means that the diameter of the target area viewed by the detector through such optics is slightly larger than the distance between the front end of the fibers and the target surface. This can be easily verified by backlighting the target, which is

	At 590 nm	At 2330 nm
Numerical Aperture	0.37	0.35
Maximum Aperture Angle 2α	43.0	41.2
Temperature Resistance Up To 500°C For Bundles, Fiber Diameter 50–70 Microns		

INFRARED FIBER OPTICS

Fiber Type T-1

% Transmissivity

Length 250 mm

Length 1000 mm

Length 3000 mm

WAVELENGTHS IN NM

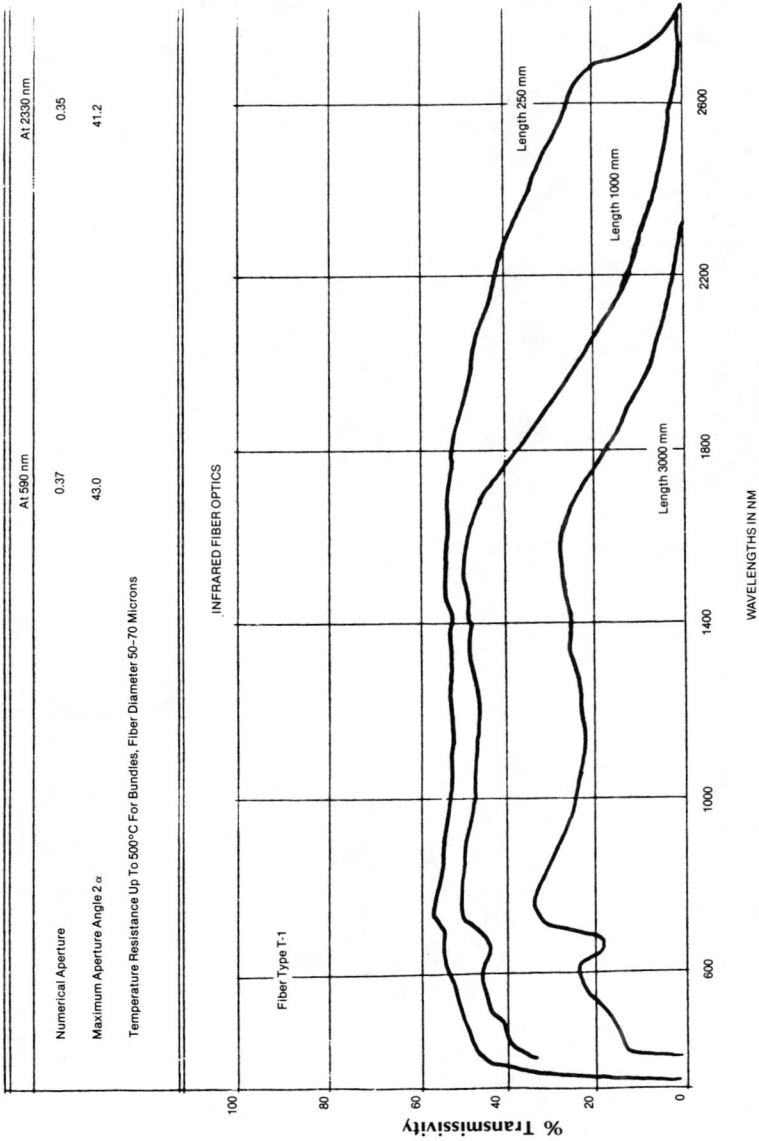

Figure 10.   Spectral transmission of infrared fiber optics.

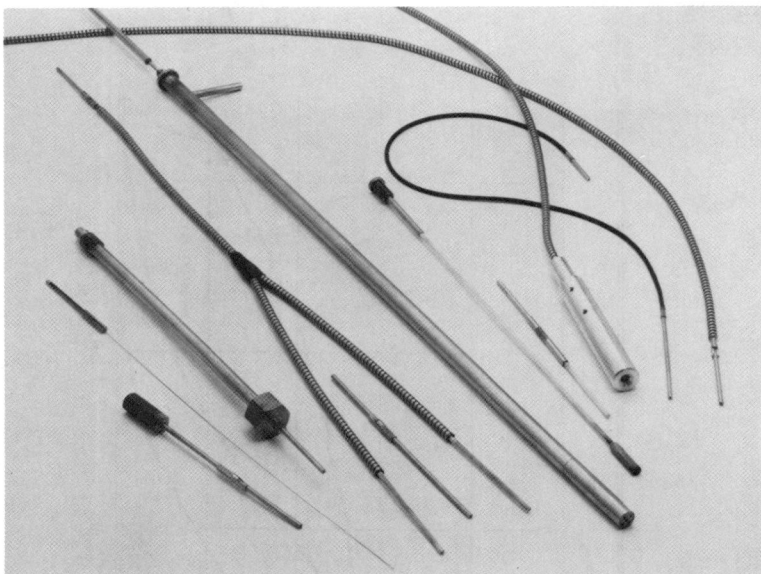

Figure 11. Various fiber optic bundles configurations.

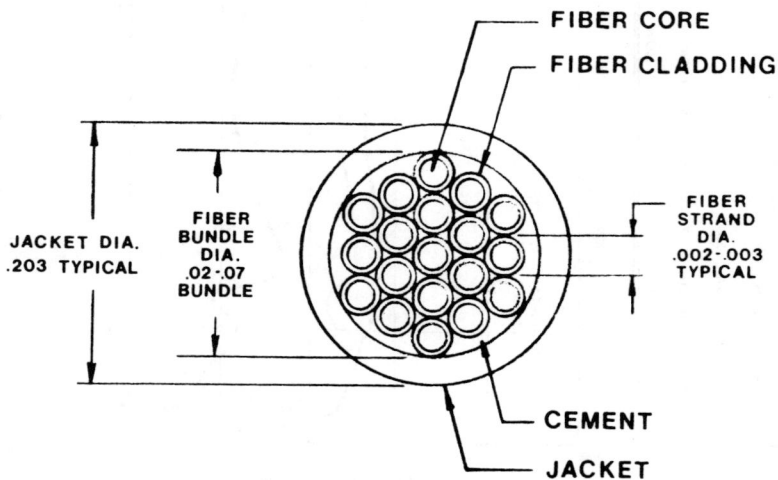

Figure 12. End view of very small fiber bundle.

achieved by injecting visible light at the opposite end of the fiber and, in this way, projecting a cone of light onto a target surface. Figure 13 compares backlighting for two different types of fibers. The one in the right hand of the operator has a 60° field of view, so it covers an area that increases with the distance between fiber front end and target. The one in the left hand of the operator is terminated with a lens, and so has a focal point at a well-defined distance from it.

Fiber optics of the first type are used when the target is large, and the average temperature of the viewed area must be measured. Fibers of the second (focused) type are used for measuring the temperature of a small area when it is not possible to place the fiber's front end close enough to the target surface. The focal point for fibers of this type can be located at almost any distance from the fiber's front end, although for practical purposes this distance is usually set between 2 cm and 3 meters.

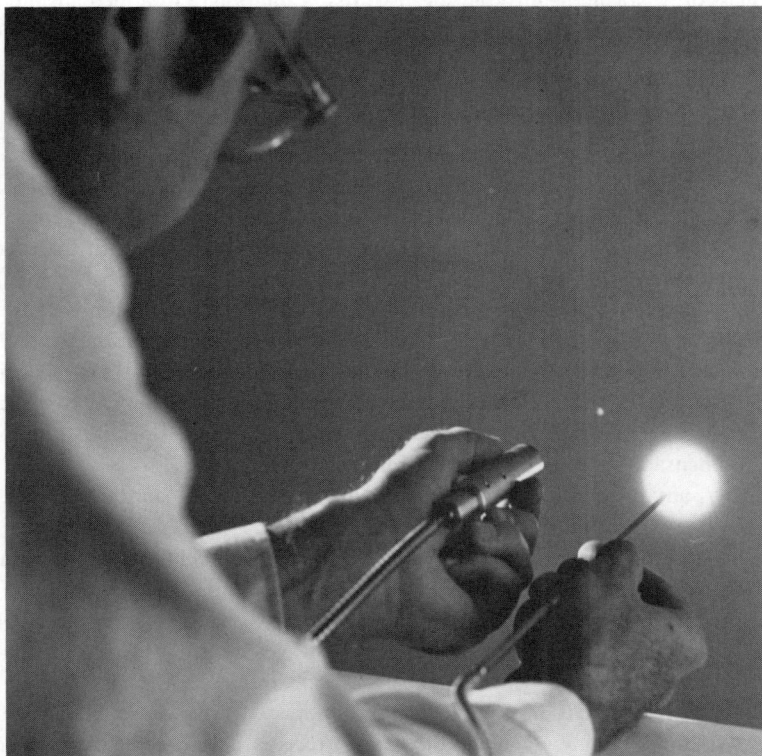

Figure 13.   Viewing areas of open and focused fiber optics.

The backlighting capability assures perfect and easy target aiming, both as location and size of the viewed area, without parallax error. In some applications it might be desirable to check the fiber's aim at any time, during the operation. In this case, a bifurcated optical fiber can be used. One branch of it is connected to a high-intensity light that will go on by pressing a switch, thus allowing the operator to verify and, if needed, to achieve a perfect aiming of the fiber. The other branch will allow the infrared detector to "see" the target at exactly the same spot that was illuminated and to measure its temperature as soon as the aiming light has been turned off.

Advantages and Disadvantages of Fiber Optics

The first and major *disadvantage* of optical fibers is the poor transmissivity of any glass or plastic in the infrared spectral area. Radiation of intermediate and far infrared (wavelengths larger than 2.5μ) is absorbed by the fibers after just a few millimeters of travel. This makes it exremely difficult to measure temperatures below 100°C, since most of the radiation emitted by a surface below 100°C is located around the 7μ region. Of course, there is some radiation in the near-infrared area, but its energy content is so low that it lies buried under the noise level.

Another disadvantage of optical fibers is the impossibility of producing two bundles of identical characteristics because a bundle is made of many hundreds, and often several thousands of individual fibers, whose final number can differ by a few percentage points at the end of the manufacturing operation because of unavoidable losses due to breakage. And even if it were possible at the beginning to assemble two perfectly identical bundles, individual fiber breakage during usage would soon create a difference in their ability to transmit infrared energy. Consequently, the signal processing electronics must have the flexibility to adjust and compensate for this lack of fiber bundles consistency.

Finally, any glass has a softening thermal threshold that cannot be exceeded without damage to the fibers. This threshold for most glasses is around 600°C. Thus, fibers that are expected to operate in hotter environments must be cooled in any of several available ways (air purge, water circulation, etc.).

Concerning *advantages,* all fibers can be looked at as low-pass filters, with gradual cutoff of transmissivity above 2.5μ in wavelength. If this is a disadvantage because it forbids to measure temperatures below circa 100°C, it is, however, an advantage because it enhances thermal resolution. This is due to the fact that the radiation power increments in the near-infrared area are larger than the corresponding power increments in the total radiation area for the same Δt of the target. The reason for it is the displacement of the peak radiation wavelength toward the visible region as the target temperature increases. Figure 14 illustrates this concept: it can be clearly seen that the area increments at the left of the vertical cutoff line are larger than the area increments of the total surface area subtended by the full radiation curves.

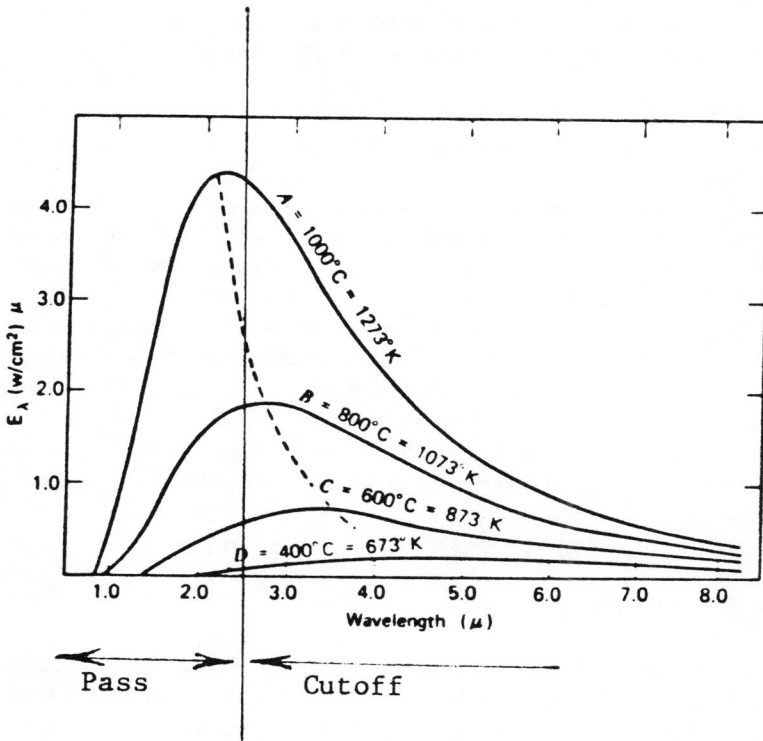

Figure 14.   The filtering action of fiber optics.

In second place, there is the advantage that optical fibers can be defined as "shielded radiation conduits" of constant transmissivity, no matter what environment they might have to cross. Smoke, water, vapors, dust, gases, etc. cannot affect the optical signal transmission through the fiber.

Among other advantages are the following:

1. *Inert, rugged construction.* In addition to being unaffected by energy fields, optical fibers contain no moving parts and are specified and constructed for use in most heat, chemical or radiation environments. For example, they can be im-

mersed in molten materials, such as polymer melts in extruders and injection molding machines, where direct, real-time response measurement has heretofore been impossible with thermocouples because of frictional heating and heat sinking of the sensing element.

2. *Easy installation.* Most fibers are flexible and of a size that permits bending around opaque obstacles for installation where ordinary instruments simply cannot fit. A direct line of sight to the target is not required.

3. *Simplified setup eliminates human error.* All fibers can be aimed with pinpoint accuracy using a source of illumination from one end to carefully line up the target area at the other end. This simple advantage overcomes most human problems associated with proper alignment of conventional noncontact temperature detectors.

4. *Optimum target viewing angle.* With the addition of small lenses, the viewing angle of the fiber can be modified from the standard 60° cone, down to a spot 1 mm in diameter. Large or small areas are precisely monitored with optimum resolution and accuracy.

5. *Wide range of temperature measurements.* With the proper selection of quick-disconnect fibers, a single system may be used to cover the full range of temperatures from 60°C to over 2000°C with extraordinarily fine resolution, accuracy and repeatability.

6. *Very large NA.* This characteristic allows one to capture from the target a very large cone of radiated energy, up to 100 times larger than it is practical with conventional optical systems.

THE CHOPPER

The majority of the IR radiometers incorporate a "chopper," that is, a mechanical device that periodically interrupts the flow of radiation from the target to the detector. In this way, the magnitude of the infrared signal can be easily measured by comparing it with the baseline of the no-signal level during the cutoff caused by the interposition of the chopper blade.

This alternating between radiation and no-radiation is reflected, at the detector output, as an alternating electrical signal, which by its own nature, is easier to produce than a dc signal. However, a minority of infrared radiometers does operate without a chopper. In this instance, the system's output is a dc signal that only varies as a function of the variations of the impinging radiation. The major difficulty of these systems is signal drift.

In its most common configuration (Figure 15), a chopper is a slotted disc rotated by an electric motor. The detector's field of view is alternatively opened and closed by the slots and the blades of the disc rotating in front of it.

Lately, tuning forks have been used as choppers. Appropriately designed vanes attached to the ends of the fork's tines perform the chopping action. Chopping

Figure 15. · Basic configuration of chopper assembly.

frequencies up to 25,000 Hz are attainable. When compared to motor-driven choppers, tuning forks have the advantage of small size, light weight, greater accuracy, long-term stability, low power drain and negligible heat dissipation.

Another class of IR radiometers operating without chopper are the scanners. In this configuration, the detector's field of view is mechanically deflected (usually by mirrors or prisms) along a preestablished repetitive pattern. The infrared signal variations occurring along this path produce at the detector's output a modulated electrical signal that can be processed by conventional means to generate, for instance, a visual display representing the distribution of infrared radiation at each point of the field of view being scanned. The instruments of this class are commonly called "infrared cameras." A better designation would be "infrared-to-visible image converters."

INFRARED DETECTORS

The transducers capable of turning the infrared radiation into an electrical signal are called infrared detectors. They are divided into several groups according to their principle of operation, as follows:

Bolometers

The operation of bolometers is based on the measurement of an electrical characteristic variation induced by the heat absorbed by a temperature-dependent element. The *metal bolometer* is based on a positive conductance variation, the *thermistor bolometer* on a negative conductance variation, the *ferroelectric bolometer* and also the *pyroelectric detector* on a dielectric constant variation. Their response time is around 1/100 of a second, with 1/1000 of a second as a typical upper limit of response.

Photocells

When infrared radiation impinges on these semiconductor devices, electrons spinning in the outer orbits of the detector's atoms can "capture" those photons that have a compatible energy content. If the so added amount of energy is sufficient to allow the electron to break its gravitational bond, the electron will escape from its orbit and will become a free charge carrier, which will immediately begin to move along the lines of the prevalent electromagnetic field towards the positive terminal of the detector. An electrical current flow is thus generated whose magnitude is directly related to the number of electrons "liberated" by the impinging photons.

However, should the energy added by the photons not be sufficient to "liberate" the electrons, the latter will merely spin faster in their orbit, thus increasing the kinetic energy content of the atoms and, consequently, their temperature. On the other hand, if the photons' energy content is excessive, they will escape capture by the electrons, so no electrical or thermal effect will be apparent at the detector terminals.

This is why the photocells, according to their chemical composition, can only operate as photon detectors in limited areas of the infrared spectrum, that is, only in those areas in which the photons of the radiation possess the "right" amount of energy—just enough to liberate electrons that, according to the characteristics of the associated electrical circuitry, can:

1. become available as current carriers, thus decreasing the dc resistance of the semiconductor;
2. accumulate at the opposite sides of a self-generated potential barrier, thus developing a voltage difference across it; and
3. move in opposite directions because of an external magnetic field, thus again generating a voltage across the semiconductor.

Case 1 deals with a conductivity effect, and the detectors of this class are called *photoconductive*. In case 2, there are the *photovoltaic* detectors. In case 3, there are the *photoelectromagnetic* detectors.

The time response of these photon detectors is on the order of microseconds or less, which is about three orders of magnitude faster than the thermal detectors and, or course, is independent from their physical mass, since no thermal effect is involved.

The infrared detectors used in fiber optics radiometric systems are mostly of the photon-detector type. Their choice is dictated by the thermal range they must cover and by the response speed required. Lead-sulfide (PbS) cells operating in the photoconductive mode are most often used, while silicon or germanium cells are used either in the photoconductive or photovoltaic mode. Since the transmissivity of optical fibers is limited to the near-infrared region of the spectrum, the detectors used in these systems do not need to be cooled. Figure 16 shows the response curves of several detectors that can operate at ambient temperature in conjunction with fiber optics. This chart plots D^* (a figure of merit related to the detector sensitivity) versus wavelength. As mentioned before, the photon detectors cover just a limited spectral area, while the bolometers and the thermocouples have a flat response throughout the whole infrared spectrum.

THE REFERENCE BLACKBODY

All the infrared radiometric systems equipped with fiber optics are using the ambient temperature as their reference. For those based on ac operation, the chopper's blades are supplying a blackbody signal to the detector in the time interval during which they interrupt the flow of radiation transmitted by the optical fibers.

For systems based on dc operation, the detector must be "zeroed" from time to time to avoid drift problems. Zeroing can be achieved either manually or automatically by instantaneously closing a shutter in front of the detector to establish a ground floor level against which to measure the radiation signal from the target.

THE COOLING SYSTEM

As mentioned previously, no cooling is needed for the detectors matching the fiber optics spectral transmission area. However, since detector performance characteristics vary with its temperature, it will be necessary to either keep its temperature constant or to compensate for the change of its electrical response whenever its temperature changes. This is achieved by varying the gain of the detector output according to its temperature variation. One or more thermistors are used for this compensation.

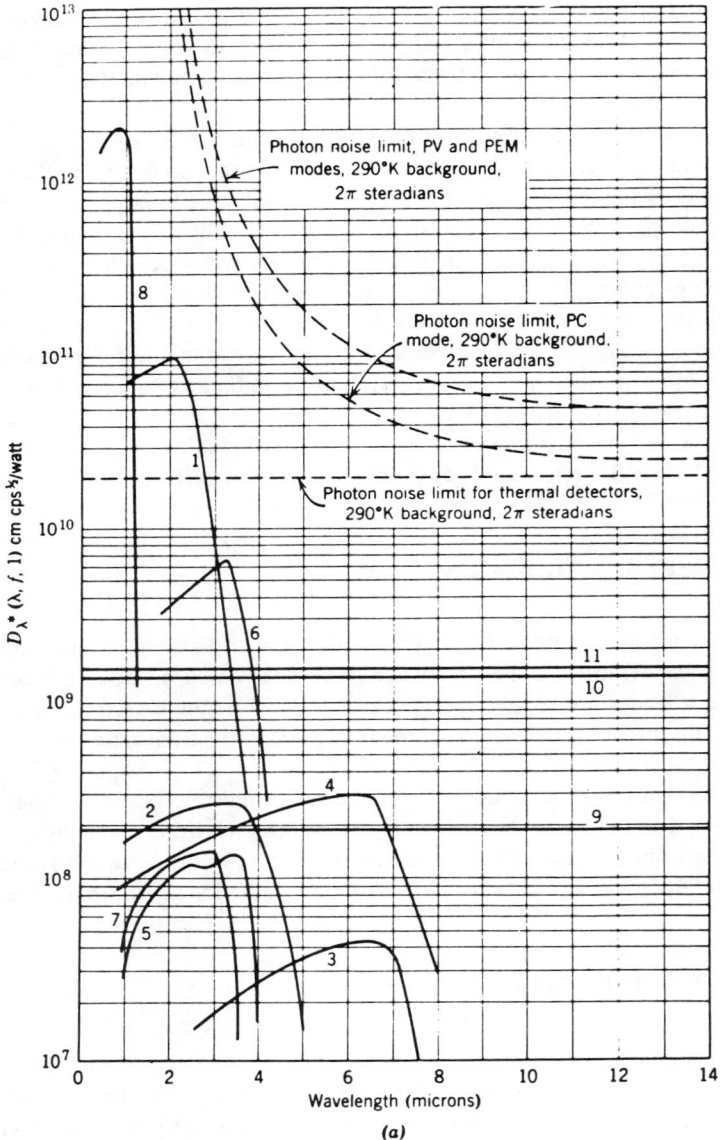

Figure 16.   Spectral D_λ^* of room-temperature detectors. (1) PbS, PC (250 usec, 90 cps); (2) PbSe, PC (90 cps); (3) InSb, PC (800 cps); (4) InSb, PEM (400 cps); (5) InAs, PC (90 cps); (6) InAs, PV (frequency unknown, sapphire immersed); (7) InAs, PEM (90 cps); (8) Tl_2S, PC (90 cps); (9) thermistor bolometer (1500 usec, 10 cps); (10) radiation thermocouple (36 msec, 5 cps); (11) Golay cell (20 msec, 10 cps).

SIGNAL PROCESSING

The electrical signal supplied by the detector is of analog nature. It is correlated with the infrared signal impinging from the target according to a nonlinear function. It is of very small magnitude, usually just below the millivolt level. According to how we want to use it, it must be amplified and recorded. Whenever a feedback function is required, dc voltages must be made available at the output terminals.

The basic block diagram depicting the signal flow was shown in Figure 6, in which the elements drawn with solid lines are essential and those with dotted lines are optional. Figure 17 is the picture of a basic industrial fiber optics radiometric system, Thermal Monitor.® It consists of three separate subassemblies: the optical fiber bundle, the infrared detector head and the signal processing and display console.

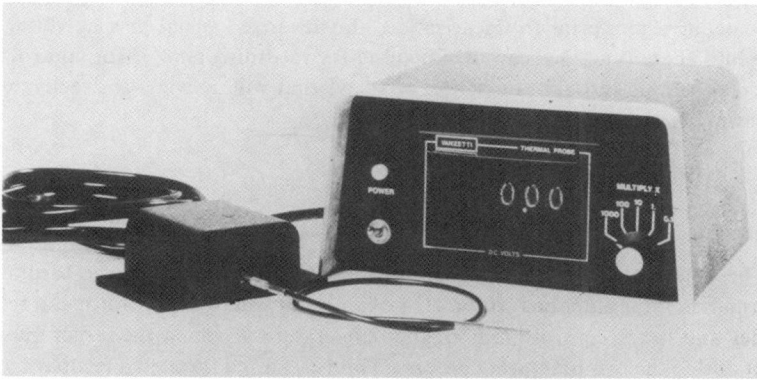

Figure 17.   Basic industrial fiber optics radiometer.

The Optical Fiber Assembly

The optical fiber assembly shown in the photograph is 0.5 meters long and its front end termination is protected in a replaceable ceramic sleeve. This makes the front end totally inert to electromagnetic energy fields. Consequently, it can be inserted without any problem between the turns of an induction coil to allow the detector to "see" and measure the temperature of the object being thermally treated inside the coil.

The Detector Head

This contains a chopper, complete with its electrical driving network, an infrared detector whose bias voltage is supplied by the display console and a very low noise preamplifier, whose output is a low-impedance analog signal that can travel along a coaxial cable (up to 100 meters long) to the display console.

The Display Console

This contains a linearizing amplifier that turns the analog exponential detector output into a linear signal directly proportional to temperature. This signal is then converted into digital and displayed on a digital panel meter (DPM) for direct visual readout of the temperature either in °F or °C. A power supply is also contained in the console.

At the back of the console, several outputs are optionally available, such as the raw analog signal prior to linearization, the linearized signal or a dc voltage or current. These functions can then be used for recording and/or for control of a thermal process through the action of relays that will activate or deactivate the necessary functions.

More Complex Systems

These have evolved from the simple unit described above. Figure 18 shows a thermal monitor equipped with a Hi-Lo logic unit where the operator can set the upper and lower control limits of the temperature at which the target must be kept during the manufacturing process. This is achieved through a feedback loop, with a high degree of precision. For instance, in semiconductor epitaxial deposition processes, where the critical temperature to be kept is 1030°C, the upper control limit can be set at 1031°C and the lower control limit at 1029°C.

Besides the Hi-Lo logic, there is an emissivity control, a gain multiplier knob, a jack for attaching a calibrator unit and, of course, a DPM for visual display of the target's temperature.

When the amount of power needed for target energization is large, a proportional controller is used instead of the Hi-Lo logic unit. In this way, the power correction is correlated to the magnitude of the deviation from the optimum level, instead of being an on-off correction.

Multichannel Thermal Monitor System

Figure 19 shows a multichannel thermal monitor system. Every channel with its own optical fiber and its own detector head is plugged into the control console,

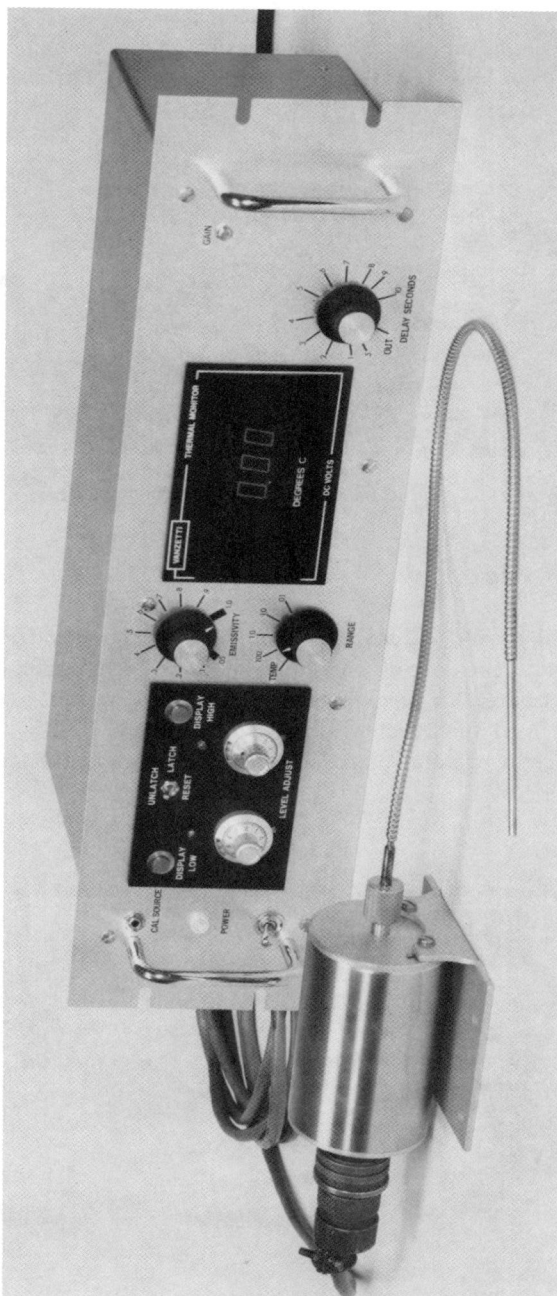

Figure 18. Hi-Lo thermal band controller.

where, according to the option, it can be processed separately all the way to its own output, or can be multiplexed through a time-shared common set of processing electronics. A knob controlling the time-sharing function is available for setting by the operator.

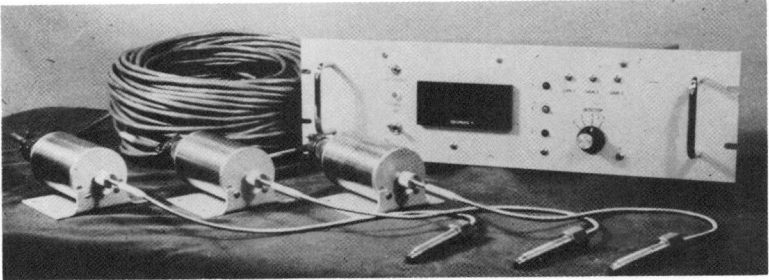

Figure 19.   Multichannel multiplexing thermal monitor.

High-Speed Thermal Monitor

A high-speed thermal monitor is shown in Figure 20. Its three DPMs display the target temperature in three different ways and at different response times, the fastest of which can reach 1 μsec. Systems of this family are ideal to control the power of lasers used for metal welding or surface heat treating, since they measure the target temperature exactly at the spot on which the laser beam is made to impinge.

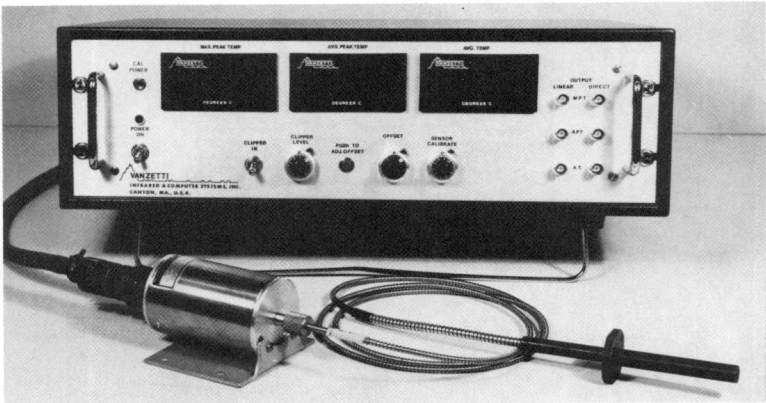

Figure 20.   High-speed thermal monitor.

Another version of these systems (Figure 37) is used to measure the temperature of the rotor blades in jet engines and other turbines. In this version, according to the setting of a control switch, a single DPM displays the average temperature of all the blades of the rotor, the average of all the hottest points of all the blades or the highest temperature of the hottest single blade. With these indications, fuel consumption can be controlled, engine temperature conditions can be observed and a catastrophic failure of the turbine from single-blade overheating can be avoided.

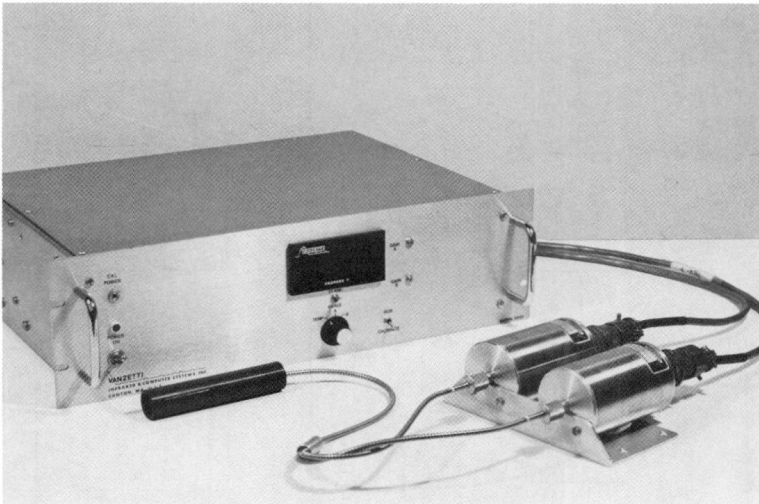

Figure 21.   Emissivity-independent thermal monitor.

Emissivity-Independent Systems

Emissivity-independent thermal monitors have been mentioned at the end of the section entitled Infrared Radiation, p. 401. Figure 21 shows one such system, whose block diagram is sketched in Figure 22. A key element is the bifurcated optical fiber, which can, according to the option, be made of different type glasses or of the same type if different detectors are used. When the emissivity characteristics of the target are those of a graybody, these systems will indicate the target's true temperature no matter what its surface emissivity is. Thanks to the fibers' large numerical aperture, temperatures as low as 200°C can be measured with 1% accuracy.

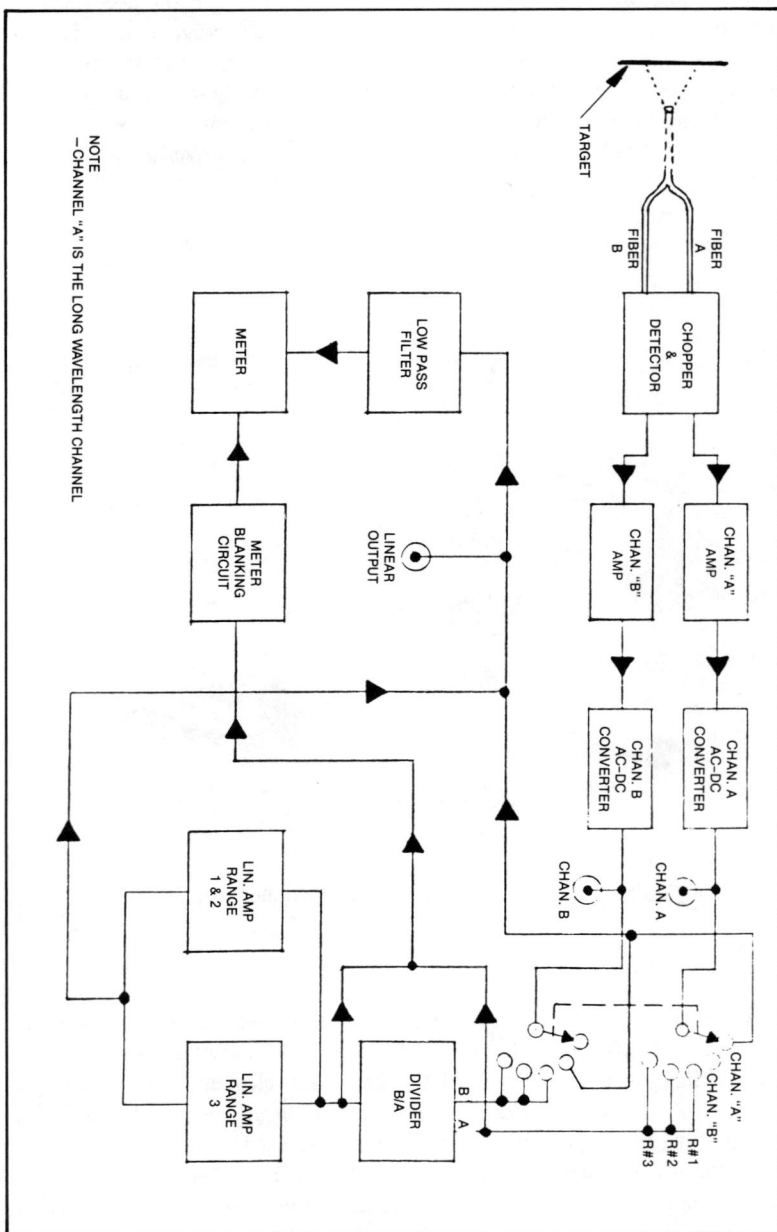

Figure 22.   Block diagram of emissivity-independent thermal monitor.

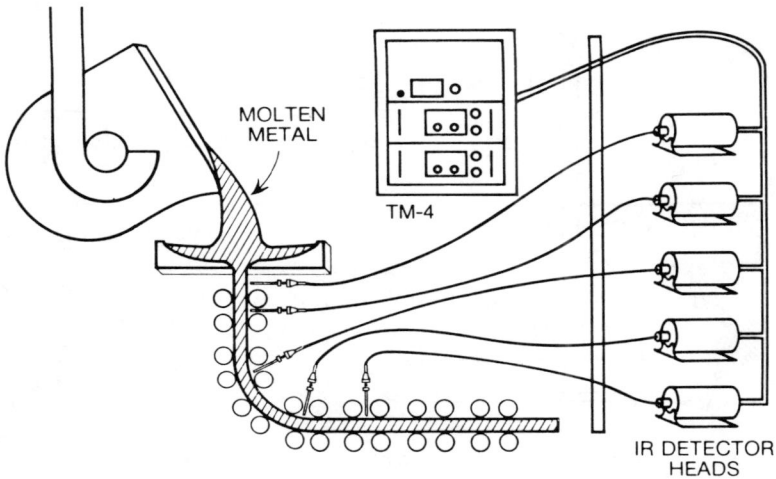

Figure 23.   Continuous casting processes.

APPLICATIONS

We have already listed, in the section entitled Advantages and Disadvantages of Fiber Optics, p. 412, the major advantages offered by optical fibers when compared with conventional optics in infrared radiometric systems. Consequently, in a host of applications in which temperature is a critical factor, it is now possible to make temperature measurements precisely where none was formerly possible, and to improve the accuracy of those measurements that until now were carried out with difficulty and lack of precision.

The main areas in which infrared fiber optics instrumentation is used to resolve difficult problems of temperature monitoring and control in industrial processes are the following:

Metal Production

Melting, continuous casting (Figure 23), annealing (Figure 24), galvanizing, roll-milling (Figure 25), etc.—in all these applications, the detector "watches" the target through smoke, fumes, vapors, water and even solid walls, thanks to the shielded path of constant transmissivity offered by the optical fibers.

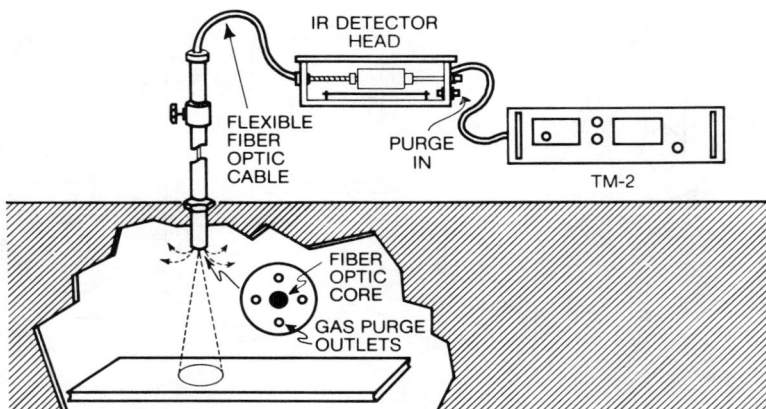

Figure 24.   Annealing/heat treating processes.

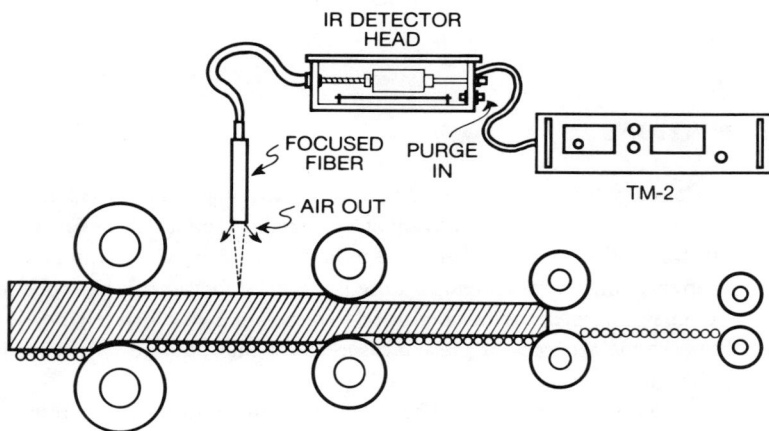

Figure 25.   Roll-milling of metals.

Metal Induction Heating

Because of the strong RF inductive energy field needed to heat the metal parts to be treated, conventional temperature-measuring devices are useless because they will be heated directly by the induction field.

Figures 26 and 27 show two typical applications of fiber optic systems to monitor and control induction treatment of metal objects either stationary in, or moving through, induction furnaces.

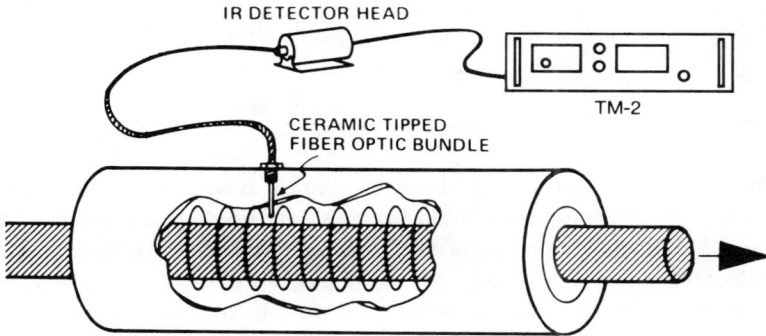

Figure 26.   Monitoring steel rod continuous induction heating.

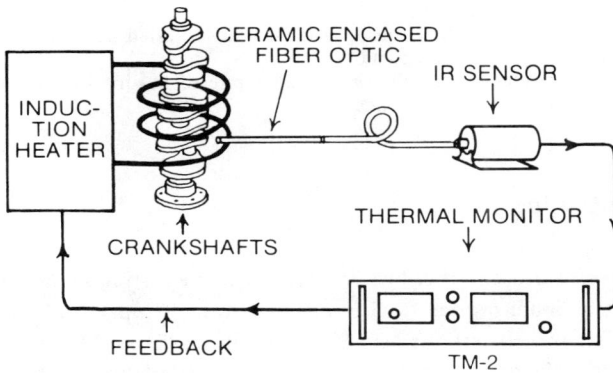

Figure 27.   Controlling induction heating of automobile crankshaft.

Precise control of the temperature needed for perfect heat treatment of metal parts (bolts, camshafts, crankshafts, axles, gears, etc.) is essential to produce the crystal structure that will ensure meeting or exceeding the mechanical characteristic specifications.

The use of fiber optics infrared control equipment:

- allows the viewing end of the fiber optic to be placed in close proximity of the target;
- saves energy by allowing only the precise amount of power to be used;
- speeds up production by controlling process by temperature instead of time and by allowing faster heat injection rate;

- prevents fiber optics from being affected by the induction energy field; and
- allows electronics to be remoted to radiation-free area.

Metal Forging, Hot Stamping, Pipe Bending

Forging of metal parts includes both rough shape as well as precision forging, which requires less material removal and waste. Pipe bending and shaping is also included in this application. These operations are carried out by heating the parts to be worked on to the optimum temperature with any of the several means available (ovens, flame, induction field, etc.). If the part temperature is below the optimum, cracks and internal tensions will develop, while if it is above the optimum, drooping will take place. The precise temperature control afforded by the use of infrared fiberoptic controllers will:

- avoid the formation of defective parts (from cracks or drooping), thus eliminating rejects and waste due to these defects;
- save thermal energy by ensuring that no heat is wasted by heating the parts beyond the optimum level; and
- speed up production by allowing a faster rate of heating the parts without danger of temperature overshoot.

Metal Die Casting

The die temperature is of critical importance in die casting of metals. Thermal cycling of aluminum products, with reference to die temperature has been successfully implemented with the help of optical fibers. Figure 28 shows schematically and in detail how the front end of the fiber is inserted through the mold frame and held in a corner of the runner plate, in contact with the aluminum flowing through it.

The major advantages offered by this solution are: (1) substantial savings of thermal energy by eliminating overheating and drastically reducing production rejects; (2) increased production due to the speedup of the casting cycle, with the operation automatically controlled by the temperature of the casting material and not solely by time, resulting in faster operation; (3) improvement in the quality of the casting due to the control of the process as a function of temperature, resulting in simpler operation and automatic compensation for a cold die startup or interrupted cycles; and (4) direct indication of the die and furnace pot temperature of the metal. Low metal level and blocked water lines are easily indicated several shots before the casting can display conditions visibly.

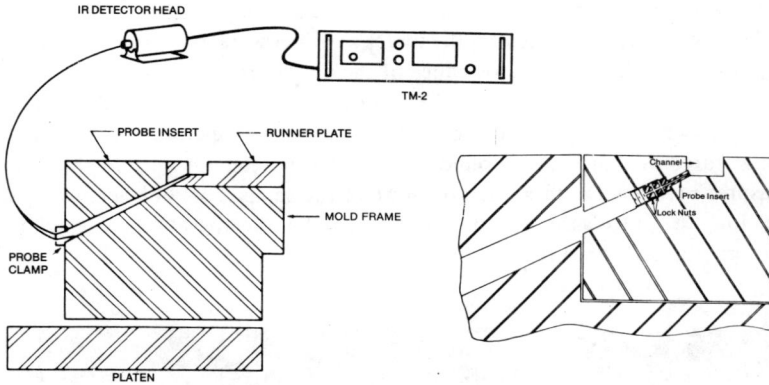

Figure 28. Metal die casting control.

Saw Production

Teeth hardening, annealing, ends forming and joining (in the case of band saws) are operations in which the optimum crystal structure of the steel is obtained only if the correct temperatures are reached, held and controlled within close tolerance limits. Infrared fiber optics systems used in these applications offer the following advantages:

1. Fast speed of response enables high production quantities to be processed.
2. Fiber optics are small and can get in close proximity of the areas under thermal treatment.
3. Very small spot size of the fiber optic enables focusing individually and sequentially on each and every tooth.

Control of Metal-Working Laser

Lasers, generally high-power CO_2 lasers, are used for welding, surface treating and finishing metals of various types. The conventional approach is periodically to sample the beam to keep its power at the desired level. This approach, however, cannot take automatically into account the emissivity variations of the target surface. These variations, in turn, affect the amount of laser power absorbed by the target and, consequently, the target's temperature, which is of paramount importance for the good performance of the operation.

This difficulty is overcome by the use of an emissivity-independent infrared fiber optics system (EITM) aimed at the spot of laser beam impact (Figure 29). The infrared system is made blind to the laser wavelength and, in this way, it measures precisely the target temperature at the same spot. Further, via a feedback loop, it controls the laser power to ensure that the operation is carried out at the optimum temperature.

Figure 29.   Control of metal-working laser.

Among the advantages offered by the fiber optics infrared approach are the following:

1. It allows noncontact temperature measurement in real time.
2. Fiber optics allow easy access to view the laser heating area because of their relatively small size.
3. EITM compensates for variations in emissivity as the part is being heated.
4. EITM response can be matched to the response speed of the laser.

Fusing Armature Windings in Electric Motors

This operation, better defined as thermocompression bonding, is carried out by heat injection and mechanical compression on the point where the wire is looped around the commutator contact hook. Ac current flowing through the heating electrode brings it up to a temperature high enough to vaporize the wire's insulation, but not enough the melt the wire. According to its mechanical configuration, the joint rises to a level between 800°F and 2000°F in a very short time (typically between 50 and 100 msec).

Figure 30 shows an infrared fiber optics system mounted on top of an automatic high-speed machine that sequentially fuses every armature winding to its corresponding hook of the commutator lug. Figure 31 shows a typical electric motor with the wire-to-hook ends prior to fusing. Note how some of the hooks are still standing up, while others already have been bent down. Figure 32 shows in detail the fusing area, with the electric motor at left and the optical focusing head of the fiber optics in the upper right corner.

Major advantages offered by the fiber optics infrared controller are: (1) increased speed of operation due to the possibility of faster heat injection and faster indexing; (2) elimination of defects from unsoldered, poorly soldered or open joints because of melted wire; and (3) ease of precise aiming at the spot of heat injection.

Crystal Growing

In semiconductor manufacture, silicon and germanium crystals must be grown from their molten state. Similar growth procedure applies for the crystals used for lasers. In all these operations the perfect lattice structure of the crystal is of paramount importance. The temperature of the meniscus between the molten material and the emerging crystal is the element controlling the diameter of the "carrot." The heat necessary to keep the molten material at optimum temperature is supplied by an RF energy field. This prevents the use of conventional temperature-measuring devices and even the proximity of infrared detectors.

Figure 30.   Electric motor commutator fusing machine.

The use of fiber optics infrared control equipment (Figure 33) solves the problem and offers the following advantages:

1. Fiber optics can be inserted inside the system and focused on the meniscus while the detector can be placed outside the energy field;
2. If other than normal atmosphere is present, fiber optics can be sealed into the meniscus area.
3. Noncontact viewing of the meniscus by the detector allows precise control of its temperature through a feedback loop acting directly on the induction energy field.

Semiconductor Epitaxial Deposition, Doping, Sputtering

Semiconductor wafer induction heating for doping, epitaxial deposition, sputtering, etc., is generally carried out in hermetically sealed quartz vessels, where vacuum or precisely controlled gas atmospheres are present. Precise temperature control of the wafers is necessary for the correct amount of doping, deposition or sputtering. Since the heating is done by an RF induction energy field, no conventional temperature monitoring devices can possibly be used.

Figure 31.   Electric motor prior to commutator fusing.

Figure 32.   Detail of fusing area.

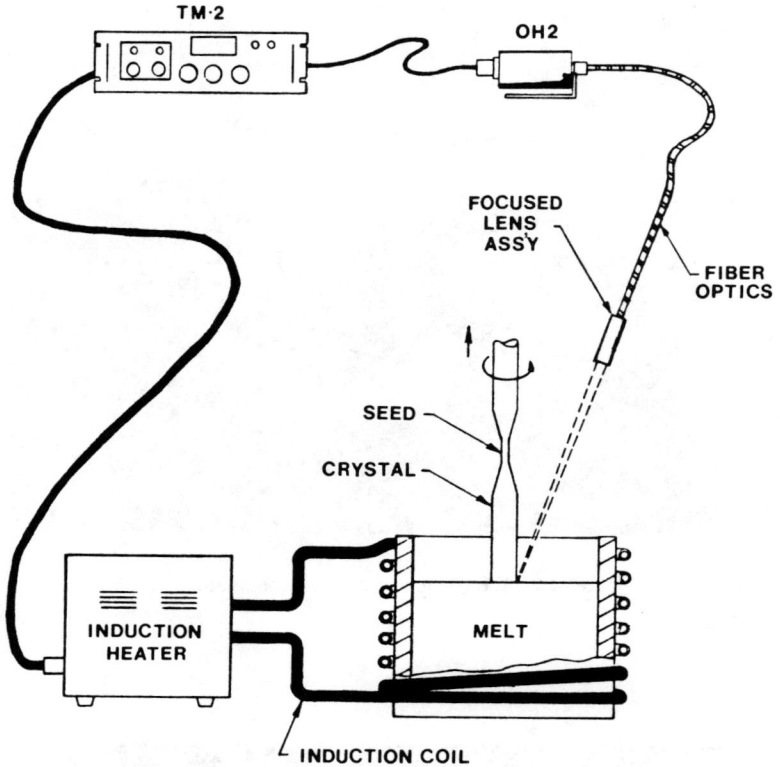

Figure 33.   Crystal growth control.

The use of fiber optics infrared monitoring and control equipment (Figure 34) solves the problem and offers the following advantages:

1.   Fiber optics enable noncontact measurement of temperature.
2.   Fiber optics can reach inside a hermetically sealed vessel and carry the radiation signal from the wafers to the remotely located detector head and electronics.
3.   Thermal process can be automatically carried out within preset optimal temperature limits.

Furthermore, the optical fiber can be manually scanned along the length of the induction oven to determine its thermal profile and to eliminate temperature gradients by adjusting the spacing between the turns of the induction coil.

IR DETECTOR HEAD

TM-2

LENS FOCUSED FIBER OPTIC

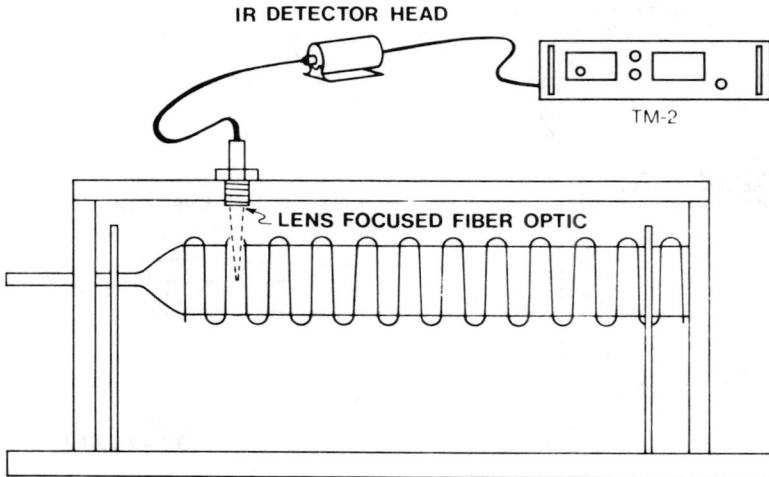

Figure 34.   Semiconductor processing inside epitaxial induction oven.

Semiconductor Eutectic Chip Bonding

Gold-silicon eutectic makes the most reliable chip-to-substrate bond in semi-conductor manufacturing. However, temperature control of the process is quite critical. The eutectic flows at 385°C and the tolerance is $-0° +20°C$. Outside said tolerances the chip might just be "tacked" or sitting above voids, or, at the high end, the gold would begin alloying into the silicon and spoil the doping.

A very thin optical fiber threaded through the collet hole needed to create the suction holding the chip to the base of the collet allows the infrared detector to "see" the chip during the bonding operation. At the instant when the eutectic flows, a large increment of the infrared radiation emitted by the chip's upper surface signals that the optimum temperature has been reached and, through a feedback loop, the process is terminated.

Advantages offered by the fiber optics infrared approach are as follows:

1. Fiber optics can be threaded into the collet in a permanent, unobtrusive setup.
2. Real-time response enables precise control of the bonding process by signaling the precise time when scrubbing must start and when a good bond is made.
3. High speed of operation allows faster rate of heat transfer from substrate to chip.
4. It ensures reliable bonds of semiconductor chips to substrate.
5. It elimintes the high-skill operator requirements and allows fully automated mass production.

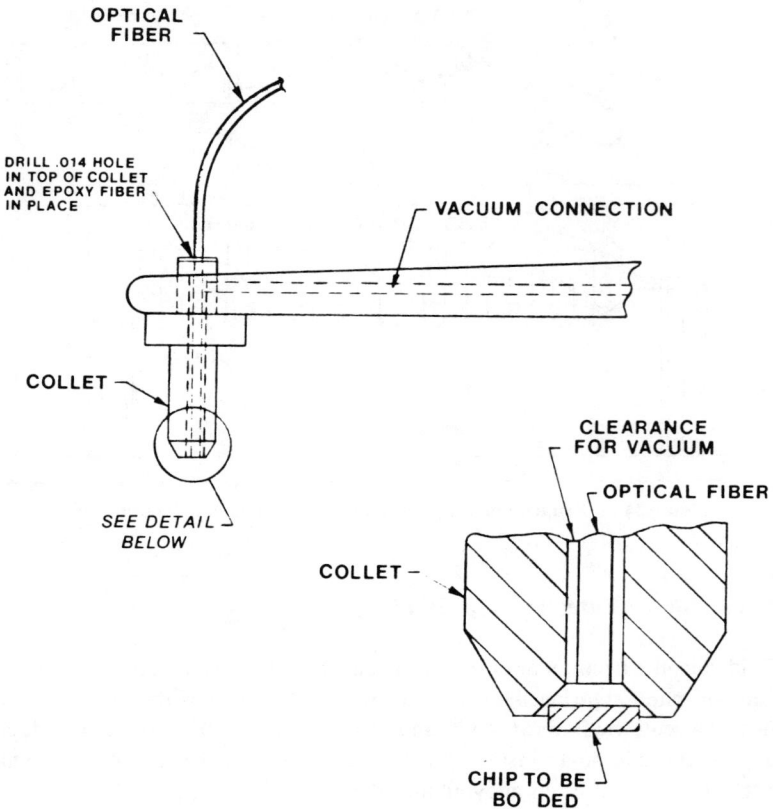

Figure 35.   Eutectic bonding of semiconductor chips onto substrate.

Figure 35 shows the detail of a typical fiber installation in the collet of a eutectic chip bonder.

Polymer Extrusion and Injection

Precise measurement and control of melt temperature is essential to ensure optimum length of the polymer molecule and, consequently, to maximize the physical properties of the product. Until now, thermocouples have been used to obtain an indication of the polymer temperature, but with the large errors due to the heat from the friction of the polymer against the thermocouple's protective capsule and the latter's thermal connection to the heated barrel. To accurately

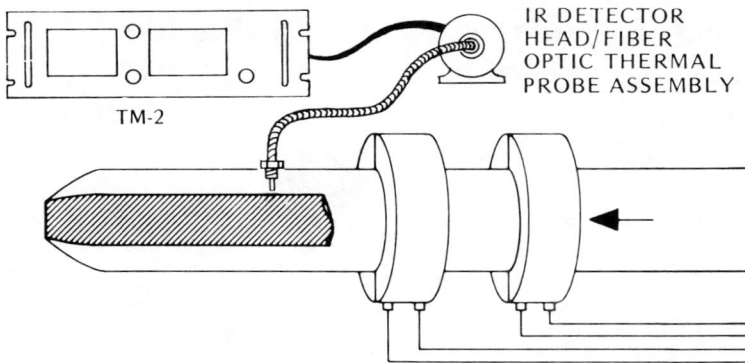

Figure 36.   Polymer extrusion and injection.

measure the melt temperature, a fiber optic probe with a high-pressure window is inserted into the barrel flush with its inside wall, so that it will measure only the temperature of the polymer flowing in front of it. If the melt temperature is at least 10% higher than the barrel's temperature, a standard polymer probe can be used. If the melt temperature is equal to, or lower than, the barrel's temperature, an air-cooled polymer probe must be used.

Advantages of the fiber optics infrared approach, which is shown schematically in Figure 36 are as follows:

1. Melt temperature measurements are accurate.
2. The response is 1000 times faster than thermocouples.
3. Since polymers are partially transparent in the infrared spectrum, the system measures average temperature to a depth between 5 and 20 mm into the melt, thereby avoiding errors caused by the barrel's temperature effect on the interface between the melt and the barrel.
4. A feedback loop ensures automatic temperature control within narrow tolerances.

Plastic Castings Ejection

Plastic material injected into molds must be ejected during its cooldown process as soon as it has solidified to the desired hardness. This is achieved by special "ejector pins" equipped with optical fibers lined along their core. Through these an infrared detector can "see" the plastic material injected into the mold. Thanks to the partial infrared transparency of the plastic, the radiation reaching

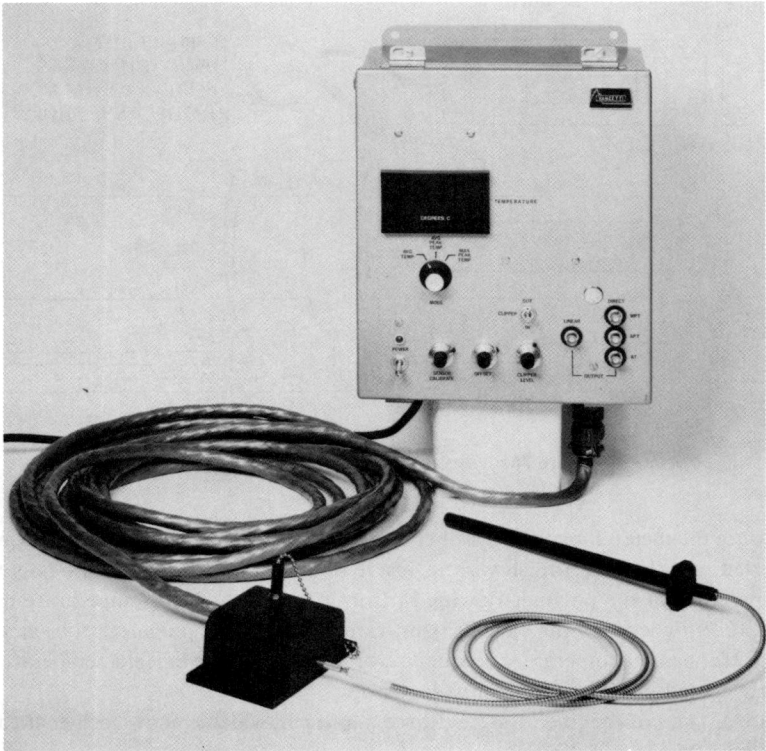

Figure 37.   Turbine blade thermal monitor.

the detector carries thermal information not only from the area contacting the fibers, but also from a certain depth within the plastic material (approximately between 5 and 20 mm).

Advantages offered by fiber optics infrared controllers are as follows:

1.   There is real-time temperature measurement in depth.
2.   The high speed of response enables material ejection at the precise moment of cooling.
3.   Fiber optics are not affected by mold temperature.
4.   They compensate for wrong temperature of the molds.

Mold Temperature Monitoring in Glass Production

The temperature of the mold into which the glass is blown is of critical importance because of the following:

1. If the mold is too hot and is opened by time sequence, the glass just molded will not have developed a hard skin and will sag.
2. If the mold is too cold, the glass will have stresses, causing it to shatter.
3. If the mold is too cold, the glass will tend to harden too quickly, thereby resulting in voids in the product.

By use of a fiber optic thermal monitor controller, the fiber optics can be focused on the inside surface of the mold that comes in contact with the molten glass during the time interval when the mold is open. The advantages of fiber optics are: (1) easy placement of the fiber optic to view the inside of the mold; and (2) long length of fiber to keep the infrared detector away from the heat of the molds.

Turbine Blade Temperature Monitoring (Figure 37)

During operation, the blades of the turbine rotors must be kept cooled constantly by circulating air through special channels located inside each blade. In the event that in one or more blades those channels should become blocked, overheating of said blades would develop, possibly reaching the softening temperature of the steel of which they are made. At this point, centrifugal force would produce deformations that could cause quick destruction of the turbine.

Optical fibers introduced through the outside "skin" of the turbine at a convenient location and with the necessary orientation will allow an infrared detector to "see" the rotor blades as they traverse, one by one, its field of view (Figure 38). In this way, thanks to its microsecond response time, the detector will be able to precisely measure the temperature of each and every blade.

In the event that even a single blade should exceed a preestablished temperature safety threshold, an alarm signal will appear at the system's output, and a feedback loop could automatically throttle down the turbine and avoid a catastrophic failure.

Spot Welding

Until now, spot welding operations were controlled by adjusting or setting pressure, time, voltage and current. However, the key parameter, temperature, could not be measured because the metal where the "nuggett" was being formed could not be reached by any temperature-measuring device.

Not any longer. Optical fibers introduced inside one of the welding electrodes are now allowing an infrared detector to "look" at the center of the area where the "nuggett" is being formed. Feedback electronics make instant-by-instant current corrections to ensure that the temperature development of the nuggett matches an optimum thermal profile stored in the controller's memory. In this

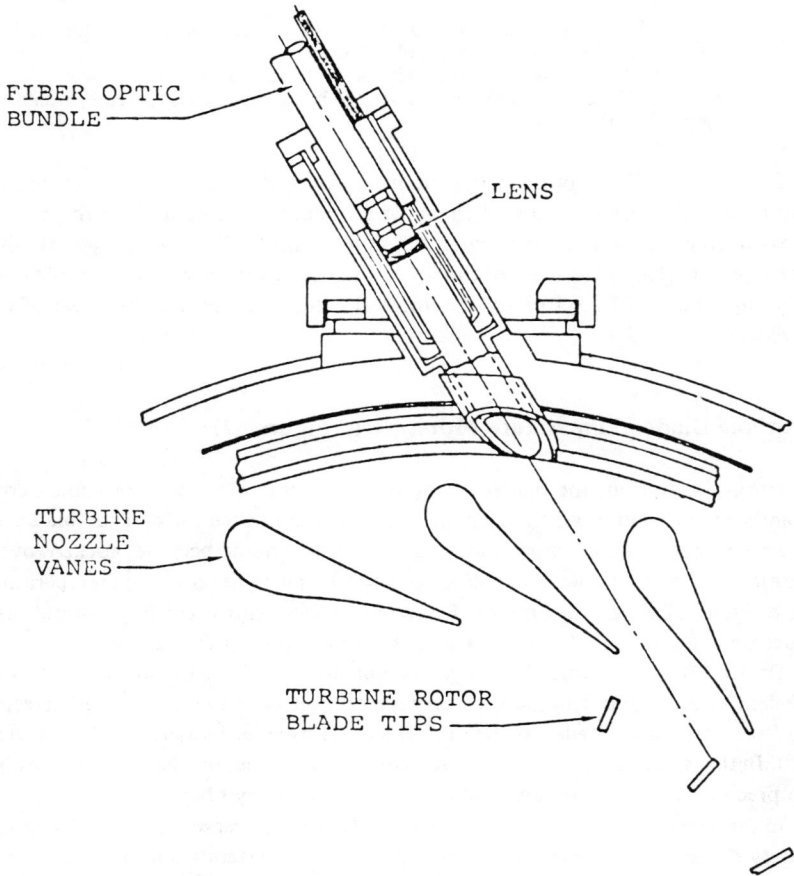

Figure 38.   Schematic of fiber optics installation in turbine.

way, a perfect spot weld is produced every time. Figure 39 is a schematic diagram of such a control system, while Figure 40 shows the welding electrodes detail and the optical fiber insertion in the upper one.

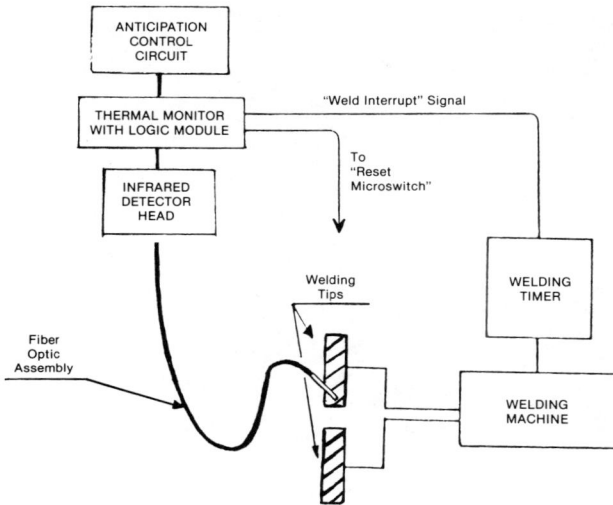

Figure 39.   Schematic of spot welder controller.

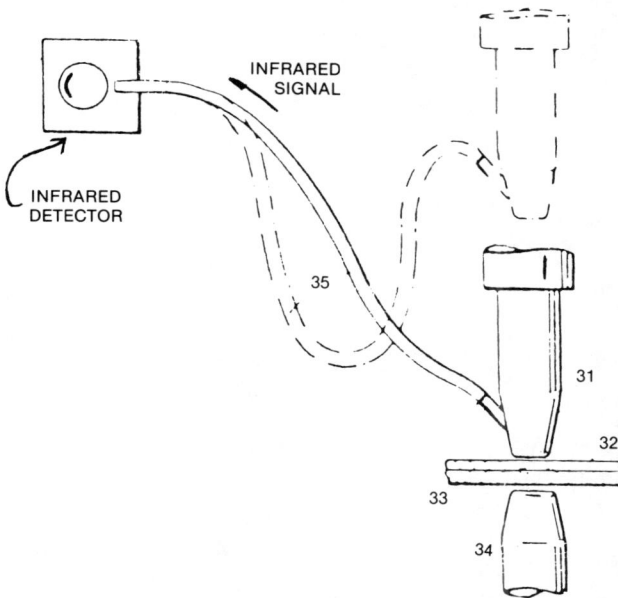

Figure 40.   Detail of spot weld electrodes with fiber optics.

CHAPTER 14

GAS PRESSURE REGULATION

Floyd D. Jury
Engineering Department
Fisher Controls Company
Marshalltown, Iowa

THE BASIC REGULATOR

The primary function of any gas regulator is to match the flow of gas through the regulator to the demand for gas placed on the system. At the same time, the regulator must maintain the system pressure within certain acceptable limits. In effect, the regulator must serve as a buffer between a relatively high pressure system and a lower pressure system.

A typical gas pressure system might be similar to that shown in Figure 1, in which the regulator is placed upstream of the valve or other load device that is varying its demand for gas from the regulator. In a system such as this, attention is too often focused only on the regulator. Actually, the term "system" includes not only the regulator, but the load device, as well as the interconnecting piping between them. These things work together as a comprehensive system.

If the flow demanded by the load decreases, then the regulator flow must decrease also. Otherwise, the regulator would put too much gas into the system and the pressure, P_2, would tend to increase. On the other hand, if the load flow increases, then the regulator flow must also increase to keep P_2 from decreasing due to a shortage of gas in the pressure system. The rate of change of P_2 in either case will depend on the characteristics of the load, the piping and the regulator.

The prime job of the regulator is to put exactly as much gas into the piping system as the load device takes out. If the regulator were capable of instantaneously matching its flow to the load flow, there would never be any major transient variation in the system pressure due to rapid load changes; however, in most real-life

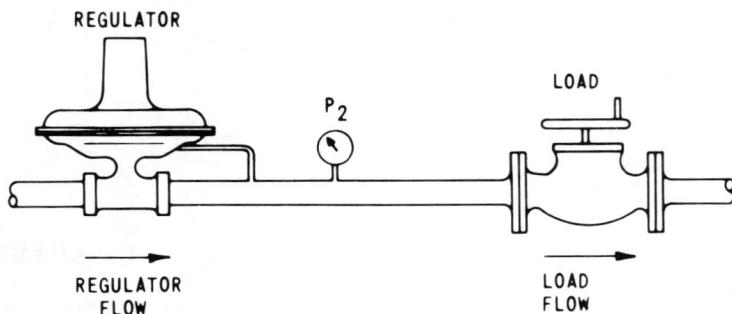

Figure 1.   Typical regulator system.

applications we would expect some fluctuations in P_2 whenever the load changes abruptly. How well the regulator is capable of performing under these dynamic situations is one of the questions to ask when selecting a regulator for a given application. Since the regulator is only part of the total system, it is imperative to know something about the characteristics of that system to obtain the best possible dynamic performance.

Transient response is the term used to describe how rapidly the regulator system responds to a load change. Transient response is not just a function of the regulator. The system behavior is greatly influenced by the process that the regulator is trying to control. A pressure process that has a large volume of gas in the piping significantly slows the response of the system to upsets. In other words, one must talk about system response, rather than regulator response. Transient response is an important characteristic of the system, especially in shock load applications; however, it is not the only important dynamic characteristic. Among others, one must be concerned about the stability of the complete system. The regulator is only part of the overall system and does not operate independently of the other components.

In addition to the dynamic performance of the regulator system, one must also be concerned about several static performance characteristics when selecting a regulator for a system. Some of the questions that should be asked are: How much droop in the system pressure will occur for a 100% load change? Is the regulator capacity sufficient to pass the maximum flow required from the system? How much noise will the system generate?

These dynamic and static performance characteristics must be augmented by desirable operational characteristics as well. For example, is the system reliable, safe and relatively easy to maintain? Is there adequate overpressure protection?

Finally, all of these characteristics must be balanced against the cost of the system. Routine operating costs must be considered as well as the initial purchase cost.

To intelligently evaluate all of these factors, it is imperative to thoroughly understand just what makes the system tick. Let us begin by taking a close look at the regulator and how it works.

Since the regulator's job is to modulate the flow of gas into the system, one can see that one of the essential elements of any regulator is a *restricting element* that will fit into the flow stream and provide a variable restriction that can modulate the flow of gas through the regulator. This restricting element is usually some type of valve arrangement. It can be a single- or double-port globe valve, a cage-style valve, rotary valve or any other type of valve capable of operating as a variable restriction to the flow. The schematic arrangement in Figure 2 shows a typical regulator restricting element.

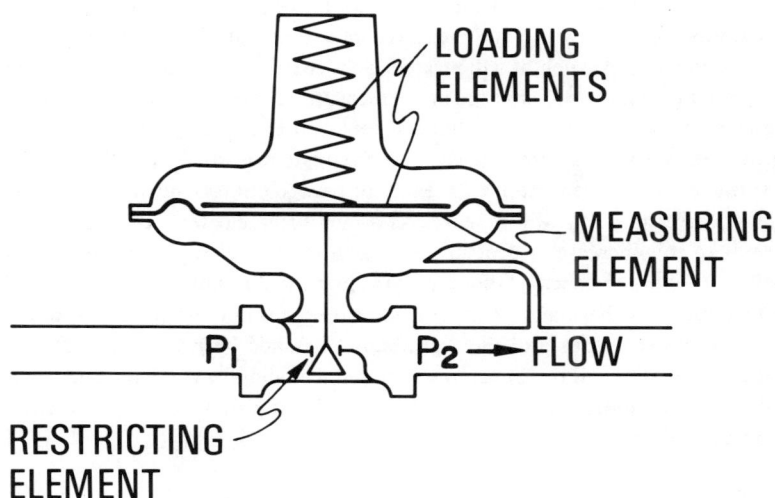

Figure 2. Regulator elements.

To cause the restricting element to vary, some type of loading force will have to be applied to it. Thus, the second essential element of a gas regulator is a *loading element,* which applies the needed force to the restricting element. The loading element can be one of any number of things, such as weights and levers, hand-jacks, springs, diaphragm actuators or piston actuators to name a few of the more common ones.

A diaphragm and a spring are frequently combined, as shown in Figure 2, to form the most common type of loading element. A loading pressure is applied to a diaphragm to produce a loading force that will act to close the restricting element. The spring provides a loading force to oppose the pressure force. Springs are popular loading elements because they provide a positive, dependable force under nearly all conditions.

The final element required to complete the regulator is a *measuring element,* which is necessary to ensure that the gas flow is being modulated correctly. If the restricting element allows too much gas into the system, P_2 will increase. If the restricting element allows too little gas into the system, P_2 will decrease. On the other hand, if the regulator flow is perfectly matched to the load flow, P_2 will remain constant. This convenient fact provides a simple means of determining whether the regulator is providing the proper flow.

Manometers, Bourdon tubes, bellows, pressure gauges and diaphragms are some of the possible measuring elements that can be used. Depending on the desired goal, some of the measuring elements would be more advantageous than others. The diaphragm, for instance, will not only act as a measuring element that will respond to changes in the measured pressure, but also simultaneously as a loading element. As such, it will produce a force to operate the restricting element that varies in response to changes in the measured pressure. If this typical measuring element is added to the loading element and the restricting element selected earlier, we will have a complete gas pressure regulator, as shown in Figure 2.

If the restricting element of this regulator tries to put too much gas into the system, the pressure, P_2, will increase. As a measuring element, the diaphragm responds to this increase in pressure and, as a loading element, produces a force that compresses the spring, thereby restricting the amount of gas going into the system. On the other hand, if the regulator does not put enough gas into the system, the pressure fails and the diaphragm responds by producing less upward force. The spring will then overcome the reduced diaphragm force and open the valve to allow more gas into the system. This type of self-correcting action is known as negative feedback.

RESTRICTING ELEMENTS

The restricting element undoubtedly will be some type of valve. There is quite a variety of valve styles to choose from, each with its own unique set of advantages and disadvantages. Regardless of the style of valve used, however, its basic purpose is to form a restriction to the flow. It forms a bottleneck that allows conversion from a high-pressure system to a low-pressure system.

The flow through the valve obviously will be a function of the flow area of the valve. This does not mean that the flow will be proportional to the valve travel, however, since the flow area may not change linearly with valve travel. By carefully shaping the internal geometry of the valve, the flow versus travel relationship can be made to assume a variety of forms. This effect is known as valve characterization.

In addition to flow being a function of the valve area, it is also a function of the valve pressure drop. The pressure differential that exists across the valve represents a difference in potential energy that causes the gas to flow just as water flows—downhill. Common sense tells us that if this pressure differential increased across the valve, one should thereby increase the flow of gas through the valve. In actual practice, this is true only up to a certain critical point.

By looking at a typical regulator installation, such as in Figure 3, one can analyze how the pressure varies along the length of the valve under steady-state flow conditions, with the valve holding steady in its wide open position.

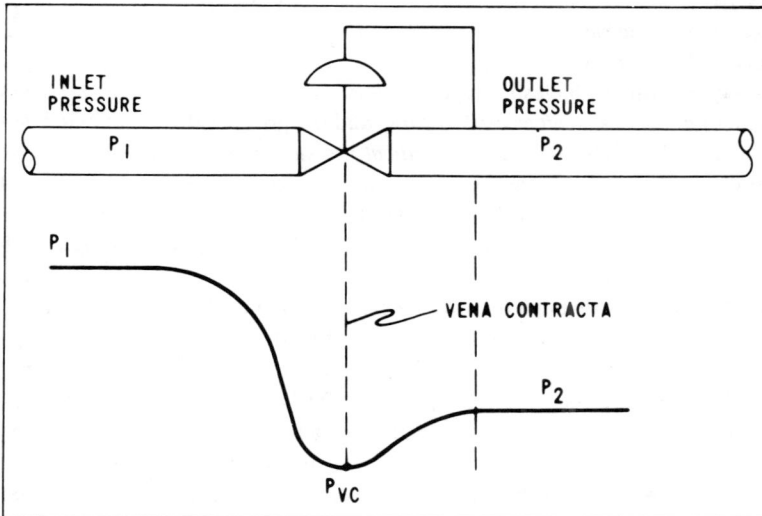

Figure 3.   Valve pressure profile.

The valve forms a restriction in the line; consequently, its flow area is smaller than that of the surrounding pipe. To maintain a steady-state flow through this system, there must be as much gas flow through the small area at the restriction as in the larger pipe area. It is obvious then that the *velocity* of the gas must be much greater at the restriction to maintain this steady flow of gas. The velocity will be the greatest at the point of maximum restriction of the flow stream. This point is called the vena contracta and is normally located a short distance downstream of the actual orifice.

This increase in velocity represents an increase in kinetic energy that must come at the expense of the potential energy, which is represented by pressure. Thus, Figure 3 shows that the pressure decreases to a minimum at the vena contracta, where the velocity is the greatest. As the gas slows down again in the larger downstream piping, some of the pressure that was lost is regained. This is called pressure recovery.

To increase the flow through the valve, one could simply open the valve and increase the flow area. If this is not possible because the valve is already wide open, then the flow can be increased only by increasing the velocity. This is normally done by lowering the downstream pressure and taking a larger pressure drop across the valve. Of course, this increases the velocity of the gas at all points in the system. As the flow continues increasing, a condition occurs in which the gas velocity reaches the speed of sound at the vena contracta. Since gas will not normally travel faster than this limiting sonic velocity, the point has been reached where the volume rate of gas flow through the valve can no longer be increased simply by lowering the outlet pressure.

This condition in which sonic velocity occurs at the vena contracta and the flow becomes limited is known as *critical flow,* and the pressure drop that exists across the valve at that point is known as *critical pressure drop.*If one were to increase the pressure drop beyond this critical point by lowering the outlet pressure of the valve, no further increase in flow through the valve would occur. The actual value of the critical pressure drop varies considerably depending on the style and flow geometry of the valve.

As the gas passes through the valve, a certain amount of turbulence occurs that results in an energy loss within the valve. Part of the kinetic energy of the gas is changed into heat energy and part of it is changed into noise energy. A valve such as a ball valve or butterfly valve, which has a fairly streamlined flow pattern, has a minimum energy loss in the valve resulting in a relatively high downstream pressure recovery. This type of valve would be called a high recovery valve and is shown in Figure 4 as the solid curve. On the other hand, a valve such as a double-ported globe valve, which has a relatively turbulent flow pattern, will have a fairly high energy loss, which means poor downstream pressure recovery. This type of valve would be called a low-recovery valve and is shown as the dashed curve in Figure 4.

The pressure differential between the inlet and the vena contracta $(P_1 - P_{vc})$ is a direct measure of flow regardless of the style of valve, whereas the pressure drop observed across the valve $(P_1 - P_2)$ is highly dependent on valve style. Thus, if the two valves plotted in Figure 4 have *equal flow areas,* they would also have *equal flow,* since the pressure drop that determines the flow $(P_1 - P_{vc})$ is the same for both valves, even though the observed pressure drops across the valves $(P_1 - P_2)$ are quite different. The observed pressure drop across the valve $(P_1 - P_2)$ is a direct measure of the energy loss in the valve.

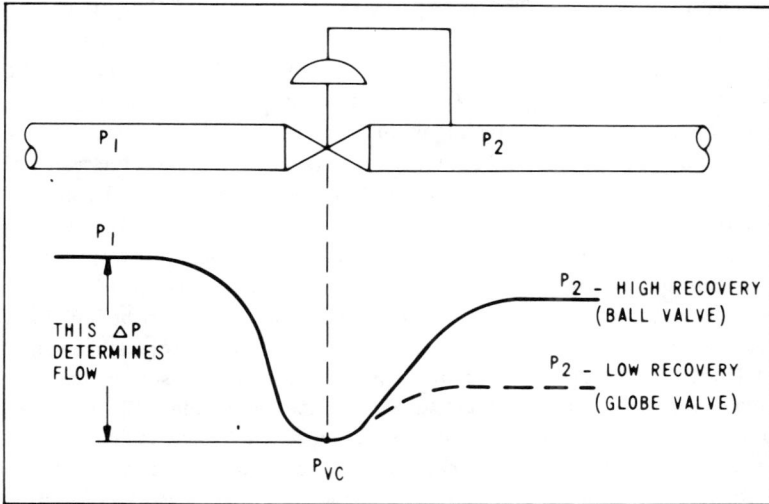

Figure 4.   High and low recovery valves.

Since it is quite difficult to measure the pressure at the vena contracta, as a practical matter we must resort to using the ΔP across the valve as a measure of the flow instead of $(P_1 - P_{vc})$. Since the ΔP across the valve is highly dependent on valve style, it is obvious that experimental means must be used to find the relationship between this pressure drop and the flow for any given style of valve. This is the reason for the many tables of valve sizing data published by the valve manufacturer.

It was indicated earlier that, in general, the flow could be increased by increasing the pressure drop across the valve; however, when this pressure drop is increased by lowering the downstream pressure, a critical condition is eventually reached where the flow no longer increases because the limiting sonic velocity has occurred at the vena contracta.

As it turns out, even after the critical flow condition has been reached, one can continue to increase the flow through the valve simply by increasing the inlet pressure to the valve. There would still be sonic velocity at the vena contracta and critical flow through the same flow area; however, by increasing P_1, the *density* of the gas entering the valve has been increased, thereby packing more *standard* cubic feet of gas into each actual cubic foot of volume passing through the regulator. In effect, the cubic feet per hour (cfh) of flow has not changed, but the standard cubic feet per hour (scfh) has increased. A standard cubic foot of gas is that amount of gas that will occupy a volume of one cubic foot under standard conditions of 60°F and 14.73 psia.

LOADING ELEMENTS

Except for a few of the older weight-loaded units, almost all gas regulators have springs. In fact, springs and diaphragms are by far the most universally used loading elements in modern gas regulators.

Spring rate is a measure of the stiffness of a spring. The greater the spring rate, the stiffer the spring. As the compression of the spring is increased, the amount of required compressive force increases a proportional amount. If the spring has been designed according to good design principles, this linear relationship will continue until the spring coils go solid. The spring normally is considered to be a linear device, but for this to be true the compression range must be limited. The spring designer specifies a maximum compression and force for each spring that should not be exceeded. Within these limits, the regulator spring rate can be treated as a constant. This spring rate, K, is defined mathematically as a number of pounds of force, F, necessary to compress a spring by one inch.

As useful as a spring is, it normally provides a loading force in one direction only. Energy is required to overcome the spring and provide a loading force in the opposite direction. This energy usually comes from the force developed by pressure acting on the area of a diaphragm loading element.

The diaphragm area for a given device usually can be obtained from the manufacturer. When using the diaphragm as a loading element, one of the main considerations is to have enough diaphragm area to develop sufficient force to compress the spring through its full travel and to properly seat the valve.

MEASURING ELEMENTS

A variety of measuring devices are available, but many of them, such as Bourdon tubes and bellows, have the disadvantages that they generate very small forces and require supplemental equipment to operate the system.

By far the most universally used gas pressure measuring element is the diaphragm. There are several reasons for this. The diaphragm is simple, very economical, highly versatile, easy to maintain and has the definite advantage of not requiring any additional equipment to supplement its action. The same diaphragm that serves as a measuring element also can serve as a loading element with no intermediate hardware being needed.

A diaphragm is simply a piece of fabric coated with a rubber-like material. The coating provides a seal to contain the pressure while the fabric gives it enough strength to withstand the pressure. Some diaphragms are molded into special shapes, but many are cut from flat sheets.

Because of its extreme simplicity, the diaphragm frequently is taken for granted and does not receive the theoretical attention it deserves. For example, it is important that the gas man be able to determine properly the diaphragm area that the pressure will be acting on.

The loading pressure on the diaphragm acts over the entire exposed surface. The diameter of the exposed diaphragm area, which the pressure acts on, is the same as the inside diameter of the upper casing. It is a mistake to conclude, however, that all of this area is useful in terms of providing a loading force. On the other hand, one should not assume that the only useful area is the area of the diaphragm plate just because it is the only place where the pressure is really pushing down on the spring assembly. The actual answer lies somewhere between these two extremes.

Consider the small portion of the diaphragm between the diaphragm plate and the case flange. Because the diaphragm is unsupported here, the pressure acting on it takes up the slack in the diaphragm and shapes it into a rounded convolution. This pressure is not only distributed uniformly over the surface, but it also acts perpendicularly to the surface at all points.

As the pressure shapes the diaphragm convolution, there is only one point—at the bottom of the cup shape—where the slope of the diaphragm is horizontal. This can be identified as the point of horizontal tangency.

By making a force analysis of the diaphragm assembly, it can be shown theoretically that the diaphragm area outside of this point of horizontal tangency is not an effective area in terms of producing thrust or motion. Thus, when using the diaphragm area in various regulator calculations, one must be sure to use the *effective area,* which is determined by using the diameter between the points of horizontal tangency on the convolution.

It is easy to assume that the point of horizontal tangency falls in the center of the convolution, but this is only true for one position in the stroke where the diaphragm plate is level with the flange. At other positions in the stroke, the effective area changes as the point of horizontal tangency moves inward or outward. It is easy to visualize how this effective area changes if the diaphragm convolution is simulated by holding a slack piece of string between outstretched hands. As one hand is moved up and down to simulate the motion of the diaphragm plate, the effective area relationship will become readily apparent.

As the diaphragm moves to *compress* the spring, the effective area *decreases.* As the diaphragm moves to *relax* the spring, the effective area *increases.* This relationship will remain at your fingertips if you remember that the effective area decreases to its least amount just when it is needed most to help compress the spring!

Because of the decrease in diaphragm area with increasing spring compression, it is important to make certain that the minimum area is large enough to compress the spring through its full travel with enough reserve to provide any necessary shutoff force. This is an important enough consideration when using the diaphragm as a loading element, but even more important is the nonlinearity that is introduced when using the diaphragm as a measuring element.

The change in effective area with travel is more pronounced with the shallow convolutions of flat-sheet diaphragms than with the more expensive molded diaphragms that have deeper convolutions. This can be illustrated with the string mentioned earlier. Molded diaphragms are frequently used where it is important to minimize the change in effective area with travel and thus improve the performance of the regulator.

Before leaving the subject of diaphragms, there are some interesting practical considerations that deserve attention. When replacing regulator diaphragms one should make certain that the material used is suitable for the application. Many quite different diaphragm materials are very difficult to tell apart by casual inspection. Diaphragm fabrics are typically made from cotton, nylon or Dacron,® each with its own physical characteristics. In addition, the type of weave employed in the fabric can significantly affect its stretch and strength.

There are also a variety of materials used to coat the diaphragm fabric. Diaphragm materials must be carefully selected to provide the proper temperature characteristics, strength, hysteresis and chemical compatibility. The coating on the diaphragm may be thinner on one side than on the other. The side with the thinner coating is usually identified by having some design or pattern printed on it. When a diaphragm of this type is used, it should be installed so that the pressure is applied to the nonpatterned side, the side with the thicker coating. The printed side goes against the diaphragm plate.

The storage of diaphragms also requires a certain amount of care. Some diaphragms will dry rot on the shelf due to ozone exposure. Ideal storage is in a plastic bag in a cool, dry location away from sunlight. Manufacturers frequently recommend that a diaphragm be placed in service within 18 months from the time it is received.

Finally, a word about installing diaphragms. On flat-sheet diaphragms the bolt circle diameter is larger than for the regulator casing flange to allow enough slack for the operation of the regulator. When installing these diaphragms, care must be taken to avoid folds or wrinkles that could cause leakage at the flange.

The clamping force holding the diaphragm is also important. The diaphragm flange should not be made as tight as a line flange or the diaphragm may be crushed or the casing warped. The pattern of bolt tightening is very important. A cross-corner tightening sequence should be used for better results. Consistency is more important than the actual tightness as long as an adequate seal is achieved.

DROOP

Figure 2 shows a typical regulator arrangement using spring-and-diaphragm loading elements. In an arrangement such as this, it is common for the spring to be installed in such a manner that some initial compression exists in the spring. With no loading pressure applied to the diaphragm, this initial spring compression holds the valve in the wide open position.

To close the valve, a loading pressure, P_2, must be applied to the diaphragm. Because of the initial compression in the spring, the loading pressure must increase to some initial value before the valve beings to move. The magnitude of this starting pressure depends entirely on the amount of initial compression that was wound into the spring. For purposes of discussion, arbitrarily assume that the system in Figure 2 requires a starting pressure of 20 psig to overcome the initial compression of the spring. If the system pressure loading the diaphragm increases above 20 psig, the valve will begin to close. The effective area of the diaphragm, the stiffness of the spring and the amount of valve travel will determine how much the loading pressure must increase to fully close the valve. Again, for purposes of discussion, arbitrarily assume that the system components in Figure 2 are such that a loading pressure increase of 5 psi is required to fully stroke the valve.

In normal operation the valve may range from its fully open to its fully closed position. Correspondingly, the system pressure, P_2, will range from 20 psig to 25 psig. In other words, when the valve is closed the system pressure will be 25 psig; when the valve is open the system pressure will be 20 psig. This may seem rather elementary, but this simple fact is the key to understanding regulator operation.

In a typical operation this regulator would be installed in a system, such as in Figure 1, so that it would supply gas to a downstream load. When the load demand is small, the regulator valve must operate in its nearly closed position. In this position, we have already seen that the system pressure loading the diaphragm must be 25 psig.

As the load demand for gas increases, the regulator valve must open up to supply the needed gas. To do this, the pressure on the diaphragm, i.e., P_2, must decrease. If the load demand increases sufficiently, the regulator valve will go wide open. The pressure, P_2, acting on the diaphragm must then be 20 psig when the valve is in the wide open position.

In summary, as the load demand changes from one extreme to the other, the controlled pressure acting on the diaphragm will range between 25 psig and 20 psig. The value of the controlled pressure, P_2, at the low load condition, i.e., 25 psig, is known as the set point pressure. As the load increases, P_2 decreases a proportionate amount from the set point pressure.

The decrease in controlled pressure that occurs as the load on a regulator increases is called droop. It is really the decrease in P_2 that must occur to allow the valve to open. Regulator droop is a direct result of spring compression and change in the effective area of the diaphragm. For this reason, it is frequently referred to as spring and diaphragm effect.

The amount of droop that occurs in the controlled pressure from a low load flow condition to a full load flow condition is sometimes called the proportional band of the regulator.

In the present example, the pressure has to decrease from 25 psig at low load to 20 psig to open the valve plug sufficiently to pass the full load flow. Furthermore, since this 20 psig is the pressure that will just hold the valve plug in the open position, this value of the pressure will have to continue as long as the high load flow condition exists. The controlled pressure, P_2, will only return to the original 25 psig set point when the load flow demand returns to its original low value at set point conditions. In other words, there is a steady-state droop of 5 psi at full load.

The 20 psig starting pressure on the diaphragm was determined solely by the amount of initial compression wound into the spring when it was installed. If we change this initial compression, we will, of course, change the starting pressure to something other than 20 psig.

On the other hand, the 5 psi pressure change required to stroke the valve is dependent on the diaphragm area, the spring stiffness and the valve travel. The amount of initial compression in the spring has nothing to do with it. In other words, for a given set of regulator components, the full load droop, or proportional band, remains essentially the same regardless of the set point pressure. For this example, the controlled pressure may range from 25 to 20 psig, 30 to 25 psig or 20 to 15 psig, but the droop will always be 5 psi.

Thus, changing the initial compression in the spring has the effect of changing the set point pressure for the regulator without affecting the amount of droop. In the illustration above, the set point could be 25, 30 or 20 psig, respectively, but the droop remains 5 psi.

Obviously, there is a limit to the adjustability range of the set point for any given regulator configuration. If one tries to put too much initial compression into the spring, the coils will be compressed solid before the valve can travel closed. This puts an upper limit on the set point pressure. At the other extreme, there is also a limit to how little initial spring compression can be tolerated. This initial compression cannot be reduced to zero without causing possible nonlinearities and problems with parts shift when the spring goes slack.

To obtain a set point pressure that is outside the range of adjustability for the regulator being used, the normal procedure is to change the spring to one of slightly different design. The manufacturer normally offers a selection of spring ranges for any particular regulator design.

As long as a spring exists in the regulator there will be some droop due to spring effect. Furthermore, the droop at any load condition will be proportional to the load flow because the size of the load flow determines how far the valve must open; this, in turn, determines the amount of droop in the pressure.

Because of this proportional relationship between droop and load flow, the spring effect is frequently referred to as proportional action. We have a proportional control system. As stated before, this leads to calling the full load droop the proportional band of the regulator.

If the effective area of the diaphragm remains constant with travel, the droop will be due strictly to spring effect; however, many regulators have flat-sheet diaphragms whose areas vary appreciably with travel, i.e., the effective area *decreases* as the diaphragm moves to *compress* the spring and close the valve. This decrease in effective area means that an even greater pressure change on the diaphragm will be required to compensate for the lost area. Otherwise, it will be unable to develop enough force to compress the spring.

The total pressure change required to stroke the valve is made up of two parts. The major part, required to compress the spring, is called spring effect. The remaining minor part required to compensate for the change in effective area of the diaphragm is known as diaphragm effect.

REDUCING DROOP

From a performance point of view, it is desirable to keep the proportional band of the regulator as narrow, or as small, as possible. Actually, there are a number of things that can be done to improve the regulator performance by reducing droop.

A first approach might be to minimize or eliminate the diaphragm effect. This can be done by using a molded diaphragm whose area is relatively unchanged with travel. The major disadvantage of this approach is the higher cost. Also, the improvement may be minimal since the spring effect is usually the major cause of droop.

The next most sophisticated approach to eliminating diaphragm effect is the use of a roll-out diaphragm. Actually, the roll-out diaphragm not only eliminates diaphragm effect, but also tends to compensate for a small part of the spring effect. The roll-out diaphragm is especially constructed so that its effective area actually *increases* with spring compression, rather than decreasing as with a standard diaphragm. The roll-out diaphragm is even more expensive than the conventional molded diaphragm, so its use tends to be limited, especially since it still leaves a major part of the spring effect.

The spring effect cannot be eliminated without eliminating the spring. Theoretically, the spring effect could be minimized and the proportional band narrowed by installing a spring with a lower spring rate, but this might introduce problems with valve plug chattering because of the lower stiffness. Another way might be to install a larger valve size so that the valve travel can be reduced. Also, a different regulator could be installed whose effective diaphragm area is greater. These steps would theoretically reduce the proportional band, although they would likely cause problems by exceeding the spring's capacity for travel or initial compression. The last two methods also have the disadvantage of requiring a major change in hardware.

Weight-loaded regulators, such as in Figure 5, provided the first attempt at eliminating the spring effect by eliminating the spring. Some regulators of this type used the weight directly on the diaphragm plate, while others used a system of levers to multiply the force developed by the weight.

Weight-loaded regulators can produce good steady-state control with very little droop, but they have very poor dynamic response and tend to be unstable. The inertia effect of the weights and levers causes the regulator to respond very sluggishly to a rapid load change, and once it begins to move it tends to overshoot, thus causing a hunting or unstable condition.

Another way to produce a constant loading force, similar to the weight yet without the inertia problems, is to use a pressure-loaded regulator similar to Figure 6.

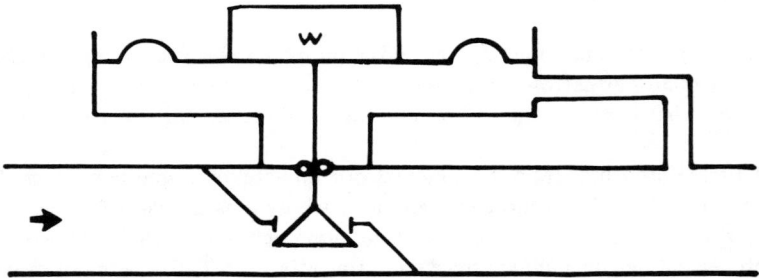

Figure 5.   Weight-loaded regulator.

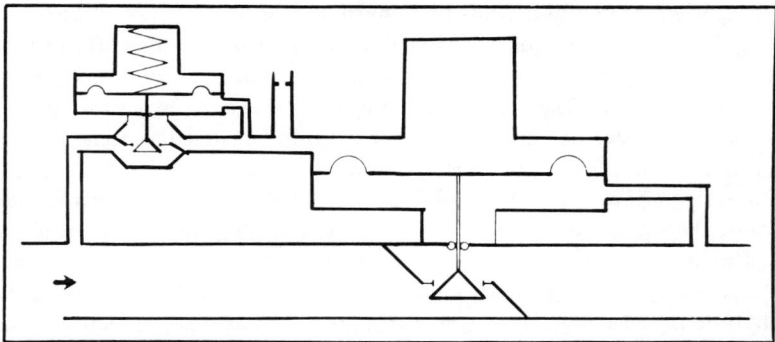

Figure 6.   Pressure-loaded regulator with atmospheric bleed.

A small auxiliary regulator maintains a constant pressure on top of the main regulator diaphragm; thus the controlled pressure, which is opposing it on the bottom of the diaphragm, must also remain constant. This type of regulator can be very effective at eliminating both spring and diaphragm effect because there is no spring and the two pressures act on the same area, thus eliminating any effect of change in this area.

Unfortunately, there are also some important limitations to this pressure-loaded regulator's usefulness. One problem is the necessity to bleed the diaphragm loading pressure to the ambient atmosphere. This not only wastes gas but can constitute a safety hazard. If a supply of instrument air is available nearby, this problem can be solved by loading the diaphragm with air pressure. Another problem is that this type of regulator is prone to instability. In most practical designs of this type a light rate spring is used to provide some stiffness to the system. This improves the stability problem but also introduces some droop due to spring effect.

Nearly any conventional self-operated regulator can be easily modified in the field to operate as a pressure-loaded regulator with atmospheric bleed. All that is required is to introduce a loading pressure on the diaphragm to produce a loading force that assists the spring. This type of device is sometimes used in batch sequencing operations, where the process must operate at one pressure for awhile and then change abruptly to another pressure. A solenoid is used to switch the loading pressure in and out as required.

The problem of atmospheric bleed can be solved by using a pressure-loaded regulator with downstream bleed, as shown in Figure 7. For the gas to bleed downstream from the diaphragm casing, the pressure in the casing must be higher than the downstream pressure. This means that a spring will have to be introduced to help the downstream pressure balance the greater pressure in the diaphragm casing. This spring introduces some spring effect, but in general fairly close control can be achieved since the spring rate can be kept low. As in the other pressure-loaded regulator, the spring also helps keep the system stable.

As a result of the small downstream orifice, the dynamic response of this regulator is rather poor. Normally the pressure differential across the small bleed ori-

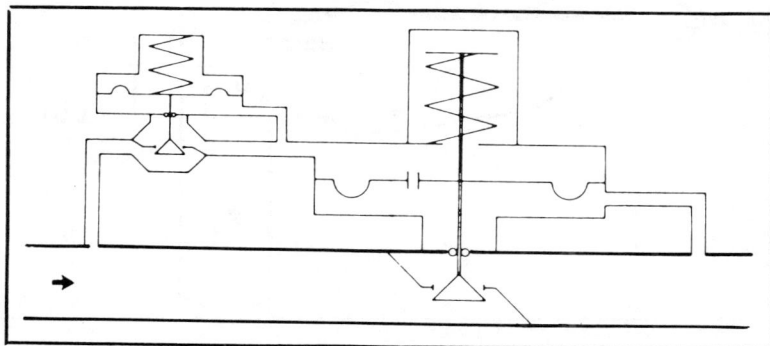

Figure 7. Pressure-loaded regulator with downstream bleed.

fice is quite low, resulting in relatively low flowrates through the bleed orifice. Thus, when the regulator experiences a shock load, it can only respond as rapidly as the pressure can equalize across this small orifice.

Another method that can be used to overcome droop is known as velocity boosting. In general, this method is not as versatile as some of the other methods, but it has the advantage of being relatively inexpensive and can be made to work very well under the proper circumstances. Velocity boosting is most frequently used in house service regulators and farm tap regulators.

Velocity boosting takes advantage of the fact that the vena contracta pressure occurs just a short distance downstream of the actual restriction in the regulator. Further downstream this pressure recovers to the value of P_2, which is the controlled pressure and the one normally applied to the diaphragm.

A regulator that employs velocity boosting is designed so that this controlled pressure is no longer applied to the diaphragm. Instead, a pitot tube, such as shown in Figure 8, or some similar design is arranged so that the lower pressure near the vena contracta acts on the diaphragm rather than using the higher controlled pressure, P_2.

As the load flow starts to increase, the sensed pressure at the pitot tube begins to droop just as P_2 does; however, since the sensed pressure is near the vena contracta and the gas velocity is greater there, this pressure, which is applied to the diaphragm, decreases more than P_2. Consequently, the valve is allowed to open

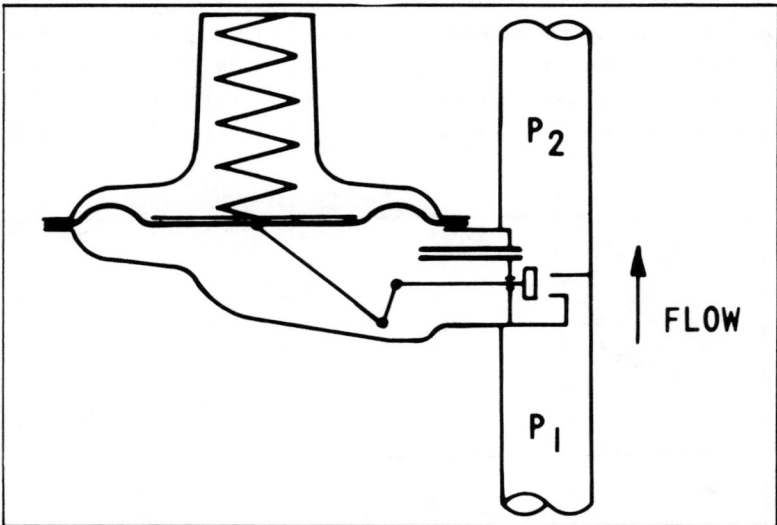

Figure 8.   Service regulator with velocity boost.

slightly wider than it would if P_2 were acting on the diaphragm. This has the effect of keeping P_2 relatively more constant and thus preventing a large droop with high load flow.

Spring and diaphragm effect still cause a drooping pressure at the vena contracta sensing point similar to the basic regulator shown as the dashed line in Figure 9, but this droop is overcome by the subsequent downstream pressure recovery of P_2. As a result, the performance curve of controlled pressure versus flow can be made relatively flat.

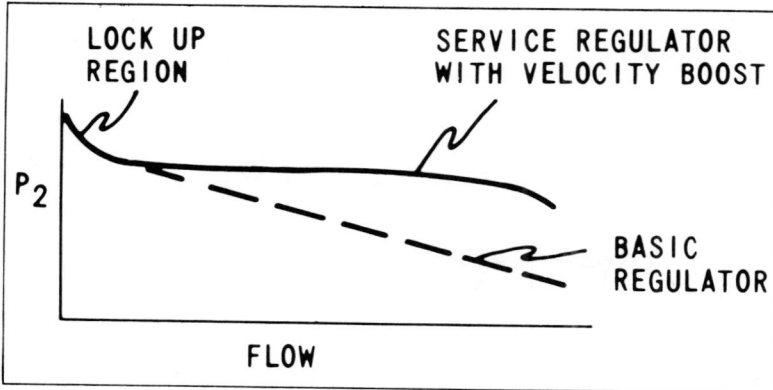

Figure 9.   Performance curves showing velocity boost effect.

All regulators discussed so far have been in the category of self-operated regulators. This means that the controlled pressure operates directly on the diaphragm to provide its own power to make the regulator respond to changes in load. In other words, the regulator operates itself using the controlled pressure directly. This common type of regulator can perform very well in many applications, but there are special situations in which the self-operated regulator has a proportional band that is too broad.

PILOT-OPERATED REGULATORS

Another important category of regulators is that group known as pilot-operated regulators. These regulators provide a more sophisticated and expensive method of reducing droop.

We have already seen that the spring rate, valve travel and effective diaphragm area are the three parameters that affect the proportional band. These factors determine how much pressure change is required on the diaphragm to fully stroke the valve. In the self-operated regulator, the controlled pressure and diaphragm pressure are the same, so the differential diaphragm pressure required to stroke the valve shows up directly as droop in the controlled pressure.

Changes can be made in these parameters to improve the proportional band, but there are practical limits to how much improvement can be obtained in this manner. The next logical step is to install a pressure amplifier in the measuring, or sensing, line, as shown in Figure 10. This pressure amplifier is frequently referred to as a pilot.

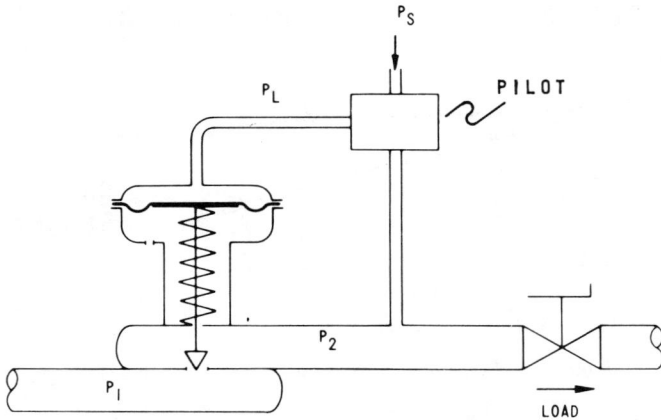

Figure 10.   Pilot-operated regulator.

The purpose of the pilot amplifier is to sense changes in the controlled pressure, P_2, and amplify them into larger changes in loading pressure, P_L, on the diaphragm. The amount of amplification obtained is called the gain of the pilot. A pilot amplifier with a gain of 20 means that a one psi change in the controlled pressure, P_2, would cause a 20 psi change in the loading pressure, P_L, on the diaphragm. To understand the significance of this device, we need to look at how the whole installation operates.

When the system load flow increases, the regulator valve must come open, just as in the case of the self-operating regulator. This is accomplished by decreasing the diaphragm pressure the same amount as would be necessary for the self-operated regulator. However, with the self-operated regulator, this entire amount of diaphragm pressure change shows up directly as droop in the controlled pressure. But with a gain of 20 in the pilot amplifier, the controlled pressure only needs to droop one-twentieth as much to get the same pressure change on the diaphragm. Thus, the proportional band of the regulator has been reduced by the amount of gain, or amplification, in the pilot.

With the rather sensational improvement in proportional band that can be achieved with this type of pilot-operated device, one might wonder why all regulators are not made this way. The answer is threefold: economics, stability and response.

Pilot-operated regulators are more expensive than similar self-operated regulators, and the improvement in proportional band may not be important enough to justify the increased cost. On the other hand, the gain of the pilot amplifier increases the gain or sensitivity of the entire pressure regulator loop. If this loop gain is increased too much, the loop can become unstable and the regulator will oscillate or hunt.

Since the feedback signal from the controlled pressure to the regulator must operate through the pilot, one can see why the response time of this type of device is much slower than the equivalent self-operated regulator. This means that its reaction to a shock load may be rather poor.

In summary, regulators employing direct, self-operated feedback tend to be stable with fast response, but also tend to have broad proportional band. On the other hand, regulators employing pilots in the feedback path tend to be less stable with poor dynamic response, but have the advantage of much better accuracy and less droop.

Although there is nothing wrong with it in principle, most practical regulators would never be constructed like the device shown in Figure 10. One of the major problems with this apparatus is that it can only be constructed with an atmospheric bleed. Since nearly all gas regulators use the process fluid as the operating medium for the pilot, it is disadvantageous to vent this gas to the atmosphere.

There are basically two different types of pilot-operated regulators that are in common use. They are the unloading system and the two-path control system. As its name implies, the two-path control system is much more sophisticated than just a simple pilot-operated device. This system's operation will be discussed in more detail later. Both the unloading and the two-path control system have the advantage of using a downstream bleed rather than venting gas to the atmosphere.

The pilot-operated, unloading regulator system, shown in Figure 11, derives its name from the manner in which the pilot unloads the pressure from the diaphragm in response to a change in the controlled pressure. The loading pressure is applied directly to the diaphragm through a fixed restriction from the upstream pressure.

A fairly accurate system, such as in Figure 11, could be constructed using conventional regulator hardware, but it would be a relatively expensive device for the performance obtained. More commonly, a special unloading-type pilot-operated regulator is constructed using the technique shown in Figure 12.

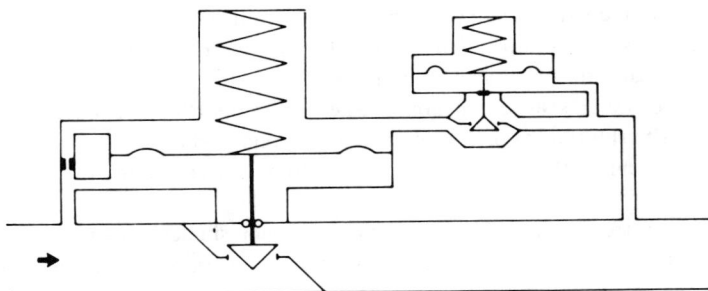

Figure 11.   Pilot-operated unloading system.

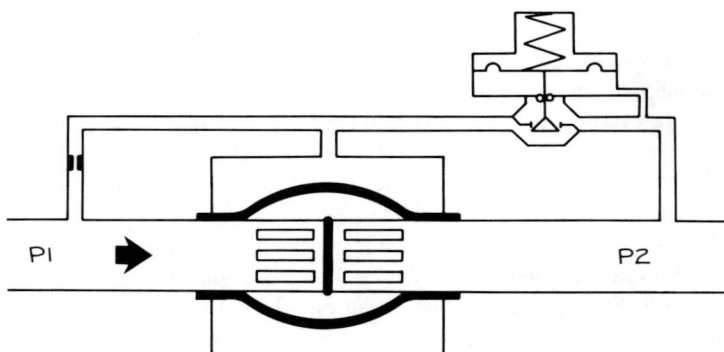

Figure 12.   Typical pilot unloading system.

A barrier is placed in the normal flow stream to prevent the flow of gas directly through the device. A series of slots is formed on each side of the barrier so that the gas can pass around the barrier by going out of the slots on one side and into the slots of the other. An elastic boot-type diaphragm covers the series of slots to regulate the amount of gas that can pass around the barrier. The unloading pilot controls the loading pressure that surrounds the exterior of the diaphragm.

This device is less expensive than the apparatus in Figure 11; however, it tends to be less accurate and requires a fairly large pressure drop across the regulator. Both of these problems stem from the fact that the pressure of the flowing gas is required to force open the rather stiff diaphragm. Also worth noting here is that this type of regulator will fail closed if the pilot freezes or plugs up.

The unloading regulator is strictly a pilot-operated system. It suffers from the problem of poor response and shock characteristics as well as some instability.

To obtain the advantages of a pilot-operated system without many of the disadvantages, a two-path control system, such as in Figure 13, was developed. The two-path control system derives its name from the direct- and pilot-controlled feedback paths that are simultaneously employed. In the direct acting path, the controlled pressure acts directly on top of the diaphragm to provide immediate, highly responsive feedback. This self-operated feedback provides good response and shock characteristics, as well as good stability.

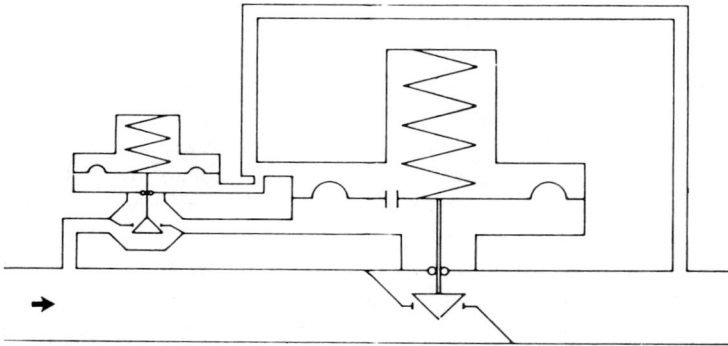

Figure 13.   Two-path control system.

In addition to the direct acting path, the controlled pressure also causes the pilot to change the loading pressure on the underside of the diaphragm. Although the pilot action is slower than the direct acting path, it gives the advantage of high accuracy and minimal droop.

THE PROPORTIONAL CONTROLLER

The two-path control system exhibits some attributes of the instrument control system, which is another important gas pressure control system. Although this type of system is truly a pressure regulation loop and is sometimes referred to as a regulator, it is really a miniature instrument system comprised of a control valve, an actuator and a controller.

The controller is sometimes called a pilot, but there are a number of significant differences between the controller and the conventional regulator pilot. In comparison, the controller has much better frequency response and general dynamic characteristics than the conventional pilot. The controller typically has very high, adjustable gain in comparison to the low, fixed gain pilot. In addition, the pilot operates strictly in the simple proportional mode, whereas the controller frequently operates with two additional modes called reset and rate.

In short, the controller has a number of adjustments that give it operating flexibility the pilot does not have. On the other hand, because of all this flexibility it is important that the controller be adjusted and used properly for optimum performance.

Many standard controllers have options for providing three common modes of operation: (1) proportional mode, (2) reset or integral mode, and (3) rate or derivative mode. These three modes of control can operate individually or in various combinations, regardless of whether the controller is pneumatic or electronic. Thus, it is possible to have a three-mode controller that combines proportional plus integral plus derivative control modes. This is frequently designated as a PID controller. A two-mode controller combining proportional plus integral control modes is called a PI controller.

In the gas industry, the proportional only controller is the most common type; however, there are many applictions in which PI controllers are used. PID controllers are not seen as frequently, but they can be important where temperature control applications are encountered.

To better understand how the various control modes function, we need to review some of the concepts that were covered in the section on droop. To close the valve in Figure 14, a loading pressure, P_2, must be applied to the diaphragm. Because of the initial compression in the spring, the loading pressure must increase to some initial value before the valve begins to move. Arbitrarily assume that this system requires a starting pressure of 25 psig to overcome the initial compression of the spring.

If the loading pressure increases above 25 psig, the valve will begin to close. The effective area of the diaphragm, the stiffness of the spring and the amount of valve travel will determine how much the loading pressure must increase to fully close the valve. Again, for purposes of discussion, arbitrarily assume that the system components in Figure 14 are such that a loading pressure increase of 5 psi is required to fully stroke the valve.

In normal operation, the valve will range from its fully open to its fully closed position. Since the controlled pressure serves as the loading pressure on the diaphragm, P_2 must range from 25 psig to 30 psig to stroke the valve through its full travel range. The value of the controlled pressure at the low load condition, i.e., $P_2 = 30$ psig, is known as the set point pressure. As the load increases, P_2 decreases a proportionate amount from the set point pressure. It is this proportional relationship between droop and the load that gives rise to the name, proportional control action.

In Figure 14, assume that the load demand is very small and that the valve plug is operating very near its seat. As long as the load demand remains low, the valve

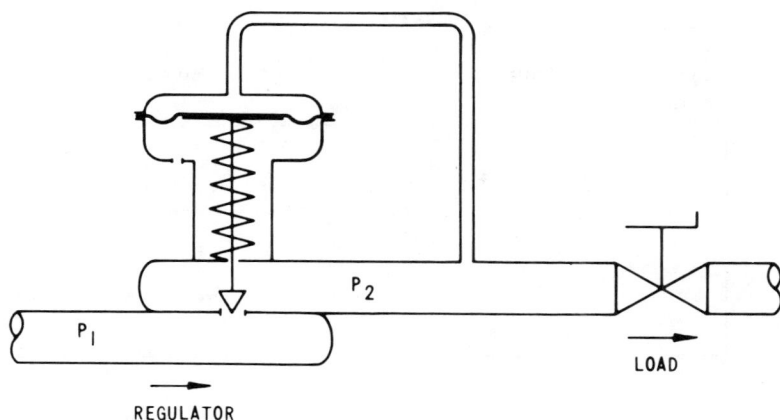

Figure 14.   Elementary pressure loop.

plug will remain in equilibrium near its seat and the downstream pressure will remain at 30 psig.

Now, suppose that the load demand abruptly increases to its maximum. To supply the needed gas to the system, the valve plug must move to its wide open position. The only way for this to happen is to decrease the pressure on the diaphragm. Since the downstream pressure is acting on the diaphragm, P_2 must decrease to 25 psig to allow the valve plug to open wide.

The amount that P_2 must decrease to fully open the valve is called droop or offset, since the controlled pressure is offset by this amount from the set point. In this example, the offset is 5 psi.

In most systems, if there were a full load offset of this amount, we would say that the system had rather low loop sensitivity, that the loop gain was low, or perhaps that the proportional band was rather broad. All these expressions are used to describe a control system that has a rather insensitive response to load changes.

Figure 15 illustrates how this example system responds to the abrupt step change in load that has been introduced. Because of inertia and other dynamic factors, P_2 will probably drop a lot more than 5 psi during the transient condition; however, the system eventually comes to equilibrium with an offset of 5 psi.

Because the offset is required to open the valve for the larger flow, it should be obvious that the offset must continue to exist as long as the load continues at the larger flowrate. The controlled pressure, P_2, will only return to the set point value when the load returns to its original set point flowrate. Thus, with proportional control, an offset exists when the load flow is something other than the set point conditions.

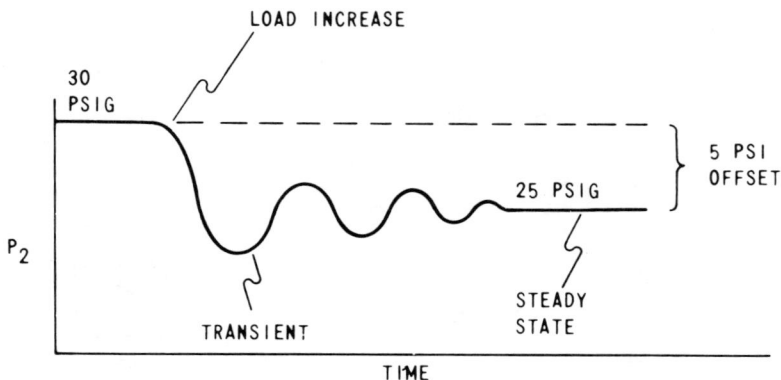

Figure 15. System response to an abrupt full-load upset.

Now revise the system slightly, as shown in Figure 16, by adding a black box device in the controlled pressure line to the diaphragm. This black box device is called a controller, but in reality is simply a pressure amplifier similar to the pilot introduced in Figure 10. The controller differs primarily from the pilot in that the controller has a knob to adjust the gain or amplification and its response is usually much better than the response of a typical pilot. Like the pilot, the controller senses changes in the controlled pressure, P_2, and amplifies them into larger changes in the loading pressure on the diaphragm, P_L. The amount of amplification is determined by the setting on the controller gain knob. For purposes of this example, assume this amplification to be 10, i.e., the change in loading pressure on the diaphragm is 10 times the change in the controlled pressure.

Returning to the previous control problem, assume that the system is operating at a very low load flow and the controlled pressure, P_2, is at the set point value of 30 psig. If an abrupt full load flow change is again experienced, the valve must open wide to pass the greater flow. As before, the loading pressure on the diaphragm must decrease by 5 psi to allow the valve to open wide; however, the controlled pressure change is only 0.5 psi because of the gain of 10 in the controller. As a result, there is a terrific improvement in system performance, as seen by comparing Figure 17 with Figure 15.

By adding the controller with its gain of 10, we can say that we have increased the loop sensitivity, raised the loop gain or perhaps narrowed the proportional band. Figure 17 shows that this increased loop gain not only improves the quality of the steady-state control, but also greatly improves the quality of the transient control.

This simple fundamental fact provides the secret of good transient and steady-state control! All that is needed is to make the loop gain as high as possible by adjusting the knob on the controller.

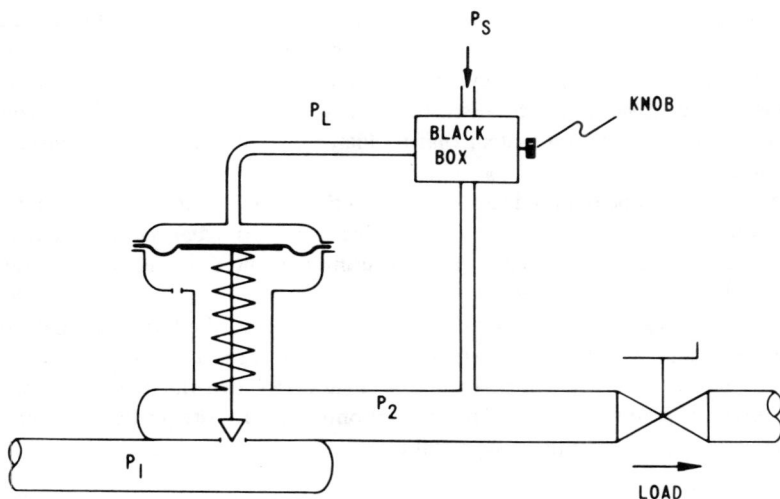

Figure 16.   Controller in the loop.

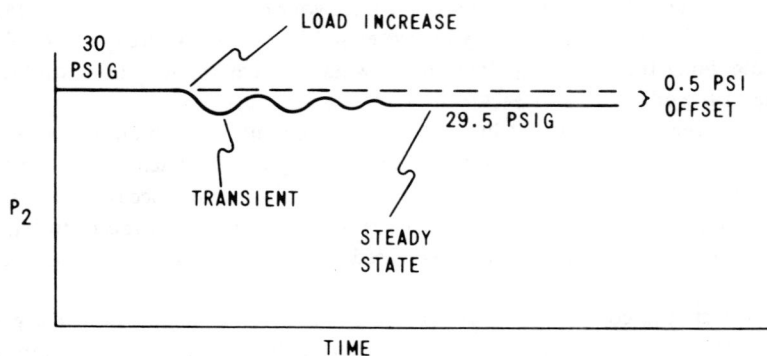

Figure 17.   Improved performance due to higher gain.

In essence, high loop gain not only improves the quality of the steady-state control, but also greatly improves the quality of the transient control. As a result, the following question must be posed: Why not simply design the loop for the highest possible gain to achieve the best possible control?

High loop gain simply means that any signal variation or disturbance will be greatly amplified as the control loop tries to correct for the disturbance. In a steady-state operating condition there is no disturbance to the system, hence there is nothing to amplify, and the high loop gain can cause no problem. In fact, it works to our advantage because it minimizes the steady-state offset.

When a disturbance does come along, high loop gain also tries to improve the transient performance by greatly amplifying the signal, which causes the system to respond quickly and accurately. If the loop gain is too high, however, the system is far too sensitive to these disturbances and it has a tendency to overcorrect severely. This results in a system that oscillates back and forth in an unstable manner.

The system can be returned to a stable condition simply by lowering the gain of the loop. This makes the system less sensitive to the disturbances and the tendency to overcorrect is eliminated. But, lowering the gain does degrade the transient and steady-state control.

There appears to be a definite control dilemma! Keep the loop gain high for good performance and risk an unstable system, or lower the loop gain for stability and suffer the poorer system performance. Figure 15 shows the type of response that might be expected from a proportional-only control system when the gain is lowered enough to keep the system stable.

CONTROL MODES

There is an old saying that "you can't have your cake and eat it too," yet the possibility for compromise nearly always exists. Let us investigate the possibility of combining these two high-gain and low-gain modes of operation to obtain some of the advantages of each.

It was seen previously that in the steady-state condition, with no disturbances to amplify, the system would tolerate the high loop gain and achieve the benefit of minimum steady-state offset; however, if a load disturbance occurs, the system will be unstable unless the gain is cut sufficiently. Finally, as the disturbance begins to die out, the gain can be increased slowly to minimize the steady-state offset.

In effect, the controller is now operating in two different modes; low gain during the transient to maintain stability and high gain in the steady state to obtain good steady-state accuracy.

Figure 15 shows the type of transient response one would expect with low gain in our system, while Figure 17 shows the improved steady-state control that can be obtained with high gain. Combining the two modes of gain operation will also combine the resulting transient and steady-state response. Slowly increasing the gain as the system approaches steady state will cause the performance to change slowly from the poorer transient response of Figure 15 to the better steady-state response of Figure 17. The combined effect of these two operational control modes is shown in Figure 18.

Note that the steady-state offset is no longer related to the low gain setting that was required during the transient state for stability. The high gain during the

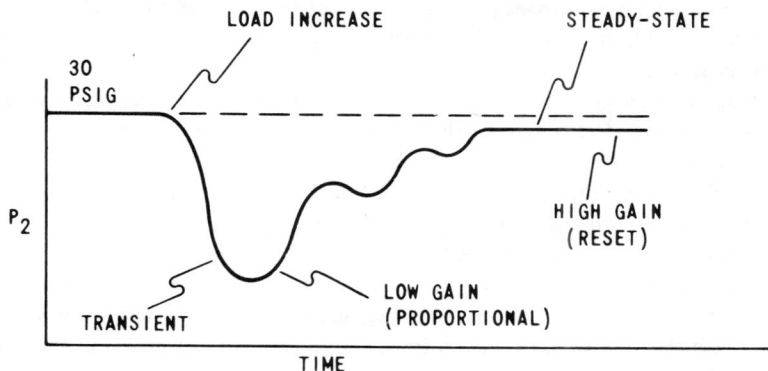

Figure 18.   Two-mode control.

steady state has, in effect, "reset" the process variable, P_2, back to the 30 psig set point after the system recovered from the load disturbance.

This phenomenon gives rise to the common name, reset action. The two modes of control action that have been combined are usually referred to as proportional-plus-reset control. The proportional action refers to the low-gain mode during the transient state to keep the system stable. Instead of remaining in this low-gain mode and suffering the poor steady-state control, we reset P_2 back to the set point by slowly making the transition back to the high-gain mode. This action is known as reset action.

A mathematical analysis of this reset action would show that the controller is adjusting its response according to the integral of the process error signal. For this reason, these two modes of control are sometimes referred to as proportional-plus-integral, or PI control.

Whether this action is called reset or integral, the simple fact remains that the loop gain is simply increasing as steady state is approached to take advantage of the more accurate steady-state control offered by the higher gain.

Theoretically, P_2 would never return completely back to the set point after an upset unless the loop had infinite gain. In actual practice, however, the steady-state gain can be made high enough so that the steady-state offset virtually disappears, i.e., it becomes so small that it is not visible on our gauges and charts.

The speed at which the gain for reset action is increased must be carefully regulated for best control. If done too slowly, it will take a long time to return to the set point; if done too quickly, the system will be unstable. The speed must be carefully tuned to the right value for the particular system being controlled.

So far we have managed to improve the quality of the steady-state control by the use of reset, but still must suffer the poorer transient control. This is because, on encountering a disturbance, the gain was cut immediately to maintain stability. However, it is again possible to effect a compromise that can also improve the transient response.

By maintaining very high gain in a system undergoing a load disturbance, the system will respond very well, but it may be far too sensitive and have a tendency to overcorrect severely. If one could only maintain the high gain for a short period of time after the load disturbance occurs, the transient response of the system could be improved greatly and yet still not endanger system stability. In other words, hold the gain high long enough to get good response to the disturbance, but then cut the gain before the system can overreact and become unstable.

It should be apparent that the timing here is again quite critical. By waiting too long to cut the gain the system will be unstable, and if the gain is cut too soon there will not be any significant improvement in transient response.

The effectiveness of this new third mode of control is so closely related to the rate at which the load disturbance occurs that it is not too surprising that it is frequently called rate control. A mathematical analysis of this rate action would show that the controller is adjusting its response according to the derivative of the process error signal. For this reason, it is also sometimes called derivative control mode.

Combining all three of these modes in a single controller would result in a proportional-plus-integral-plus-derivative (PID) controller. It could also be called a proportional-plus-reset-plus-rate controller. Figure 19 shows the improved transient response and improved steady-state control that can result from the use of a three-mode PID controller.

It is important to realize that the reset or integral mode is used to improve the quality of steady-state control; rate or derivative mode is used to improve the quality of transient control; and proportional mode is used to provide system stability.

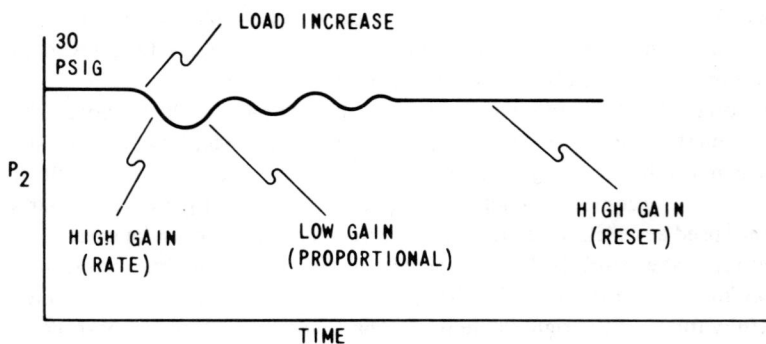

Figure 19.   Three-mode control.

HOW IT WORKS

Whether electronic or pneumatic, all conventional three-mode controllers operate on the same principle, i.e., the principle of automatically varying the gain of the system when a disturbance occurs to obtain stable control with the best possible transient and steady-state control.

Using a three-mode controller, this is accomplished by momentarily keeping the gain high when a load disturbance occurs to get good initial transient response. The gain is then quickly lowered enough to keep the system stable and, finally, as the disturbance dies out, slowly increased again to decrease the steady-state offset. We can easily illustrate how a controller performs this function by looking at a typical pneumatic controller. The basic nozzle-flapper amplifier, shown in Figure 20, is a key element in the operation of the controller.

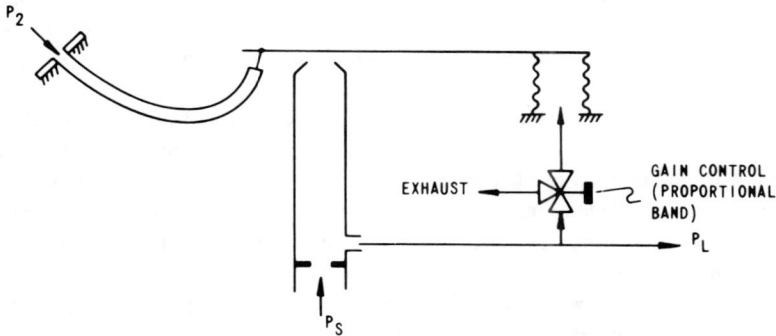

Figure 20. Pneumatic proportional controller.

A source of supply pressure, P_s, is fed into the nozzle chamber through a fixed restriction. This is usually obtained from a compressed air system, which provides a source of energy to many different instruments at the same time. A nozzle opening, which is larger than the fixed orifice, allows the air to escape. As the movable end of the pivoted flapper undergoes an input motion, the flow area of the nozzle changes and varies the pressure in the nozzle chamber accordingly. This nozzle pressure can then be used as an output loading pressure, P_L, to some other device, such as an actuator.

The movable end of the curved Bourdon tube is attached to the flapper through a mechanical linkage. As the input pressure, P_2, increases, the Bourdon tube tries to straighten itself and results in a flapper motion that tends to restrict the flow out of the nozzle. This causes a resulting increase in the output loading pressure, P_L. A typical system such as this is designed so that the output pressure, P_L, is rather sensitive to changes in the input pressure, P_2. In other words, it is a high-gain pneumatic amplifier. To make this unit useful as a controller, some type of adjustment must be provided for the gain.

The output loading pressure, P_L, is tapped and fed to an adjustable three-way valve. By adjusting this valve, the percentage of pressure signal, P_L, that is fed back to the bellows can be varied. In essence, the three-way valve acts as a splitter. Part of the output signal going to the three-way valve is fed to the bellows and the remainder is exhausted to the atmosphere.

When a change occurs in the input pressure, P_2, the action of the nozzle-flapper changes the output pressure, P_L. Part of this output pressure change is fed to the bellows in such a way as to decrease the effect that P_2 has on P_L. This is technically known as negative feedback. The more negative feedback used, the more the sensitivity or gain of the device is decreased.

If the gain control is adjusted so that none of the change in P_L is registered in the bellows, then the system will respond with high gain. As more of the changes in P_L are fed back to the bellows, the gain of the unit is cut accordingly. The negative feedback adjustment on the controller is called the gain, or proportional band, control.

The unit shown in Figure 20 is a single-mode pneumatic proportional controller. This device is the black-box amplifier that was added to the system in Figure 16. When it is installed in an operating system, it must always have some offset in the controlled pressure, P_2, to accommodate changes in the load flow. It is called a proportional controller because the offset required is proportional to the amount of change in load flow. The amount of offset required in any case can be decreased by increasing the gain of the system. If the gain is sufficiently high, the steady-state offset can be reduced to a negligible amount. The only catch in this approach is that problems with stability are encountered.

The way to solve the instability problem with the proportional controller is to adjust the gain to a lower value, i.e., broaden the proportional band setting on the controller by feeding more pressure back to the proportional bellows in Figure 20. When this is done, we must accept the larger steady-state offset that results from the lower gain.

The main reason for having the proportional bellows in Figure 20 is to provide a means for cutting the gain of the controller. Earlier, we suggested the desirability of being able to automatically vary the gain of the controller when a load disturbance occurs. The second bellows (Figure 21) provides a means of accomplishing this.

This second bellows, known as the reset bellows, is located opposite the proportional bellows. Its operating characteristics match those of the proportional bellows. If the pressures in these two bellows are the same, one cancels the effect of the other and the system has high gain. Indeed, this is the real secret behind the action of a proportional-plus-reset controller.

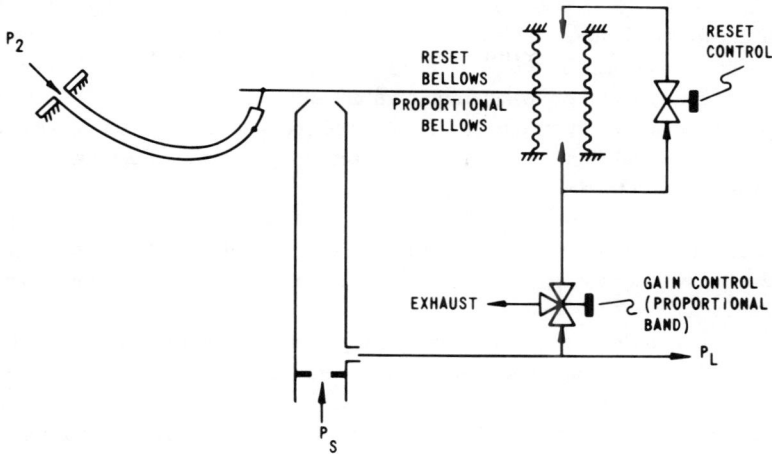

Figure 21.   Proportional-plus-reset controller.

Figure 21 shows that the pressure being sent to the proportional bellows is also sent to the reset bellows. Before this pressure can reach the reset bellows, however, it must first pass through an adjustable needle valve known as the reset control. It should be noted that this reset control is not a three-way valve like the gain control. The sole purpose of this reset valve is to provide an adjustable time delay; the more resistance adjusted into the valve, the longer it will take a change in pressure to register in the reset bellows.

The signal that goes to the reset bellows is technically known as positive feedback. The more positive feedback the greater the increase in the sensitivity or gain of the controller. In the present example, the positive and negative feedback signals are canceling each other out in the steady-state condition. In the transient condition, the reset restriction temporarily delays the positive feedback signal so that the negative feedback can achieve its gain cutting effect.

A simple illustration will help explain how this system works. Assume a steady-state operating condition with everything in equilibrium. In this situation, the pressures in both bellows are equal and are canceling each other. A high gain exists to provide the best possible steady-state accuracy.

If a step decrease occurs in the load flow of the pressure control system, the controlled pressure, P_2, increases. This increase in P_2 registers in the Bourdon tube of the controller and results in an increase in diaphragm loading pressure, P_L. Part of this increase in P_L passes through the gain control three-way valve and registers in the proportional bellows. The restriction of the reset valve prevents this pressure increase from immediately reaching the reset bellows. At this stage of the operation, the pressure in the proportional bellows is much greater than that in the reset bellows. This has the effect of immediately reducing the gain of the controller. This gain reduction during the transient portion of the disturbance prevents the system from oscillating in an unstable manner.

As the transient response to the disturbance decays and the system once again approaches equilibrium, the pressure signal to the proportional bellows has time to bleed through the reset restriction to the reset bellows. As a result, the pressure in the reset bellows slowly increases and cancels the gain cutting effect of the proportional bellows. When a steady-state condition is finally reached, the pressures in the two bellows are again equal and the controller is back to the same high gain as before.

If the gain in the steady state is high enough, the amount of steady-state offset is so small that, for all practical purposes, P_2 returns to the original set point. This type of controller action has earned the name reset action because after a load disturbance occurs it appears that the controller resets the controlled pressure, P_2, back to the set point. Those who are more theoretically oriented may refer to this same controller action as integral control because of the mathematical relationship that exists.

Regardless of the name used, the net result is the same. The high gain of the controller is sacrificed during the transient condition to keep the system stable. As a result, a relatively large deviation may occur. As the transient condition decays, the reset action slowly increases the controller gain. By the time a steady-state condition is reached, the high gain results in the good steady-state performance illustrated in Figure 18.

To get the best performance from the system and still keep it stable, the restriction of the reset control must be adjusted properly. The reset signal must be delayed just enough to match the recovery characteristics of the process under control. If the gain increases too rapidly, the system will be unstable. In this case, one would say that the reset is too fast, i.e., too many repeats-per-minute (RPM)! On the other hand, if the gain increases too slowly, the system will be very sluggish and will not achieve the best possible control because the reset is too slow, i.e., not enough RPM.

It should be noted that the reset adjustment determines the reset time! It does not influence the size of the steady-state offset; it merely determines how quickly P_2 is returned to the set point. In addition to the reset adjustment, the proportional control also must be set properly. The gain during the transient condition must be sufficiently low to maintain stability, but not so low that the transient deviation in P_2 is unnecessarily large. The proper adjustment of these controls for a given process is called controller tuning and will be discussed in more detail later.

The relatively poor transient control that results from the immediate gain reduction in a two-mode controller can be significantly improved by delaying the gain reduction just long enough to start the system responding to the load disturbance, but not so long that the system becomes unstable. The amount of delay needed will have to be adjustable since it depends on the dynamics of the particular system under control.

Figure 22 shows the addition of another valve restriction in the pressure line feeding both the controller bellows. This valve is known as the rate control valve. Like the reset control, the rate control restriction can be varied to provide just the proper amount of time delay before the gain cutting effect begins. Once the pressure change has time to bleed across the rate valve restriction, the controller acts just as explained for the two-mode control action.

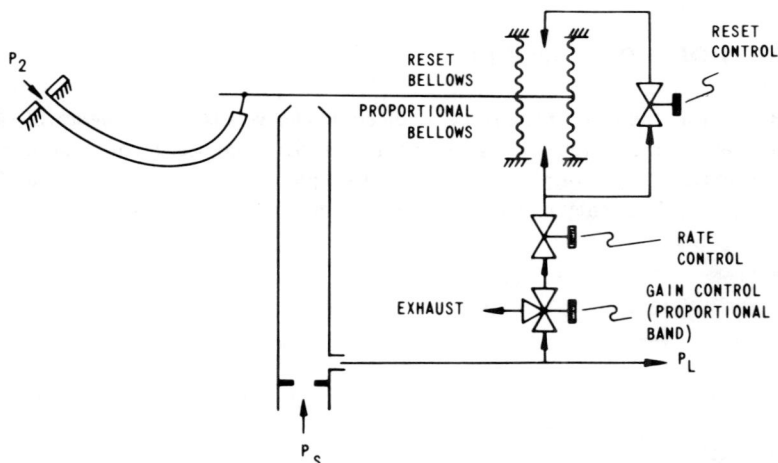

Figure 22.   Proportional-plus-reset-plus-rate controller.

The control action just described is known as rate action because its effect is usually influenced by the rate at which the load disturbance occurs. If the disturbance occurs rather slowly, the rate valve offers no significant delay to the pressure signal. On the other hand, if the load disturbance is very rapid, the delay caused by the rate valve is comparatively more significant. The theorists refer to this controller action as derivative control.

How does the gain of this three-mode controller vary when it encounters a step change in the load? Initially, in the steady state, the gain is high. When the step change in load occurs, the rate control momentarily maintains the high gain to improve the initial transient response of the system. Before this high gain has an opportunity to make the system unstable, the pressure signal to the proportional bellows cuts the gain during the major portion of the transient condition. Finally, as the system again approaches steady state, the reset control slowly increases the gain back to its high steady-state value. This improves the steady-state control. Figure 19 illustrates how this gain variation of the three-mode controller improves both the transient and steady-state control, yet maintains system stability.

To get the best performance from the system and still keep it stable, the restriction of the rate control must be properly adjusted. The feedback signal must be delayed just enough to match the dynamic characteristics of the process under control. If we wait too long before reducing the gain, the system will be unstable. In this case, the rate is said to be too slow, i.e., too many minutes. Conversely, if we cut the gain too soon, no significant improvement in transient control will be obtained. In this case, the rate is said to be too fast, i.e., not enough minutes.

CONTROLLER RESPONSE

By plotting the gain of the three-mode controller versus the *frequency* of signals to which it must respond, some rather interesting and useful conclusions can be derived. This type of plot is known as a frequency response curve. Figure 23 shows such a frequency response plot for an ideal three-mode controller.

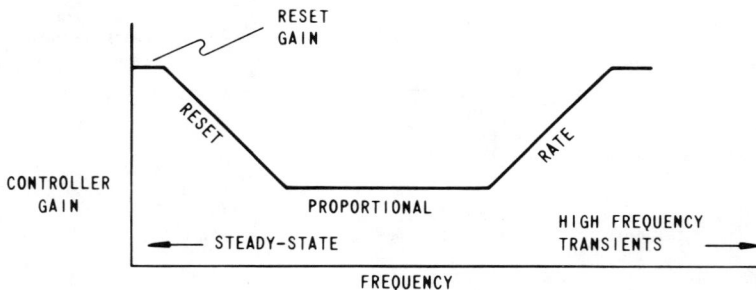

Figure 23.   Frequency response of an ideal three-mode controller.

At this stage, it is not particularly important to understand what this curve represents or where it comes from. Rather, it should be thought of as a visual memory aid that will be very useful later when discussing the proper tuning of a controller. It will suffice to know a few facts about the general shape of the curve.

The low end of the frequency scale represents what has been referred to as steady state. Previously, it was shown that as steady state is approached, the reset action of the controller steadily increases the gain. This can be seen quite clearly in Figure 23. This increase in gain continues until a maximum is reached where the two bellows are completely balancing each other. This maximum gain is known as the reset gain. The higher the reset gain, the closer the controlled variable, P_2, will be held to the desired set point in the steady state.

At the other end of the frequency scale occur the rather rapid, or high-frequency, transient changes in the load disturbance. The resulting high-frequency changes in nozzle pressure are effectively isolated from the proportional bellows by the rate control valve. This keeps the gain of the controller at its maximum value. As the frequency of the signal decreases or slows down, the rate valve restriction becomes less and less significant and the gain gradually decreases. This is also shown in Figure 23.

The lowest gain that the controller will reach is determined by the setting of the proportional band or gain control valve. This occurs in the middle range of frequencies.

As shown in Figure 24, the gain versus frequency curve for a real, practical controller approximates that of an ideal controller. At low frequencies, the gain is limited by the reset gain designed into the controller. At high frequencies, gain of the real controller is affected by inertia, capacitance and resistance. The frequency where the gain is attenuated by these parameters is referred to as the controller frequency response. This is analogous to the limitation on the high-frequency response of a sound stereo system. Above the frequency response of the system, the signal is attenuated because the system can no longer faithfully respond to the input. A major difference, of course, between the controller and the stereo system is the breadth of the frequency response range. At frequencies above the controller frequency response, the gain is determined by physical limitations and is independent of controller adjustments.

Figure 25 shows how the depth of the notch in the gain curve can be varied with the gain or proportional band control. By making the notch deeper, we say that we have cut the gain or broadened the proportional band. This is shown as the solid line at the bottom of the curve. By narrowing the proportional band or increasing the gain, the notch is raised as indicated by the broken line in the figure. In either case, it is worth noting that the width at the bottom of the notch remains the same when adjusting the proportional band or gain control. The corner frequencies of the notch remain unaffected. Their values depend on the reset and derivative settings on the controller.

Figure 26 shows that increasing the reset setting of the controller in RPMs moves the left side of the notch to the right and, therefore, narrows the notch. A common set of measurement units on the reset control is RPM. Increasing the RPM means faster reset. This brings the reset action into play sooner, i.e., at a higher frequency, and the left side of the notch moves to the right. This faster reset is indicated by a decrease in the integral time constant. This faster reset also means that the process variable is being returned more quickly to the set point.

Increasing the derivative setting of the controller moves the right side of the notch to the left and again narrows the notch as shown in Figure 27. A common

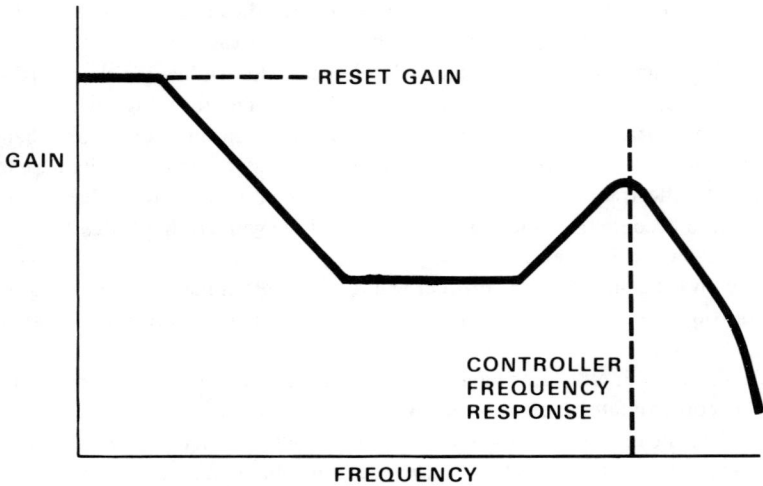

Figure 24.   Frequency response of a practical controller.

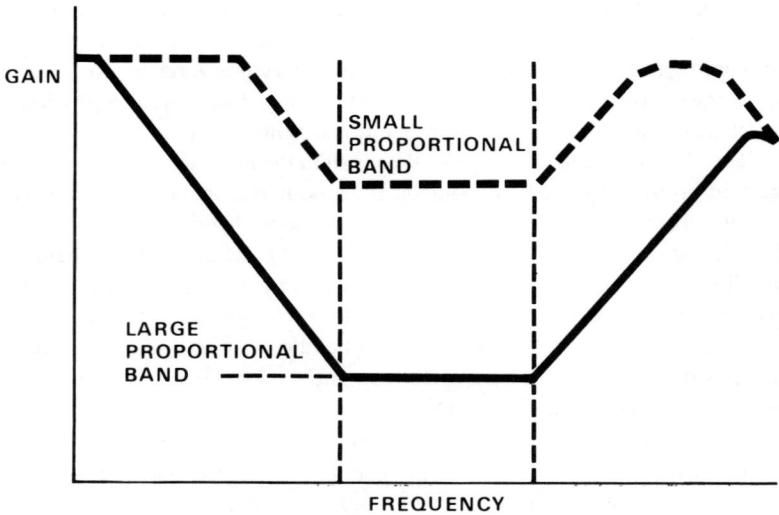

Figure 25.   Effect of proportional band adjustment.

set of measurement units on the rate control is minutes (MIN). Increasing the MIN means slower rate. This means that the rate action will maintain the gain longer, i.e., down to a lower frequency, and the right side of the notch moves to the left. This slower derivative action is indicated by an increase in the derivative

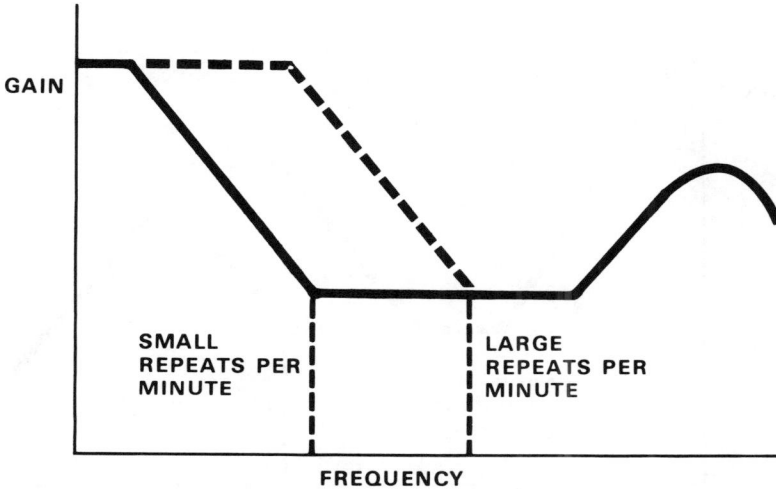

Figure 26.   Effect of reset adjustment.

time constant. This also means that the transient response of the system is being improved.

It is natural for one to wonder about the meaning behind such terms as RPM and MIN. Sometimes such knowledge can even be useful in the application of such terms. This is not one of those situations, however. An explanation of the meaning of these units would be lengthy and would not contribute significantly to understanding their use. If we simply concentrate our attention on how to use these terms it will greatly simplify the situation.

Increasing both the reset (RPM) and rate (MIN) values narrows the notch, and decreasing these values broadens the notch. Broadening or increasing the proportional band (% PB) lowers the bottom of the notch; narrowing or decreasing the proportional band (% PB) raises the bottom of the notch. These three controls theoretically give a method of adjusting the notch and moving it around in virtually any manner desired. This ability allows the controller to be tuned to match the dynamic characteristics of the system.

LOOP STABILITY

In every control loop there is a critical frequency, f_c, where disturbances tend to reinforce themselves. If conditions are not satisfactory when a disturbance occurs, the loop will be unstable and will cycle at this critical frequency. This frequency is sometimes referred to as the system cycling frequency and can be thought of as loosely analogous to the resonant frequency of a simple mechanical

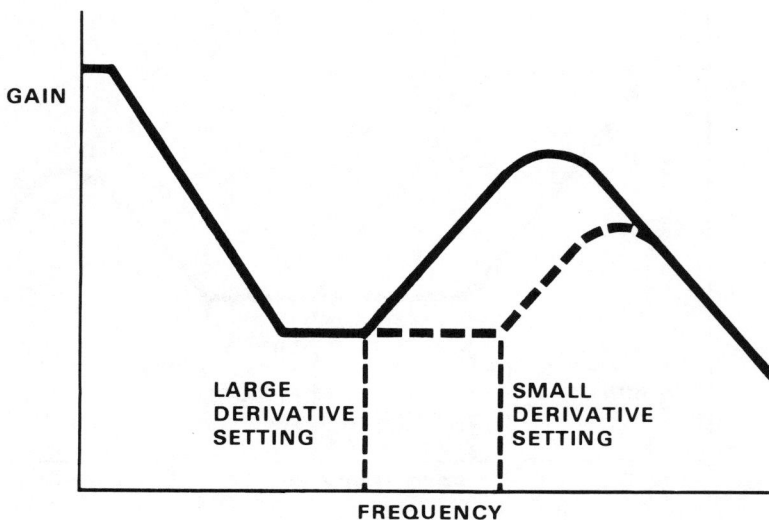

Figure 27.   Effect of derivative adjustment.

device. To keep the loop from being unstable, the system loop gain must be sufficiently low at this critical frequency. At other frequencies, the gain can be maintained at much higher values to obtain the best possible control performance. It is
not necessary to know this critical frequency; it is only necessary to realize that at
this frequency there is some value of the controller gain that will keep the loop
stable. The objective is to tune the controller for maximum performance, yet
keep the gain at the critical frequency below this limiting value.

The three-mode controller is simply a notch filter with low gain at the system
cycling frequency and high gain everywhere else. The purpose of tuning such a
controller is to make certain that this gain notch is centered about the critical frequency of the particular system. Since the characteristic cycling frequency varies
from system to system, the adjustability provided by the three-mode controller
gives the flexibility needed to control many different types of systems.

Adjusting the controller knobs makes the notch wide or narrow and one can
raise or lower the bottom, as well as move the whole notch to the right or left.
When the best place for the notch is found, the controller will be tuned!

CONTROLLER TUNING

There have been many theoretical studies to determine formulas and techniques for calculating settings for proper tuning. These studies certainly have

their usefulness in the general investigation of controller theory, but they have definite limitations when applying them to real processes. These studies usually make a number of simplifying assumptions that render results of limited value in real systems. In addition, the criteria for defining a properly tuned controller are very subjective and vary significantly from one instance to the next.

Experienced control people agree that the most realistic method of tuning a controller is for the operator to tune it around the actual system being controlled. This has the advantage of not requiring any assumptions about the dynamic nature of the process. Furthermore, this technique allows the operator to tune the system according to established company policy or his own subjective tastes, whichever is the determining factor.

First, consider why you want to tune your controller. You want a stable system that recovers rapidly after a disturbance with a minimum amount of overshoot. You want the system to settle to equilibrium as soon as possible. At equilibrium, the controlled variable should be as close as possible to the set point; however, as discovered earlier, that is governed by the use of reset, not by tuning. Of course the system must meet these goals in spite of varying operating conditons, changing throughput and imprecise process information.

For a system that is properly tuned, the notch will be as narrow and as high as possible, and will be approximately centered on the characteristic frequency, f_c.

This condition means that the system will have high gain in the low-frequency, or steady-state, region for better steady-state accuracy, and will also have high gain in the high-frequency region for better transient response. On the other hand, the gain at the characteristic frequency will be low enough to keep the system stable, yet not so low as to be unduly sluggish.

If the notch is too high or narrow, the system will be oscillatory. If it is too low or wide, the system will be sluggish and very slow to recover after a disturbance. Remember that the reset setting does not affect the size of the steady-state error, but it does affect how fast the system gets to steady state. The size of the steady-state error is determined by the magnitude of the disturbance and the reset gain designed into the controller.

Before actually beginning to tune the controller, one very important point should be noted. If the controller is already in service, it is usually wise to record the current controller settings before beginning the tuning process. If difficulty is encountered during the tuning process, one will at least be able to return to the previous working condition.

In the process of tuning the controller, the proportional band, reset and derivative knobs are adjusted to various appropriate settings, but what actually happens is that the frequency response notch is moved to an optimum position for best performance.

The best place for the notch to achieve a properly tuned system depends on all elements in the loop because they determine the value of the cycling frequency, f_c, which is characteristic of the system. This frequency is one index to system behavior. It indicates how fast the system responds to disturbances.

The first requirement for tuning is that the characteristic frequency be within the controller notch. The controller adjustments move the notch, and to be effective in stabilizing the system they must be able to change the gain at the characteristic frequency. Thus, for effective adjustments, f_c must be within the notch.

If the system cycling frequency, f_c, is larger than the controller frequency response, the adjustments cannot affect controller gain at this frequency. The adjustments can move the notch, but not to a frequency high enough to affect system performance. This situation frequently occurs when a pneumatic controller is directly driving a large actuator.

This does not necessarily mean the system will be unstable. It just means that there is no adjustability, or tunability, in the system. This situation can usually be corrected by inserting a volume-booster relay or a positioner between the controller and the actuator. This decreases the load on the controller and increases its frequency response to a value larger than the characteristic frequency of the system. Then the notch can be moved enough to change the controller gain at the characteristic frequency.

To ensure that the system will be initially stable, one should start with a wide, low notch. Wide means a small value for the reset setting in RPM and a small value for the derivative setting in MIN. Low means a large value for the proportional band setting.

How small and how large should these initial settings be? The answer is, small enough and large enough to stabilize the system! There is no one set of values that is the universal place to start tuning all systems. However, two sets of values are helpful as beginning points. For ordinary systems, a reset value of one RPM and zero derivative setting should give a notch wide enough to encompass the system characteristic frequency. A proportional band of about 200% should normally be sufficient to ensure stability.

Once the system is stabilized, one is ready to start tuning. It must be possible to introduce a small disturbance into the system to check stability. The system can be disturbed by changing the set point about 4% or 5%. If the process variable settles out at the new set point, the system is stable.

For an electronic controller, a second way of disturbing the system begins by switching the controller to the manual mode of operation. Next, manually drive the control valve until the controlled process variable is 2–4% off the set point. Switching back to the automatic mode of operation closes the loop. A stable system should settle out with the process variable at its set point.

For a pneumatic controller, a third way of disturbing the system is to bump the flapper of the nozzle-flapper amplifier. Regardless of the method used to disturb the system, the stability can be easily determined; however, the system may occasionally respond in an unexpected manner. For example, one may change the set point a small amount and find that, while proportional action has moved the process variable in the right direction, reset has apparently failed to bring it to the new set point. At first, it might appear as though the reset action were faulty. But remember you are starting with slow reset to get a wide notch. The reset speed should be increased until the controlled variable visibly begins to approach its set point. This new value then should be used as the initial reset setting.

The first step in the actual tuning procedure is to raise the bottom of the notch, as shown in Figure 28. Suppose, as an example, tuning is begun at the recommended 200% proportional band setting. The proportional band should be decreased in increments and the system disturbed at each step to check for stability. This should be repeated until the proportional band is found where the system begins to cycle; then that proportional band setting should be doubled. Thus, if cycling occurs at 40% proportional band, the setting should be changed to 80%. Figure 29 illustrates the system response at the three different proportional band settings of Figure 28. The controller gain at the bottom of the notch is considerably higher than its initial value, and the system is still stable.

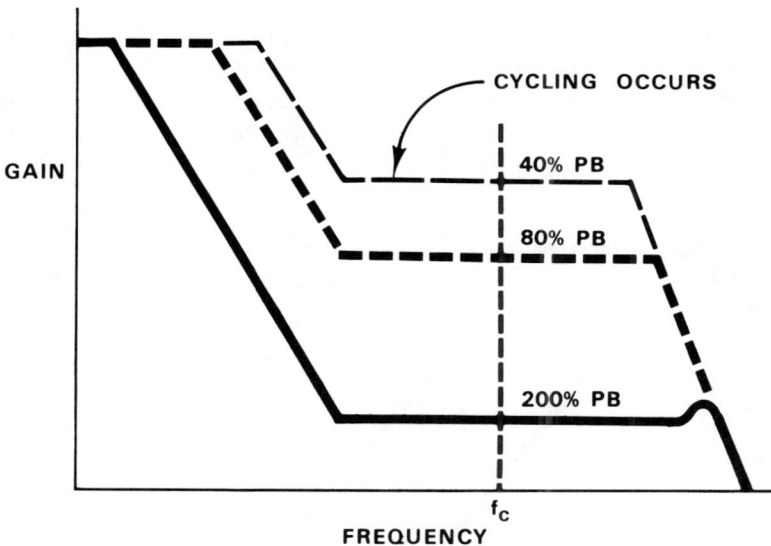

Figure 28.   Adjusting proportional band.

The next step is to narrow the notch to get the best control. Reset action will ensure that eventually, any steady-state error will be small, but you want to move the left corner of the notch toward f_c so that the system will get to the steady-state condition as soon as possible.

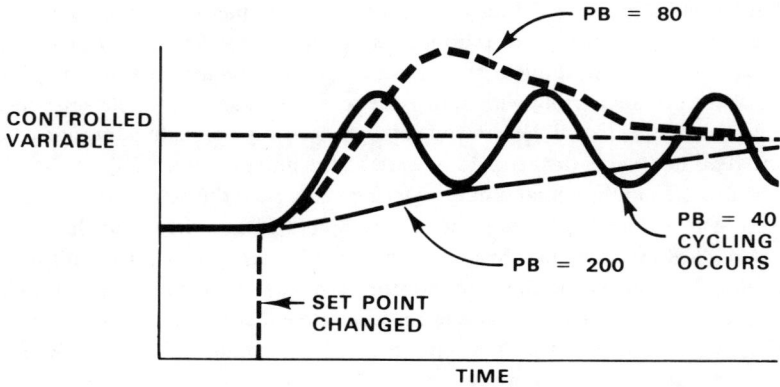

Figure 29.   System response to proportional band adjustment.

During this part of the tuning procedure, the reset should be speeded up a reasonable amount, the system disturbed and a check made for stability. This should be repeated until the reset setting that causes cycling is determined. Then the reset should be slowed down by about a factor of three.

For the example shown in Figure 30, cycling occurred at two RPM and the reset was slowed to 0.7 RPM. It is instructive to note here that the fast reset of 2 RPM

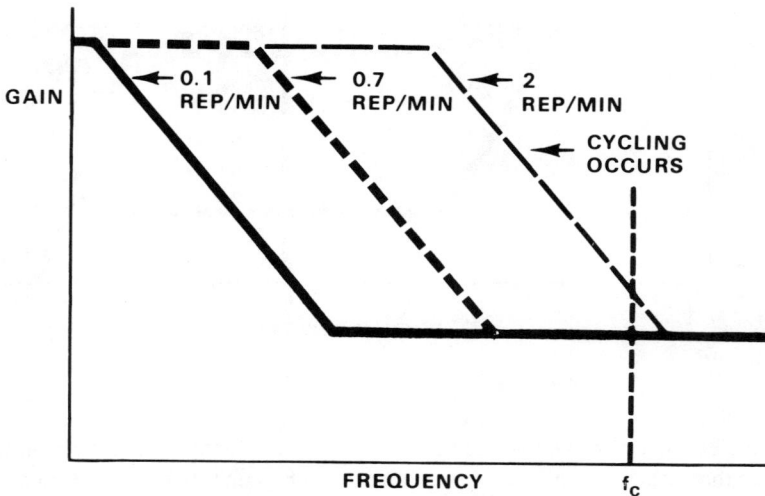

Figure 30.   Adjusting reset.

caused the system to cycle because the gain was increased at the characteristic frequency.

The object in tuning the reset function is to get the reset as fast as we dare without causing the system to cycle. In other words, we want to bring the left side of the notch in close to f_c, but not so close that we raise the gain at the characteristic frequency.

The system response curves at three different reset settings in Figure 31 show that the faster reset setting drives the controlled variable to the set point sooner. The system cycles if the reset is too fast, but the notch can then be widened slightly to achieve stability and still establish equilibrium quickly.

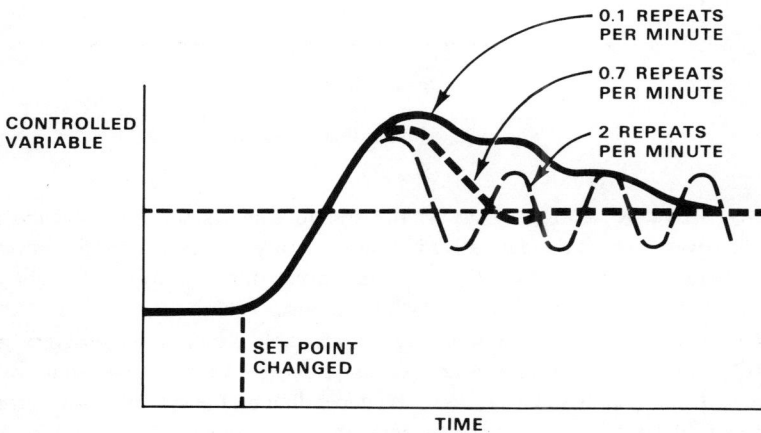

Figure 31. System response to reset adjustment.

To adjust the derivative setting for this hypothetical system, begin at zero minutes. Increase the setting in reasonable increments, disturb the system, and check for stability. Repeat this until the derivative setting that causes cycling is determined, then speed up the derivative by a factor of three.

If the system became unstable at a three-minute derivative setting, as shown in Figure 32, speed the derivative up to a one-minute setting. As with the case of reset, the slow derivative of 3 minutes caused the system to cycle because the gain was increased at the characteristic frequency.

The object in tuning the derivative function is to get the derivative as slow as we dare without causing the system to cycle. In other words, we want to bring the right side of the notch in close to f_c, but not so close that we raise the gain at the characteristic frequency.

The system response curves in Figure 33 show improved transient response with

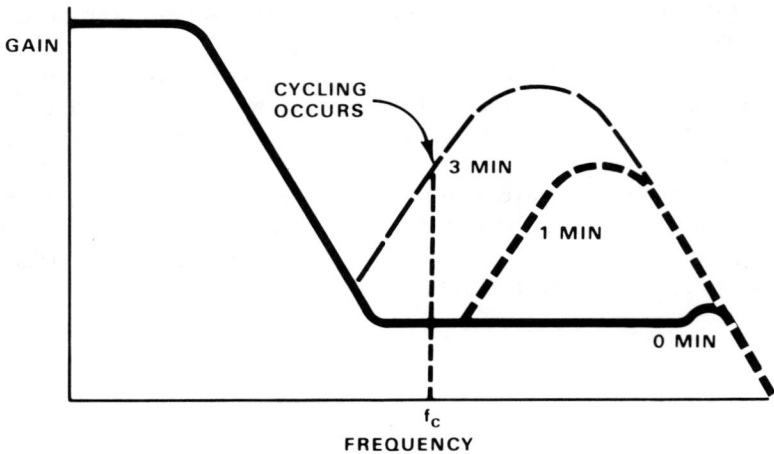

Figure 32.   Adjusting derivative.

slower derivative settings. That is, the system exhibits smaller rise time and smaller overshoot as the derivative is slowed. However, the system does cycle if the derivative becomes too slow. The notch can then be widened slightly to achieve both stability and good transient response.

You now have controller settings that yield a stable system which recovers fairly rapidly from small disturbances and which quickly settles to a steady-state value. These settings are satisfactory; however, depending on the system, various trade-offs can be made between the three controller adjustments. The purpose of the control system may dictate that the notch should be lower and narrower, or higher and wider.

Consider, for example, the settings on a proportional-plus-reset controller used to control the pressure of gas supplied to a large industrial user. Reasonable stability can be achieved with either broad proportional band and fast reset, or with small proportional band and slow reset. The differences show up in the transient performance of the system in response to load changes. Figure 34 illustrates how this trade-off in controller settings can result in the same relative system stability by maintaining the same relative gain at the characteristic frequency.

An example of where this trade-off technique might be useful is when there is a pressure relief valve which you don't want to be popping because of frequent load changes. A small proportional band setting can be used to prevent the transient pressure rise from popping the relief valve. The system response will be rather oscillatory and you will have to go to a lower reset setting, in repeats per minute, to achieve stability.

In another situation, you may be metering flow and want a smooth pressure instead of an extended series of small oscillations. In this case, a large propor-

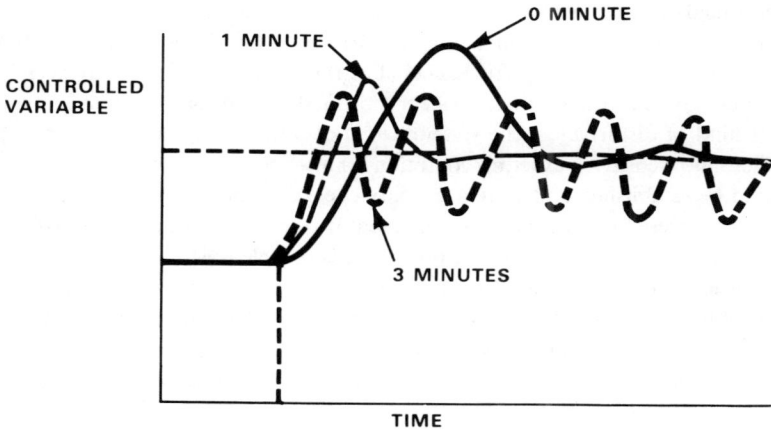

Figure 33.   System response to derivative adjustment.

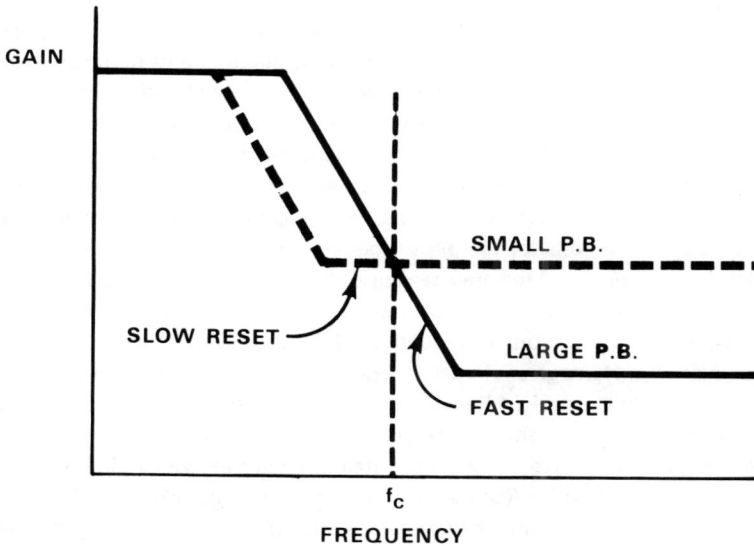

Figure 34.   Controller settings trade-off.

tional band can be used to achieve a very stable loop. The system will respond to appreciable load changes with a relatively large transient deviation. A fast reset setting can be used to get the system back to the set point rapidly while the large proportional band ensures stability.

Following the tuning technique outlined here will ensure that the system has been tuned for satisfactory response to small disturbances, but in actual operation the system may experience large disturbances. In addition, you have tuned for response to set point disturbances, but the system usually experiences load changes. To be confident that a system is properly tuned, one needs to know just what kind of disturbances the system will encounter during normal operations.

The most sensitive, exacting conditions that occur during normal operations should be established and system stability checked. The worst throughput value and most severe load change should be used. If system performance is satisfactory under these conditions, the tuning procedure is complete. If, however, the system response to these large load changes is unsatisfactory, or if it is impractical to establish these worst conditions, then one should take a more conservative approach. Lower the notch a little by increasing the proportional band, widen it by decreasing the RPM of the reset and the MIN of the derivative settings.

TWO-PATH CONTROL SYSTEMS

Earlier the different types of regulator devices were discussed and divided into the three basic categories of self-operated regulators, pilot-operated regulators and instrument control systems. Some of the advantages and disadvantages of each type were discussed. In addition, some definite conclusions were drawn about the relative advantages between self-operated and pilot-operated regulators. It would be helpful to review briefly the highlights of these comparisons.

In general, self-operated regulators, such as shown in Figure 2, are simple in design and construction, easy to operate and maintain, and are usually stable devices. Except for some of the pitot tube-type service regulators, self-operated regulators have very good dynamic response characteristics in general because any changes in the controlled pressure operates directly and immediately on the main diaphragm to produce a quick response to the disturbance.

The main disadvantage of the self-operated regulator is that it is not generally capable of drawing a straight-line pressure curve. In other words, the droop that results from both spring and diaphragm effect must be tolerated.

Another category of regulators, illustrated in Figure 10, uses a pilot to provide the loading pressure on the main diaphragm. This pilot, also called a relay, amplifier, multiplier or some similar terminology, has the ability to multiply a small change in downstream pressure into a large change in pressure applied to the regulator diaphragm. Due to this high gain feature of the pilot, these pilot-operated regulators can achieve a dramatic improvement in steady-state accuracy.

Pilot-operated regulators are more expensive than similar self-operated regulators, and the improvement in proportional band may not be important enough to justify the increased cost. In addition, the gain of the pilot amplifier increases the gain or sensitivity of the entire pressure regulator loop. If this loop gain is increased too much, the loop can become unstable and, when a disturbance occurs, the regulator will continue to oscillate or hunt.

Since the feedback signal from the controlled pressure to the regulator must operate through the pilot, we can see that the response time of this type of device is much slower than the equivalent self-operated regulator. This means that its reaction to a shock load may be rather poor.

In summary, regulators employing direct, self-operated feedback tend to be stable with fast response, but also tend to have broad proportional band. On the other hand, regulators employing pilots in the feedback path tend to be less stable with poor dynamic response, but have the advantage of much better accuracy and less droop.

To take advantage of the most desirable features of both the self-operated and pilot-operated devices, the two-path control system, shown in Figure 13, was developed. This is called two-path control because the feedback action to the regulator diaphragm follows two paths.

The primary path provides immediate feedback directly to the regulator diaphragm. This provides us with the advantages already listed for self-operated regulators. In addition to this primary self-operated feedback path, a secondary feedback path is provided to the regulator diaphragm through the pilot. The response of the pilot feedback path is much slower than the self-operated path, but it eventually provides a high-gain corrective action, which gives the performance benefits associated with pilot-operated regulators.

A close inspection of Figure 13 allows one to analyze how this system operates. The two-path control system is equipped with a spring that works to close the main regulator valve. If the controlled pressure downstream of the regulator tries to increase, immediate feedback is applied to the top of the regulator diaphragm. This pressure produces a force that helps the spring try to close the valve. At the same time, this increased pressure feedback signal is applied to the underside of the diaphragm in the pilot. This tends to close the valve in the pilot, therefore reducing the loading pressure that the pilot is applying to the underside of the regulator diaphragm.

Note that the decrease in pilot loading pressure also acts to close the main regulator valve. In other words, both feedback paths work together to close the main regulator valve when the controlled pressure tries to increase. The low-gain primary path provides a fast, stable response, but one that may have less than desirable accuracy. A short time later, the secondary path provides a high-gain response through the pilot that supplements the response of the primary path. The high-gain action provides a touch-up correction that greatly improves the steady-state accuracy.

The main spring in the two-path control system is positioned so that it tries to close the main valve. As the controlled pressure downstream of the regulator tries to increase, it produces a force on the main diaphragm, which also acts to close the main valve. The force of the loading gas from the pilot must therefore overcome both the force of downstream gas and the closing force of the spring. This convenient relationship allows the higher pressure loading gas to be bled off to the lower pressure downstream system. Note that when the main valve locks up, the force of the downstream gas also closes the pilot.

By judiciously combining the concepts of both the self-operated and pilot-operated devices one can obtain most of the advantages of both systems. The two-path pilot-operated regulator provides fast feedback action through the self-operated path. This produces excellent response characteristics from the system and the low gain of this path assures that the system will be stable. After a short, finite amount of time, the feedback through the pilot-operated path comes into play raising the gain of the system and improving the steady-state accuracy.

This sequence of events should begin to sound familiar by now. This two-step action should be recognized as being the same as the proportional-plus-reset control of the instrument control system. In other words, the two-path control regulator operates in two distinct modes: low gain during the beginning of the transient to maintain stability, then increasing gain as we approach the steady state to obtain good steady-state accuracy.

Although the physical construction of the two-path control regulator is quite different from the proportional-plus-reset controller, functionally they operate in a similar manner. A mathematical analysis of the two-path regulator would also verify the proportional-plus-reset action of this device.

When studying the proportional-plus-reset controller it was seen that the speed at which the gain for reset action is increased is very important. If it is too slow, it will take a long time to return to the set point. If it is too fast, the system will be unstable. The speed must be carefully tuned to the right value for the particular system to obtain the best control.

On the proportional-plus-reset controller there was a knob for adjusting the reset time according to the process being controlled. The adjustment of this knob is called tuning. However, the two-path control regulator has no such knob. This means that we must satisfy ourselves with a fixed reset time that must work on all systems where we intend to apply this device.

Thus, of necessity, this fixed reset time must be slightly more conservative than that which we might be able to use with the controller, but in most practical applications the two-path system can do an outstanding job of control on a wide variety of pressure systems, and it is widely used on systems in which both fast response and steady-state accuracy are important.

INDEX